THERMAL STRESSES

THERMAL STRESSES

2nd Edition

Naotake Noda
Department of Mechanical Engineering
Shizuoka University
Hamamatsu, Japan

Richard B. Hetnarski
Department of Mechanical Engineering
Rochester Institute of Technology
Rochester, New York, USA

Yoshinobu Tanigawa
Department of Mechanical Systems Engineering
Osaka Prefecture University
Sakai, Japan

Taylor & Francis
Taylor & Francis Group

NEW YORK • LONDON

Denise T. Schanck, *Vice President*
Robert H. Bedford, *Editor*
Thomas Hastings, *Marketing Manager*

Dennis P. Teston, *Production Director*
Anthony Mancini Jr., *Production Manager*
Brandy Mui, *STM Production Editor*
Jennifer Crisp, *Art Manager*
Daniel Sierra, *Cover Designer*

Published in 2003 by
Taylor & Francis
29 West 35th Street
New York, NY 10001

Published in Great Britain by
Taylor & Francis
11 New Fetter Lane
London EC4P 4EE

Printed in the United States of America on acid-free paper.

10 9 8 7 6 5 4 3 2 1

Library of Congress Cataloging-in-Publication Data

Noda, Naotake.
Thermal stresses/edited by Naotake Noda, Richard B. Hetnarski, & Yoshinobu Tanigawa.–2nd ed.
p. cm.
Includes bibliographical references and index.
ISBN 1-56032-971-8
1. Thermal stresses. I. Hetnarski, Richard B. II. Tanigawa, Yoshinobu. III. Title.
TA418.58.N63 2002
620.1′121—dc21

2002066209

CONTENTS

PREFACE TO THE SECOND EDITION

The first edition of this book was published in June 2000. As it was stated in the Preface to the first edition, the book has been planned as a textbook that should satisfy the needs of students, scientists, and practicing engineers. Following the comments received from the readers, we felt the need to upgrade and enlarge the contents of the book. We have exercised such an opportunity with the publication of the second edition.

First, we have carefully corrected various errors and misprints found in the first edition. Then, we have added a large number of illustrative examples in Chapters 1 through 9. Moreover, in Chapter 9, which is concerned with Thermally Induced Instability, we added a section on Thermally Induced Instability of Plates. Finally, in order to make it more convenient for instructors using this text, we have prepared a solutions manual for problems given at the end of each chapter. We expect that these improvements and additions will be well received by the users of the book.

We wish to express our gratitude to Research Associate Masayuki Ishihara at Shizuoka University for proofreading and typing the manuscript, and for the preparation of the figures and tables. Our thanks go also to the people of Taylor & Francis, in particular to Robert H. Bedford, our editor, and Brandy Mui, the STM production editor for taking care of the publication of the book.

N. Noda
R. B. Hetnarski
Y. Tanigawa

June 2002

PREFACE TO THE FIRST EDITION

With the development of new industries and modern processes, many machines and structures experience extreme thermal environments, resulting in various types of thermal loads. This situation has created a need for a text that is focused on the analysis of thermal stresses. Most existing publications on the subject, however, address the needs of researchers and senior level engineers working in laboratories, rather than those of engineering students who are exposed to this very rich and diverse field for the first time.

This book is a textbook, but it may also satisfy the needs of practicing scientists and engineers. The development of the material is such that a student is guided gradually from elementary methods and simple problems to sophisticated methods and difficult problems. Most of the book's content is suitable for senior undergraduate or first year graduate students who have a basic knowledge of engineering mechanics.

While writing the book, we were especially careful to present the material in a simple and consistent manner. Chapters 1 and 2 present an elementary treatment of thermal stresses in bars and beams. Since the theory of thermal stresses relies on the methods of heat conduction analysis, Chapter 3 develops heat conduction theory to the extent it is needed in subsequent chapters. Chapter 4 presents the basic equations of thermoelasticity. In Chapters 5 through 8, we apply the theory developed in Chapter 4 to solutions of plane thermoelastic problems, problems of circular cylinders, problems of spherical bodies, and problems of rectangular and circular plates. Chapter 9 introduces thermally induced instability, and Chapter 10 deals with thermodynamics of thermoelasticity.

In most of the chapters we incorporated a number of solved problems which demonstrate the methods employed. We took pains to ensure that these problems are easy to understand.

We wish to express our gratitude to Research Associate Masayuki Ishihara who contributed many ideas during the preparation of the book. To him and to Takao Fujimoto and Masaki Makino, both graduate students at Shizuoka University, we express our thanks for proofreading and typing the manuscript. In addition, we convey our thanks to Dr. Louis G. Hector, Jr., of Alcoa Technical Center, for reading critically the entire text and for suggestions for various improvements.

N. Noda
R. B. Hetnarski
Y. Tanigawa

June 2000

Chapter 1

THERMAL STRESSES IN BARS

In this chapter, the concept of thermal stresses is introduced for the simple case of homogeneous bars subjected to either uniform or non-uniform thermal loads. The stresses, strains, free thermal expansion, and free thermal strain are defined, and the Hooke's law is stated. Thermal stress fields are discussed for both uniform and non-uniform temperature changes in clamped bars, a circular bar, a trapezoidal plate, a circular frustum with and without a circular hole, and a hollow truncated cone. Thermal stresses are considered in partially restrained bars, as they appear in machine elements subjected to thermal loading. Discussed are problems of two parallel bars, a bar with a small gap, two bars attached to each other, and two parallel bars attached in a series to a spring. Deformation in partially restrained bars due to free thermal expansion and thermal stress, and the equilibrium between internal and external forces are considered in order to find stresses and strains.

1.1 Stress and Strain

1.1.1 Stress and Strain

Consider a body which deforms under the action of external forces. The force produced between the parts of the body by the external forces and distributed in the body is called an internal force. The force divided by the area on which it acts is called stress. We define the stress p by

$$p = \lim_{\Delta A \to 0} \frac{\Delta P}{\Delta A} \tag{1.1}$$

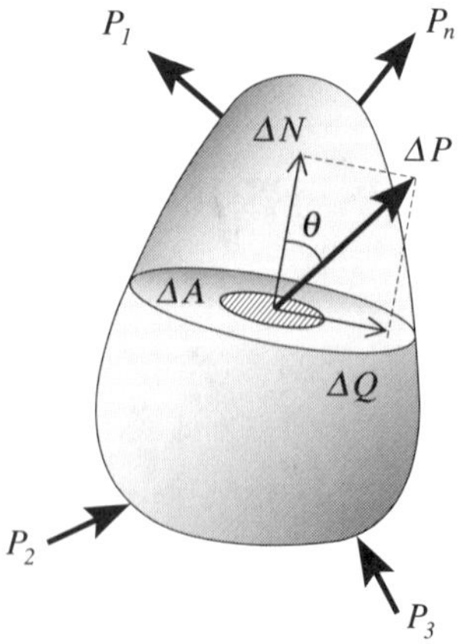

Figure 1.1 An internal force in the body.

in which ΔP is the internal force with dimension [N] acting on a small area ΔA in the body shown in Fig. 1.1.

In the general case, the direction of the force is inclined to the surface normal to the small area. Denoting the angle between the direction of the internal force and the surface normal to the small area ΔA by θ, the normal component ΔN, and tangential component ΔQ of the internal force ΔP are given by

$$\Delta N = \Delta P \cos\theta, \qquad \Delta Q = \Delta P \sin\theta \tag{1.2}$$

Dividing the normal component ΔN and the tangential component ΔQ of the internal force ΔP by the small area ΔA, and taking the limit of the ratios, we have

$$\sigma = \lim_{\Delta A \to 0} \frac{\Delta N}{\Delta A}, \qquad \tau = \lim_{\Delta A \to 0} \frac{\Delta Q}{\Delta A} \tag{1.3}$$

where σ and τ are called the normal and the shearing stresses, respectively. The stresses are measured in Newtons per square meter, that is, Pascals [Pa].

When the body is deformed under the action of external forces, the strain is defined by the elongation per unit length. When a circular bar of length l and diameter d is extended in the axial direction by a tensile force, the bar undergoes an elongation λ in the axial direction and a contraction δ in the radial direction. Two normal strains, the longitudinal strain ϵ and the lateral strain ϵ', are defined by the equations

$$\epsilon = \frac{(l+\lambda) - l}{l} = \frac{\lambda}{l} \tag{1.4}$$

$$\epsilon' = \frac{(d+\delta) - d}{d} = \frac{\delta}{d} \tag{1.5}$$

The strain is measured in meters per meter, which is a non-dimensional quantity. The ratio of the unit lateral contraction to the unit axial elongation, taken with the negative

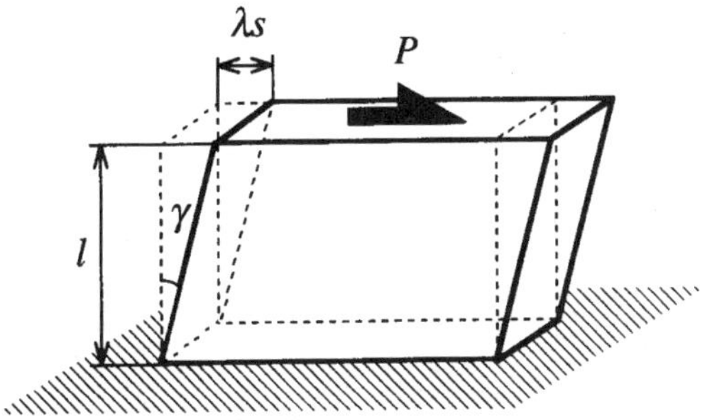

Figure 1.2 A shearing force.

sign, is known as the Poisson's ratio ν

$$\nu = -\frac{\epsilon'}{\epsilon} \tag{1.6}$$

Since ϵ' is a negative quantity whenever ϵ is positive, and vice versa, the Poisson's ratio is always positive.

When the distortion is produced by a shearing force P shown in Fig. 1.2, a shearing strain γ occurs; it is defined by the ratio of a small horizontal displacement λ_s to the length l, and is approximately equal to a small distortional angle γ

$$\frac{\lambda_s}{l} = \tan\gamma \cong \gamma \tag{1.7}$$

1.1.2 Hooke's Law

When an external tensile force acts on tensile test specimen with a length l and a cross-sectional area A made of mild steel, a corresponding elongation in the specimen is produced. The relationship between the tensile force and the elongation is shown in Fig. 1.3. Dividing the tensile force and the elongation by the initial cross-sectional area A and the initial length l of the specimen gives the tensile stress and tensile strain, respectively, so that Fig. 1.3 also gives the relationship between tensile stress and tensile strain. From O to A the stress and the strain are proportional. The stress at point A is called the *proportional limit.* The deformation before the proportional limit at point A is called a linear elastic deformation. Beyond the proportional limit at point A, the strain rapidly increases and is not proportional to the stress. When the deformation is within the *elastic limit* at point B, the strain is not proportional to the stress, and the deformation (between A and B on the curve) is called a nonlinear elastic deformation. The distance between points A and B is very small for mild steel so that the distinction between points A and B may be negligible. When the specimen is subjected to unloading from an arbitrary point on the stress–strain curve within the elastic deformation, the stress–strain curve in unloading is identical to that in loading.

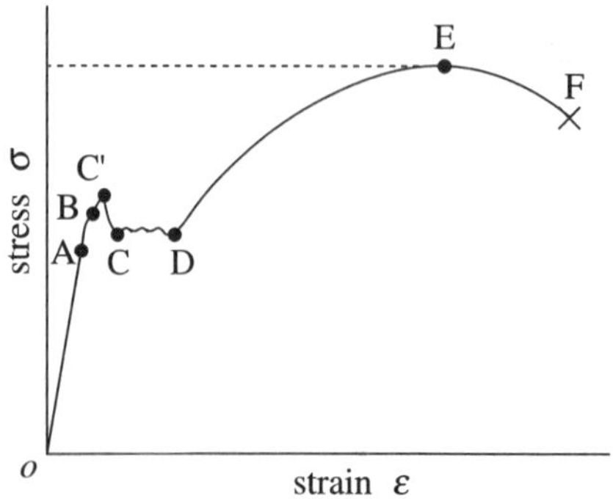

Figure 1.3 Relation between stress and strain.

After the strain reaches point C′, the stress immediately falls down to point C without increment of the strain. The stresses at points C′ and C are called the *upper yield stress* and *lower yield stress*, respectively. Beyond point C, the strain increases along CD under constant stress, that is, the *yield stress*. The deformation after the yield stress has been reached is called *plastic deformation*. Beyond point D, the material recovers its resistance and the stress increases again with the strain. The stress has a maximum value at a point E: stress at point E is called *tensile strength*. After point E, the strain increases with a decrease in stress and the specimen is finally destroyed at point F.

When the deformation in the material is produced by the action of external forces, within certain limits the elongation of the material λ is proportional to the tensile force as shown in Fig. 1.3 and the displacement λ_s is proportional to the shearing force. That is, the normal stress is proportional to the normal strain and the shearing stress is proportional to the shearing strain. These relations are known as Hooke's law which is expressed by

$$\sigma = E\epsilon \tag{1.8}$$

$$\tau = G\gamma \tag{1.9}$$

The material constant E is called the Young's modulus, and the material constant G is called the shear modulus. These material constants are measured in Pascals [Pa].

1.1.3 Free Thermal Expansion and Free Thermal Strain

When the temperature of a circular bar of length l and diameter d rises from an initial temperature T_0 to a final temperature T_1, this temperature rise produces an elongation of the bar, and the length of the bar and its diameter become l' and d', respectively. The free thermal expansions λ and δ of the bar are given by

$$\lambda = l' - l = \alpha\,(T_1 - T_0)\,l, \qquad \delta = d' - d = \alpha\,(T_1 - T_0)\,d \tag{1.10}$$

where α is another material constant called the coefficient of linear thermal expansion. It denotes the elongation of the bar per unit length per unit temperature change, and is measured in one per one degree of the temperature [1/K]. The normal strain ϵ in the longitudinal direction and the normal strain ϵ' in the radial direction are given by

$$\epsilon = \frac{l'-l}{l} = \frac{\lambda}{l} = \frac{\alpha(T_1 - T_0)\,l}{l} = \alpha(T_1 - T_0) \tag{1.11}$$

$$\epsilon' = \frac{d'-d}{d} = \frac{\delta}{d} = \frac{\alpha(T_1 - T_0)\,d}{d} = \alpha(T_1 - T_0) \tag{1.12}$$

and are called the free thermal strains.

1.1.4 Action of External Force and Temperature Change

The strains in the body due to simultaneous action of the external force and the temperature change are given by

$$\epsilon = \epsilon_s + \epsilon_T, \qquad \gamma = \gamma_s \tag{1.13}$$

The subscript s denotes the strain produced by the external force, and the subscript T denotes the strain produced by the temperature change. Changing ϵ in Eq. (1.8) to ϵ_s and ϵ in Eq. (1.11) to ϵ_T, substitution of strains due to the external force from Eq. (1.8) and due to the temperature change from Eq. (1.11) into Eq. (1.13) will result in

$$\epsilon = \frac{\sigma}{E} + \alpha(T_1 - T_0) = \frac{\sigma}{E} + \alpha T \tag{1.14}$$

$$\gamma = \frac{\tau}{G} \tag{1.15}$$

where T is given by

$$T = T_1 - T_0 \tag{1.16}$$

and denotes the temperature change or the temperature difference. The stress–strain relations, Eqs. (1.14) and (1.15), represent the Hooke's law with the temperature change.

1.2 Thermal Stresses in Clamped Bars

1.2.1 Constant Temperature Change

Consider thermal stress in a perfectly clamped bar, Fig. 1.4, with length l and cross-sectional area A subjected to a temperature change from initial temperature T_0 to final temperature T_1. If the bar is unconstrained, it will expand freely. The free thermal elongation λ is

$$\lambda = l' - l = \alpha(T_1 - T_0)\,l = \alpha T l \tag{a}$$

Such a free thermal elongation, however, cannot occur in a perfectly clamped bar. A thermal stress is produced in the bar in order to eliminate the free thermal elongation λ.

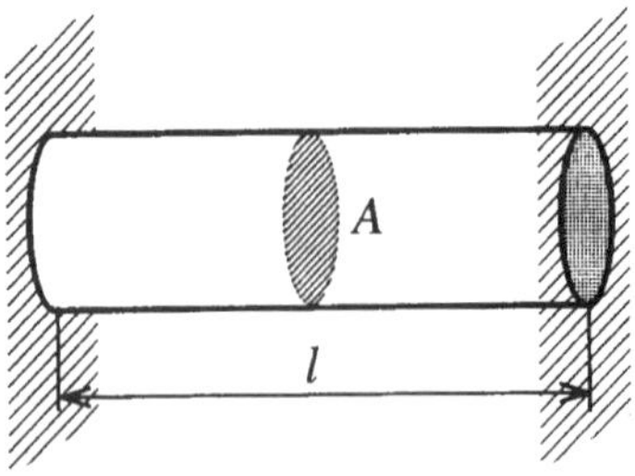

Figure 1.4 A perfectly clamped bar.

If the bar is perfectly restrained, the free thermal elongation λ must be eliminated by the contraction of the bar due to the action of the external force P. The contraction λ' of the bar produced by the external force P is

$$\lambda' = \epsilon_s l = \frac{\sigma}{E} l = \frac{Pl}{AE} \tag{b}$$

Therefore, from the condition $\lambda' = -\lambda$ we obtain

$$P = -\alpha E (T_1 - T_0) A \tag{c}$$

Thus, the thermal stress in the perfectly clamped bar is

$$\sigma = \frac{P}{A} = -\alpha E (T_1 - T_0) = -\alpha E T \tag{1.17}$$

An alternative way is to consider the final deformation as the sum of the free thermal expansion and the deformation by the stress. If the bar is perfectly clamped, the final deformation of the bar is zero

$$\epsilon = \frac{\sigma}{E} + \alpha (T_1 - T_0) = 0 \tag{d}$$

Therefore, the thermal stress σ is in the form of Eq. (1.17).

As an example, consider a perfectly clamped and initially unstressed bar made of mild steel. The coefficient of linear thermal expansion is $\alpha = 11.2 \times 10^{-6}$ 1/K and the Young's modulus is $E = 206$ GPa. If the temperature is now raised by 100 K, the thermal stress in the bar is $\sigma = -\alpha E T = -230.7$ MPa. The absolute value of the thermal stress per unit temperature change is 2.31 MPa/K.

Table 1.1 shows the values of the coefficient of linear thermal expansion α, the Young's modulus E, and the absolute value of thermal stress per unit temperature change $|\sigma/T|$ or αE for some materials. Observe that the range of thermal stresses per unit temperature change for these materials is approximately 1 to 5 MPa.

Example 1.1 A trapezoidal plate of thickness t shown in Fig. 1.5 is subjected to a temperature change T. Calculate the thermal stress produced in the plate.

Table 1.1 The coefficient of linear thermal expansion α, the Young's modulus E, and the absolute value of the thermal stress per unit temperature change $|\sigma/T|$ or αE

Material	Coeffcient of linear thermal expansion α $\alpha \times 10^6$ (1/K)	Young's modulus E (GPa)	$\lvert\sigma/T\rvert$ (MPa/K)
Aluminum (Al)	23.1	72	1.66
Copper (Cu)	16.5	120	1.98
Gold (Au)	14.2	80	1.14
Iron (Fe)	11.8	206	2.43
Lead (Pb)	28.9	170	4.91
Platinum (Pt)	8.8	160	1.41
Silver (Ag)	18.9	80	1.51
Zinc (Zn)	30.2	120	3.62
Mild steel	11.2	206	2.31
Hard steel	10.7	206	2.20
High-carbon steel	10.7	200	2.14
Gray pig iron	11.9	60–150	0.714–1.79
18-8 stainless steel	14.7	206	3.03
Cr14 stainless steel	11.1	206	2.29
Duralumin	21.6	70	1.51
Brass	19.0	110	2.09
Inconel	11.6	210	2.44
Ti + 6Al + 4V	8.9	106	0.94

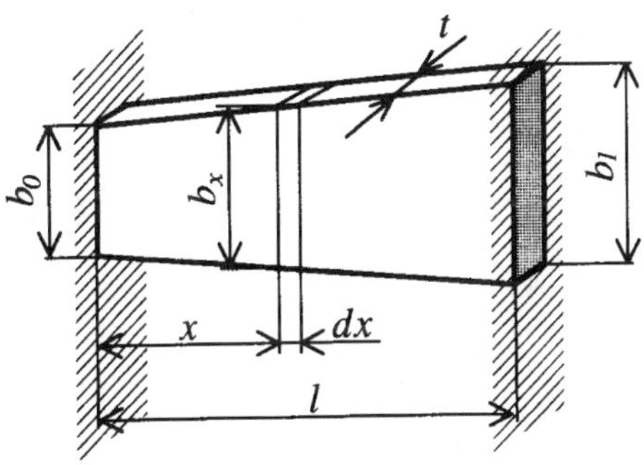

Figure 1.5 A trapezoidal plate.

Solution The free thermal elongation λ is

$$\lambda = \alpha T l \tag{a}$$

The width b_x of a plate at a distance x from the left end is

$$b_x = b_0 + (b_1 - b_0)\frac{x}{l} \tag{b}$$

The cross-sectional area A_x at a distance x is given by

$$A_x = b_x t = \left[b_0 + (b_1 - b_0)\frac{x}{l}\right] t \tag{c}$$

The strain ϵ_x at a distance x from the left end due to a virtual force P is

$$\epsilon_x = \frac{\sigma_x}{E} = \frac{P}{EA_x} = \frac{P}{E\left[b_0 + (b_1 - b_0)\dfrac{x}{l}\right]t} \tag{d}$$

The elongation λ' of the plate due to the virtual force P is

$$\lambda' = \int d\lambda' = \int_0^l \epsilon_x\, dx = \int_0^l \frac{P}{E\left[b_0 + (b_1 - b_0)\dfrac{x}{l}\right]t}\, dx$$

$$= \frac{Pl}{E(b_1 - b_0)t} \ln\left[b_0 + (b_1 - b_0)\frac{x}{l}\right]\Bigg|_0^l = \frac{Pl}{E(b_1 - b_0)t} \ln\left(\frac{b_1}{b_0}\right) \tag{e}$$

The plate is perfectly constrained in the x direction, so that the combined deformation of the free thermal expansion λ and the elongation λ' due to the force P is zero

$$\lambda + \lambda' = 0 \tag{f}$$

From Eqs. (a), (e) and (f) the force P is

$$P = -\alpha ET\frac{(b_1 - b_0)t}{\ln(b_1/b_0)} \tag{g}$$

and the thermal stress is calculated to be:

$$\sigma_x = \frac{P}{A_x} = -\alpha ET\frac{(b_1 - b_0)}{\left[b_0 + (b_1 - b_0)\dfrac{x}{l}\right]\ln(b_1/b_0)} \tag{1.18}$$

If $b_1 > b_0$, the maximum thermal stress $(\sigma_x)_{max}$ occurs at the minimum cross-sectional area of the plate and the minimum thermal stress $(\sigma_x)_{min}$ occurs at the maximum cross-sectional area

$$(\sigma_x)_{max} = -\alpha ET\frac{(b_1 - b_0)}{b_0\ln(b_1/b_0)}, \qquad (\sigma_x)_{min} = -\alpha ET\frac{(b_1 - b_0)}{b_1\ln(b_1/b_0)} \tag{1.19}$$

Example 1.2 A clamped circular frustum shown in Fig. 1.6 is subjected to a temperature change T. Calculate the thermal stress produced in the frustum.

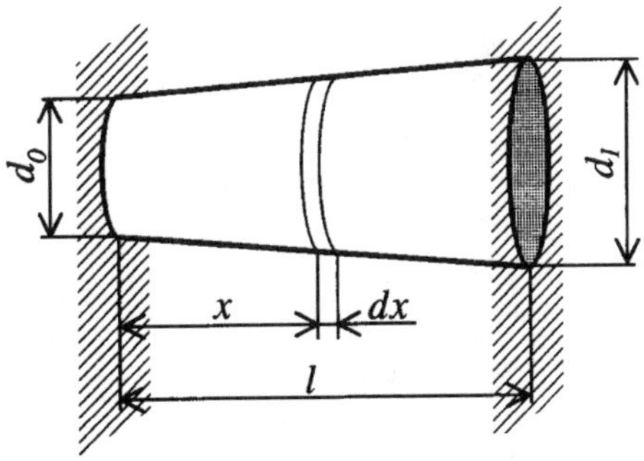

Figure 1.6 A clamped circular frustum.

Solution The free thermal expansion λ is

$$\lambda = \alpha T l \tag{a}$$

and the diameter d_x of the frustum at a position x from the left end is

$$d_x = d_0 + (d_1 - d_0)\frac{x}{l} \tag{b}$$

The cross-sectional area A_x at the position x is given by

$$A_x = \frac{\pi}{4}d_x^2 = \frac{\pi}{4}\left[d_0 + (d_1 - d_0)\frac{x}{l}\right]^2 \tag{c}$$

Thus, the strain ϵ_x of the frustum at x due to a virtual force P becomes

$$\epsilon_x = \frac{\sigma_x}{E} = \frac{P}{EA_x} = \frac{4P}{E\pi\left[d_0 + (d_1 - d_0)\frac{x}{l}\right]^2} \tag{d}$$

and the elongation λ' of the frustum due to the virtual force P equals

$$\lambda' = \int d\lambda' = \int_0^l \epsilon_x\,dx = \int_0^l \frac{4P}{E\pi\left[d_0 + (d_1 - d_0)\frac{x}{l}\right]^2}dx$$

$$= -\frac{4Pl}{E\pi(d_1 - d_0)}\left[\frac{1}{d_0 + (d_1 - d_0)\frac{x}{l}}\right]_0^l = \frac{4Pl}{E\pi d_1 d_0} \tag{e}$$

Because the frustum is perfectly constrained in the x direction, the combined deformation of the free thermal expansion λ and the elongation λ' due to the force P must be zero

$$\lambda + \lambda' = 0 \tag{f}$$

From Eqs. (a), (e), and (f) the force P is

$$P = -\alpha ET \frac{\pi}{4} d_1 d_0 \tag{g}$$

and the thermal stress is calculated to be:

$$\sigma_x = \frac{P}{A_x} = -\alpha ET \frac{d_1 d_0}{\left[d_0 + (d_1 - d_0)\frac{x}{l}\right]^2} \tag{1.5}$$

If $d_1 > d_0$, the maximum thermal stress $(\sigma_x)_{max}$ occurs at the minimum cross-sectional area and the minimum thermal stress $(\sigma_x)_{min}$ occurs at the maximum cross-sectional area

$$(\sigma_x)_{max} = -\alpha ET \frac{d_1}{d_0}, \qquad (\sigma_x)_{min} = -\alpha ET \frac{d_0}{d_1} \tag{1.6}$$

Example 1.3 The clamped circular frustum with a cylindrical hole of diameter d_i shown in Fig. 1.7 is subjected to a temperature change T. Calculate the thermal stress produced in the frustum.

Solution The free thermal expansion λ is

$$\lambda = \alpha T l \tag{a}$$

and the outer diameter d_x of the frustum at a distance x from the left end is

$$d_x = d_0 + (d_1 - d_0)\frac{x}{l} \tag{b}$$

The cross-sectional area A_x at the position x is given by

$$A_x = \frac{\pi}{4}(d_x^2 - d_i^2) = \frac{\pi}{4}\left\{\left[d_0 + (d_1 - d_0)\frac{x}{l}\right]^2 - d_i^2\right\} \tag{c}$$

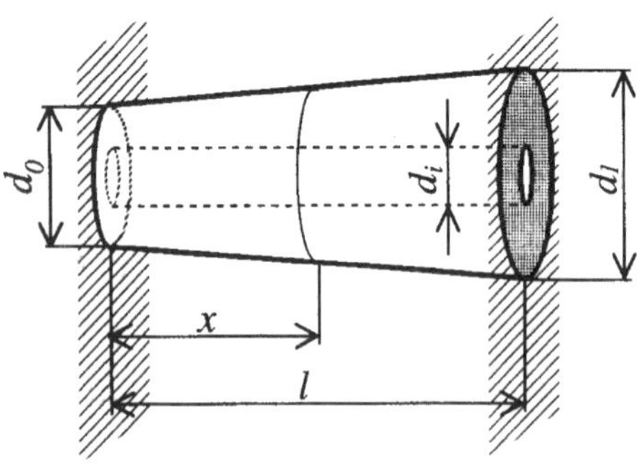

Figure 1.7 A clamped circular frustum with a cylindrical hole.

Thus, the strain ϵ_x of the frustum at x due to the virtual force P becomes

$$\epsilon_x = \frac{\sigma_x}{E} = \frac{P}{EA_x} = \frac{4P}{E\pi\left\{\left[d_0 + (d_1 - d_0)\frac{x}{l}\right]^2 - d_i^2\right\}} \tag{d}$$

and the elongation λ' of the frustum due to the force P equals

$$\lambda' = \int d\lambda' = \int_0^l \epsilon_x \, dx = \int_0^l \frac{4P}{E\pi\left\{\left[d_0 + (d_1 - d_0)\frac{x}{l}\right]^2 - d_i^2\right\}} dx$$

$$= \frac{2Pl}{E\pi d_i(d_1 - d_0)} \ln\left[\frac{(d_1 - d_i)(d_0 + d_i)}{(d_1 + d_i)(d_0 - d_i)}\right] \tag{e}$$

Since the frustum is perfectly constrained in the x direction, the combined deformation of the free thermal expansion λ and the elongation λ' due to the force P is zero

$$\lambda + \lambda' = 0 \tag{f}$$

From Eqs. (a), (e), and (f) the force P is calculated

$$P = -\frac{1}{2}\alpha ET \frac{\pi d_i(d_1 - d_0)}{\ln\left[\frac{(d_1 - d_i)(d_0 + d_i)}{(d_1 + d_i)(d_0 - d_i)}\right]} \tag{g}$$

and the thermal stress becomes

$$\sigma_x = \frac{P}{A_x} = -\alpha ET \frac{2d_i(d_1 - d_0)}{\left\{\left[d_0 + (d_1 - d_0)\frac{x}{l}\right]^2 - d_i^2\right\} \ln\left[\frac{(d_1 - d_i)(d_0 + d_i)}{(d_1 + d_i)(d_0 - d_i)}\right]} \tag{1.22}$$

If $d_1 > d_0$, the maximum thermal stress $(\sigma_x)_{max}$ occurs at the minimum cross-sectional area and the minimum thermal stress $(\sigma_x)_{min}$ occurs at the maximum cross-sectional area

$$\begin{aligned} (\sigma_x)_{max} &= -\alpha ET \frac{2d_i(d_1 - d_0)}{(d_0^2 - d_i^2)\ln\left[\frac{(d_1 - d_i)(d_0 + d_i)}{(d_1 + d_i)(d_0 - d_i)}\right]} \\ (\sigma_x)_{min} &= -\alpha ET \frac{2d_i(d_1 - d_0)}{(d_1^2 - d_i^2)\ln\left[\frac{(d_1 - d_i)(d_0 + d_i)}{(d_1 + d_i)(d_0 - d_i)}\right]} \end{aligned} \tag{1.23}$$

Example 1.4 A clamped, hollow, truncated cone of thickness t shown in Fig. 1.8 is subjected to a temperature change T. Calculate the thermal stress in the cone.

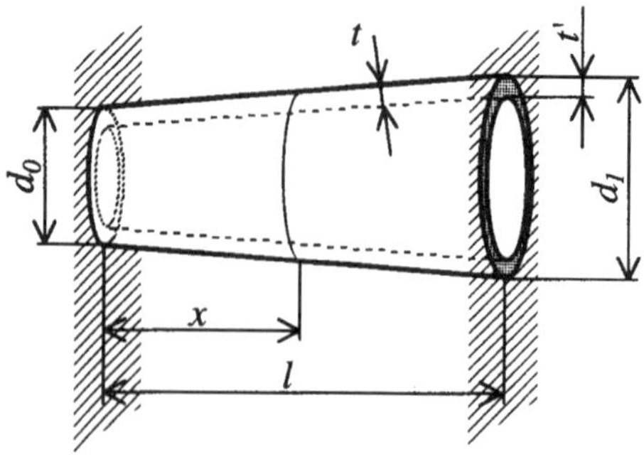

Figure 1.8 A clamped hollow truncated cone.

Solution The free thermal expansion λ is

$$\lambda = \alpha T l \tag{a}$$

and the outer diameter d_x of the conical surface at a distance x from the left end is

$$d_x = d_0 + (d_1 - d_0)\frac{x}{l} \tag{b}$$

The difference t' of the outer radius and the inner radius is

$$t' = t\sqrt{1 + \left(\frac{d_1 - d_0}{2l}\right)^2} \tag{c}$$

The cross-sectional area A_x at the position x equals

$$A_x = \frac{\pi}{4}\left[d_x^2 - (d_x - 2t')^2\right] = \pi t'\left[d_0 - t' + (d_1 - d_0)\frac{x}{l}\right] \tag{d}$$

Thus, the strain ϵ_x at x due to the force P equals

$$\epsilon_x = \frac{\sigma_x}{E} = \frac{P}{EA_x} = \frac{P}{E\pi t'\left[d_0 - t' + (d_1 - d_0)\frac{x}{l}\right]} \tag{e}$$

and the elongation λ' of the hollow truncated cone due to the virtual force P is given by

$$\begin{aligned}\lambda' &= \int d\lambda' = \int_0^l \epsilon_x \, dx = \int_0^l \frac{P}{E\pi t'\left[d_0 - t' + (d_1 - d_0)\frac{x}{l}\right]} dx \\ &= \frac{Pl}{E\pi t'(d_1 - d_0)} \ln\left[d_0 - t' + (d_1 - d_0)\frac{x}{l}\right]\Bigg|_0^l \\ &= \frac{Pl}{E\pi t'(d_1 - d_0)} \ln\left(\frac{d_1 - t'}{d_0 - t'}\right)\end{aligned} \tag{f}$$

Because of the perfect constraint in the x direction, the combined deformation of the free thermal expansion λ and the elongation λ' due to the force P must be zero

$$\lambda + \lambda' = 0 \tag{g}$$

From Eqs. (a), (f), and (g) the force P is

$$P = -\alpha E T \frac{\pi t'(d_1 - d_0)}{\ln\left(\dfrac{d_1 - t'}{d_0 - t'}\right)} \tag{h}$$

and the thermal stress is

$$\sigma_x = \frac{P}{A_x} = -\alpha E T \frac{1}{\left(\dfrac{d_0 - t'}{d_1 - d_0} + \dfrac{x}{l}\right) \ln\left(\dfrac{d_1 - t'}{d_0 - t'}\right)} \tag{1.24}$$

If $d_1 > d_0$, the maximum thermal stress $(\sigma_x)_{max}$ occurs at the minimum cross-sectional area and the minimum thermal stress $(\sigma_x)_{min}$ occurs at the maximum cross-sectional area

$$\begin{aligned} (\sigma_x)_{max} &= -\alpha E T \frac{1}{\left(\dfrac{d_0 - t'}{d_1 - d_0}\right) \ln\left(\dfrac{d_1 - t'}{d_0 - t'}\right)} \\ (\sigma_x)_{min} &= -\alpha E T \frac{1}{\left(\dfrac{d_1 - t'}{d_1 - d_0}\right) \ln\left(\dfrac{d_1 - t'}{d_0 - t'}\right)} \end{aligned} \tag{1.25}$$

1.2.2 Non-Uniform Temperature Change

In the proceeding discussion, the temperature change in the bar was assumed constant. Consider now the thermal stress in a perfectly clamped bar shown in Fig. 1.9, when the temperature change $T(x)$ is a function of the position x. The free thermal expansion $d\lambda$ of the small element dx may be written as

$$d\lambda = \alpha T(x)\, dx \tag{1.26}$$

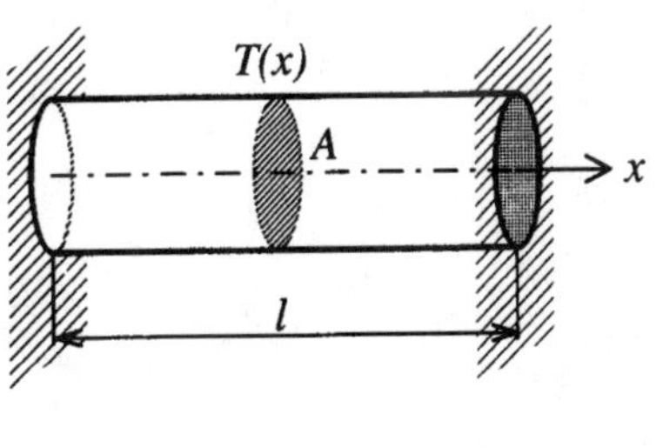

Figure 1.9 A perfectly clamped bar.

The free thermal expansion λ of a bar of length l is therefore

$$\lambda = \int d\lambda = \int_0^l \alpha T(x)\,dx = \alpha \int_0^l T(x)\,dx \tag{1.27}$$

However, this free thermal expansion cannot occur if the bar is perfectly clamped. Thermal stress is produced in the bar when the free thermal expansion is eliminated. As the bar is perfectly restrained, the free thermal expansion λ must be eliminated by the contraction of the bar induced by the virtual force P. The contraction λ' of the bar produced by the compressive virtual force P is

$$\lambda' = \frac{\sigma}{E} l = \frac{Pl}{EA} \tag{a}$$

The combined deformation of the free thermal expansion λ and the contraction by the force P is zero, or $\lambda' + \lambda = 0$. Therefore,

$$P = -\frac{\lambda E A}{l} \tag{b}$$

and the resulting thermal stress in the perfectly clamped bar is

$$\sigma = \frac{P}{A} = -\frac{\lambda E}{l} = -\frac{\alpha E}{l} \int_0^l T(x)\,dx \tag{1.28}$$

From a comparison with the case of the constant temperature change and the case with the temperature change as a function of the position, it may be observed that the thermal stresses in the bar subjected to the temperature change as the function of the position is obtained from the thermal stress with the constant temperature change by substitution of $\alpha \int_0^l T(x)\,dx$ for $\alpha T l$. Therefore, from Examples 1.1–1.4 it follows that for a temperature change in the form $T = T(x)$, the thermal stresses are:

[A] In a clamped trapezoidal plate:

$$\sigma = -\alpha E \frac{1}{l\left(\dfrac{b_0}{b_1 - b_0} + \dfrac{x}{l}\right) \ln\left(\dfrac{b_1}{b_0}\right)} \int_0^l T(\eta)\,d\eta \tag{1.29}$$

[B] In a clamped circular frustum:

$$\sigma_x = -\alpha E \frac{d_1 d_0}{l\left[d_0 + (d_1 - d_0)\dfrac{x}{l}\right]^2} \int_0^l T(\eta)\,d\eta \tag{1.30}$$

[C] In a clamped circular frustum with a cylindrical hole with a diameter d_i:

$$\sigma_x = -\alpha E \frac{2d_i(d_1 - d_0)}{l\left\{\left[d_0 + (d_1 - d_0)\frac{x}{l}\right]^2 - d_i^2\right\} \ln\left[\frac{(d_1 - d_i)(d_0 + d_i)}{(d_1 + d_i)(d_0 - d_i)}\right]} \times \int_0^l T(\eta)\, d\eta \tag{1.31}$$

[D] In a clamped, hollow, truncated cone:

$$\sigma_x = -\alpha E \frac{1}{l\left(\frac{d_0 - t'}{d_1 - d_0} + \frac{x}{l}\right) \ln\left(\frac{d_1 - t'}{d_0 - t'}\right)} \int_0^l T(\eta)\, d\eta \tag{1.32}$$

1.3 Thermal Stresses in Partially Restrained Bars

Consider the thermal stresses in two parallel bars of the same length l, shown in Fig. 1.10, subjected to different temperature changes $T_i, i = 1, 2$. The two bars are connected to two rigid plates of which one is fixed and the other may translate but cannot rotate. Therefore, the change in length of both bars is always the same. Let A_i, E_i, and α_i denote cross-sectional area, the Young's modulus, and the coefficient of linear thermal expansion of the i-th bar, respectively. The deformations λ_i due to both the free thermal expansion and the thermal stress are

$$\lambda_1 = \alpha_1 T_1 l + \frac{\sigma_1}{E_1} l, \qquad \lambda_2 = \alpha_2 T_2 l + \frac{\sigma_2}{E_2} l \tag{a}$$

Since the final length of both bars after deformation is the same, the following relation holds

$$l + \alpha_1 T_1 l + \frac{\sigma_1}{E_1} l = l + \alpha_2 T_2 l + \frac{\sigma_2}{E_2} l \tag{b}$$

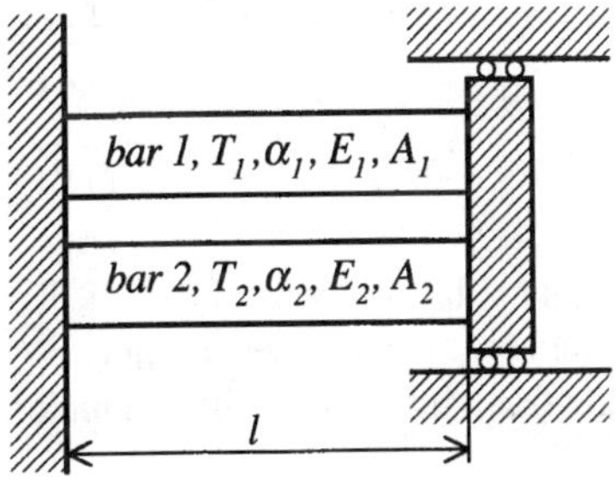

Figure 1.10 Two parallel bars of the same length.

The internal forces in the first and the second bar are, respectively,

$$\sigma_1 A_1 \quad \text{and} \quad \sigma_2 A_2 \tag{c}$$

and since there is no external force applied to the bars, the equilibrium of forces is described by

$$\sigma_1 A_1 + \sigma_2 A_2 = 0 \tag{d}$$

Solving Eqs. (b) and (d) gives the stresses

$$\sigma_1 = -K\alpha_1 E_1 T_1, \qquad \sigma_2 = -\sigma_1 \frac{A_1}{A_2} \tag{1.33}$$

where

$$K = \frac{1 - \dfrac{\alpha_2 T_2}{\alpha_1 T_1}}{1 + \dfrac{A_1 E_1}{A_2 E_2}} \tag{1.34}$$

The coefficient K is called the restraint coefficient. If $K > 0$, thermal stress σ_1 is compressive and thermal stress σ_2 is tensile, and when $K < 0$, σ_1 is tensile, and σ_2 is compressive. From Eqs. (1.33) one bar undergoes tensile stress and the other undergoes compressive stress.

Substitution of Eqs. (1.33) into Eq. (a) leads to formulae for the deformation of the bars

$$\lambda_1 = \lambda_2 = \frac{(\alpha_1 T_1 E_1 A_1 + \alpha_2 T_2 E_2 A_2)l}{A_1 E_1 + A_2 E_2} \tag{1.35}$$

When the temperature changes T_i are positive, the deformation λ $(= \lambda_1 = \lambda_2)$ is positive and the bars become elongated. The two-bar system with positive T_i undergoes an elongation given by Eq. (1.35) independently of whether K is positive or negative. This discussion leads to a more general conclusion for thermal stress problems. Namely, the fact that the bars in a system are elongated does not produce a sufficient basis to conclude which of the bars is under tension and which is under compression.

By subtracting the free thermal expansion $\alpha_i T_i l$ from the final deformation λ of the bar given by Eq. (1.35) the following relations result for the first and the second bar, respectively,

$$\lambda - \alpha_1 T_1 l = \frac{\sigma_1}{E_1} l, \qquad \lambda - \alpha_2 T_2 l = \frac{\sigma_2}{E_2} l \tag{1.36}$$

Thus, $\lambda - \alpha_i T_i l$ is proportional to the thermal stress in bar i. It is clear from Eqs. (1.36) that when $\lambda - \alpha_i T_i l$ is positive, the bar is under tension and vice versa, when this quantity is negative, bar i is under compression. As already mentioned, one of the bars is under tension while the other is under compression.

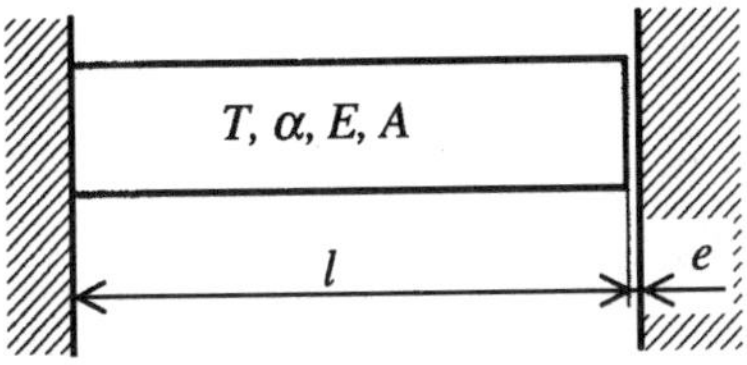

Figure 1.11 A bar with a small gap.

Example 1.5 If a bar with a small gap e between its free end and a rigid wall shown in Fig. 1.11 is subjected to the positive temperature change T, how large a thermal stress is produced?

Solution When free thermal expansion $\alpha T l$ is larger than the small gap e, the combined elongation due to the free thermal expansion and the deformation due to the stress must be equal to the small gap e

$$\alpha T l + \frac{\sigma}{E} l = e \tag{a}$$

Solving this equation for thermal stress σ we obtain

$$\sigma = -\alpha E T \left(1 - \frac{e}{\alpha T l}\right) \tag{1.37}$$

When the free thermal expansion $\alpha T l$ is smaller than e, no thermal stress σ results.

Example 1.6 Two bars are attached to each other in series, placed between rigid walls, and are subjected to the different temperature changes T_i, Fig. 1.12. What thermal stresses are produced?

Solution The elongations of bar 1 and bar 2, respectively, are

$$\alpha_1 T_1 l_1 + \frac{\sigma_1}{E_1} l_1, \qquad \alpha_2 T_2 l_2 + \frac{\sigma_2}{E_2} l_2 \tag{a}$$

The combined elongation of the bars is zero because of the rigid walls. Thus,

$$\alpha_1 T_1 l_1 + \frac{\sigma_1}{E_1} l_1 + \alpha_2 T_2 l_2 + \frac{\sigma_2}{E_2} l_2 = 0 \tag{b}$$

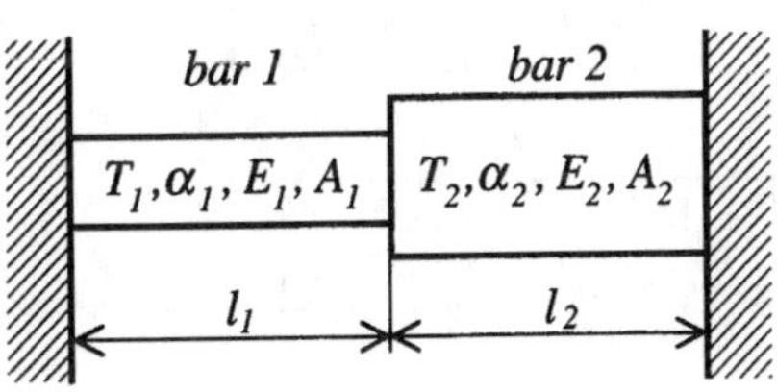

Figure 1.12 Two bars attached to each other.

The internal forces in bar 1 and bar 2, respectively, are

$$\sigma_1 A_1, \qquad \sigma_2 A_2 \tag{c}$$

From equilibrium, the internal force in bar 1 is equal to the internal force in bar 2

$$\sigma_1 A_1 = \sigma_2 A_2 \tag{d}$$

Solution of Eqs. (b) and (d) gives the thermal stresses σ_1 and σ_2 as

$$\sigma_1 = -\alpha_1 E_1 T_1 \left[\frac{1 + \dfrac{\alpha_2 T_2 l_2}{\alpha_1 T_1 l_1}}{1 + \dfrac{A_1 E_1 l_2}{A_2 E_2 l_1}} \right], \qquad \sigma_2 = \sigma_1 \frac{A_1}{A_2} \tag{1.38}$$

Example 1.7 A bar is attached to a spring in series at one end and restrained by a rigid wall at another, Fig. 1.13. When the bar is subjected to a temperature change T, what thermal stress is produced?

Solution The sum of the elongation λ_1 of the bar and the contraction λ_2 of the spring is zero

$$\lambda_1 + \lambda_2 = \alpha T l + \frac{\sigma}{E} l + \lambda_2 = 0 \tag{a}$$

The internal force in the bar equals the force $k\lambda_2$ in the spring

$$\sigma A = k\lambda_2 \tag{b}$$

where k is the spring constant. Solving Eqs. (a) and (b) gives the following thermal stress:

$$\sigma = -\alpha E T \left[\frac{1}{1 + \dfrac{AE}{kl}} \right] \tag{1.39}$$

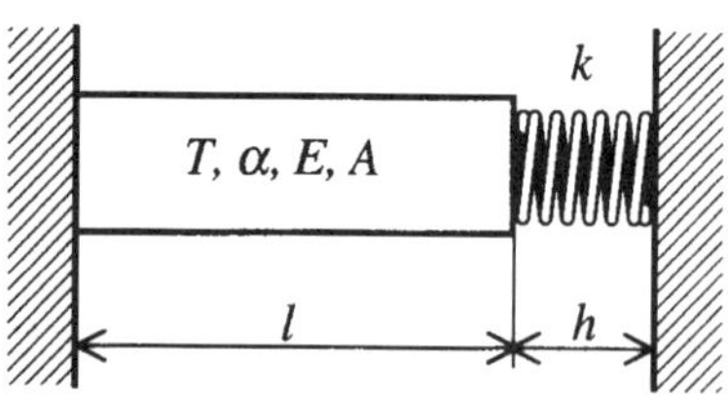

Figure 1.13 A bar attached to the spring.

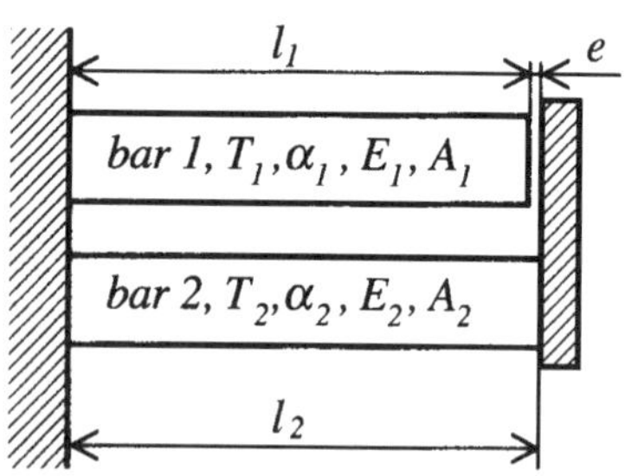

Figure 1.14 Two parallel bars attached to walls.

Example 1.8 Two parallel bars are attached to a fixed wall at one side and at the other side the end of bar 1 is a small distance e from the rigid but movable plate, while bar 2 is attached to that plate. The bars, shown in Fig. 1.14, are subjected to different temperature changes T_i. Calculate the resulting thermal stresses.

Solution Assume the free thermal expansion in bar 1 to be larger than the free thermal expansion in bar 2. After the right end of bar 1 makes contact with the rigid plate, thermal stresses are produced. The elongations of bar 1 and bar 2 are, respectively,

$$\alpha_1 T_1 l_1 + \frac{\sigma_1}{E_1} l_1, \qquad \alpha_2 T_2 l_2 + \frac{\sigma_2}{E_2} l_2 \tag{a}$$

The elongation of bar 1 is larger by the distance of the gap e, where $e = l_2 - l_1$, compared to the elongation of bar 2

$$\alpha_1 T_1 l_1 + \frac{\sigma_1}{E_1} l_1 = \alpha_2 T_2 l_2 + \frac{\sigma_2}{E_2} l_2 + e \tag{b}$$

The internal forces in bar 1 and in bar 2 are, respectively,

$$\sigma_1 A_1, \qquad \sigma_2 A_2 \tag{c}$$

From the equilibrium of the internal force and the external force we receive

$$\sigma_1 A_1 + \sigma_2 A_2 = 0 \tag{d}$$

By solving Eqs. (b) and (d), the thermal stresses σ_1 and σ_2 are determined to be

$$\sigma_1 = -\alpha_1 E_1 T_1 \left[\frac{1 - \dfrac{\alpha_2 T_2 l_2 + e}{\alpha_1 T_1 l_1}}{1 + \dfrac{A_1 E_1 l_2}{A_2 E_2 l_1}} \right], \qquad \sigma_2 = -\sigma_1 \frac{A_1}{A_2} \tag{1.40}$$

If the gap e is very small compared with the lengths of the bars, then the length of each bar is approximately the same, $l_1 \cong l_2 \equiv l$. In this case, Eqs. (1.40) reduce to

$$\sigma_1 = -\alpha_1 E_1 T_1 \left[\frac{1 - \dfrac{\alpha_2 T_2 l + e}{\alpha_1 T_1 l}}{1 + \dfrac{A_1 E_1}{A_2 E_2}} \right], \qquad \sigma_2 = -\sigma_1 \frac{A_1}{A_2} \tag{1.41}$$

Example 1.9 There are two parallel bars attached to a third bar in series, see Fig. 1.15. The ends of the bars are completely restrained by fixed walls. If the bars are subjected to different temperature changes, T_i, calculate the thermal stresses produced in the bars.

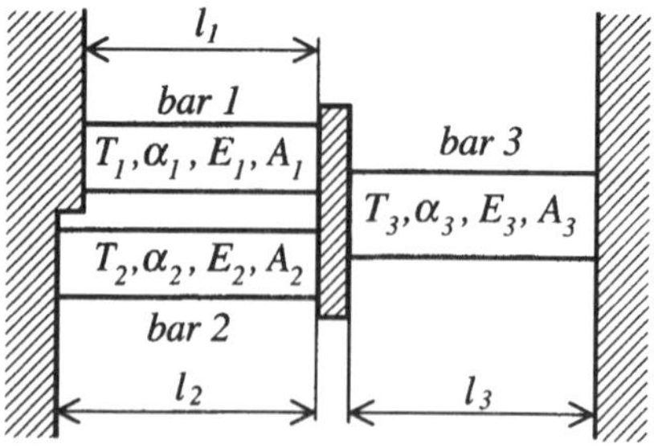

Figure 1.15 Two parallel bars attached to the third bar.

Solution The elongations of the bars are

$$\alpha_i T_i l_i + \frac{\sigma_i}{E_i} l_i \qquad (i = 1, 2, 3) \tag{a}$$

The elongation of bar 1 equals the elongation of bar 2

$$\alpha_1 T_1 l_1 + \frac{\sigma_1}{E_1} l_1 = \alpha_2 T_2 l_2 + \frac{\sigma_2}{E_2} l_2 \tag{b}$$

and the combined elongations of bar 1 (or bar 2) and bar 3 are zero

$$\left(\alpha_1 T_1 l_1 + \frac{\sigma_1}{E_1} l_1\right) + \left(\alpha_3 T_3 l_3 + \frac{\sigma_3}{E_3} l_3\right) = 0 \tag{c}$$

From the equilibrium of forces produced in each bar, it follows that

$$\sigma_1 A_1 + \sigma_2 A_2 = \sigma_3 A_3 \tag{d}$$

By solving Eqs. (b), (c), and (d), the following thermal stresses are obtained

$$\begin{aligned}
\sigma_1 &= -\alpha_1 E_1 T_1 + \left[\frac{\alpha_1 E_1 T_1 + \alpha_2 E_2 T_2 A_2/A_1 - \alpha_3 E_3 T_3 A_3/A_1}{1 + l_1 A_2 E_2/l_2 A_1 E_1 + l_1 A_3 E_3/l_3 A_1 E_1}\right] \\
\sigma_2 &= -\alpha_2 E_2 T_2 + \left[\frac{\alpha_2 E_2 T_2 + \alpha_1 E_1 T_1 A_1/A_2 - \alpha_3 E_3 T_3 A_3/A_2}{1 + l_2 A_1 E_1/l_1 A_2 E_2 + l_2 A_3 E_3/l_3 A_2 E_2}\right] \\
\sigma_3 &= -\alpha_3 E_3 T_3 + \left[\frac{\alpha_3 E_3 T_3 - \alpha_1 E_1 T_1 A_1/A_3 - \alpha_2 E_2 T_2 A_2/A_3}{1 + l_3 A_1 E_1/l_1 A_3 E_3 + l_3 A_2 E_2/l_2 A_3 E_3}\right]
\end{aligned} \tag{1.42}$$

Example 1.10 Two parallel bars are attached to a spring in series, see Fig. 1.16. The walls to which the bars and the spring are attached, are fixed. If the bars are subjected to different temperature changes T_i, calculate the thermal stresses produced in the bars.

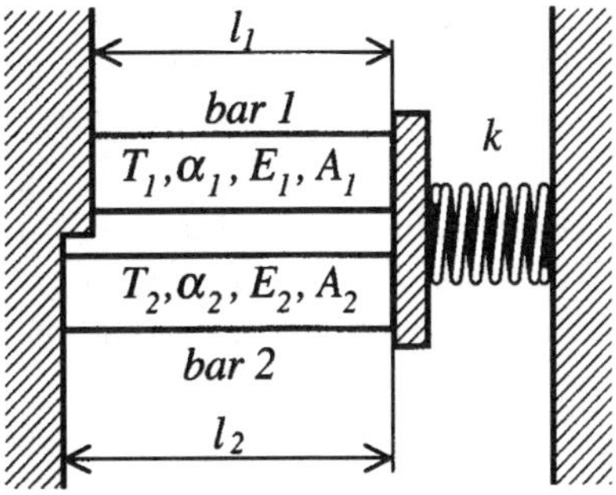

Figure 1.16 Two parallel bars attached to a spring.

Solution The elongations of the bars are

$$\alpha_i T_i l_i + \frac{\sigma_i}{E_i} l_i \qquad (i = 1, 2) \tag{a}$$

The elongation of bar 1 is equal to the elongation of bar 2

$$\alpha_1 T_1 l_1 + \frac{\sigma_1}{E_1} l_1 = \alpha_2 T_2 l_2 + \frac{\sigma_2}{E_2} l_2 \tag{b}$$

and the combined elongation of bar 1 (or bar 2) and the elongation λ_3 of the spring is zero

$$\left(\alpha_1 T_1 l_1 + \frac{\sigma_1}{E_1} l_1\right) + \lambda_3 = 0 \tag{c}$$

while the equilibrium condition of the forces produced in each bar and the spring requires that

$$\sigma_1 A_1 + \sigma_2 A_2 = k\lambda_3 \tag{d}$$

where k is the spring constant. By solving Eqs. (b), (c), and (d) the thermal stresses and elongation λ_3 of the spring are obtained

$$\begin{aligned}
\sigma_1 &= -\alpha_1 E_1 T_1 \left[\frac{1 - \alpha_2 T_2 l_2/\alpha_1 T_1 l_1 + l_2 k/A_2 E_2}{1 + l_2 A_1 E_1/l_1 A_2 E_2 + l_2 k/A_2 E_2}\right] \\
\sigma_2 &= -\alpha_2 E_2 T_2 \left[\frac{1 - \alpha_1 T_1 l_1/\alpha_2 T_2 l_2 + l_1 k/A_1 E_1}{1 + l_1 A_2 E_2/l_2 A_1 E_1 + l_1 k/A_1 E_1}\right] \\
\lambda_3 &= -\alpha_1 T_1 l_1 \left[\frac{1 + \alpha_2 T_2 A_2 E_2/\alpha_1 T_1 A_1 E_1}{1 + l_1 A_2 E_2/l_2 A_1 E_1 + l_1 k/A_1 E_1}\right]
\end{aligned} \tag{1.43}$$

1.4 Thermal Stresses in Bars Under Bending

When two parallel bars that are clamped to a rigid plate are subjected to different temperature changes T_i (Fig. 1.17), the bars may elongate and bend. Consider the thermal stresses in the bar with elongation and bending. The coordinate system is shown in Fig. 1.17 and e_i denotes the distance from $y = 0$ to the centroid of the cross section of each bar, and ρ is the radius of curvature. If the stresses in each bar are constant, the final lengths of bar 1 and bar 2 are, respectively,

$$l + \alpha_1 T_1 l + \frac{\sigma_1}{E_1} l = (\rho - e_1)\theta, \qquad l + \alpha_2 T_2 l + \frac{\sigma_2}{E_2} l = (\rho + e_2)\theta \tag{a}$$

As external forces and bending moments do not act on the bars, equilibrium of forces and of moments requires that

$$\sigma_1 A_1 + \sigma_2 A_2 = 0 \tag{b}$$

$$\sigma_1 A_1 e_1 - \sigma_2 A_2 e_2 = 0 \tag{c}$$

From Eqs. (a), (b), and (c) it follows that the thermal stresses become zero,

$$\sigma_1 = \sigma_2 = 0 \tag{1.44}$$

while the angle of rotation θ of the plate and the radius of curvature ρ are

$$\theta = (\alpha_2 T_2 - \alpha_1 T_1) l/(e_1 + e_2) \tag{1.45}$$

$$\rho = \frac{e_1 + e_2 + \alpha_1 T_1 e_2 + \alpha_2 T_2 e_1}{\alpha_2 T_2 - \alpha_1 T_1} \tag{1.46}$$

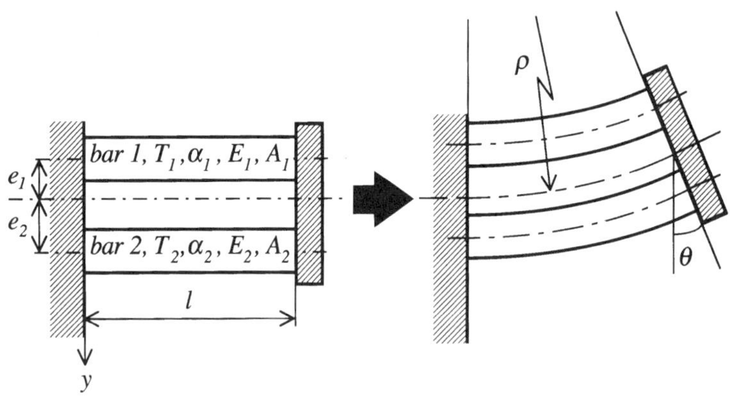

Figure 1.17 Two parallel bars attached to walls.

The radii of curvature ρ_i at the centroidal axis of the cross section of each bar and the elongations λ_i are calculated to be

$$\rho_i = \frac{(e_1 + e_2)(1 + \alpha_i T_i)}{\alpha_2 T_2 - \alpha_1 T_1} \tag{1.47}$$

$$\lambda_i = \alpha_i T_i l \tag{1.48}$$

As was to be expected, because the two bars may freely elongate and bend, thermal stresses are not produced in the bars.

Example 1.11 In the foregoing problem calculate the thermal stresses if the elongation of the bars is restrained while bending is not restrained.

Solution Since the elongation is restrained, we have $l = \rho\theta$, and Eqs. (a) in Section 1.4 yield the following relations:

$$\alpha_1 T_1 l + \frac{\sigma_1}{E_1} l = -e_1\theta, \qquad \alpha_2 T_2 l + \frac{\sigma_2}{E_2} l = e_2\theta \tag{a}$$

The bars may bend freely, which means that there is no bending moment in the bars

$$\sigma_1 A_1 e_1 - \sigma_2 A_2 e_2 = 0 \tag{b}$$

By solving Eqs. (a) and (b) the stresses σ_1 and σ_2 and the angle of rotation θ of the bar are

$$\sigma_1 = -\alpha_1 E_1 T_1 \left[\frac{1 + \alpha_2 T_2 e_1/\alpha_1 T_1 e_2}{1 + E_1 A_1 e_1^2/E_2 A_2 e_2^2}\right], \qquad \sigma_2 = \frac{e_1 A_1}{e_2 A_2}\sigma_1 \tag{1.49}$$

$$\theta = \alpha_1 T_1 \left[\frac{l(\alpha_2 T_2 E_2 A_2 e_2/\alpha_1 T_1 E_1 A_1 e_1 - 1)}{e_1(1 + E_2 A_2 e_2^2/E_1 A_1 e_1^2)}\right] \tag{1.50}$$

while $\rho = l/\theta$.

Problems

1.1 If the temperature in a mild steel rail with length 25 m is raised by 50 K, and the coefficient of linear thermal expansion for mild steel is 11.2×10^{-6} 1/K, what elongation is produced in the rail?

Answer 14 mm

1.2 A bar of length 1 m of mild steel is kept at 300 K. If the temperature at one end of the bar is raised to 380 K and at the other end to 480 K, and the temperature distribution is linear along the bar, what elongation is produced in the bar? The coefficient of linear thermal expansion for mild steel is 11.2×10^{-6} 1/K.

Answer 1.46 mm

1.3 A bar of mild steel at 300 K is clamped between two walls. Calculate the thermal stress produced in the bar when the bar is heated to 360 K. See Table 1.1 for material properties.

Answer −138 MPa

1.4 A bar of mild steel at 300 K is clamped between two walls in such a way that the initial stress is approximately zero. At what temperature will the thermal stress in the bar reach the compressive strength ($\sigma_{BC} = 400$ MPa)? See Table 1.1 for material properties.

Answer 473 K

1.5 In Problem 1.2, calculate the thermal stress produced in the bar if it is clamped between two walls. See Table 1.1 for material properties.

Answer −300 MPa

1.6 If a clamped circular frustum of mild steel with $d_0 = 1$ cm, $d_1 = 2$ cm, and $l = 2$ m in Fig. 1.6 is subjected to a temperature change −50 K, calculate the resulting thermal stress.

Answer

$$\sigma_x = 231 \Big/ \left(1 + \frac{x}{2}\right)^2 \text{ MPa}, \qquad (\sigma_x)_{max} = 231 \text{ MPa},$$

$$(\sigma_x)_{min} = 57.8 \text{ MPa}$$

1.7 If the temperature in the circular frustum in Problem 1.6 changes linearly from 0 K at one end to −50 K at the other end, calculate the resulting thermal stresses.

Answer

$$115\Big/\left(1+\frac{x}{2}\right)^2 \text{ MPa}, \qquad (\sigma_x)_{max} = 115\,\text{MPa},$$

$$(\sigma_x)_{min} = 28.8\,\text{MPa}$$

1.8 If a bar with a small gap e between its free end and a rigid wall shown in Fig. 1.11 is subjected to a positive temperature change $T(x)$, calculate the resulting thermal stress.

Answer

$$\sigma = -\frac{E}{l}\left[\alpha\int_0^l T(x)\,dx - e\right]$$

1.9 If a bar with a small gap e between its free end and a rigid wall shown in Fig. 1.11 is subjected to the positive temperature change $T(x)$, and the cross-sectional area of the bar is given by $A(x)$, what thermal stress is produced in the bar?

Answer

$$\sigma_x = -\frac{E}{A(x)\displaystyle\int_0^l \frac{1}{A(x)}\,dx}\left[\alpha\int_0^l T(x)\,dx - e\right]$$

$$(\sigma_x)_{max} = -\frac{E}{[A(x)]_{min}\displaystyle\int_0^l \frac{1}{A(x)}\,dx}\left[\alpha\int_0^l T(x)\,dx - e\right]$$

$$(\sigma_x)_{min} = -\frac{E}{[A(x)]_{max}\displaystyle\int_0^l \frac{1}{A(x)}\,dx}\left[\alpha\int_0^l T(x)\,dx - e\right]$$

1.10 Two circular bars, one of mild steel of length 50 cm and diameter 1 cm, and the other of aluminum of length 25 cm and diameter 2 cm, are attached to each other in series, placed between rigid walls, and subjected to different temperature changes: 50 K for the bar of mild steel and 100 K for the bar of aluminum, as shown in Fig. 1.12. Calculate the thermal stresses produced in each bar.

Answer $\sigma_s = -260\,\text{MPa}$, $\sigma_a = -65\,\text{MPa}$

1.11 Two circular bars, one of mild steel of length 50 cm and diameter 1 cm, and the other of aluminum of length 25 cm and diameter 2 cm, are attached to each other in series, placed between rigid walls, and subjected to a temperature change T, as shown in Fig. 1.12. What temperature rise is needed for the thermal stresses in the bars to reach the compressive strength? The compressive strengths for mild steel and for aluminum are 400 MPa and 70 MPa, respectively.

Answer $T = 81\,\mathrm{K}$

1.12 A bar of mild steel of cross-sectional area A_s is placed between two parallel bars of copper of cross-sectional area A_c. All three bars are of the same length l. If the bars are bonded together and are then subjected to a temperature change of T_s in the bar of mild steel and T_c in the bar of copper, calculate the thermal stresses produced in each bar.

Answer

$$\sigma_s = E_s\left[\frac{\alpha_c T_c - \alpha_s T_s}{1 + A_s E_s/2A_c E_c}\right], \qquad \sigma_c = -\sigma_s \frac{A_s}{2A_c}$$

1.13 A copper tube is fastened by a mild steel bolt, as shown in Fig. 1.18. The length of the tube is 50 cm, and the cross-sectional areas of the bolt and the tube are 1 cm^2 and 2 cm^2, respectively. Calculate the thermal stresses produced if the system is subjected to the temperature change of 80 K.

Answer $\sigma_s = 47.0\,\mathrm{MPa}$, $\sigma_c = -23.5\,\mathrm{MPa}$

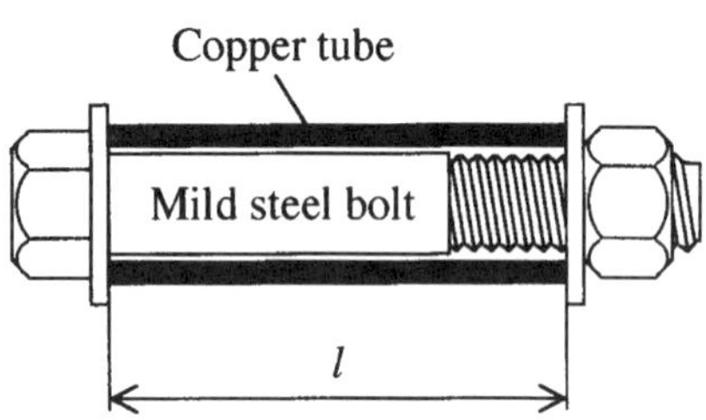

Figure 1.18 A copper tube fastened by a mild steel bolt.

1.14 A copper tube is fastened by a mild steel bolt, as shown in Fig. 1.18. The length of the tube is 50 cm, and the cross-sectional areas of the bolt and the tube are 1 cm^2 and 2 cm^2, respectively. What is the maximum tolerable temperature rise such that stresses in the system do not exceed the compressive or the tensile strength? We assume the compressive strength has the same magnitude as the tensile strength, and the tensile strengths for mild steel and for copper are 400 MPa and 300 MPa, respectively. The safety factor (defined by the ratio of yield stress or the tensile strength and the allowable stress) is $f = 3$.

Answer $T = 227\,\mathrm{K}$

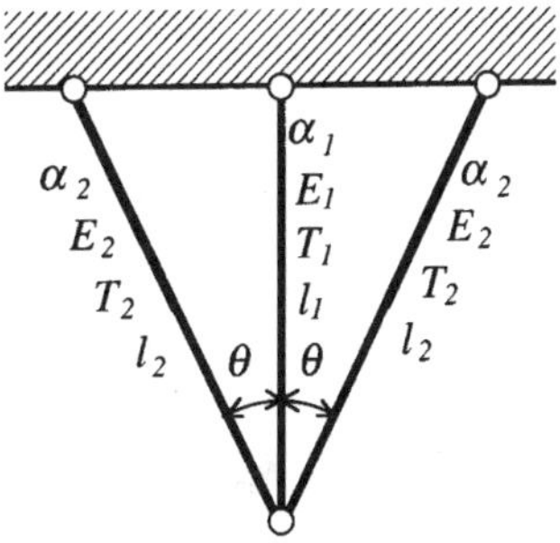

Figure 1.19 Bars of the truss.

1.15 Calculate the thermal stresses produced in the bars of the truss shown in Fig. 1.19 if the temperature changes of the bars are T_i.

Answer

$$\sigma_1 = -\alpha_1 E_1 T_1 \left[\frac{1 - \alpha_2 T_2/\alpha_1 T_1 \cos^2\theta}{1 + A_1 E_1/2A_2 E_2 \cos^3\theta} \right], \qquad \sigma_2 = -\sigma_1 \frac{A_1}{2A_2 \cos\theta}$$

Chapter 2

THERMAL STRESSES IN BEAMS

In this chapter, the Bernoulli-Euler hypothesis is used to calculate thermal stresses in beams. Thermal stresses in rectangular beams are presented for four cases: (1) built-in ends; (2) free expansion and restrained bending; (3) restrained expansion and free bending; (4) free expansion and free bending. Thermal stresses are considered in composite beams with various boundary conditions, including composite multi-layered beams and non-homogeneous beams. Solutions are developed for an arbitrary steady temperature change $T(x, y, z)$. Thermal deflections in beams subjected to a symmetrical thermal load, $T(x, y)$, are derived for both simply supported and built-in edges. Finally, solutions for stresses and strains in curved beams and in beams on an elastic foundation are presented.

2.1 Thermal Stresses in Beams

2.1.1 Stresses in Beams Subjected to Mechanical Loads

We will first discuss the strain and the stress in beams subjected not to thermal loads but to mechanical loads. When the beam is subjected to a bending moment, it will be deformed. We assume that the Bernoulli-Euler hypothesis is valid, which means that a plane section which is perpendicular to the longitudinal axis of the beam before loading remains plane and perpendicular to the axis after loading. An element ABCD changes after deformation to A′B′C′D′, Fig. 2.1. Longitudinal fibers on the convex side are extended and the fibers on the concave side are compressed so that there exists a plane in which the fibers do not undergo the elongation in the beam. This plane is called the neutral plane (plane $m'n'$), and the intersection between this plane and any cross section

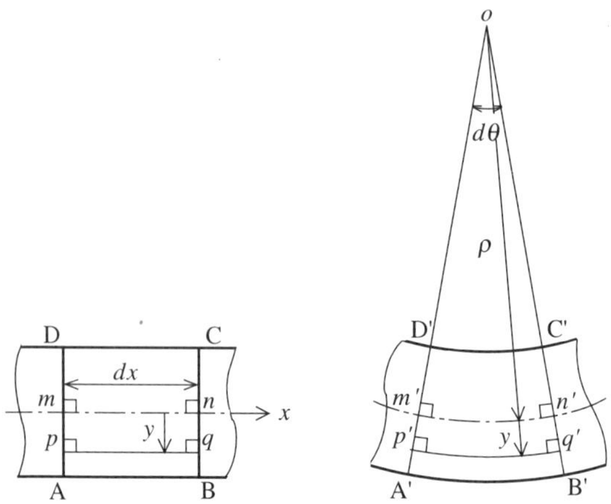

Figure 2.1 Deformation of an element subjected to bending.

is called the neutral axis. Any fibers pq at distance y from the neutral plane in the element change to the fibers $p'q'$. The bending strain ϵ at the distance y from the neutral plane after bending is

$$\epsilon = \frac{\overset{\frown}{p'q'} - \overline{pq}}{\overline{pq}} = \frac{\overset{\frown}{p'q'} - \overline{mn}}{\overline{mn}} = \frac{\overset{\frown}{p'q'} - \overset{\frown}{m'n'}}{\overset{\frown}{m'n'}} = \frac{(\rho + y)\,d\theta - \rho\,d\theta}{\rho\,d\theta} = \frac{y}{\rho} \tag{2.1}$$

where ρ denotes the radius of curvature at the neutral plane. From Hooke's law (1.8), the normal stress σ is

$$\sigma = E\epsilon = E\frac{y}{\rho} \tag{a}$$

The normal stress σ is called the bending stress. The distribution of the bending stress is proportional to the distance y from the neutral plane. If the axial force is not applied to the beam, the resultant force due to the bending stress must be zero

$$\int_A \sigma\,dA = 0 \tag{b}$$

where dA denotes a small elemental area of cross section at distance y from the neutral plane. When the material is homogeneous in the cross section, the Young's modulus is independent of the integral with respect to the area, so that after substitution of Eq. (a), the integral (b) reduces to

$$\int_A \sigma\,dA = \int_A E\epsilon\,dA = \frac{E}{\rho}\int_A y\,dA = 0 \tag{c}$$

We find that the neutral axis passes through the centroid of the cross section.

The resultant moment due to the bending stress with respect to the neutral axis equals the bending moment M applied to the beam, so that

$$M = \int_A y\sigma\, dA = \int_A E\epsilon y\, dA = \frac{E}{\rho}\int_A y^2\, dA \tag{d}$$

or

$$\frac{1}{\rho} = \frac{M}{EI} \tag{2.2}$$

where I is called the moment of inertia of the cross section with respect to the neutral axis and is defined by

$$I = \int_A y^2\, dA \tag{2.3}$$

Eliminating the radius of curvature ρ from Eqs. (a) and (2.2), the bending stress is expressed by

$$\sigma = \frac{My}{I} \tag{2.4}$$

When the bending moment M is known in the beam, the bending stress σ is received from Eq. (2.4).

2.1.2 Thermal Stresses in Clamped Beams

There is given a beam of length l and a rectangular cross section $b \times h$ whose both ends are clamped at the rigid walls shown in Fig. 2.2 (a). If the initial temperature of the beam is constant, let us consider the thermal stress in the beam after the temperature rises by

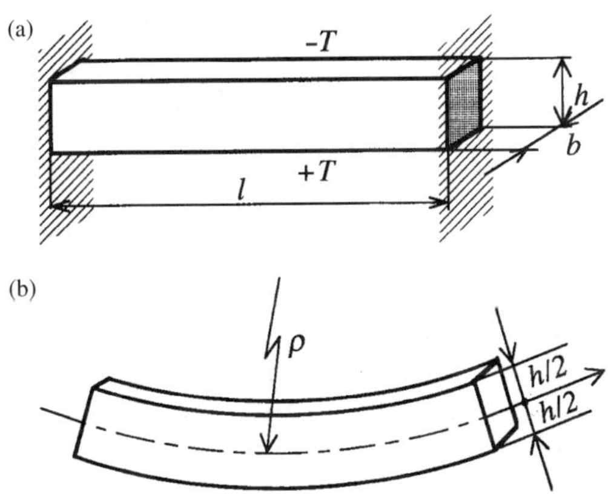

Figure 2.2 A beam of rectangular cross section clamped at rigid walls.

T at the lower surface and decreases by T at the upper surface of the beam. As the beam is clamped at both ends, the length of the beam does not change after the temperature change. If the beam is not clamped, both ends are free and the beam may deform to a circular arc shown in Fig. 2.2 (b). Comparing the length of the lower surface and of the neutral surface, we have

$$\frac{\left(\rho + \frac{h}{2}\right) d\theta}{\rho\, d\theta} = \frac{l + \alpha T l}{l}$$

Then,

$$\frac{1}{\rho} = \frac{\alpha T}{\frac{h}{2}} \tag{a}$$

As the beam is perfectly clamped and is kept straight, the curvature $1/\rho$ due to a free expansion must be eliminated by a virtual bending moment M. The beam will not bend when such a virtual bending moment M is applied. Eliminating ρ from Eqs. (a) and (2.2), the virtual bending moment M is

$$M = -\alpha E T \frac{I}{h/2} \tag{b}$$

Substitution of Eq. (b) into Eq. (2.4) gives the thermal stress

$$\sigma = -\alpha E T \frac{y}{h/2} \tag{2.5}$$

The thermal stress is expressed as a linear function of the distance y from the neutral axis. The thermal stress at the lower surface, which was subjected to the temperature rise T, is negative and is $\sigma = -\alpha E T$. On the other hand, the thermal stress at the upper surface, which is subjected to a temperature decrease T, is positive and is $\sigma = \alpha E T$. The stress is zero at the neutral plane $y = 0$ since the temperature at the axis of the beam has not changed and, therefore, the length of the beam at the axis remained constant.

2.1.3 Thermal Stresses in Rectangular Beams

Consider a beam of rectangular cross section $b \times h$ with no restraints at both end surfaces. We wish to determine the thermal stress in the beam subjected to any temperature change $T(y)$ along the y direction, as shown in Fig. 2.3. This means that the temperature changes from the initial temperature T_0 to the final temperature $T_1(y)$ or $T(y) = T_1(y) - T_0$. If both ends of the beam are perfectly clamped, the thermal stress σ_{x1} in the beam is

$$\sigma_{x1} = -\alpha E T(y) \tag{2.6}$$

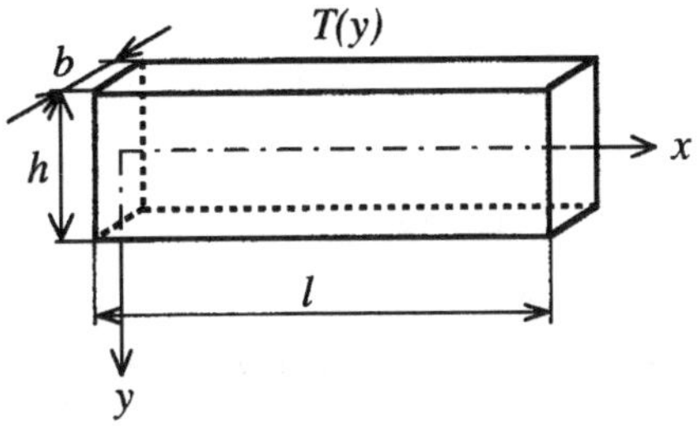

Figure 2.3 A beam of rectangular cross section with no restraint.

This stress prevents both extension of the beam and bending of the beam. Since both ends are clamped, the thermal stress (2.6) produces an internal force P

$$P = \int_{-h/2}^{h/2} \sigma_{x1} b\, dy = -\int_{-h/2}^{h/2} \alpha E T(y) b\, dy \tag{a}$$

If the beam has no restraints against extension, this internal force P at the ends must be eliminated by a virtual force P_T at the ends:

$$P_T = -P = \int_{-h/2}^{h/2} \alpha E T(y) b\, dy \tag{2.7}$$

The thermal stress corresponding to the virtual force P_T is

$$\sigma_{xP} = \frac{P_T}{A} = \frac{1}{A}\int_{-h/2}^{h/2} \alpha E T(y) b\, dy = \frac{1}{h}\int_{-h/2}^{h/2} \alpha E T(y)\, dy \tag{b}$$

Therefore, the summation of Eqs. (2.6) and (b) gives the thermal stress σ_{x2} for the case where the beam may freely extend in the longitudinal direction but is restrained from bending:

$$\sigma_{x2} = -\alpha E T(y) + \frac{1}{h}\int_{-h/2}^{h/2} \alpha E T(y)\, dy \tag{2.8}$$

From the thermal stress σ_{x1}, the bending moment M in the beam is

$$M = \int_{-h/2}^{h/2} \sigma_{x1} y b\, dy = -\int_{-h/2}^{h/2} \alpha E T(y) y b\, dy \tag{c}$$

If the ends of the beam are free to rotate, and they are not subjected to moment M, the bending moment M at the ends must be eliminated by a virtual bending moment M_T at the ends:

$$M_T = -M = \int_{-h/2}^{h/2} \alpha E T(y) y b\, dy \tag{2.9}$$

Substituting Eq. (2.9) into Eq. (2.4), we obtain the thermal bending stress σ_{xM} produced by the virtual bending moment M_T:

$$\sigma_{xM} = \frac{M_T y}{I} = \frac{12y}{h^3}\int_{-h/2}^{h/2} \alpha E T(y) y\, dy \tag{d}$$

where the moment of inertia I for the rectangular cross section $b \times h$ is given by $I = bh^3/12$. The thermal stress σ_{x3} in the beam with free bending and with the restrained extension in the longitudinal direction can be determined by the summation of Eqs. (2.6) and (d)

$$\sigma_{x3} = -\alpha E T(y) + \frac{12y}{h^3}\int_{-h/2}^{h/2} \alpha E T(y) y\, dy \tag{2.10}$$

Therefore, the thermal stress σ_x in the unclamped beam which is free from external force is given by

$$\sigma_x = -\alpha E T(y) + \frac{1}{h}\int_{-h/2}^{h/2} \alpha E T(y)\, dy + \frac{12y}{h^3}\int_{-h/2}^{h/2} \alpha E T(y) y\, dy \tag{2.11}$$

Thus, we have derived equations for thermal stresses in a beam subjected to a temperature change $T(y)$ corresponding to various boundary conditions at the ends.

[1]	Beam with perfectly clamped ends	Eq. (2.6)
[2]	Beam with free extension and restrained bending	Eq. (2.8)
[3]	Beam with restrained extension and free bending	Eq. (2.10)
[4]	Beam with free extension and free bending	Eq. (2.11)

Here, numerical results for [1] to [4] are shown in Fig. 2.4. The material of the beam is mild steel and thickness h is 1 cm. Constants α and E for mild steel are given in Table 1.1. Figure 2.4 (b) shows the distribution of thermal stress for the prescribed linear temperature shown in Fig. 2.4 (a), and Fig. 2.4 (d) shows the distribution of thermal stress for the prescribed quadratic temperature shown in Fig. 2.4 (c).

2.2 General Technique for Thermal Stresses in Beams

We will now discuss the strain and stress in beams subjected to thermal loads. When the beam is subjected to thermal loads, it will be deformed. The deformation will consist of axial elongation and of bending. We assume that the Bernoulli-Euler assumption is valid. An element ABCD may be deformed to A′B′C′D′ by thermal loads shown in Fig. 2.1. We take the neutral axis to pass through the centroid of the cross section of the beam. A fiber pq at a distance y from the neutral axis in the element elongates to $p'q'$. As a fiber mn at the neutral axis elongates to $m'n'$ due to thermal loads, the strain ϵ at the

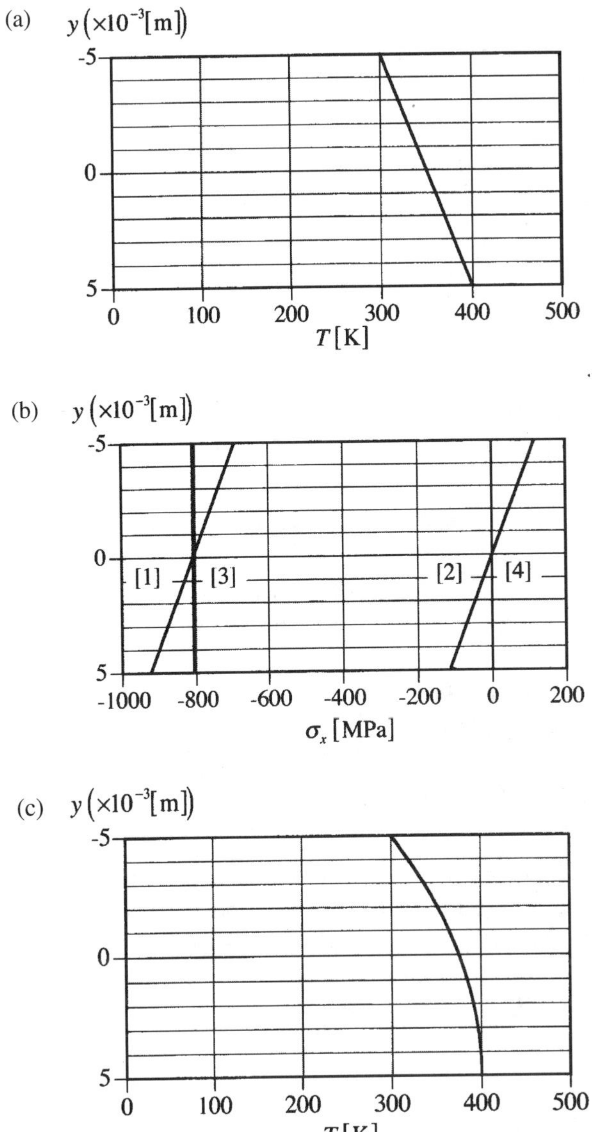

Figure 2.4 Temperatures and thermal stresses in a beam of rectangular cross section: (a) linear temperature distribution; (b) thermal stresses due to temperature shown in (a); (c) quadratic temperature distribution; and (d) thermal stresses due to temperature shown in (c).

(d)

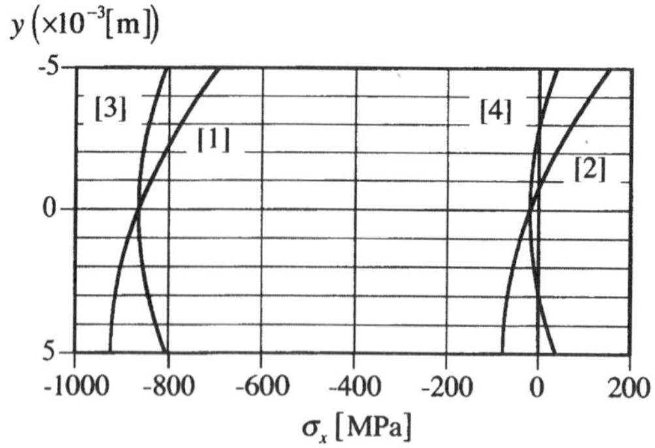

Figure 2.4 continued.

distance y from the neutral axis after thermal loads are applied is

$$\epsilon = \frac{\widehat{p'q'} - \overline{pq}}{\overline{pq}} = \frac{\widehat{p'q'} - \overline{mn}}{\overline{mn}} = \frac{\left(\widehat{p'q'} - \widehat{m'n'}\right) + \left(\widehat{m'n'} - \overline{mn}\right)}{\overline{mn}}$$

$$= \frac{\widehat{m'n'} - \overline{mn}}{\overline{mn}} + \frac{\widehat{p'q'} - \widehat{m'n'}}{\overline{mn}} = \epsilon_0 + \frac{(\rho + y)\,d\theta - \rho\,d\theta}{\rho\,d\theta}\frac{\rho\,d\theta}{\overline{mn}}$$

$$= \epsilon_0 + \frac{y}{\rho}\left(1 + \frac{\rho\,d\theta - \overline{mn}}{\overline{mn}}\right) = \epsilon_0 + \frac{y}{\rho}\left(1 + \frac{\widehat{m'n'} - \overline{mn}}{\overline{mn}}\right)$$

$$= \epsilon_0 + \frac{y}{\rho}(1 + \epsilon_0) \cong \epsilon_0 + \frac{y}{\rho} \tag{2.12}$$

where ρ denotes, as before, the radius of curvature at the neutral plane and ϵ_0 denotes the axial strain at the neutral plane. The strain ϵ at a distance y from the neutral axis in the beam is expressed as the sum of free thermal strain and the strain due to stress σ_x:

$$\epsilon = \alpha T + \frac{\sigma_x}{E} = \epsilon_0 + \frac{y}{\rho} \tag{2.13}$$

Solving this equation for σ_x, we find

$$\sigma_x = -\alpha E T + \epsilon_0 E + E\frac{y}{\rho} \tag{2.14}$$

Since the beam is free from external forces, the external force and bending moment must each be zero:

$$\int_A \sigma_x\,dA = 0 \tag{a}$$

$$\int_A \sigma_x y\,dA = 0 \tag{b}$$

where dA denotes a small element area of the cross section at a distance y from the neutral plane. Substitution of Eq. (2.14) into Eqs. (a) and (b) gives the axial strain ϵ_0 and the curvature $1/\rho$ at the neutral plane $y = 0$ as follows:

$$\epsilon_0 = \frac{1}{EA}\int_A \alpha E T(y)\, dA \tag{c}$$

$$\frac{1}{\rho} = \frac{1}{EI}\int_A \alpha E T(y) y\, dA \tag{d}$$

where I denotes the moment of inertia of the cross section, and the moment of the area of the cross section $\int_A y\, dA$ is zero because the neutral plane passes through the centroid of the cross section. Then, the substitution of Eqs. (c) and (d) into Eqs. (2.14) gives the thermal stress

$$\sigma_x(y) = -\alpha E T(y) + \frac{1}{A}\int_A \alpha E T(y)\, dA + \frac{y}{I}\int_A \alpha E T(y) y\, dA \tag{2.15}$$

Equation (2.15) gives the general solution for the thermal stress in a beam under thermal loads. For the beam with rectangular cross section of the height h and the width b, the areas dA and A, and the moment of inertia I are

$$dA = b\, dy, \qquad A = bh, \qquad I = \frac{bh^3}{12} \tag{e}$$

Thus, from Eqs. (2.15) and (e), the thermal stress σ_x for the beam with the rectangular cross section is

$$\sigma_x(y) = -\alpha E T(y) + \frac{1}{h}\int_{-h/2}^{h/2} \alpha E T(y)\, dy + \frac{12y}{h^3}\int_{-h/2}^{h/2} \alpha E T(y) y\, dy \tag{2.16}$$

This result agrees with Eq. (2.11).

Example 2.1 Calculate the thermal stress σ_x in a beam of a circular cross section with radius a when the beam is subjected to a temperature change $T(y)$.

Solution For the beam of circular cross section with radius a, a small area element dA, cross section area A, and the moment of inertia I are, respectively,

$$dA = 2\sqrt{a^2 - y^2}\, dy, \qquad A = \pi a^2, \qquad I = \frac{\pi}{4}a^4 \tag{a}$$

Substitution of Eqs. (a) into Eq. (2.15) gives

$$\sigma_x = -\alpha E T(y) + \frac{2}{\pi a^2}\int_A \alpha E T(y)\sqrt{a^2 - y^2}\, dy + \frac{8y}{\pi a^4}\int_A \alpha E T(y) y\sqrt{a^2 - y^2}\, dy \tag{b}$$

Example 2.2 Calculate the thermal stress σ_x in a beam of a circular cross section with radius a, when the beam is subjected to a temperature change $T(y) = C_1 y + C_0$.

Solution The thermal stress σ_x in the beam of the circular cross section with radius a is given by

$$\sigma_x = -\alpha E T(y) + \frac{2}{\pi a^2}\int_A \alpha E T(y)\sqrt{a^2 - y^2}\, dy + \frac{8y}{\pi a^4}\int_A \alpha E T(y) y \sqrt{a^2 - y^2}\, dy \tag{a}$$

We change the variable y to θ as follows:

$$y = a\cos\theta \tag{b}$$

Therefore,

$$\sqrt{a^2 - y^2} = a\sin\theta, \qquad dy = -a\sin\theta\, d\theta, \qquad T(y) = C_1 y + C_0 = C_1 a\cos\theta + C_0 \tag{c}$$

From Eqs. (a), (b), and (c), the thermal stress σ_x is

$$\sigma_x = -\alpha E(C_1 a\cos\theta + C_0) - \frac{2}{\pi a^2}\int_\pi^0 \alpha E(C_1 a\cos\theta + C_0) a^2 \sin^2\theta\, d\theta - \frac{8a\cos\theta}{\pi a^4}\int_\pi^0 \alpha E(C_1 a\cos\theta + C_0) a^3 \cos\theta\sin^2\theta\, d\theta \tag{d}$$

However,

$$\int_0^\pi \sin^2\theta\, d\theta = \frac{1}{2}\int_0^\pi (1 - \cos 2\theta)\, d\theta = \frac{1}{2}\left[\theta - \frac{1}{2}\sin 2\theta\right]_0^\pi = \frac{\pi}{2}$$

$$\int_0^\pi \cos\theta\sin^2\theta\, d\theta = \frac{1}{3}[\sin^3\theta]_0^\pi = 0$$

$$\int_0^\pi \cos^2\theta\sin^2\theta\, d\theta = \frac{1}{4}\int_0^\pi \sin^2 2\theta\, d\theta = \frac{1}{8}\int_0^\pi (1 - \cos 4\theta)\, d\theta = \frac{1}{8}\left[\theta - \frac{1}{4}\sin 4\theta\right]_0^\pi = \frac{\pi}{8}$$

Using these relations, Eq. (d) reduces to

$$\sigma_x = -\alpha E(C_1 a\cos\theta + C_0) + \frac{2}{\pi a^2}\frac{\pi}{2} C_0 \alpha E a^2 + \frac{8a\cos\theta}{\pi a^4}\frac{\pi}{8} C_1 \alpha E a^4 = 0 \tag{e}$$

The thermal stress is not produced in the beam.

2.3 Thermal Stresses in Composite Beams

2.3.1 Thermal Stresses in a Two-Layered Beam

We consider a beam consisting of two layers bonded together and subjected to temperature changes $T_i(y)$ shown in Fig. 2.5. Let the subscript i $(i = 1, 2)$ refer to the i-th layer, and b_i and h_i denote the width and height of the layers, respectively. We take the origin $(y = 0)$ at the bonding surface of the composite beam. The strains after bending are expressed by

$$\epsilon_{xi} = \alpha_i T_i(y) + \frac{\sigma_{xi}}{E_i} = \epsilon_0 + \frac{y}{\rho} \qquad (i = 1, 2) \tag{2.17}$$

where ϵ_0 and $1/\rho$ denote the strain at the bonding surface $y = 0$ and the curvature, respectively. Since external forces do not act on the beam, then

$$\int_{-h_1}^{0} \sigma_{x1} b_1 \, dy + \int_{0}^{h_2} \sigma_{x2} b_2 \, dy = 0 \tag{a}$$

$$\int_{-h_1}^{0} \sigma_{x1} y b_1 \, dy + \int_{0}^{h_2} \sigma_{x2} y b_2 \, dy = 0 \tag{b}$$

Substituting stress σ_{xi} from Eq. (2.17) into Eqs. (a) and (b), we get

$$\begin{aligned} &2(E_2 h_2 b_2 + E_1 h_1 b_1)\epsilon_0 + \left(E_2 h_2^2 b_2 - E_1 h_1^2 b_1\right)\frac{1}{\rho} \\ &\quad = 2\int_{-h_1}^{0} \alpha_1 E_1 T_1(y) b_1 \, dy + 2\int_{0}^{h_2} \alpha_2 E_2 T_2(y) b_2 \, dy \\ &3\left(E_2 h_2^2 b_2 - E_1 h_1^2 b_1\right)\epsilon_0 + 2\left(E_2 h_2^3 b_2 + E_1 h_1^3 b_1\right)\frac{1}{\rho} \\ &\quad = 6\int_{-h_1}^{0} \alpha_1 E_1 T_1(y) b_1 y \, dy + 6\int_{0}^{h_2} \alpha_2 E_2 T_2(y) b_2 y \, dy \end{aligned} \tag{c}$$

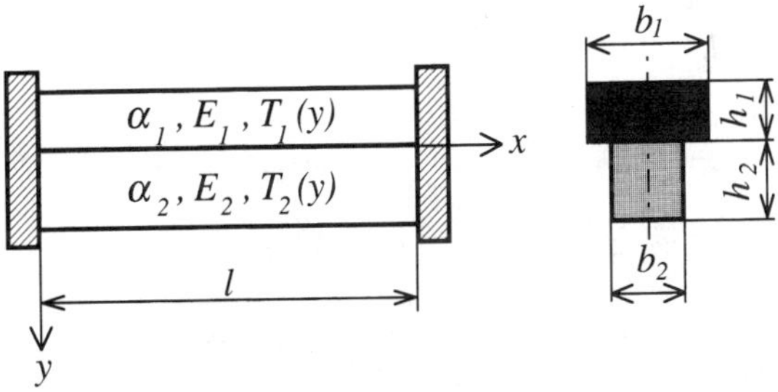

Figure 2.5 A beam consisting of two layers bonded together.

From the above equations, the strain ϵ_0 and the curvature $1/\rho$ at the bonding surface $y = 0$ is obtained

$$\epsilon_0 = \frac{2}{D}\left\{2\left[\int_{-h_1}^{0} \alpha_1 E_1 T_1(y) b_1\,dy + \int_0^{h_2} \alpha_2 E_2 T_2(y) b_2\,dy\right]\left(E_2 h_2^3 b_2 + E_1 h_1^3 b_1\right)\right.$$
$$\left. - 3\left[\int_{-h_1}^{0} \alpha_1 E_1 T_1(y) b_1 y\,dy + \int_0^{h_2} \alpha_2 E_2 T_2(y) b_2 y\,dy\right]\left(E_2 h_2^2 b_2 - E_1 h_1^2 b_1\right)\right\}$$
$$\frac{1}{\rho} = \frac{6}{D}\left\{2\left[\int_{-h_1}^{0} \alpha_1 E_1 T_1(y) b_1 y\,dy + \int_0^{h_2} \alpha_2 E_2 T_2(y) b_2 y\,dy\right](E_2 h_2 b_2 + E_1 h_1 b_1)\right.$$
$$\left. - \left[\int_{-h_1}^{0} \alpha_1 E_1 T_1(y) b_1\,dy + \int_0^{h_2} \alpha_2 E_2 T_2(y) b_2\,dy\right]\left(E_2 h_2^2 b_2 - E_1 h_1^2 b_1\right)\right\} \qquad \text{(d)}$$

where

$$D = \left(E_2 h_2^2 b_2 - E_1 h_1^2 b_1\right)^2 + 4E_1 E_2 h_1 h_2 (h_1 + h_2)^2 b_1 b_2 \qquad \text{(e)}$$

Thus, the thermal stresses in a two-layered beam are given by

$$\sigma_{xi}(y) = -\alpha_i E_i T_i(y)$$
$$+ \frac{2E_i}{D}\left\{2\left[\int_{-h_1}^{0} \alpha_1 E_1 T_1(y) b_1\,dy + \int_0^{h_2} \alpha_2 E_2 T_2(y) b_2\,dy\right]\right.$$
$$\times \left(E_2 h_2^3 b_2 + E_1 h_1^3 b_1\right)$$
$$\left. - 3\left[\int_{-h_1}^{0} \alpha_1 E_1 T_1(y) b_1 y\,dy + \int_0^{h_2} \alpha_2 E_2 T_2(y) b_2 y\,dy\right]\left(E_2 h_2^2 b_2 - E_1 h_1^2 b_1\right)\right\}$$
$$+ \frac{6E_i y}{D}\left\{2\left[\int_{-h_1}^{0} \alpha_1 E_1 T_1(y) b_1 y\,dy + \int_0^{h_2} \alpha_2 E_2 T_2(y) b_2 y\,dy\right]\right.$$
$$\times (E_2 h_2 b_2 + E_1 h_1 b_1)$$
$$\left. - \left[\int_{-h_1}^{0} \alpha_1 E_1 T_1(y) b_1\,dy + \int_0^{h_2} \alpha_2 E_2 T_2(y) b_2\,dy\right]\left(E_2 h_2^2 b_2 - E_1 h_1^2 b_1\right)\right\}$$
$$(i = 1, 2) \qquad (2.18)$$

Let us consider some special cases that can be derived from Eq. (2.18).

[1] The upper layer and the lower layer are of the same cross sections. If both cross sections of each layer are of the same dimensions

$$h_1 = h_2 \equiv h, \qquad b_1 = b_2 \equiv b \qquad \text{(f)}$$

substitution of Eq. (f) into (2.18) gives the thermal stresses

$$\begin{aligned}\sigma_{xi}(y) &= -\alpha_i E_i T_i(y)\\ &+ \frac{2E_i}{D}\left\{2\left[\int_{-h}^{0}\alpha_1 E_1 T_1(y)\,dy + \int_{0}^{h}\alpha_2 E_2 T_2(y)\,dy\right](E_2+E_1)h\right.\\ &\qquad\left. -3\left[\int_{-h}^{0}\alpha_1 E_1 T_1(y)y\,dy + \int_{0}^{h}\alpha_2 E_2 T_2(y)y\,dy\right](E_2-E_1)\right\}\\ &+ \frac{6E_i}{D}\frac{y}{h}\left\{2\left[\int_{-h}^{0}\alpha_1 E_1 T_1(y)y\,dy + \int_{0}^{h}\alpha_2 E_2 T_2(y)y\,dy\right](E_2+E_1)\right.\\ &\qquad\left. -\left[\int_{-h}^{0}\alpha_1 E_1 T_1(y)\,dy + \int_{0}^{h}\alpha_2 E_2 T_2(y)\,dy\right](E_2-E_1)h\right\}\\ &\qquad (i=1,2)\end{aligned} \tag{2.19}$$

where

$$D = \left[(E_2+E_1)^2 + 12E_1E_2\right]h^2 \tag{g}$$

[2] The temperature changes in the layers are constant but different. Taking the temperature changes as

$$T_1(y) = T_1, \qquad T_2(y) = T_2 \tag{h}$$

the thermal stresses are obtained as

$$\begin{aligned}\sigma_{xi}(y) &= -\alpha_i E_i T_i\\ &+\frac{E_i}{D}\left[4(\alpha_1 E_1 T_1 h_1 b_1 + \alpha_2 E_2 T_2 h_2 b_2)\left(E_2 h_2^3 b_2 + E_1 h_1^3 b_1\right)\right.\\ &\left.+3\left(\alpha_1 E_1 T_1 h_1^2 b_1 - \alpha_2 E_2 T_2 h_2^2 b_2\right)\left(E_2 h_2^2 b_2 - E_1 h_1^2 b_1\right)\right]\\ &-\frac{6E_i y}{D}\left[\left(\alpha_1 E_1 T_1 h_1^2 b_1 - \alpha_2 E_2 T_2 h_2^2 b_2\right)(E_2 h_2 b_2 + E_1 h_1 b_1)\right.\\ &\left.+(\alpha_1 E_1 T_1 h_1 b_1 + \alpha_2 E_2 T_2 h_2 b_2)\left(E_2 h_2^2 b_2 - E_1 h_1^2 b_1\right)\right] \quad (i=1,2)\end{aligned} \tag{2.20}$$

where

$$D = \left(E_2 h_2^2 b_2 - E_1 h_1^2 b_1\right)^2 + 4E_1E_2h_1h_2(h_1+h_2)^2 b_1 b_2 \tag{i}$$

[3] Both cross sections of each layer are of the same dimensions. The temperature changes in the layers are constant but different. This is a special case of [2]. In this case, taking

$$T_1(y) = T_1, \qquad T_2(y) = T_2, \qquad h_1 = h_2 \equiv h, \qquad b_1 = b_2 \equiv b \tag{j}$$

we can obtain the thermal stresses from Eqs. (2.20) and (j)

$$\sigma_{xi}(y) = -\alpha_i E_i T_i + \frac{E_i}{D}\big[\alpha_1 E_1 T_1(E_1 + 7E_2) + \alpha_2 E_2 T_2(7E_1 + E_2) - 12\frac{y}{h}E_1 E_2(\alpha_1 T_1 - \alpha_2 T_2)\big] \qquad (i = 1, 2) \tag{2.21}$$

An alternative and more convenient form of Eq. (2.21) is

$$\sigma_{xi}(y) = \frac{E_i}{D}\big[(\alpha_1 T_1 - \alpha_i T_i)E_1^2 + (\alpha_2 T_2 - \alpha_i T_i)E_2^2 + 7(\alpha_1 T_1 + \alpha_2 T_2 - 2\alpha_i T_i)E_1 E_2 - 12\frac{y}{h}E_1 E_2(\alpha_1 T_1 - \alpha_2 T_2)\big] \qquad (i = 1, 2) \tag{2.21'}$$

where

$$D = (E_1 + E_2)^2 + 12E_1 E_2 \tag{k}$$

[4] The temperature changes in the layers of the same width b are expressed by a linear function of the variable y.
In this case, taking

$$T_1(y) = A_1 + B_1 y, \qquad T_2(y) = A_2 + B_2 y, \qquad b_1 = b_2 \equiv b \tag{l}$$

we can obtain the thermal stresses from Eqs. (2.18) and (l)

$$\begin{aligned}\sigma_{xi}(y) = &-\alpha_i E_i(A_i + B_i y)\\ &+ \frac{E_i}{D}\Big\{2\left[\alpha_1 E_1 h_1(2A_1 - B_1 h_1) + \alpha_2 E_2 h_2(2A_2 + B_2 h_2)\right]\left(E_2 h_2^3 + E_1 h_1^3\right)\\ &- \left[\alpha_1 E_1 h_1^2(-3A_1 + 2B_1 h_1) + \alpha_2 E_2 h_2^2(3A_2 + 2B_2 h_2)\right]\left(E_2 h_2^2 - E_1 h_1^2\right)\Big\}\\ &+ \frac{E_i y}{D}\Big\{2\left[\alpha_1 E_1 h_1^2(-3A_1 + 2B_1 h_1) + \alpha_2 E_2 h_2^2(3A_2 + 2B_2 h_2)\right]\\ &\times (E_2 h_2 + E_1 h_1)\\ &- 3\left[\alpha_1 E_1 h_1(2A_1 - B_1 h_1) + \alpha_2 E_2 h_2(2A_2 + B_2 h_2)\right]\left(E_2 h_2^2 - E_1 h_1^2\right)\Big\}\\ &(i = 1, 2)\end{aligned} \tag{2.22}$$

where

$$D = \left(E_2 h_2^2 - E_1 h_1^2\right)^2 + 4E_1 E_2 h_1 h_2(h_1 + h_2)^2 \tag{m}$$

Here, numerical results for Eqs. (2.21) and (2.22) are shown in Fig. 2.6. Partially stabilized zirconia (PSZ) and Ti+6Al+4V are assumed for beams 1 and 2, respectively, and thickness h_i and length l of each beam are 1 cm and 1 m, respectively. Values for α and E of Ti + 6Al + 4V are given in Table 1.1 and those for PSZ are 10×10^{-6} 1/K and 151 GPa, respectively. Figure 2.6 (b) shows the distribution of thermal stress due to the prescribed temperature shown in Fig. 2.6 (a). Figure 2.6 (d) shows the thermal stress distribution due to the prescribed temperature in Fig. 2.6 (c). Continuity of heat flux (i.e., perfect thermal contact) is assumed at the interface between the two layers. Note that the thermal conductivities of PSZ and Ti+6Al+4V are 2.09 W/(mK) and 7.5 W/(mK), respectively.

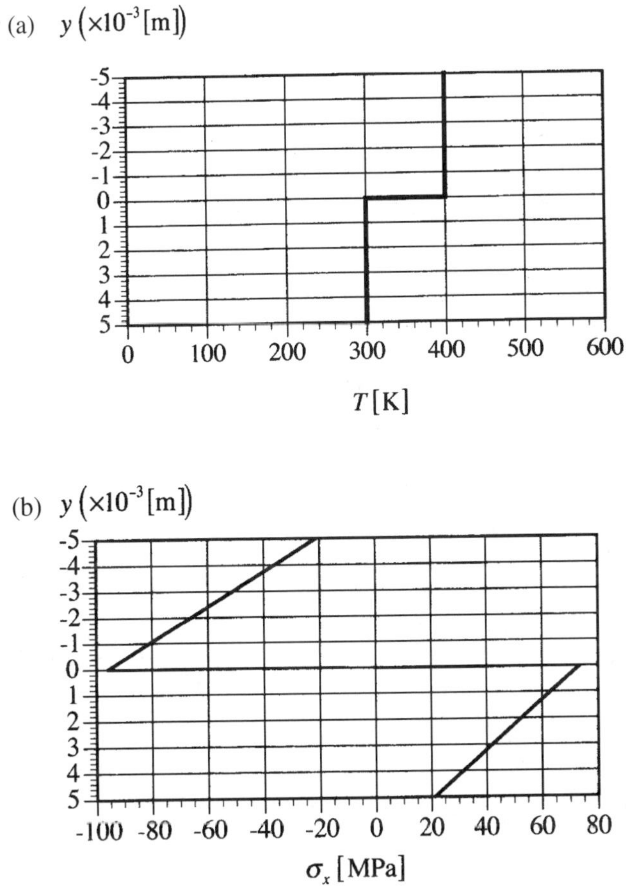

Figure 2.6 Temperatures and thermal stresses in a two-layered beam: (a) uniform temperature distribution; (b) thermal stress due to temperature shown in (a); (c) linear temperature distribution; and (d) thermal stress due to temperature shown in (c).

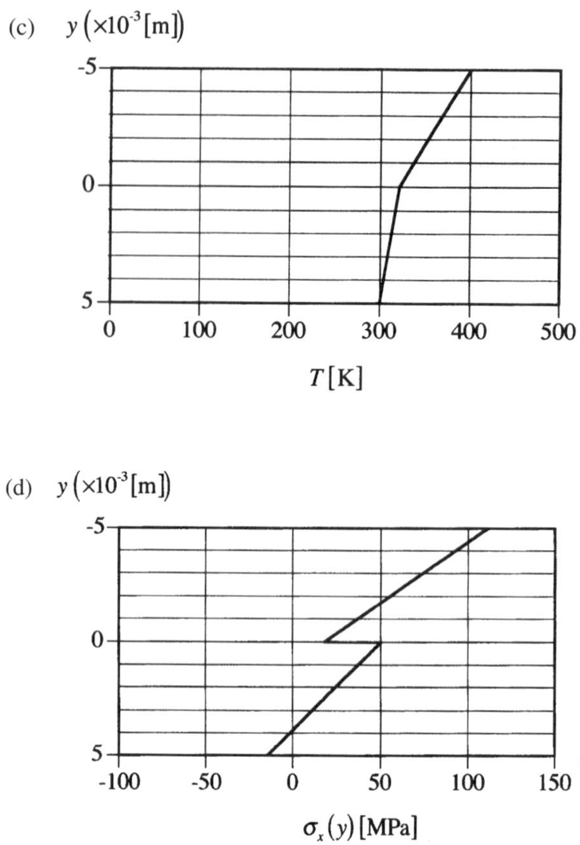

Figure 2.6 continued.

A more complex case will now be considered.

Example 2.3 Calculate the strain ϵ_0 at the bonding surface and the curvature $1/\rho$ for cases **[1]** through **[4]**.

Solution

[1] From Eqs. (d), (e), and (f) we get

$$\epsilon_0 = \frac{2}{D}\left\{2\left[\int_{-h}^{0} \alpha_1 E_1 T_1(y)\,dy + \int_{0}^{h} \alpha_2 E_2 T_2(y)\,dy\right](E_2 + E_1)h \right.$$
$$\left. - 3\left[\int_{-h}^{0} \alpha_1 E_1 T_1(y) y\,dy + \int_{0}^{h} \alpha_2 E_2 T_2(y) y\,dy\right](E_2 - E_1)\right\}$$

$$\frac{1}{\rho} = \frac{6}{D}\frac{1}{h}\left\{2\left[\int_{-h}^{0} \alpha_1 E_1 T_1(y) y\, dy + \int_{0}^{h} \alpha_2 E_2 T_2(y) y\, dy\right](E_2 + E_1) - \left[\int_{-h}^{0} \alpha_1 E_1 T_1(y)\, dy + \int_{0}^{h} \alpha_2 E_2 T_2(y)\, dy\right](E_2 - E_1)h\right\}$$

in which D is given by Eq. (g).

[2] Substitution of Eqs. (h) into Eqs. (d) and (e) gives

$$\epsilon_0 = \frac{1}{D}\Big[4(\alpha_1 E_1 T_1 h_1 b_1 + \alpha_2 E_2 T_2 h_2 b_2)\big(E_2 h_2^3 b_2 + E_1 h_1^3 b_1\big) + 3\big(\alpha_1 E_1 T_1 h_1^2 b_1 - \alpha_2 E_2 T_2 h_2^2 b_2\big)\big(E_2 h_2^2 b_2 - E_1 h_1^2 b_1\big)\Big]$$

$$\frac{1}{\rho} = -\frac{6}{D}\Big[\big(\alpha_1 E_1 T_1 h_1^2 b_1 - \alpha_2 E_2 T_2 h_2^2 b_2\big)(E_2 h_2 b_2 + E_1 h_1 b_1) + (\alpha_1 E_1 T_1 h_1 b_1 + \alpha_2 E_2 T_2 h_2 b_2)\big(E_2 h_2^2 b_2 - E_1 h_1^2 b_1\big)\Big]$$

in which D is given by Eq. (i).

[3] Using Eqs. (d), (e), and (j) we obtain

$$\epsilon_0 = \frac{1}{D}[\alpha_1 E_1 T_1 (E_1 + 7E_2) + \alpha_2 E_2 T_2 (7E_1 + E_2)]$$

$$\frac{1}{\rho} = -\frac{12}{D}\frac{1}{h} E_1 E_2 (\alpha_1 T_1 - \alpha_2 T_2)$$

in which D is given by Eq. (k).

[4] From Eqs. (d), (e), and (l) we get

$$\epsilon_0 = \frac{1}{D}\Big\{2\,[\alpha_1 E_1 h_1 (2A_1 - B_1 h_1) + \alpha_2 E_2 h_2 (2A_2 + B_2 h_2)]\,\big(E_2 h_2^3 + E_1 h_1^3\big) - \Big[\alpha_1 E_1 h_1^2(-3A_1 + 2B_1 h_1) + \alpha_2 E_2 h_2^2 (3A_2 + 2B_2 h_2)\Big]\,\big(E_2 h_2^2 - E_1 h_1^2\big)\Big\}$$

$$\frac{1}{\rho} = \frac{1}{D}\Big\{2\Big[\alpha_1 E_1 h_1^2(-3A_1 + 2B_1 h_1) + \alpha_2 E_2 h_2^2 (3A_2 + 2B_2 h_2)\Big] \times (E_2 h_2 + E_1 h_1) - 3\,[\alpha_1 E_1 h_1 (2A_1 - B_1 h_1) + \alpha_2 E_2 h_2 (2A_2 + B_2 h_2)]\,\big(E_2 h_2^2 - E_1 h_1^2\big)\Big\}$$

in which D is given by Eq. (m).

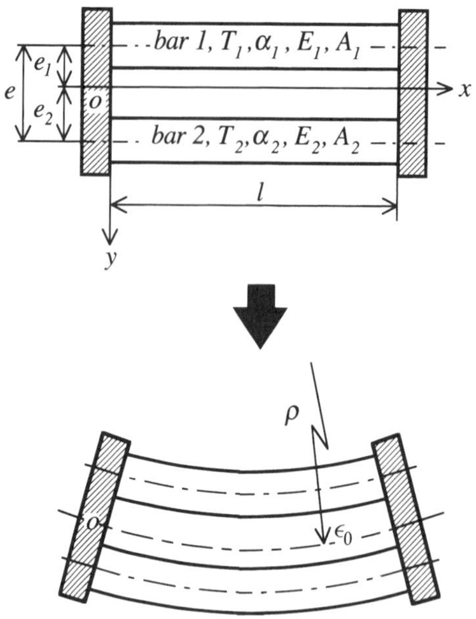

Figure 2.7 Two parallel beams clamped to rigid plates.

2.3.2 Thermal Stresses in Two Beams Clamped at Each End

We again consider the thermal stresses in the beams treated in Section 1.4. When two parallel beams clamped at each end to rigid, non-conducting plates are subjected to different temperature changes T_i, Fig. 2.7, they may elongate and bend. We wish to determine the thermal stresses in the beams with elongation and bending. The coordinate system is shown in Fig. 2.7. The origin of the coordinate y is taken at an arbitrary position between two beams, the origin of the local coordinate y_i is taken at the centroid of the cross section of each beam, e denotes the distance between the centroids of cross sections of both beams, and e_i denotes the distance from $y = 0$ to the centroid of the cross section of each bar. We do not adopt here the assumption that the stresses in each beam are constant, the assumption which we used in Section 1.4, since the thermal stresses in each beam will be functions of a position. The moments of the area of the cross section for each beam are zero:

$$\int_{A_i} y_i \, dA_i = 0 \qquad (i = 1, 2) \tag{a}$$

because the origin of the coordinate system passes through the centroid of the section of each beam and A_i denotes the area of the cross section.

When ϵ_0 and ρ denote the axial strain at $y = 0$ and the radius of curvature at $y = 0$, respectively, the stress σ_x is

$$\sigma_x = E\left(\epsilon_0 + \frac{y}{\rho} - \alpha T\right) \tag{b}$$

Since external forces are not applied to the beam,

$$\int_A \sigma_x \, dA = 0, \qquad \int_A \sigma_x y \, dA = 0 \tag{c}$$

where the integration extends from the top of the upper beam to the bottom of the lower beam. Using the relationship $y_2 = y - e_2$ and $y_1 = y + e_1$, we obtain from Eqs. (b) and (c)

$$\begin{aligned}
&(E_1 A_1 + E_2 A_2)\epsilon_0 + (e_2 E_2 A_2 - e_1 E_1 A_1)\frac{1}{\rho} \\
&\quad = \sum_{i=1}^{2} \int_{A_i} \alpha_i E_i T_i(y_i) \, dA_i \\
&(e_2 E_2 A_2 - e_1 E_1 A_1)\epsilon_0 + \left[E_1\left(I_1 + A_1 e_1^2\right) + E_2\left(I_2 + A_2 e_2^2\right)\right]\frac{1}{\rho} \\
&\quad = \sum_{i=1}^{2} \int_{A_i} \alpha_i E_i T_i(y_i) y_i \, dA_i + e_2 \int_{A_2} \alpha_2 E_2 T_2(y_2) \, dA_2 \\
&\quad\quad - e_1 \int_{A_1} \alpha_1 E_1 T_1(y_1) \, dA_1
\end{aligned} \tag{d}$$

where $I_i = \int_{A_i} y_i^2 \, dA_i$ denotes the moment of inertia of the cross section of each beam with respect to its neutral axis. By solving Eq. (d), the normal strain ϵ_0 and the curvature $1/\rho$ at $y = 0$ can be obtained as

$$\epsilon_0 = \frac{P_T I_{E2} - M_T I_{E1}}{I_{E0} I_{E2} - I_{E1}^2}, \qquad \frac{1}{\rho} = \frac{M_T I_{E0} - P_T I_{E1}}{I_{E0} I_{E2} - I_{E1}^2} \tag{e}$$

where

$$\begin{aligned}
&I_{E0} = (E_1 A_1 + E_2 A_2) \\
&I_{E1} = (e_2 E_2 A_2 - e_1 E_1 A_1) \\
&I_{E2} = E_1\left(I_1 + A_1 e_1^2\right) + E_2\left(I_2 + A_2 e_2^2\right) \\
&P_T = P_{T1} + P_{T2} \\
&P_{T1} = \int_{A_1} \alpha_1 E_1 T_1(y_1) \, dA_1, \qquad P_{T2} = \int_{A_2} \alpha_2 E_2 T_2(y_2) \, dA_2 \\
&M_T = M_{T1} + M_{T2} + e_2 P_{T2} - e_1 P_{T1} \\
&M_{T1} = \int_{A_1} \alpha_1 E_1 T_1(y_1) y_1 \, dA_1, \qquad M_{T2} = \int_{A_2} \alpha_2 E_2 T_2(y_2) y_2 \, dA_2
\end{aligned} \tag{f}$$

Thus, the thermal stresses may be expressed by

$$
\begin{aligned}
\sigma_{x1}(y_1) =& -\alpha_1 E_1 T_1(y_1) + E_1 \left[\frac{P_T I_{E2} - M_T I_{E1}}{I_{E0} I_{E2} - I_{E1}^2} \right] \\
&+ (y_1 - e_1) E_1 \left[\frac{M_T I_{E0} - P_T I_{E1}}{I_{E0} I_{E2} - I_{E1}^2} \right] \\
\sigma_{x2}(y_2) =& -\alpha_2 E_2 T_2(y_2) + E_2 \left[\frac{P_T I_{E2} - M_T I_{E1}}{I_{E0} I_{E2} - I_{E1}^2} \right] \\
&+ (y_2 + e_2) E_2 \left[\frac{M_T I_{E0} - P_T I_{E1}}{I_{E0} I_{E2} - I_{E1}^2} \right]
\end{aligned} \tag{2.23}
$$

or in an alternative form

$$
\begin{aligned}
\sigma_{x1}(y_1) =& -\alpha_1 E_1 T_1(y_1) \\
&+ \frac{E_1}{D}\big[(P_{T1} + P_{T2})(E_1 I_1 + E_2 I_2) + P_{T1} E_2 A_2 e^2 \\
&\qquad - (M_{T1} + M_{T2}) E_2 A_2 e\big] \\
&+ \frac{E_1 y_1}{D}[(M_{T1} + M_{T2})(E_1 A_1 + E_2 A_2) + (P_{T2} E_1 A_1 - P_{T1} E_2 A_2) e]
\end{aligned} \tag{2.23$'$}
$$

$$
\begin{aligned}
\sigma_{x2}(y_2) =& -\alpha_2 E_2 T_2(y_2) \\
&+ \frac{E_2}{D}\big[(P_{T1} + P_{T2})(E_1 I_1 + E_2 I_2) + P_{T2} E_1 A_1 e^2 \\
&\qquad + (M_{T1} + M_{T2}) E_1 A_1 e\big] \\
&+ \frac{E_2 y_2}{D}[(M_{T1} + M_{T2})(E_1 A_1 + E_2 A_2) + (P_{T2} E_1 A_1 - P_{T1} E_2 A_2) e]
\end{aligned}
$$

where

$$
D = (E_1 A_1 + E_2 A_2)(E_1 I_1 + E_2 I_2) + E_1 E_2 A_1 A_2 e^2 \tag{g}
$$

Here, a numerical result from Eq. (2.23) is shown in Fig. 2.8. Partially stabilized zirconia (PSZ) and Ti + 6Al + 4V are assumed for beams 1 and 2, respectively, thickness h_i and length l of each beam are 1 cm and 1 m, respectively, and the gap e between two centroids of beams is taken to be 2 cm. Temperatures in two beams are assumed to be uniform; $T_1 = 400$ K and $T_2 = 300$ K. α and E of Ti + 6Al + 4V are given in Table 1.1 and those of PSZ are 10×10^{-6} 1/K and 151 MPa, respectively.

Example 2.4 Derive Eq. (2.23) using the same technique as in Section 1.4.

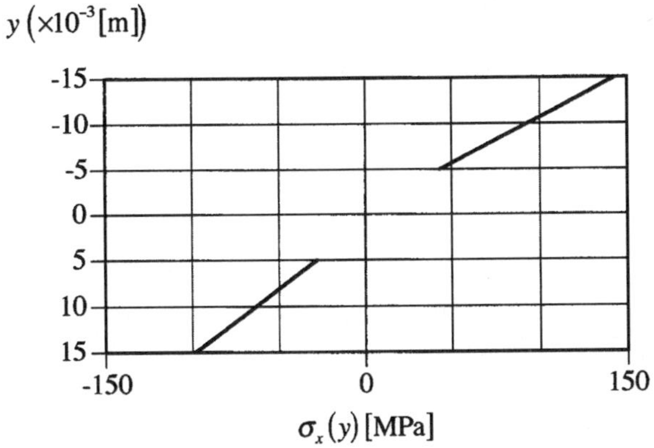

Figure 2.8 Thermal stresses in two parallel beams clamped to rigid plates.

Solution The elongations of beam 1 and beam 2 are given by

$$
\begin{aligned}
l + \alpha_1 T_1(y_1) l + \frac{\sigma_{x1}(y_1)}{E_1} l &= (\rho - e_1 + y_1)\theta \\
l + \alpha_2 T_2(y_2) l + \frac{\sigma_{x2}(y_2)}{E_2} l &= (\rho + e_2 + y_2)\theta
\end{aligned}
\tag{h}
$$

From Eqs. (h), the thermal stresses are

$$
\begin{aligned}
\sigma_{x1} &= E_1 \left[\left(\frac{\rho\theta}{l} - 1 \right) + (y_1 - e_1)\frac{\theta}{l} \right] - \alpha_1 E_1 T_1(y_1) \\
\sigma_{x2} &= E_2 \left[\left(\frac{\rho\theta}{l} - 1 \right) + (y_2 + e_2)\frac{\theta}{l} \right] - \alpha_2 E_2 T_2(y_2)
\end{aligned}
\tag{i}
$$

Substitution of the above equations into Eqs. (c) yields

$$
\begin{aligned}
&(E_1 A_1 + E_2 A_2)\left(\frac{\rho\theta}{l} - 1 \right) + (e_2 E_2 A_2 - e_1 E_1 A_1)\frac{\theta}{l} \\
&\quad = \sum_{i=1}^{2} \int_{A_i} \alpha_i E_i T_i(y_i)\, dA_i \\
&(e_2 E_2 A_2 - e_1 E_1 A_1)\left(\frac{\rho\theta}{l} - 1 \right) + \left[E_1\left(I_1 + A_1 e_1^2\right) + E_2\left(I_2 + A_2 e_2^2\right) \right] \frac{\theta}{l} \\
&\quad = \sum_{i=1}^{2} \int_{A_i} \alpha_i E_i T_i(y_i) y_i\, dA_i + e_2 \int_{A_2} \alpha_2 E_2 T_2(y_2)\, dA_2 \\
&\quad\quad - e_1 \int_{A_1} \alpha_1 E_1 T_1(y_1)\, dA_1
\end{aligned}
\tag{j}
$$

$(\rho\theta/l - 1)$ and θ/l are

$$\left(\frac{\rho\theta}{l} - 1\right) = \frac{P_T I_{E2} - M_T I_{E1}}{I_{E0} I_{E2} - I_{E1}^2}, \qquad \frac{\theta}{l} = \frac{M_T I_{E0} - P_T I_{E1}}{I_{E0} I_{E2} - I_{E1}^2} \tag{k}$$

Substitution of Eq. (k) into Eq. (i) gives the thermal stresses expressed by Eq. (2.23).

Example 2.5 When the distance between the two beams is very large compared to the height in each beam in Example 2.1, show that the mean stresses in each beam subjected to the different constant temperature change of each beam are very small compared to the stresses expressed by Eq. (1.33).

Solution Using Eqs. (2.23′) and (g), the mean stresses of each beam are given by

$$\begin{aligned}
\overline{\sigma}_{x1} &= \frac{1}{A_1}\int_{A_1} \sigma_{x1}(y_1)\,dA_1 \\
&= -\alpha_1 E_1 T_1 + \frac{E_1}{D}\big[(P_{T_1} + P_{T_2})(E_1 I_1 + E_2 I_2) \\
&\quad + P_{T_1} E_2 A_2 e^2 - (M_{T_1} + M_{T_2}) E_2 A_2 e\big] \\
\overline{\sigma}_{x2} &= \frac{1}{A_2}\int_{A_2} \sigma_{x2}(y_2)\,dA_2 \\
&= -\alpha_2 E_2 T_2 + \frac{E_2}{D}\big[(P_{T_1} + P_{T_2})(E_1 I_1 + E_2 I_2) \\
&\quad + P_{T_2} E_1 A_1 e^2 + (M_{T_1} + M_{T_2}) E_1 A_1 e\big]
\end{aligned} \tag{l}$$

where

$$\begin{aligned}
P_{T1} &= \int_{A_1} \alpha_1 E_1 T_1(y_1)\,dA_1 = \alpha_1 E_1 T_1 A_1 \\
P_{T2} &= \int_{A_2} \alpha_2 E_2 T_2(y_2)\,dA_2 = \alpha_2 E_2 T_2 A_2 \\
M_{T1} &= \int_{A_1} \alpha_1 E_1 T_1(y_1) y_1\,dA_1 = \alpha_1 E_1 T_1 \int_{A_1} y_1\,dA_1 = 0 \\
M_{T2} &= \int_{A_2} \alpha_2 E_2 T_2(y_2) y_2\,dA_2 = \alpha_2 E_2 T_2 \int_{A_2} y_2\,dA_2 = 0 \\
D &= (E_1 A_1 + E_2 A_2)(E_1 I_1 + E_2 I_2) + E_1 E_2 A_1 A_2 e^2
\end{aligned} \tag{m}$$

Using Eqs. (m), Eqs. (l) reduce to

$$\begin{aligned}
\overline{\sigma}_{x1} &= \frac{(\alpha_2 T_2 - \alpha_1 T_1) E_1 E_2 A_2 (E_1 I_1 + E_2 I_2)}{(E_1 A_1 + E_2 A_2)(E_1 I_1 + E_2 I_2) + E_1 E_2 A_1 A_2 e^2} \\
\overline{\sigma}_{x2} &= -\frac{(\alpha_2 T_2 - \alpha_1 T_1) E_1 E_2 A_1 (E_1 I_1 + E_2 I_2)}{(E_1 A_1 + E_2 A_2)(E_1 I_1 + E_2 I_2) + E_1 E_2 A_1 A_2 e^2}
\end{aligned} \tag{n}$$

and the curvature is given by

$$\frac{1}{\rho} = \frac{(\alpha_2 T_2 - \alpha_1 T_1)E_1 E_2 A_1 A_2 e}{(E_1 A_1 + E_2 A_2)(E_1 I_1 + E_2 I_2) + E_1 E_2 A_1 A_2 e^2} \tag{o}$$

where the moments of inertia I_1 and I_2 are rewritten as

$$I_1 = A_1 k_1^2, \qquad I_2 = A_2 k_2^2 \tag{p}$$

and k_i $(i = 1, 2)$ denotes the radius of inertia. When the dimension of both beams and the Young's moduli of both beams are of the same order, Eqs. (n) reduce to

$$\begin{aligned}
\overline{\sigma}_{x1} &= \frac{(\alpha_2 T_2 - \alpha_1 T_1)E_1 E_2 A_2 \left(E_1 A_1 k_1^2 + E_2 A_2 k_2^2\right)}{(E_1 A_1 + E_2 A_2)\left(E_1 A_1 k_1^2 + E_2 A_2 k_2^2\right) + E_1 E_2 A_1 A_2 e^2} \\
&= (\alpha_2 T_2 - \alpha_1 T_1)E_1 \frac{1}{1 + A_1 E_1 / A_2 E_2} \left(\frac{k_1}{e}\right)^2 \\
&\quad \times \frac{(1 + A_1 E_1 / A_2 E_2)\left(1 + E_2 A_2 k_2^2 / E_1 A_1 k_1^2\right)}{1 + (1 + E_1 A_1 / E_2 A_2)\left(1 + E_2 A_2 k_2^2 / E_1 A_1 k_1^2\right)\left(k_1^2 / e^2\right)} \\
&\ll (\alpha_2 T_2 - \alpha_1 T_1)\frac{1}{1 + A_1 E_1 / A_2 E_2} E_1
\end{aligned} \tag{q}$$

$$\begin{aligned}
\overline{\sigma}_{x2} &= \frac{-(\alpha_2 T_2 - \alpha_1 T_1)E_1 E_2 A_1 \left(E_1 A_1 k_1^2 + E_2 A_2 k_2^2\right)}{(E_1 A_1 + E_2 A_2)\left(E_1 A_1 k_1^2 + E_2 A_2 k_2^2\right) + E_1 E_2 A_1 A_2 e^2} \\
&= -(\alpha_2 T_2 - \alpha_1 T_1)E_1 \frac{1}{1 + A_1 E_1 / A_2 E_2} \frac{A_1}{A_2} \left(\frac{k_1}{e}\right)^2 \\
&\quad \times \frac{(1 + A_1 E_1 / A_2 E_2)\left(1 + E_2 A_2 k_2^2 / E_1 A_1 k_1^2\right)}{1 + (1 + E_1 A_1 / E_2 A_2)\left(1 + E_2 A_2 k_2^2 / E_1 A_1 k_1^2\right)\left(k_1^2 / e^2\right)} \\
&\ll -(\alpha_2 T_2 - \alpha_1 T_1)\frac{1}{1 + A_1 E_1 / A_2 E_2} E_1 \frac{A_1}{A_2}
\end{aligned} \tag{r}$$

because $k_1/e \ll 1$. These stresses are very small compared to the stresses expressed by Eq. (1.33). Eq. (o) approximately equals

$$\frac{1}{\rho} \cong \frac{\alpha_2 T_2 - \alpha_1 T_1}{e} \tag{s}$$

This curvature nearly equals that of Eq. (1.47).

2.3.3 Thermal Stresses in Multi-Layered Composite Beams

Consider an n-layered composite beam, Fig. 2.9, subjected to temperature change $T_i(y)$. We take the origin $y = 0$ of the coordinate system y at the upper surface of the composite beam. Assuming that the normal strain at the upper surface $y = 0$ is ϵ_0 and the curvature at $y = 0$ is $1/\rho$, the bending strains at an arbitrary position y are expressed by

$$\alpha_i T_i(y) + \frac{\sigma_{xi}}{E_i} = \epsilon_0 + \frac{y}{\rho} \qquad (i = 1, 2, \ldots, n) \tag{2.24}$$

Solving this equation for σ_{xi} yields

$$\sigma_{xi} = -\alpha_i E_i T_i(y) + \epsilon_0 E_i + \frac{E_i y}{\rho} \tag{2.25}$$

Since external forces do not act on the layers,

$$\sum_{i=1}^{n} \int_{y_{i-1}}^{y_i} \sigma_{xi} b_i \, dy = 0 \tag{a}$$

$$\sum_{i=1}^{n} \int_{y_{i-1}}^{y_i} \sigma_{xi} b_i y \, dy = 0 \tag{b}$$

where b_i and $y_i - y_{i-1} = h_i$ denote the width and the height of each layer, respectively, and y_i $(i = 1, 2, \ldots, n)$ means the lower surface of the i-th beam, $y_0 = 0$. Substitution

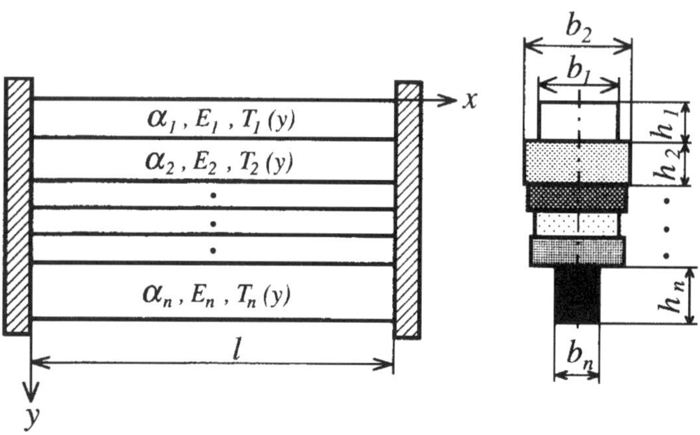

Figure 2.9 Multi-layered composite beam.

of Eq. (2.25) into (a) and (b) yields

$$\sum_{i=1}^{n} E_i b_i (y_i - y_{i-1})\epsilon_0 + \sum_{i=1}^{n} E_i b_i \left(y_i^2 - y_{i-1}^2\right)\frac{1}{2\rho} = \sum_{i=1}^{n} \int_{y_{i-1}}^{y_i} \alpha_i E_i T_i(y) b_i \, dy$$

$$\sum_{i=1}^{n} E_i b_i \left(y_i^2 - y_{i-1}^2\right)\frac{\epsilon_0}{2} + \sum_{i=1}^{n} E_i b_i \left(y_i^3 - y_{i-1}^3\right)\frac{1}{3\rho} = \sum_{i=1}^{n} \int_{y_{i-1}}^{y_i} \alpha_i E_i T_i(y) b_i y \, dy \qquad \text{(c)}$$

The normal strain ϵ_0 and the curvature $1/\rho$ at $y = 0$ are obtained as

$$\epsilon_0 = \frac{P_T I_{E2} - M_T I_{E1}}{I_{E0} I_{E2} - I_{E1}^2}, \qquad \frac{1}{\rho} = \frac{M_T I_{E0} - P_T I_{E1}}{I_{E0} I_{E2} - I_{E1}^2} \qquad \text{(d)}$$

where

$$I_{E0} = \sum_{i=1}^{n} E_i b_i (y_i - y_{i-1})$$

$$I_{E1} = \frac{1}{2}\sum_{i=1}^{n} E_i b_i \left(y_i^2 - y_{i-1}^2\right), \qquad I_{E2} = \frac{1}{3}\sum_{i=1}^{n} E_i b_i \left(y_i^3 - y_{i-1}^3\right) \qquad \text{(e)}$$

$$P_T = \sum_{i=1}^{n} \int_{y_{i-1}}^{y_i} \alpha_i E_i T_i(y) b_i \, dy, \qquad M_T = \sum_{i=1}^{n} \int_{y_{i-1}}^{y_i} \alpha_i E_i T_i(y) y b_i \, dy$$

The thermal stresses may be expressed from Eq. (2.25) as

$$\sigma_{xi}(y) = -\alpha_i E_i T_i(y) + E_i \frac{P_T I_{E2} - M_T I_{E1}}{I_{E0} I_{E2} - I_{E1}^2} + y E_i \frac{M_T I_{E0} - P_T I_{E1}}{I_{E0} I_{E2} - I_{E1}^2} \qquad (2.26)$$

Example 2.6 When an n-layered composite beam with restrained extension at $y = 0$ and free bending is subjected to a temperature change $T_i(y)$, what thermal stress is produced in the beam?

Solution When the normal strain ϵ_0 is zero at $y = 0$, Eq. (2.25) gives

$$\sigma_{xi} = -\alpha_i E_i T_i(y) + \frac{E_i y}{\rho} \qquad \text{(a)}$$

Since an external bending moment does not act on the beam, we receive

$$\sum_{i=1}^{n} \int_{y_{i-1}}^{y_i} \sigma_{xi} b_i y \, dy = 0 \tag{b}$$

Substitution of Eq. (a) into Eq. (b) gives the curvature $1/\rho$ at $y = 0$

$$\frac{1}{\rho} = \frac{M_T}{I_{E2}} \tag{c}$$

where M_T and I_{E2} are given by Eqs. (e) in Subsection 2.3.3. Thermal stress is

$$\sigma_{xi} = -\alpha_i E_i T_i(y) + y E_i \frac{M_T}{I_{E2}} \tag{d}$$

In the same way, thermal stress in a composite beam with free extension at $y = 0$ and restrained bending is obtained as:

$$\sigma_{xi} = -\alpha_i E_i T_i(y) + E_i \frac{P_T}{I_{E0}} \tag{e}$$

where P_T and I_{E0} are given by Eqs. (e) in Subsection 2.3.3.

2.4 Thermal Stresses in Nonhomogeneous Beams

Materials in which properties such as the coefficient of linear thermal expansion α and the Young's modulus E depend on position, are known as nonhomogeneous materials. Consider the thermal stress in a nonhomogeneous beam, Fig. 2.10, in which the coefficient of linear thermal expansion and the Young's modulus are functions of y.

Take the origin $y = 0$ of the coordinate system (x, y) at the centroid of the cross section of the beam. Thus, the bending strain is

$$\alpha(y)T(y) + \frac{\sigma_x}{E(y)} = \epsilon_0 + \frac{y}{\rho} \tag{2.27}$$

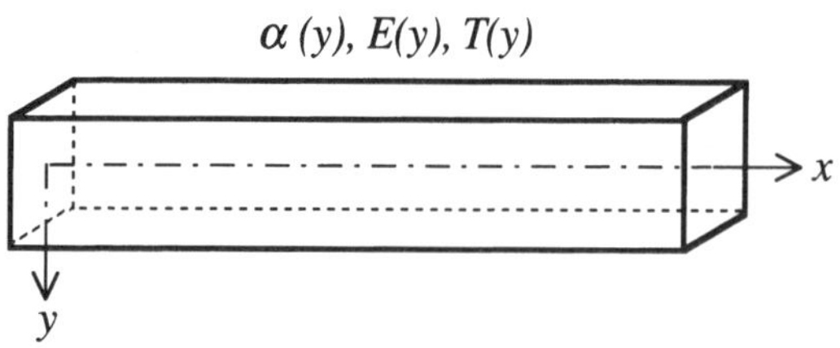

Figure 2.10 Nonhomogeneous beam.

Solving this equation for σ_x yields

$$\sigma_x = -\alpha(y)E(y)T(y) + \epsilon_0 E(y) + \frac{E(y)y}{\rho} \tag{2.28}$$

Since the external forces and moments do not act on the beam, then

$$\int_A \sigma_x \, dA = 0 \tag{a}$$

$$\int_A \sigma_x y \, dA = 0 \tag{b}$$

Substituting Eq. (2.28) into Eqs. (a) and (b), we can obtain the normal strain ϵ_0 at $y = 0$ and the curvature $1/\rho$ as

$$\epsilon_0 = \frac{P_T I_{E2} - M_T I_{E1}}{I_{E0} I_{E2} - I_{E1}^2}, \qquad \frac{1}{\rho} = \frac{M_T I_{E0} - P_T I_{E1}}{I_{E0} I_{E2} - I_{E1}^2} \tag{c}$$

where

$$\begin{aligned} I_{E0} &= \int_A E(y) \, dA, \qquad I_{E1} = \int_A E(y) y \, dA, \qquad I_{E2} = \int_A E(y) y^2 \, dA \\ P_T &= \int_A \alpha(y) E(y) T(y) \, dA, \qquad M_T = \int_A \alpha(y) E(y) T(y) y \, dA \end{aligned} \tag{d}$$

The thermal stress σ_x in the nonhomogeneous beam may be expressed from Eq. (2.28) as

$$\begin{aligned} \sigma_x(y) = &- \alpha(y)E(y)T(y) + E(y)\frac{P_T I_{E2} - M_T I_{E1}}{I_{E0} I_{E2} - I_{E1}^2} \\ &+ yE(y)\frac{M_T I_{E0} - P_T I_{E1}}{I_{E0} I_{E2} - I_{E1}^2} \end{aligned} \tag{2.29}$$

Example 2.7 When the mechanical boundary condition of a nonhomogeneous beam subjected to a temperature change $T(y)$ is given by two cases:

[1] with restrained extension at $y = 0$ and free bending
[2] with free extension at $y = 0$ and restrained bending

what thermal stress is produced in the beam?

Solution

[1] When the normal strain ϵ_0 is zero at $y = 0$ in case 1, Eq. (2.28) gives

$$\sigma_x(y) = -\alpha(y)E(y)T(y) + \frac{E(y)y}{\rho} \tag{a}$$

Since an external bending moment does not act on the beam, we get

$$\int_A \sigma_x(y) y \, dA = 0 \tag{b}$$

Substitution of Eq. (a) into Eq. (b) gives the curvature $1/\rho$ at $y = 0$

$$\frac{1}{\rho} = \frac{M_T}{I_{E2}} \tag{c}$$

where M_T and I_{E2} are given by Eqs. (d) in Section 2.4. The thermal stress is

$$\sigma_x(y) = -\alpha(y)E(y)T(y) + yE(y)\frac{M_T}{I_{E2}} \tag{d}$$

[2] When the curvature $1/\rho$ is zero at $y = 0$ in case 2, Eq. (2.28) gives

$$\sigma_x(y) = -\alpha(y)E(y)T(y) + \epsilon_0 E(y) \tag{e}$$

Since an external force does not act on the beam, we have

$$\int_A \sigma_x(y) \, dA = 0 \tag{f}$$

The normal strain ϵ_0 at $y = 0$ is obtained as:

$$\epsilon_0 = \frac{P_T}{I_{E0}} \tag{g}$$

where P_T and I_{E0} are given by Eqs. (d) in Section 2.4. The thermal stress is

$$\sigma_x(y) = -\alpha(y)E(y)T(y) + E(y)\frac{P_T}{I_{E0}} \tag{h}$$

2.5 Thermal Stresses at Arbitrary Temperature

2.5.1 Homogeneous Beams

Consider a cantilever beam of homogeneous material subjected to an arbitrary temperature change $T(x, y, z)$, shown in Fig. 2.11.We take the origin of the coordinate system (x, y, z) at the centroid of the cross section. The total strain is the sum of the free thermal strain and the strain due to stress σ_x and is equal to the strain after bending

$$\alpha T(x, y, z) + \frac{\sigma_x}{E} = \epsilon_0 + \frac{y}{\rho_y} + \frac{z}{\rho_z} \tag{2.30}$$

where ϵ_0 and ρ_y, ρ_z denote the axial strain and the radii of curvature in y and z directions at $y = z = 0$, respectively. Solving Eq. (2.30) for σ_x yields

$$\sigma_x = -\alpha E T(x, y, z) + \epsilon_0 E + \frac{Ey}{\rho_y} + \frac{Ez}{\rho_z} \tag{2.31}$$

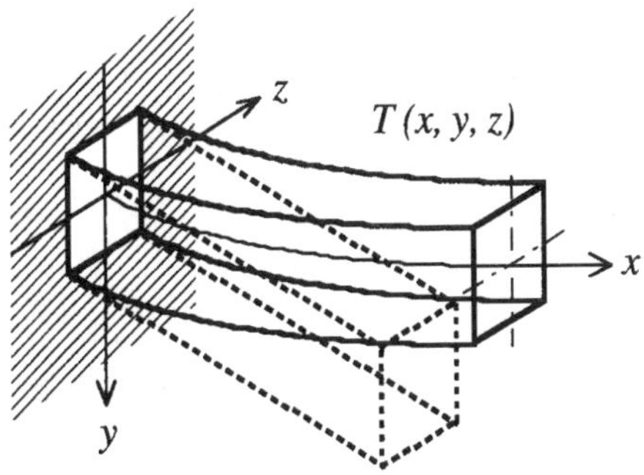

Figure 2.11 Homogeneous beam subjected to an arbitrary temperature change.

Since external forces and moments do not act on the beam

$$\int_A \sigma_x \, dA = 0 \tag{a}$$

$$\int_A \sigma_x y \, dA = 0 \tag{b}$$

$$\int_A \sigma_x z \, dA = 0 \tag{c}$$

Substituting Eq. (2.31) into Eqs. (a), (b), and (c), we find

$$\epsilon_0 \int_A E \, dA + \frac{1}{\rho_y} \int_A E y \, dA + \frac{1}{\rho_z} \int_A E z \, dA = \int_A \alpha E T(x, y, z) \, dA \tag{a$'$}$$

$$\epsilon_0 \int_A E y \, dA + \frac{1}{\rho_y} \int_A E y^2 \, dA + \frac{1}{\rho_z} \int_A E y z \, dA = \int_A \alpha E T(x, y, z) y \, dA \tag{b$'$}$$

$$\epsilon_0 \int_A E z \, dA + \frac{1}{\rho_y} \int_A E y z \, dA + \frac{1}{\rho_z} \int_A E z^2 \, dA = \int_A \alpha E T(x, y, z) z \, dA \tag{c$'$}$$

By solving the above equations, the axial strain ϵ_0 and the curvatures $1/\rho_y$ and $1/\rho_z$ at $y = z = 0$ are expressed as

$$\epsilon_0 = \frac{1}{EA} \int_A \alpha E T(x, y, z) \, dA = \frac{P_T}{EA} \tag{d}$$

$$\frac{1}{\rho_y} = \frac{I_y \int_A \alpha E T(x, y, z) y \, dA - I_{yz} \int_A \alpha E T(x, y, z) z \, dA}{E\left(I_y I_z - I_{yz}^2\right)}$$

$$= \frac{I_y M_{Tz} - I_{yz} M_{Ty}}{E\left(I_y I_z - I_{yz}^2\right)} \tag{e}$$

$$\frac{1}{\rho_z} = \frac{I_z \int_A \alpha E T(x, y, z) z \, dA - I_{yz} \int_A \alpha E T(x, y, z) y \, dA}{E\left(I_y I_z - I_{yz}^2\right)}$$
$$= \frac{I_z M_{Ty} - I_{yz} M_{Tz}}{E\left(I_y I_z - I_{yz}^2\right)} \tag{f}$$

where I_y and I_z are the moments of inertia about the y and z axes, respectively, and I_{yz} is the product of inertia about these axes:

$$I_y = \int_A z^2 \, dA, \qquad I_z = \int_A y^2 \, dA, \qquad I_{yz} = \int_A yz \, dA \tag{2.32}$$

If P_T denotes the force due to the temperature change, and M_{Ty} and M_{Tz} denote the moments about y and z axes due to the temperature change, then

$$P_T = \int_A \alpha E T(x, y, z) \, dA$$
$$M_{Ty} = \int_A \alpha E T(x, y, z) z \, dA, \qquad M_{Tz} = \int_A \alpha E T(x, y, z) y \, dA \tag{g}$$

Therefore, the thermal stress is given from Eq. (2.31) by

$$\sigma_x(x, y, z) = -\alpha E T + \frac{P_T}{A} + \frac{I_y M_{Tz} - I_{yz} M_{Ty}}{I_y I_z - I_{yz}^2} y + \frac{I_z M_{Ty} - I_{yz} M_{Tz}}{I_y I_z - I_{yz}^2} z \tag{2.33}$$

From Eq. (2.33), we may obtain the neutral plane made up of the line

$$y = -\frac{I_z M_{Ty} - I_{yz} M_{Tz}}{I_y M_{Tz} - I_{yz} M_{Ty}} z \tag{2.34}$$

Therefore, the beam may be bent to the normal direction of the neutral plane:

$$y = \frac{I_y M_{Tz} - I_{yz} M_{Ty}}{I_z M_{Ty} - I_{yz} M_{Tz}} z \tag{2.35}$$

The thermal stress in the neutral plane is then given by

$$\sigma_x(x, y, z) = -\alpha E T + \frac{P_T}{A} \tag{2.36}$$

When a cross section is symmetry about y axis, the product of inertia I_{yz} reduces to zero

$$I_{yz} = \int_A yz \, dA = \int_{-y_1}^{y_2} y \left(\int_{-z_1(y)}^{z_1(y)} z \, dz \right) dy$$
$$= \int_{-y_1}^{y_2} y \left[\frac{1}{2} z_1^2(y) - \frac{1}{2} \{-z_1(y)\}^2 \right] dy = 0 \tag{h}$$

The thermal stress, Eq. (2.33), in the beam with a symmetric cross section is simplified to the form

$$\sigma_x(x, y, z) = -\alpha ET + \frac{P_T}{A} + \frac{M_{Tz}}{I_z}y + \frac{M_{Ty}}{I_y}z \tag{2.37}$$

Example 2.8 When a beam of a rectangular cross section $b \times h$ is subjected to a temperature change $T(y, z) = C_0 + C_1 y + C_2 z$, what thermal stress is produced in the beam?

Solution The substitution of the temperature change $T(y, z) = C_0 + C_1 y + C_2 z$ into Eqs. (g) in Subsection 2.5.1 gives

$$\begin{aligned}
P_T &= \int_A \alpha ET(y, z)\, dA \\
&= \int_{-b/2}^{b/2}\int_{-h/2}^{h/2} \alpha E(C_0 + C_1 y + C_2 z)\, dy\, dz = \alpha EbhC_0 \\
M_{Ty} &= \int_A \alpha ET(y, z) z\, dA \\
&= \int_{-b/2}^{b/2}\int_{-h/2}^{h/2} \alpha E\left(C_0 z + C_1 yz + C_2 z^2\right) dy\, dz = \frac{1}{12}\alpha Eb^3hC_2 \\
M_{Tz} &= \int_A \alpha ET(y, z) y\, dA \\
&= \int_{-b/2}^{b/2}\int_{-h/2}^{h/2} \alpha E\left(C_0 y + C_1 y^2 + C_2 yz\right) dy\, dz = \frac{1}{12}\alpha Ebh^3C_1
\end{aligned}$$

However,

$$A = bh, \qquad I_z = \frac{1}{12}bh^3, \qquad I_y = \frac{1}{12}b^3h, \qquad I_{yz} = 0$$

Therefore, the thermal stress (2.37) reduces to

$$\sigma_x = -\alpha E(C_0 + C_1 y + C_2 z) + \frac{\alpha EbhC_0}{bh} + \frac{\alpha Ebh^3C_1/12}{bh^3/12}y + \frac{\alpha Eb^3hC_2/12}{b^3h/12}z = 0$$

Thermal stress is not produced in the beam.

2.5.2 Nonhomogeneous Beams

Next, we consider the stress in a nonhomogeneous beam due to an arbitrary temperature field. We take the coordinate system (x', y', z') in which the origin is located at an arbitrary point in the cross section of the beam. But it is important how the origin of the coordinate system is selected for the nonhomogeneous beam as well as the

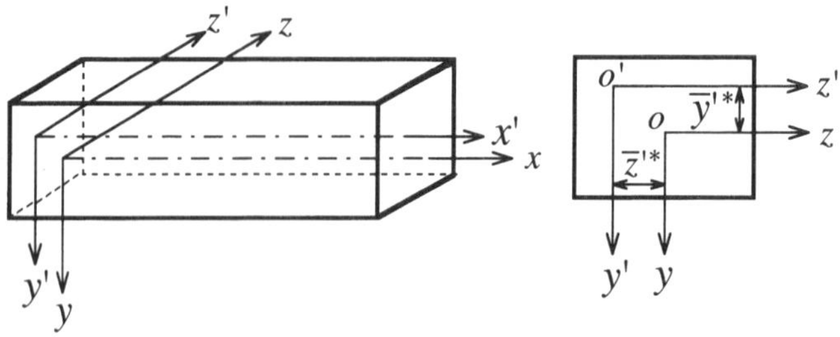

Figure 2.12 New coordinate system.

homogeneous beam. The origin of the coordinate system is generally selected at the centroid of the cross section of a homogeneous beam. If the origin is appropriately selected in the cross section of a nonhomogenous beam, the nonhomogenous beam can be easily treated. Therefore, we define the modulus weighted section quantities under the coordinate system (x', y', z') as follows:

$$A^* = \int_A \frac{E}{E_0}\, dA, \qquad \overline{y}'^* = \frac{1}{A^*}\int_A \frac{E}{E_0} y'\, dA, \qquad \overline{z}'^* = \frac{1}{A^*}\int_A \frac{E}{E_0} z'\, dA \tag{a}$$

in which E_0 is a reference Young's modulus and may be taken equal to the Young's modulus of the actual material for the homogeneous case. We select the new coordinate system (x, y, z) shown in Fig. 2.12:

$$x = x', \qquad y = y' - \overline{y}'^*, \qquad z = z' - \overline{z}'^* \tag{b}$$

Substitution of Eqs. (b) into Eq. (a) gives

$$\begin{aligned}
\overline{y}'^* &= \frac{1}{A^*}\int_A \frac{E}{E_0}(y + \overline{y}'^*)\, dA = \frac{1}{A^*}\int_A \frac{E}{E_0} y\, dA + \frac{\overline{y}'^*}{A^*}\int_A \frac{E}{E_0}\, dA \\
&= \frac{1}{A^*}\int_A \frac{E}{E_0} y\, dA + \overline{y}'^* \\
\overline{z}'^* &= \frac{1}{A^*}\int_A \frac{E}{E_0} z\, dA + \overline{z}'^*
\end{aligned}$$

Therefore,

$$\int_A \frac{E}{E_0} y\, dA = 0, \qquad \int_A \frac{E}{E_0} z\, dA = 0 \tag{2.38}$$

The stress–strain relation in the new coordinate system (x, y, z) in which Eqs. (2.38) must be satisfied is given by

$$\epsilon_x = \frac{\sigma_x}{E} + \alpha T(x, y, z) = \epsilon_0 + \frac{y}{\rho_y} + \frac{z}{\rho_z} \tag{2.39}$$

where ϵ_0 denotes a normal strain in the direction of the x axis, and ρ_y and ρ_z are the radii of the curvature in y and z directions at $y = z = 0$. Since the external forces and moments do not act on the beam

$$\int_A \sigma_x \, dA = 0, \qquad \int_A \sigma_x y \, dA = 0, \qquad \int_A \sigma_x z \, dA = 0 \tag{c}$$

Solving for σ_x in Eq. (2.39) and substituting into Eq. (c), we find

$$\begin{aligned}
\epsilon_0 \int_A E \, dA + \frac{1}{\rho_y} \int_A E y \, dA + \frac{1}{\rho_z} \int_A E z \, dA &= \int_A \alpha E T \, dA \\
\epsilon_0 \int_A E y \, dA + \frac{1}{\rho_y} \int_A E y^2 \, dA + \frac{1}{\rho_z} \int_A E y z \, dA &= \int_A \alpha E T y \, dA \\
\epsilon_0 \int_A E z \, dA + \frac{1}{\rho_y} \int_A E y z \, dA + \frac{1}{\rho_z} \int_A E z^2 \, dA &= \int_A \alpha E T z \, dA
\end{aligned} \tag{d}$$

By consideration of Eqs. (2.38), Eqs. (d) reduce to

$$\begin{aligned}
\epsilon_0 \int_A E \, dA &= \int_A \alpha E T \, dA \\
\frac{1}{\rho_y} \int_A E y^2 \, dA + \frac{1}{\rho_z} \int_A E y z \, dA &= \int_A \alpha E T y \, dA \\
\frac{1}{\rho_y} \int_A E y z \, dA + \frac{1}{\rho_z} \int_A E z^2 \, dA &= \int_A \alpha E T z \, dA
\end{aligned} \tag{e}$$

By solving Eqs. (e), the normal strain ϵ_0 and the curvatures $1/\rho_y$ and $1/\rho_z$ at $y = z = 0$ are expressed as

$$\epsilon_0 = \frac{P_T}{A^* E_0}, \qquad \frac{1}{\rho_y} = \frac{M_{Tz} I_y^* - M_{Ty} I_{yz}^*}{E_0 \left(I_y^* I_z^* - I_{yz}^{*2} \right)}, \qquad \frac{1}{\rho_z} = \frac{M_{Ty} I_z^* - M_{Tz} I_{yz}^*}{E_0 \left(I_y^* I_z^* - I_{yz}^{*2} \right)} \tag{2.40}$$

where

$$\begin{aligned}
A^* &= \int_A \frac{E}{E_0} \, dA \\
I_y^* &= \int_A \frac{E}{E_0} z^2 \, dA, \qquad I_z^* = \int_A \frac{E}{E_0} y^2 \, dA, \qquad I_{yz}^* = \int_A \frac{E}{E_0} y z \, dA \\
P_T &= \int_A \alpha E T \, dA, \qquad M_{Ty} = \int_A \alpha E T z \, dA, \qquad M_{Tz} = \int_A \alpha E T y \, dA
\end{aligned} \tag{2.41}$$

Therefore, the thermal stress is given from Eqs. (2.39) and (2.40) by

$$\begin{aligned}
\sigma_x = -\alpha E T &+ \frac{E}{E_0} \frac{P_T}{A^*} \\
&+ \frac{E}{E_0} \left[\frac{M_{Tz} I_y^* - M_{Ty} I_{yz}^*}{I_y^* I_z^* - I_{yz}^{*2}} y \right] + \frac{E}{E_0} \left[\frac{M_{Ty} I_z^* - M_{Tz} I_{yz}^*}{I_y^* I_z^* - I_{yz}^{*2}} z \right]
\end{aligned} \tag{2.42}$$

2.6 Thermal Deflection in Beams

When a homogeneous beam shown in Fig. 2.11 is subjected to the symmetrical thermal load $T(x, y)$, the total strain ϵ_x, which is the sum of the free thermal strain and the strain due to stress σ_x, equals the sum of the strain ϵ_0 at the centroid of the cross section and the bending strain:

$$\epsilon_x = \alpha T(x, y) + \frac{\sigma_x}{E} = \epsilon_0 + \frac{y}{\rho} \tag{a}$$

Using the equations

$$\epsilon_0 = \frac{1}{EA}\int_A \alpha E T(x, y)\, dA, \qquad \frac{1}{\rho} = \frac{1}{EI}\int_A \alpha E T(x, y) y\, dA \tag{2.43}$$

the strain ϵ_x is

$$\epsilon_x = \frac{1}{EA}\int_A \alpha E T(x, y)\, dA + \frac{y}{EI}\int_A \alpha E T(x, y) y\, dA \tag{b}$$

The axial displacement u may be obtained by integrating the strain $\epsilon_x = \dfrac{\partial u}{\partial x}$.

$$u = u_0 + \int_0^x \left[\frac{1}{EA}\int_A \alpha E T(x, y)\, dA + \frac{y}{EI}\int_A \alpha E T(x, y) y\, dA\right] dx \tag{2.44}$$

where u_0 is the axial displacement at $x = 0$. The average axial displacement for the cross section u_{av} is

$$\begin{aligned} u_{av} &= \frac{1}{A}\int_A u\, dA \\ &= \frac{1}{A}\int_A \left\{u_0 + \int_0^x \left[\frac{1}{EA}\int_A \alpha E T(x, y)\, dA \right.\right. \\ &\qquad \left.\left. + \frac{y}{EI}\int_A \alpha E T(x, y) y\, dA\right] dx\right\} dA \end{aligned} \tag{c}$$

The neutral axis passes through the centroid of the cross section, so the moment of the area of the cross section becomes zero, and the last integral in Eq. (c) becomes zero. Indeed, we observe that the last integral is

$$\frac{1}{A}\int_A \left[\int_0^x \left(\frac{y}{EI}\int_A \alpha E T(x, y) y\, dA\right) dx\right] dA \tag{d}$$

Taking $f(x) = \int_A \alpha E T(x, y) y\, dA$, integral (d) is equal to

$$\frac{1}{A}\int_A \left[\int_0^x \frac{y}{EI} f(x)\, dx\right] dA = \frac{1}{A}\int_0^x \frac{f(x)}{EI}\, dx \int_A y\, dA = 0 \tag{e}$$

because $\int_A y\,dA = 0$. Therefore, we can obtain the average axial displacement u_{av} as

$$u_{av} = u_{0av} + \frac{1}{A}\int_A \left\{ \int_0^x \frac{1}{EA}\left[\int_A \alpha E T(x, y)\,dA\right] dx \right\} dA \qquad (2.45)$$

where

$$u_{0av} = \frac{1}{A}\int_A u_0\,dA \qquad \text{(f)}$$

Next, consider the deflection v of the beam. Using the relationship between the deflection and the curvature

$$\frac{1}{\rho} = -\frac{\dfrac{d^2v}{dx^2}}{\left\{1 + \left(\dfrac{dv}{dx}\right)^2\right\}^{3/2}} \cong -\frac{d^2v}{dx^2} \qquad (2.46)$$

and substituting Eq. (2.43) into the approximate form of Eq. (2.46), the governing equation for the deflection v can be obtained:

$$\frac{d^2v}{dx^2} = -\frac{M_T}{EI} \qquad (2.47)$$

where

$$M_T = \int_A \alpha E T(x, y) y\,dA \qquad (2.48)$$

Integrating Eq. (2.47), the deflection v is given by

$$v = -\int \left(\int \frac{M_T}{EI} dx \right) dx + C_1 x + C_2 \qquad (2.49)$$

The unknown constants C_1, C_2 will be determined from the boundary conditions. There are two types of boundary conditions shown in Fig. 2.13, that is, the simply

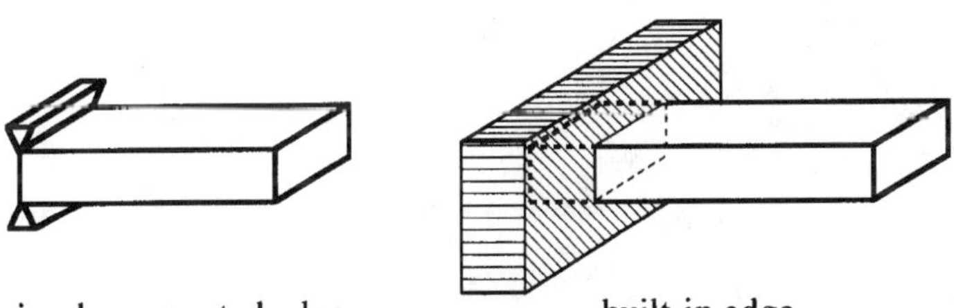

Figure 2.13 Boundary conditions.

supported edge and the built-in edge:

$$\left.\begin{array}{lll} \text{simply supported edge} & v = 0, & \dfrac{d^2 v}{dx^2} + \dfrac{M_T}{EI} = 0 \\ \text{built-in edge} & v = 0, & \dfrac{dv}{dx} = 0 \end{array}\right\} \quad \text{at } x = 0 \tag{2.50}$$

Example 2.9 When a cantilever beam with length l, height h and width b is subjected to the temperature change

$$T = \sum_{i=0}^{n} T_{2i} y^{2i} + \sum_{i=0}^{n} T_{2i+1} y^{2i+1} \tag{a}$$

what are the thermal stress and the thermal deflection in the beam?

Solution The thermal stress in the beam is obtained from Eq. (2.11)

$$\sigma_x = -\alpha E T(y) + \frac{1}{h} \int_{-h/2}^{h/2} \alpha E T(y)\, dy + \frac{12y}{h^3} \int_{-h/2}^{h/2} \alpha E T(y) y\, dy \tag{b}$$

Substitution of Eq. (a) into Eq. (b) yields the thermal stress:

$$\sigma_x = -\alpha E \sum_{i=0}^{n} T_{2i} \left[y^{2i} - \frac{1}{2i+1} \left(\frac{h}{2}\right)^{2i} \right] - \alpha E y \sum_{i=0}^{n} T_{2i+1} \left[y^{2i} - \frac{3}{2i+3} \left(\frac{h}{2}\right)^{2i} \right]$$

[*Answer*]

From Eqs. (a) and (2.48), we can calculate the bending moment due to the temperature change

$$M_T = 2\alpha E b \sum_{i=0}^{n} T_{2i+1} \frac{1}{2i+3} \left(\frac{h}{2}\right)^{2i+3} \tag{c}$$

The bending moment does not contain even terms in the temperature change, i.e., the even terms do not influence the bending moment. A beam with a symmetric cross section subjected to a temperature change $T(x, y)$ which is an even function of the variable y does not deflect but is elongated, since the bending moment becomes zero:

$$M_T = \int_A \alpha E T(x, y) y\, dA = 0$$

From Eqs. (2.49) and (c), we obtain

$$v = -\frac{M_T}{2EI} x^2 + C_1 x + C_2 \tag{d}$$

Considering the boundary conditions of the built-in edge

$$v = 0, \quad \frac{dv}{dx} = 0 \quad \text{at } x = 0 \tag{e}$$

the constants of integration are equal to zero

$$C_1 = 0, \qquad C_2 = 0 \tag{f}$$

The deflection of a beam with $I = bh^3/12$ is then expressed by

$$v = -\frac{3\alpha}{2}\sum_{i=0}^{n} T_{2i+1}\frac{1}{2i+3}\left(\frac{h}{2}\right)^{2i} x^2 \qquad \text{[Answer]}$$

Beam deflection occurs when the temperature change is an odd function of y.

2.7 Curved Beams

We will consider curved beams using the same assumptions of the theory of straight beams. The origin of the coordinate system will be taken at the center line of the curved beam, as shown in Fig. 2.14. The center line lies in one of the centroidal principal planes. If the element of the curved beam with curvature $1/R$ and angle $d\theta$ before thermal loading deforms to the curvature $1/\rho$ and angle $d\theta + \Delta d\theta$ after the temperature change T, the strain ϵ_0 at the center line is given by

$$\begin{aligned}\epsilon_0 &= \frac{\overset{\frown}{n'm'} - \overset{\frown}{nm}}{\overset{\frown}{nm}} = \frac{\rho(d\theta + \Delta d\theta) - R d\theta}{R d\theta} = \frac{\rho - R}{R} + \frac{\rho}{R}\frac{\Delta d\theta}{d\theta} \\ &= \frac{\rho - R}{R} + \frac{\rho}{R}\omega_0\end{aligned} \tag{2.51}$$

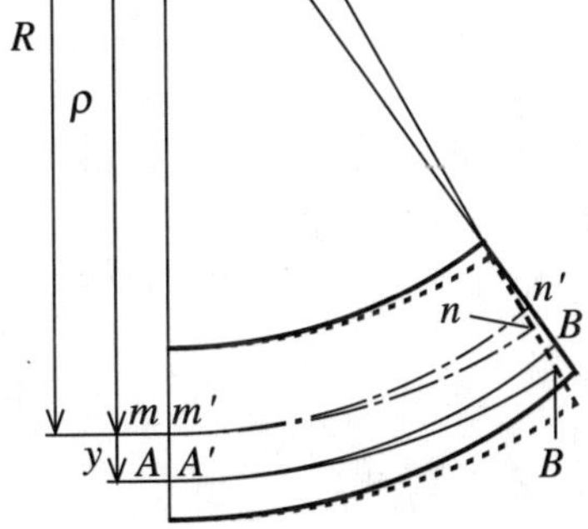

Figure 2.14 Curved beam.

where

$$\omega_0 = \frac{\Delta d\theta}{d\theta} \tag{2.52}$$

The strain ϵ at a distance y from the center line is given by

$$\begin{aligned}\epsilon &= \frac{\widehat{A'B'} - \widehat{AB}}{\widehat{AB}} = \frac{\left(\rho + y + \int_0^y \alpha T\, dy\right)(d\theta + \Delta d\theta) - (R + y)\, d\theta}{(R + y)\, d\theta} \\ &\cong \frac{1}{R + y}\left(\epsilon_0 R + \omega_0 y + \int_0^y \alpha T\, dy\right)\end{aligned} \tag{2.53}$$

In the last expression we neglected the term $\omega_0 \int_0^y \alpha T dy$ as being small compared to $\int_0^y \alpha T dy$. The final strain ϵ in the beam is expressed as the sum of the free thermal strain and the strain due to stress $\sigma_{\theta\theta}$:

$$\epsilon = \alpha T + \frac{\sigma_{\theta\theta}}{E} = \frac{1}{R + y}\left(\epsilon_0 R + \omega_0 y + \int_0^y \alpha T\, dy\right) \tag{2.54}$$

The stress can, therefore, be expressed by

$$\sigma_{\theta\theta} = -\alpha E T + \frac{E}{R + y}\left(\epsilon_0 R + \omega_0 y + \int_0^y \alpha T\, dy\right) \tag{2.55}$$

Since the curved beam is not subjected to external forces and moments, we have

$$\int_A \sigma_{\theta\theta}\, dA = 0 \tag{a}$$

$$\int_A \sigma_{\theta\theta}\, y\, dA = 0 \tag{b}$$

We can obtain the normal strain ϵ_0 and the curvature ω_0 at the center surface $y = 0$ as follows:

$$\epsilon_0 = \frac{N}{EA} + \frac{M}{EAR}, \qquad \omega_0 = \frac{N}{EA} + \frac{M}{EAR}\left(1 + \frac{1}{\kappa}\right) \tag{2.56}$$

where

$$\kappa = -\frac{1}{A}\int_A \frac{y}{R + y}\, dA \tag{2.57}$$

$$\begin{aligned} N &= \int_A \alpha E T dA - \int_A \frac{E}{R + y}\left(\int_0^y \alpha T dy\right) dA \\ M &= \int_A \alpha E T y\, dA - \int_A \frac{Ey}{R + y}\left(\int_0^y \alpha T dy\right) dA \end{aligned} \tag{c}$$

We obtain the curvature after thermal loading at the center surface $y = 0$ from Eqs. (2.51) and (2.56)

$$\rho = \left(\frac{1+\epsilon_0}{1+\omega_0}\right) R = \left(1 - \frac{\dfrac{M}{EAR}\dfrac{1}{\kappa}}{1 + \dfrac{N}{EA} + \dfrac{M}{EAR}\left(1+\dfrac{1}{\kappa}\right)}\right) R \tag{2.58}$$

The substitution of Eq. (2.56) into Eq. (2.55) gives the thermal stress $\sigma_{\theta\theta}$

$$\begin{aligned}
\sigma_{\theta\theta} &= -\alpha E T + \frac{1}{A}\left[N + \frac{M}{R}\left(1 + \frac{y}{\kappa(R+y)}\right)\right] + \frac{E}{R+y}\int_0^y \alpha T\, dy \\
&= -\alpha E T + \frac{1}{A}\left[\int_A \alpha E T\, dA + \frac{1}{R}\left(1 + \frac{y}{\kappa(R+y)}\right)\int_A \alpha E T y\, dA\right] \\
&\quad - \frac{1}{A}\left[\int_A \frac{E}{R+y}\int_0^y \alpha T\, dy\, dA + \frac{1}{R}\left(1 + \frac{y}{\kappa(R+y)}\right)\right. \\
&\quad \left.\times \int_A \frac{Ey}{R+y}\int_0^y \alpha T\, dy\, dA\right] + \frac{E}{R+y}\int_0^y \alpha T\, dy
\end{aligned} \tag{2.59}$$

Example 2.10 When a curved beam with a rectangular cross section $b \times h$ is subjected to a temperature change $T(y, z) = C_0 + C_1 y$, derive the thermal stress in the curved beam.

Solution First, we calculate integrals:

$$\begin{aligned}
\kappa &= -\frac{1}{A}\int_A \frac{y}{R+y}\, dA = -\frac{1}{A}\int_A \left(1 - \frac{R}{R+y}\right) dA \\
&= -\left(1 - \frac{R}{h}\int_{-h/2}^{h/2} \frac{1}{R+y}\, dy\right) = -\left(1 - \frac{R}{h}\ln\frac{R+h/2}{R-h/2}\right)
\end{aligned}$$

$$\int_0^y \alpha T(y)\, dy = \int_0^y \alpha(C_0 + C_1 y)\, dy = \alpha\left(C_0 y + \frac{1}{2}C_1 y^2\right)$$

$$\int_A \alpha E T(y)\, dA = b\int_{-h/2}^{h/2} \alpha E(C_0 + C_1 y)\, dy = \alpha E C_0 b h$$

$$\int_A \alpha E T(y) y\, dA = b\int_{-h/2}^{h/2} \alpha E(C_0 + C_1 y) y\, dy = \alpha E C_1 \frac{bh^3}{12}$$

$$\begin{aligned}
\int_A \frac{E}{R+y}\left(\int_0^y \alpha T\, dy\right) dA &= \alpha E b \int_{-h/2}^{h/2} \frac{1}{R+y}\left(C_0 y + \frac{1}{2}C_1 y^2\right) dy \\
&= \alpha E b \int_{-h/2}^{h/2}\left[\frac{1}{2}C_1 y + C_0 - \frac{1}{2}C_1 R + \left(\frac{1}{2}C_1 R^2 - C_0 R\right)\frac{1}{R+y}\right] dy \\
&= \alpha E b\left(C_0 - \frac{1}{2}C_1 R\right)\left(h - R\ln\frac{R+h/2}{R-h/2}\right)
\end{aligned}$$

$$\int_A \frac{Ey}{R+y}\left(\int_0^y \alpha T\,dy\right)dA$$

$$= \alpha Eb\int_{-h/2}^{h/2}\frac{1}{R+y}\left(C_0y^2+\frac{1}{2}C_1y^3\right)dy$$

$$= \alpha Eb\int_{-h/2}^{h/2}\left[\frac{1}{2}C_1y^2+\left(C_0-\frac{1}{2}C_1R\right)y\right.$$

$$\left.-\left(C_0-\frac{1}{2}C_1R\right)R+\left(C_0-\frac{1}{2}C_1R\right)R^2\frac{1}{R+y}\right]dy$$

$$= \alpha Eb\left[\frac{1}{24}C_1h^3-\left(C_0-\frac{1}{2}C_1R\right)R\left(h-R\ln\frac{R+h/2}{R-h/2}\right)\right]$$

Substitution of these results into Eqs. (c) in Section 2.7 gives

$$N = \alpha Eb\left[C_0h-\left(C_0-\frac{1}{2}C_1R\right)\left(h-R\ln\frac{R+h/2}{R-h/2}\right)\right]$$
$$M = \alpha Eb\left[C_1\frac{h^3}{24}+\left(C_0-\frac{1}{2}C_1R\right)R\left(h-R\ln\frac{R+h/2}{R-h/2}\right)\right] \tag{a}$$

We now receive the thermal stress:

$$\sigma_{\theta\theta} = -\alpha E(C_0+C_1y)$$
$$+\frac{\alpha Eb}{bh}\left\{\left[C_0h-\left(C_0-\frac{1}{2}C_1R\right)\left(h-R\ln\frac{R+h/2}{R-h/2}\right)\right]\right.$$
$$+\frac{1}{R}\left[C_1\frac{h^3}{24}+\left(C_0-\frac{1}{2}C_1R\right)R\left(h-R\ln\frac{R+h/2}{R-h/2}\right)\right]$$
$$\left.\times\left[1+\frac{y}{\kappa(R+y)}\right]\right\}+\frac{\alpha E}{R+y}\left(C_0y+\frac{1}{2}C_1y^2\right)$$
$$= -\frac{1}{2}\alpha EC_1\left\{y-\left[1+\frac{y}{\kappa(R+y)}\right]\frac{h^2}{12R}\right\} \tag{b}$$

2.8 Thermal Shearing Stresses in Thin-Walled Beams

Up to this point, we have focused on the normal stress, the normal strain, and the thermal deflection due to the normal stress that occurred in beams subjected to specific temperature changes. The shearing stress in beams due to thermal loads is usually very small in comparison with the normal stress, so that it is customary to neglect it in most cases. The shearing stress, however, should be considered in special cases such as a *thin-walled* open beam shown in Fig. 2.15. The term *thin-walled* is typically used when the thickness is less than or equal to 5% of the average radius. In Fig. 2.15, s is the length measured from one of its free ends and $t(s)$ is the thickness of the

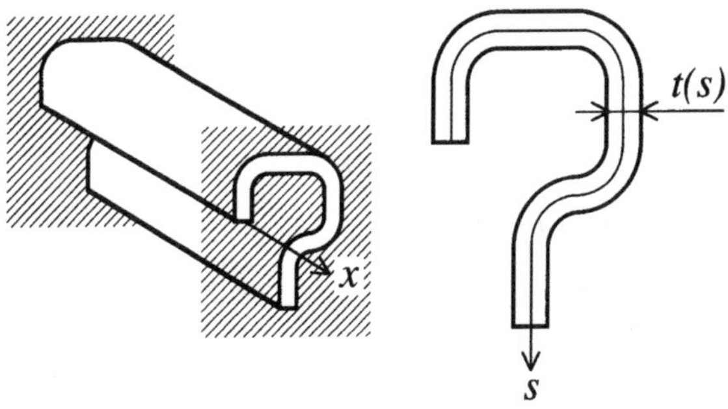

Figure 2.15 A thin-walled open beam.

thin-walled open beam. We will consider the deformation of the beam under the following assumptions:

1. The stress does not vary through the thickness.
2. The direction of the shearing stress is tangent to the median line.

We introduce the shearing force per unit length q, called the shear flow, defined by

$$q = \sigma_{xs} t(s) \tag{2.60}$$

where the symbol σ_{xs} denotes the shearing stress along the s axis on the x plane.

Let us consider the equilibrium of forces in the x direction in a small element of the beam shown in Fig. 2.15. As the forces in the x direction due to stresses acting on the x planes

$$\left(\sigma_x + \frac{\partial \sigma_x}{\partial x}\,dx\right) t\,ds - \sigma_x t\,ds$$

must balance the forces in the x direction due to stresses acting on the s planes:

$$\left(\sigma_{xs} + \frac{\partial \sigma_{xs}}{\partial s}\,ds\right)\left(t + \frac{dt}{ds}\,ds\right) dx - \sigma_{xs} t\,dx$$

we get

$$\left(\sigma_x + \frac{\partial \sigma_x}{\partial x}\,dx\right) t\,ds - \sigma_x t\,ds + \left(\sigma_{xs} + \frac{\partial \sigma_{xs}}{\partial s}\,ds\right)\left(t + \frac{dt}{ds}\,ds\right) dx - \sigma_{xs} t\,dx = 0 \tag{a}$$

Simplification of Eq. (a) gives

$$\frac{\partial \sigma_x}{\partial x} t + \frac{\partial \sigma_{xs}}{\partial s} t + \sigma_{xs} \frac{dt}{ds} = 0 \tag{2.61}$$

By use of Eq. (2.60), Eq. (2.61) reduces to

$$\frac{\partial \sigma_x}{\partial x} + \frac{1}{t}\frac{\partial q}{\partial s} = 0 \tag{2.62}$$

Integration of Eq. (2.62) from $s = 0$ to $s = s$ yields

$$q = q_0 - \int_0^s \frac{\partial \sigma_x}{\partial x} t\, ds \tag{2.63}$$

in which q_0 is the shearing force per unit length at the starting point. The starting point is usually chosen as a free edge in order to ensure that the shearing force per unit length is zero. Substituting Eq. (2.42) into Eq. (2.63), the shearing force per unit length for the nonhomogeneous beam is given by

$$\begin{aligned} q(x,s) = -\int_0^s \frac{\partial}{\partial x}\Bigg\{ & -\alpha E T + \frac{E}{E_0}\frac{P_T}{A^*} + \frac{E}{E_0}\left[\frac{M_{Tz}I_y^* - M_{Ty}I_{yz}^*}{I_y^* I_z^* - I_{yz}^{*2}}\right] y \\ & + \frac{E}{E_0}\left[\frac{M_{Ty}I_z^* - M_{Tz}I_{yz}^*}{I_y^* I_z^* - I_{yz}^{*2}}\right] z \Bigg\} t\, ds \end{aligned} \tag{2.64}$$

When the material properties do not vary in x direction, Eq. (2.64) reduces to

$$\begin{aligned} q(x,s) = & \int_0^s \alpha E \frac{\partial T}{\partial x} t\, ds - \frac{1}{A^*}\frac{dP_T}{dx}\int_0^s \frac{E}{E_0} t\, ds \\ & - \left[\frac{\dfrac{dM_{Tz}}{dx} I_y^* - \dfrac{dM_{Ty}}{dx} I_{yz}^*}{I_y^* I_z^* - I_{yz}^{*2}}\right] \int_0^s \frac{E}{E_0} y t\, ds \\ & - \left[\frac{\dfrac{dM_{Ty}}{dx} I_z^* - \dfrac{dM_{Tz}}{dx} I_{yz}^*}{I_y^* I_z^* - I_{yz}^{*2}}\right] \int_0^s \frac{E}{E_0} z t\, ds \end{aligned} \tag{2.65}$$

Example 2.11 When thickness of a homogeneous thin-walled open beam is constant, determine the shear flow in the beam.

Solution From Eqs. (2.33) and (2.63) we get

$$\begin{aligned} q(x,s) = q_0 - \int_0^s \frac{\partial}{\partial x}\Bigg\{ & -\alpha E T + \frac{P_T}{A} + \frac{M_{Tz}I_y - M_{Ty}I_{yz}}{I_y I_z - I_{yz}^2} y \\ & + \frac{M_{Ty}I_z - M_{Tz}I_{yz}}{I_y I_z - I_{yz}^2} z \Bigg\} t\, ds \end{aligned}$$

$$= q_0 + \alpha E t \int_0^s \frac{\partial T}{\partial x} ds - \frac{dP_T}{dx}\frac{ts}{A} - \frac{\dfrac{dM_{Tz}}{dx} I_y - \dfrac{dM_{Ty}}{dx} I_{yz}}{I_y I_z - I_{yz}^2} t \int_0^s y\, ds$$

$$- \frac{\dfrac{dM_{Ty}}{dx} I_z - \dfrac{dM_{Tz}}{dx} I_{yz}}{I_y I_z - I_{yz}^2} t \int_0^s z\, ds \tag{a}$$

2.9 Thermal Stresses in Beams on an Elastic Foundation

Let us consider a homogeneous beam supported by an elastic foundation along its whole length, shown in Fig. 2.16. When the beam is deflected, the reaction from the elastic foundation which is proportional to the deflection v supports the beam. The reaction per unit length along the beam can be expressed by kv. The coefficient k is called the modulus of the foundation. When the beam is subjected to thermal and mechanical loads, the equilibrium equations of the external force and the bending moment M are

$$\int_A \sigma_x \, dA = 0 \tag{a}$$

$$\int_A \sigma_x y \, dA = M \tag{b}$$

The relation between stress and strain is given by Eq. (2.14)

$$\sigma_x = -\alpha E T + \epsilon_0 E + E\frac{y}{\rho} \tag{c}$$

From Eqs. (a), (b), and (c), the axial strain ϵ_0 and the curvature $1/\rho$ at the neutral plane $y = 0$ are

$$\epsilon_0 = \frac{\alpha}{A}\int_A T(x, y)\, dA \tag{d}$$

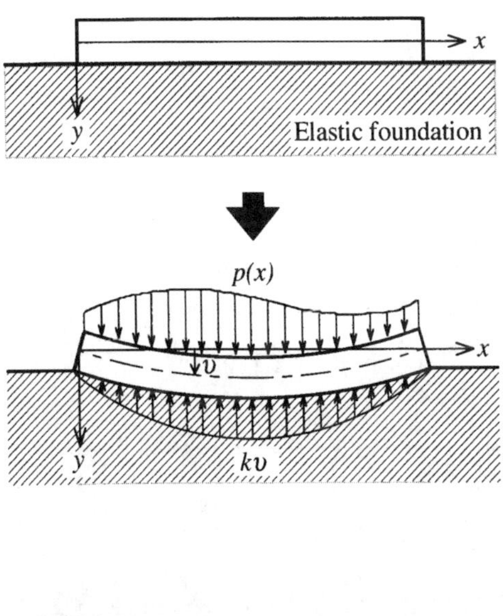

Figure 2.16 A beam on an elastic foundation.

$$\frac{1}{\rho} = \frac{M}{EI} + \frac{\alpha}{I}\int_A T(x, y)y\,dA \tag{e}$$

We recall that when the beam is subjected to a load per unit length q, the shearing force F, the bending moment M, and the load q are given by*

$$\frac{dM}{dx} = F, \qquad \frac{dF}{dx} = -q, \qquad \frac{d^2M}{dx^2} = -q \tag{f}$$

In the case of the beam on an elastic foundation, the load per unit length along the beam q consists of two parts, one is lateral load per unit length p, the other is reaction from the foundation, kv, since it is assumed that the beam does not separate from the elastic foundation. Hence $q = p - kv$ and Eq. (f) reduces to

$$\frac{d^2M}{dx^2} = -p + kv \tag{g}$$

Using the relation (2.46)

$$\frac{1}{\rho} = -\frac{\dfrac{d^2v}{dx^2}}{\left\{1 + \left(\dfrac{dv}{dx}\right)^2\right\}^{3/2}} \cong -\frac{d^2v}{dx^2} \tag{h}$$

the substitution of Eq. (e) into Eq. (g) gives the differential equation for deflection in beam on an elastic foundation subjected to mechanical and thermal loads

$$\frac{d^4v}{dx^4} + \frac{k}{EI}v = \frac{p}{EI} - \frac{\alpha}{I}\frac{d^2}{dx^2}\int_A T(x, y)y\,dA \tag{2.66}$$

When lateral load p is removed, the differential equation for deflection reduces to

$$\frac{d^4v}{dx^4} + \frac{k}{EI}v = -\frac{\alpha}{I}\frac{d^2}{dx^2}\int_A T(x, y)y\,dA \tag{2.67}$$

Using the notation

$$\sqrt[4]{\frac{k}{4EI}} = \beta \tag{2.68}$$

the general solution of Eq. (2.67) can be expressed by

$$v(x) = e^{\beta x}(C_1 \cos\beta x + C_2 \sin\beta x) + e^{-\beta x}(C_3 \cos\beta x + C_4 \sin\beta x) + v_p \tag{2.69}$$

*See: S. Timoshenko, Strength of Materials, Part 1 Elementary, Van Nostrand Reinhold, Third Edition, Berkshire (1995), Eqs. (50) and (51), pp. 77–78.

where v_p is a particular solution of Eq. (2.67). The unknown constants C_1, C_2, C_3, C_4 can be determined from the boundary conditions. There are three types of the boundary conditions, that is, the free edge, the simply supported edge, and the built-in edge:

Free edge

$$F = EI\left[-\frac{d^3v}{dx^3} - \frac{\alpha}{I}\frac{d}{dx}\int_A T(x, y)y\,dA\right] = 0$$
$$M = EI\left[-\frac{d^2v}{dx^2} - \frac{\alpha}{I}\int_A T(x, y)y\,dA\right] = 0 \tag{2.70a}$$

Simply supported edge

$$v = 0, \qquad M = EI\left[-\frac{d^2v}{dx^2} - \frac{\alpha}{I}\int_A T(x, y)y\,dA\right] = 0 \tag{2.70b}$$

Built-in edge

$$v = 0, \qquad \frac{dv}{dx} = 0 \tag{2.70c}$$

Moreover, the thermal stress is determined from Eqs. (c), (d), and (h)

$$\sigma_x = -\alpha E T(x, y) + \frac{\alpha E}{A}\int_A T(x, y)\,dA - Ey\frac{d^2v}{dx^2} \tag{2.71}$$

Let us consider a beam with a rectangular cross section $b \times h$ on an elastic foundation with simply supported edges subjected to the temperature distribution $T = [(T_1 + T_0)/2] + [(T_1 - T_0)/h]y$. In this case, the differential equation for deflection (2.67) reduces to

$$\frac{d^4v}{dx^4} + \frac{k}{EI}v = 0 \tag{i}$$

The boundary conditions are

$$v = 0, \quad M = EI\left[-\frac{d^2v}{dx^2} - \frac{\alpha}{I}\int_A T(x, y)y\,dA\right] = 0 \qquad \text{at } x = 0, l \tag{j}$$

The general solution of Eq. (i) is

$$v(x) = e^{\beta x}(C_1 \cos\beta x + C_2 \sin\beta x) + e^{-\beta x}(C_3 \cos\beta x + C_4 \sin\beta x) \tag{k}$$

From the boundary conditions, we get

$$C_1 + C_3 = 0$$
$$2\beta^2(C_2 - C_4) = -\alpha\frac{T_1 - T_0}{h}$$

$$e^{\beta l}(C_1 \cos \beta l + C_2 \sin \beta l) + e^{-\beta l}(C_3 \cos \beta l + C_4 \sin \beta l) = 0$$

$$2\beta^2 \left[e^{\beta l}(C_2 \cos \beta l - C_1 \sin \beta l) + e^{-\beta l}(-C_4 \cos \beta l + C_3 \sin \beta l) \right] = -\alpha \frac{T_1 - T_0}{h} \tag{l}$$

By solving the system of Eqs. (l), the unknown constants are obtained

$$\begin{aligned}
C_1 &= \alpha \frac{T_1 - T_0}{h} \frac{\sin \beta l (\cosh \beta l - \cos \beta l)}{4\beta^2 (\sinh^2 \beta l + \sin^2 \beta l)} \\
C_2 &= \alpha \frac{T_1 - T_0}{h} \frac{(e^{-\beta l} \sinh \beta l - \sin^2 \beta l - \sinh \beta l \cos \beta l)}{4\beta^2 (\sinh^2 \beta l + \sin^2 \beta l)} \\
C_3 &= -\alpha \frac{T_1 - T_0}{h} \frac{\sin \beta l (\cosh \beta l - \cos \beta l)}{4\beta^2 (\sinh^2 \beta l + \sin^2 \beta l)} \\
C_4 &= \alpha \frac{T_1 - T_0}{h} \frac{(e^{\beta l} \sinh \beta l + \sin^2 \beta l - \sinh \beta l \cos \beta l)}{4\beta^2 (\sinh^2 \beta l + \sin^2 \beta l)}
\end{aligned} \tag{m}$$

Then, the deflection is

$$v = \alpha \frac{T_1 - T_0}{h} \frac{\cosh \beta l - \cos \beta l}{2\beta^2 (\sinh^2 \beta l + \sin^2 \beta l)} \times [\sinh \beta x \sin \beta (l - x) + \sinh \beta (l - x) \sin \beta x] \tag{2.72}$$

Substitution of Eq. (2.72) into Eq. (2.71) gives the thermal stress in the beam:

$$\sigma_x = -\alpha E \frac{T_1 - T_0}{h} y \times \left\{ 1 - \frac{\cosh \beta l - \cos \beta l}{\sinh^2 \beta l + \sin^2 \beta l} \times [\cosh \beta x \cos \beta (l - x) + \cosh \beta (l - x) \cos \beta x] \right\} \tag{2.73}$$

Example 2.12 A beam of a rectangular cross section $b \times h$ supported by an elastic foundation is subjected to a temperature change $T = [(T_1 + T_0)/2] + [(T_1 - T_0)/h]y$. When one end of the beam is clamped and the other end is free, calculate the deflection and the thermal stress in the beam.

Solution Because of the form of the temperature change, the governing equation for the deflection (2.67) is reduced to

$$\frac{d^4 v}{dx^4} + \frac{k}{EI} v = 0 \tag{a}$$

The boundary conditions are

$$v = 0, \quad \frac{dv}{dx} = 0 \qquad \text{at } x = 0 \tag{b}$$

$$\left.\begin{aligned} \frac{d^3v}{dx^3} + \frac{\alpha}{I}\frac{d}{dx}\int_A T(x,y)y\,dA = 0 \\ \frac{d^2v}{dx^2} + \frac{\alpha}{I}\int_A T(x,y)y\,dA = 0 \end{aligned}\right\} \qquad \text{at } x = l \tag{c}$$

The general solution of Eq. (a) is

$$v(x) = e^{\beta x}(C_1 \cos\beta x + C_2 \sin\beta x) + e^{-\beta x}(C_3 \cos\beta x + C_4 \sin\beta x) \tag{d}$$

Substitution of Eq. (d) into the boundary conditions yields

$$\begin{aligned}
&C_1 + C_3 = 0 \\
&C_1 + C_2 - C_3 + C_4 = 0 \\
&-e^{-\beta l}(\cos\beta l + \sin\beta l)C_1 + e^{\beta l}(\cos\beta l - \sin\beta l)C_2 \\
&\quad + e^{-\beta l}(\cos\beta l - \sin\beta l)C_3 + e^{-\beta l}(\cos\beta l + \sin\beta l)C_4 = 0 \\
&e^{\beta l}\sin\beta l C_1 - e^{\beta l}\cos\beta l C_2 - e^{-\beta l}\sin\beta l C_3 + e^{-\beta l}\cos\beta l C_4 = \alpha\frac{T_1 - T_0}{h}\frac{1}{2\beta^2}
\end{aligned} \tag{e}$$

By solving the system of equations (e), the unknown constants are obtained

$$\begin{aligned}
C_1 &= \alpha\frac{T_1 - T_0}{h}\frac{1}{4\beta^2(\cosh^2\beta l + \cos^2\beta l)}(\cosh\beta l \sin\beta l - \sinh\beta l \cos\beta l) \\
C_2 &= -\alpha\frac{T_1 - T_0}{h}\frac{1}{4\beta^2(\cosh^2\beta l + \cos^2\beta l)}[\cosh\beta l(\cos\beta l + \sin\beta l) + e^{-\beta l}\cos\beta l] \\
C_3 &= \alpha\frac{T_1 - T_0}{h}\frac{1}{4\beta^2(\cosh^2\beta l + \cos^2\beta l)}(\sinh\beta l \cos\beta l - \cosh\beta l \sin\beta l) \\
C_4 &= \alpha\frac{T_1 - T_0}{h}\frac{1}{4\beta^2(\cosh^2\beta l + \cos^2\beta l)}[\cosh\beta l(\cos\beta l - \sin\beta l) + e^{\beta l}\cos\beta l]
\end{aligned}$$

Then, the deflection is

$$\begin{aligned}
v = \alpha\frac{T_1 - T_0}{h}&\frac{1}{4\beta^2(\cosh^2\beta l + \cos^2\beta l)} \\
&\times [\cosh\beta(l - x)\cos\beta(l + x) - \cosh\beta(l + x)\cos\beta(l - x) \\
&+ \sinh\beta(l + x)\sin\beta(l - x) - \sinh\beta(l - x)\sin\beta(l + x)]
\end{aligned} \tag{f}$$

Substitution of Eq. (f) into Eq. (2.71) gives the thermal stress:

$$\sigma_x = -\alpha E \frac{T_1 - T_0}{h} y \left\{ 1 + \frac{1}{2(\cosh^2 \beta l + \cos^2 \beta l)} \right.$$
$$\times [\cosh \beta(l-x) \cos \beta(l+x) - \cosh \beta(l+x) \cos \beta(l-x)$$
$$\left. - \sinh \beta(l+x) \sin \beta(l-x) + \sinh \beta(l-x) \sin \beta(l+x)] \right\} \quad \text{(g)}$$

Problems

2.1 A rectangular beam with a cross section $b \times h$ is subjected to the temperature change $T(y) = C_1 y + C_0$. Calculate the thermal stresses in the beam in the following cases:

[1] Beam with perfectly clamped end.
[2] Beam with free extension and restrained bending.
[3] Beam with restrained extension and free bending.
[4] Beam with free extension and free bending.

Answer

$$[1]\quad \sigma_x = -\alpha E(C_1 y + C_0) \qquad [2]\quad \sigma_x = -\alpha E C_1 y$$
$$[3]\quad \sigma_x = -\alpha E C_0 \qquad [4]\quad \sigma_x = 0, \quad \frac{1}{\rho} = \alpha C_1$$

2.2 When a rectangular beam with a cross section $b \times h$ is subjected to the temperature change $T(y) = \sum_{n=0}^{\infty}(C_{2n} y^{2n} + C_{2n+1} y^{2n+1})$, what is the thermal stress and what curvature is produced in the beam assuming a stress free boundary condition?

Answer

$$\sigma_x = -\alpha E \sum_{n=0}^{\infty} \left\{ C_{2n} \left[y^{2n} - \frac{1}{2n+1} \left(\frac{h}{2} \right)^{2n} \right] \right.$$
$$\left. + C_{2n+1} \left[y^{2n+1} - \frac{3}{2n+3} \left(\frac{h}{2} \right)^{2n} y \right] \right\}$$
$$(\sigma_x)_{y=\pm h/2} = -\alpha E \sum_{n=0}^{\infty} n \left(\frac{h}{2} \right)^{2n} \left(\frac{2}{2n+1} C_{2n} \pm \frac{h}{2n+3} C_{2n+1} \right)$$
$$\frac{1}{\rho} = 3\alpha \sum_{n=0}^{\infty} \frac{C_{2n+1}}{2n+3} \left(\frac{h}{2} \right)^{2n}$$

2.3 Prove that the thermal stress in an unrestrained beam is zero when the temperature rise in the beam is given by a linear function.

2.4 When the origin of the coordinate system does not lie at the centroid of the section, find the thermal stresses in the beam with temperature rise T.

Answer

$$\sigma_x = -\alpha E T(y) + \frac{1}{(AI_2 - I_1^2)}\left[I_2 \int_A \alpha E T(y)\, dA - I_1 \int_A \alpha E T(y) y\, dA\right]$$
$$+ \frac{y}{(AI_2 - I_1^2)}\left[A \int_A \alpha E T(y) y\, dA - I_1 \int_A \alpha E T(y)\, dA\right]$$

where

$$I_2 = \int_A y^2\, dA, \qquad I_1 = \int_A y\, dA$$

2.5 When a two-layered beam in case [3] in Subsection 2.3.1 is subjected to the same temperature change T, what are the thermal stress and curvature produced? The upper beam and the lower beam are made of mild steel and of aluminum, respectively, and the temperature change is $T = 100$ K. See Table 1.1 for material properties.

Answer

$$\sigma_{xs} = \left(104.7 + 170.9\frac{y}{h}\right) \text{MPa} \quad -h \le y \le 0$$
$$\sigma_{xa} = -\left(49.1 - 59.7\frac{y}{h}\right) \text{MPa} \quad 0 \le y \le h$$

2.6 When two parallel bars with a rectangular cross section are clamped to a rigid plate and are subjected to the different constant temperature rises T_i in Fig. 2.7, what thermal stresses are produced?

Answer

$$\sigma_{x1} = \frac{1}{D}(\alpha_2 T_2 - \alpha_1 T_1) E_1 E_2 A_2 (E_1 I_1 + E_2 I_2 + y_1 e E_1 A_1)$$
$$\sigma_{x2} = \frac{1}{D}(\alpha_2 T_2 - \alpha_1 T_1) E_1 E_2 A_1 [-(E_1 I_1 + E_2 I_2) + y_2 e E_2 A_2]$$

where

$$D = (E_1 A_1 + E_2 A_2)(E_1 I_1 + E_2 I_2) + E_1 E_2 A_2 A_1 e^2$$
$$A_1 = b_1 h_1, \quad A_2 = b_2 h_2, \quad I_1 = \frac{1}{12} b_1 h_1^3 = \frac{1}{12} A_1 h_1^2, \quad I_2 = \frac{1}{12} b_2 h_2^3 = \frac{1}{12} A_2 h_2^2$$

2.7 Calculate the thermal stress and curvature produced in a beam if the elongation at $y = 0$ is restrained to zero, as in Example 2.1.

Answer

$$\sigma_1(y_1) = -\alpha_1 E_1 T_1(y_1) + E_1(y_1 - e_1)\frac{M_T}{I_{E2}}$$

$$\sigma_2(y_2) = -\alpha_2 E_2 T_2(y_2) + E_2(y_2 + e_2)\frac{M_T}{I_{E2}}$$

where

$$I_{E2} = E_1\left(I_1 + A_1 e_1^2\right) + E_2\left(I_2 + A_2 e_2^2\right)$$

$$M_T = \sum_{i=1}^{2} \int_{A_i} \alpha_i E_i T_i(y_i) y_i \, dA_i$$

$$+ e_2 \int_{A_2} \alpha_2 E_2 T_2(y_2)\, dA_2 - e_1 \int_{A_1} \alpha_1 E_1 T_1(y_1)\, dA_1$$

2.8 When the temperature on the upper surface and the lower surface of the two-layered beam with rectangular cross section shown in Fig. 2.5 is raised from an initial temperature T_0 to T_a and T_b, respectively, the temperature rises from initial temperature are given by

$$\begin{aligned} T_1^* &= T_a - (T_a - T_b)\frac{1 + y/h_1}{1 + \lambda_1 h_2/\lambda_2 h_1} \qquad -h_1 \le y \le 0 \\ T_2^* &= T_a - (T_a - T_b)\frac{1 + \lambda_1 y/\lambda_2 h_1}{1 + \lambda_1 h_2/\lambda_2 h_1} \qquad 0 \le y \le h_2 \end{aligned} \tag{a}$$

in which λ_1 and λ_2 denote the thermal conductivities of the upper and lower beams, respectively. Calculate the resulting thermal stress and curvature if the two-layered beam is subjected to temperature rises (a).

Answer

$$\sigma_{xi} = -\alpha_i E_i (T_a - T_0)$$

$$\times \left\{1 - \frac{1}{D\alpha_i}\left[4(\alpha_1 E_1 b_1 h_1 + \alpha_2 E_2 b_2 h_2)\left(E_1 b_1 h_1^3 + E_2 b_2 h_2^3\right)\right.\right.$$

$$\left. - 3\left(\alpha_1 E_1 b_1 h_1^2 - \alpha_2 E_2 b_2 h_2^2\right)\left(E_1 b_1 h_1^2 - E_2 b_2 h_2^2\right)\right]$$

$$+ \frac{6y}{D\alpha_i}\left[\left(\alpha_1 E_1 b_1 h_1^2 - \alpha_2 E_2 b_2 h_2^2\right)(E_1 h_1 b_1 + E_2 h_2 b_2)\right.$$

$$\left.\left. -(\alpha_1 E_1 b_1 h_1 + \alpha_2 E_2 b_2 h_2)\left(E_1 b_1 h_1^2 - E_2 b_2 h_2^2\right)\right]\right\}$$

$$
\begin{aligned}
&- \alpha_i E_i (T_b - T_a) K \\
&\times \left[(1 + C_i y) - \frac{1}{D\alpha_i} \left\{ 2[\alpha_1 E_1 b_1 h_1 (2 - C_1 h_1) + \alpha_2 E_2 b_2 h_2 (2 + C_2 h_2)] \right. \right. \\
&\times \left(E_1 b_1 h_1^3 + E_2 b_2 h_2^3 \right) - \left[\alpha_1 E_1 b_1 h_1^2 (3 - 2C_1 h_1) - \alpha_2 E_2 b_2 h_2^2 (3 + 2C_2 h_2) \right] \\
&\times \left. \left(E_1 b_1 h_1^2 - E_2 b_2 h_2^2 \right) \right\} \\
&+ \frac{y}{D\alpha_i} \left\{ 2\left[\alpha_1 E_1 b_1 h_1^2 (3 - 2C_1 h_1) - \alpha_2 E_2 b_2 h_2^2 (3 + 2C_2 h_2) \right] \right. \\
&\times (E_1 h_1 b_1 + E_2 h_2 b_2) - 3\left[\alpha_1 E_1 b_1 h_1 (2 - C_1 h_1) + \alpha_2 E_2 b_2 h_2 (2 + C_2 h_2) \right] \\
&\times \left. \left. \left(E_1 b_1 h_1^2 - E_2 b_2 h_2^2 \right) \right\} \right] \qquad (i = 1, 2)
\end{aligned}
$$

where

$$
K = 1/(1 + \lambda_1 h_2 / \lambda_2 h_1), \qquad C_1 = 1/h_1, \qquad C_2 = \lambda_1 / \lambda_2 h_1
$$

$$
D = \left(E_2 b_2 h_2^2 - E_1 b_1 h_1^2 \right)^2 + 4 E_1 E_2 h_1 h_2 (h_1 + h_2)^2 b_1 b_2
$$

2.9 Prove that the boundary condition of simply supported edge of the beam is given by Eqs. (2.50).

2.10 When a simply supported beam with length l, height h, and width b is subjected to the temperature change

$$
T = \sum_{i=0}^{n} T_{2i} y^{2i} + \sum_{i=0}^{n} T_{2i+1} y^{2i+1} \tag{a}
$$

what is the thermal deflection in the beam?

Answer

$$
v = \frac{3}{2} \alpha \sum_{i=0}^{n} \frac{T_{2i+1}}{2i + 3} \left(\frac{h}{2} \right)^{2i} x(l - x)
$$

2.11 When a curved beam of uniform rectangular cross section, height h, and width b, whose center line is a quarter of a circle in Fig. 2.14, is subjected to uniform temperature rise T, what are the thermal stress, the radius of curvature, and the change of θ in the curved beam?

Answer

$$
\sigma_{\theta\theta} = 0, \qquad \rho = (1 + \kappa T) R, \qquad \Delta\theta = 0
$$

Chapter 3

HEAT CONDUCTION

The Fourier heat conduction equation is derived and stated in various coordinate systems. Six sets of boundary conditions, along with the appropriate initial condition, are introduced: (1) prescribed surface temperature; (2) the adiabatic surface; (3) prescribed heat flux; (4) heat transfer by convection; (5) heat transfer by radiation; (6) heat transfer by convection and radiation. Detailed solution methodologies for one-dimensional temperature variations are provided by the separation of variables method and the Laplace transform technique in Cartesian, cylindrical, and spherical coordinate systems. The temperature distributions in a strip, a solid cylinder, a hollow cylinder, a solid sphere and a hollow sphere are described in detail.

3.1 Heat Conduction Equation

It was noted that it was necessary to know the temperature distribution in a bar and in a beam in order to obtain the thermal stresses considered in Chapters 1 and 2. We will here discuss means for calculating the temperature distribution in a solid body. There are three kinds of the transfer of heat, that is (1) conduction, (2) convection, and (3) radiation. The transfer of heat between two particles of the solid body takes place by conduction in which heat flows through the particles of the body. When over-all transfer of the heat in the solid body is considered, the transfer due to radiation is usually negligible. There is also no convection. Therefore, the temperature distribution in the solid body is governed mainly by heat conduction. The amount of heat transferred from the high temperature point in the solid body to the low temperature point through the isothermal surface per unit area per unit time is called the heat flux q with dimension $[\mathrm{W/m^2}]$ and is equal to

$$q = -\lambda \frac{\partial T}{\partial n} \tag{3.1}$$

where λ is the thermal conductivity of the solid which is measured in [W/(m · K)]. Here, $\partial/\partial n$ denotes differentiation along out-drawn normal n to the isothermal surface. Equation (3.1) is referred to as the Fourier law of heat conduction.

We will derive the Fourier heat conduction equation for the solid based on the Fourier law of the heat conduction (3.1). When the temperature at an arbitrary point $A(x, y, z)$ and at time t is denoted by $T(x, y, z, t)$, consider the energy balance in an element of volume $(dx\,dy\,dz)$ of the solid with thermal conductivity λ, shown in Fig. 3.1. If the thermal conductivity is different in all directions, the components of heat flux along the x, y, and z directions are expressed, respectively, by

$$q_x = -\lambda_x \frac{\partial T}{\partial x}, \qquad q_y = -\lambda_y \frac{\partial T}{\partial y}, \qquad q_z = -\lambda_z \frac{\partial T}{\partial z} \tag{a}$$

where λ_x, λ_y, and λ_z denote the thermal conductivities along the x, y, and z directions, respectively. The amount of heat $(dQ_x)_{in}$ with dimension [J] which flows into the element of volume through the surface ABCD in the time interval dt is

$$(dQ_x)_{in} = q_x\,dy\,dz\,dt = -\lambda_x \frac{\partial T}{\partial x}\,dy\,dz\,dt \tag{b}$$

The heat $(dQ_x)_{out}$ which flows out of the element of volume through the surface EFGH in the time interval dt is

$$\begin{aligned}(dQ_x)_{out} &= \left(q_x + \frac{\partial q_x}{\partial x}\,dx\right) dy\,dz\,dt \\ &= -\left[\lambda_x \frac{\partial T}{\partial x} + \frac{\partial}{\partial x}\left(\lambda_x \frac{\partial T}{\partial x}\right) dx\right] dy\,dz\,dt\end{aligned} \tag{c}$$

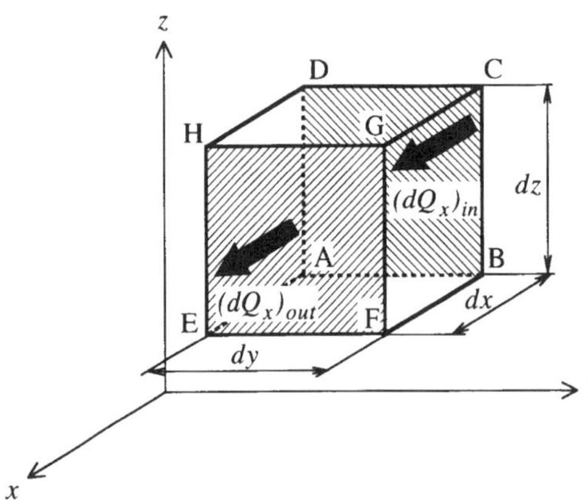

Figure 3.1 An element of volume.

The amount of heat which remains in the element of volume due to the heat flow along the x direction is

$$(dQ_x)_{in} - (dQ_x)_{out} = -\frac{\partial q_x}{\partial x}\,dx\,dy\,dz\,dt = \frac{\partial}{\partial x}\left(\lambda_x \frac{\partial T}{\partial x}\right) dx\,dy\,dz\,dt \tag{d}$$

Similar expressions hold for the directions of y and z:

$$(dQ_y)_{in} - (dQ_y)_{out} = -\frac{\partial q_y}{\partial y}\,dx\,dy\,dz\,dt = \frac{\partial}{\partial y}\left(\lambda_y \frac{\partial T}{\partial y}\right) dx\,dy\,dz\,dt \tag{e}$$

$$(dQ_z)_{in} - (dQ_z)_{out} = -\frac{\partial q_z}{\partial z}\,dx\,dy\,dz\,dt = \frac{\partial}{\partial z}\left(\lambda_z \frac{\partial T}{\partial z}\right) dx\,dy\,dz\,dt \tag{f}$$

The amount of heat dQ_1 which remains in the element of volume due to the heat flow through the surfaces is

$$\begin{aligned} dQ_1 &= -\left(\frac{\partial q_x}{\partial x} + \frac{\partial q_y}{\partial y} + \frac{\partial q_z}{\partial z}\right) dx\,dy\,dz\,dt = -\operatorname{div}\boldsymbol{q}\,dx\,dy\,dz\,dt \\ &= \left[\frac{\partial}{\partial x}\left(\lambda_x \frac{\partial T}{\partial x}\right) + \frac{\partial}{\partial y}\left(\lambda_y \frac{\partial T}{\partial y}\right) + \frac{\partial}{\partial z}\left(\lambda_z \frac{\partial T}{\partial z}\right)\right] dx\,dy\,dz\,dt \end{aligned} \tag{g}$$

If there is in the solid an internal heat generation Q per unit volume per unit time, the amount of heat dQ_2 generated in the element of volume in the time interval dt is

$$dQ_2 = Q\,dx\,dy\,dz\,dt \tag{h}$$

The temperature rise dT in the element of volume occurs due to the amount of heat dQ_1 and dQ_2. On the other hand, the amount of heat dQ_3 which accumulates due to the temperature rise dT in the element of volume is

$$dQ_3 = c\rho\,dT\,dx\,dy\,dz \tag{i}$$

where c is the specific heat with dimension [J/(kg · K)], and ρ is the density with dimension [kg/m^3] of the solid. The energy balance $dQ_3 = dQ_1 + dQ_2$ gives

$$c\rho \frac{\partial T}{\partial t} = \frac{\partial}{\partial x}\left(\lambda_x \frac{\partial T}{\partial x}\right) + \frac{\partial}{\partial y}\left(\lambda_y \frac{\partial T}{\partial y}\right) + \frac{\partial}{\partial z}\left(\lambda_z \frac{\partial T}{\partial z}\right) + Q \tag{3.2}$$

Equation (3.2) is a partial differential equation of the second order and it is called the Fourier heat conduction equation. This equation holds for a nonhomogeneous anisotropic solid.

A nonhomogeneous body is a body in which material constants, such as the Young's modulus E, and the thermal conductivity λ, may change from point to point. In a homogeneous body these constants are the same at all points. An anisotropic body is a body in which material constants at a point may change in each direction. In an isotropic body these constants at a point in a body are the same in all directions.

Therefore, concentrating our attention on thermal conductivities, we note that when the thermal conductivities have different values at various points, but have the same value in all directions at each particular point,

$$\lambda_x = \lambda_y = \lambda_z \equiv \lambda(x, y, z) \tag{j}$$

this solid is a nonhomogeneous isotropic solid. For the nonhomogeneous isotropic solid Eq. (3.2) reduces to

$$c\rho\frac{\partial T}{\partial t} = \frac{\partial}{\partial x}\left(\lambda\frac{\partial T}{\partial x}\right) + \frac{\partial}{\partial y}\left(\lambda\frac{\partial T}{\partial y}\right) + \frac{\partial}{\partial z}\left(\lambda\frac{\partial T}{\partial z}\right) + Q \tag{3.3}$$

Similarly, if the thermal conductivities have same value at any point, but have different value in each direction

$$\lambda_x\text{: constant,} \qquad \lambda_y\text{: constant,} \qquad \lambda_z\text{: constant} \tag{k}$$

this solid is an homogeneous anisotropic solid. For the homogeneous anisotropic solid Eq. (3.2) reduces to

$$c\rho\frac{\partial T}{\partial t} = \lambda_x\frac{\partial^2 T}{\partial x^2} + \lambda_y\frac{\partial^2 T}{\partial y^2} + \lambda_z\frac{\partial^2 T}{\partial z^2} + Q \tag{3.4}$$

For the homogeneous isotropic solid in which the thermal conductivity λ in Eq. (3.3) is constant, or the thermal conductivities λ_x, λ_y, and λ_z in Eq. (3.4) have the same value in all directions, the heat conduction equation reduces to

$$c\rho\frac{\partial T}{\partial t} = \lambda\left(\frac{\partial^2 T}{\partial x^2} + \frac{\partial^2 T}{\partial y^2} + \frac{\partial^2 T}{\partial z^2}\right) + Q \tag{3.5}$$

or

$$c\rho\frac{\partial T}{\partial t} = \lambda\nabla^2 T + Q \tag{3.5$'$}$$

or

$$\frac{1}{\kappa}\frac{\partial T}{\partial t} = \nabla^2 T + \frac{Q}{\lambda} \tag{3.5$''$}$$

where

$$\nabla^2 = \frac{\partial^2}{\partial x^2} + \frac{\partial^2}{\partial y^2} + \frac{\partial^2}{\partial z^2} \tag{3.6}$$

$$\kappa = \frac{\lambda}{c\rho} \tag{3.7}$$

The constant κ is called the thermal diffusivity and its dimension is [m^2/s], and ∇^2 is the Laplacian operator. If there is no heat generation in the solid, Eq. (3.5″) reduces to

$$\frac{1}{\kappa}\frac{\partial T}{\partial t} = \nabla^2 T \tag{3.8}$$

As the temperature expressed by Eqs. (3.5) and (3.8) varies with the time, these equations are often called transient heat conduction equations.

It is often the case that the temperature does not vary with time. The temperature field has thus reached a steady state. From Eq. (3.5″), the steady state heat conduction equation is

$$\nabla^2 T + \frac{Q}{\lambda} = 0 \tag{3.9}$$

Furthermore, if there is no heat generation in the solid, Eq. (3.9) reduces to the Laplace equation:

$$\nabla^2 T = 0 \tag{3.10}$$

We have discussed the heat conduction equation in a Cartesian coordinate system (x, y, z). We will now consider how to transform the heat conduction equation into two other coordinate systems, namely, cylindrical coordinates and spherical coordinates. The heat conduction equation in Cartesian coordinate system (x, y, z) can be transformed into other systems as well.

Cylindrical coordinates (r, θ, z)

Using the relationship $x = r\cos\theta$, and $y = r\sin\theta$, Eq. (3.5′) retains its form in cylindrical coordinate system (r, θ, z) shown in Fig. 3.2

$$c\rho\frac{\partial T}{\partial t} = \lambda\nabla^2 T + Q \tag{3.11}$$

where the Laplacian operator ∇^2 is

$$\nabla^2 = \frac{\partial^2}{\partial r^2} + \frac{1}{r}\frac{\partial}{\partial r} + \frac{1}{r^2}\frac{\partial^2}{\partial \theta^2} + \frac{\partial^2}{\partial z^2} \tag{3.12}$$

Spherical coordinates (r, θ, ϕ)

Using the relationship $x = r\sin\theta\cos\phi$, $y = r\sin\theta\sin\phi$, and $z = r\cos\theta$, Eq. (3.5′) is again of the form in spherical coordinate system (r, θ, ϕ) shown in Fig. 3.3

$$c\rho\frac{\partial T}{\partial t} = \lambda\nabla^2 T + Q \tag{3.13}$$

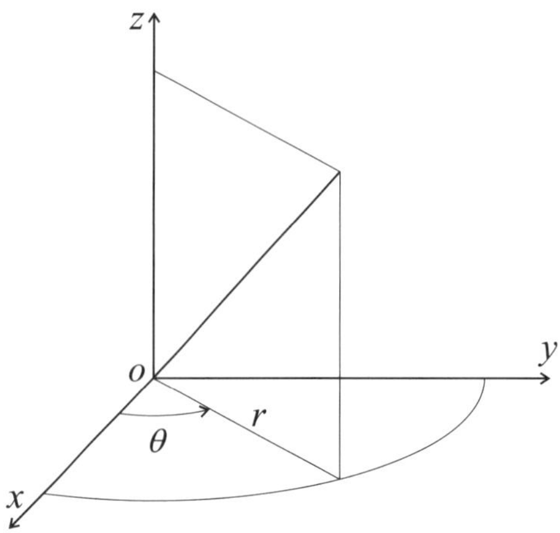

Figure 3.2 Cylindrical coordinates.

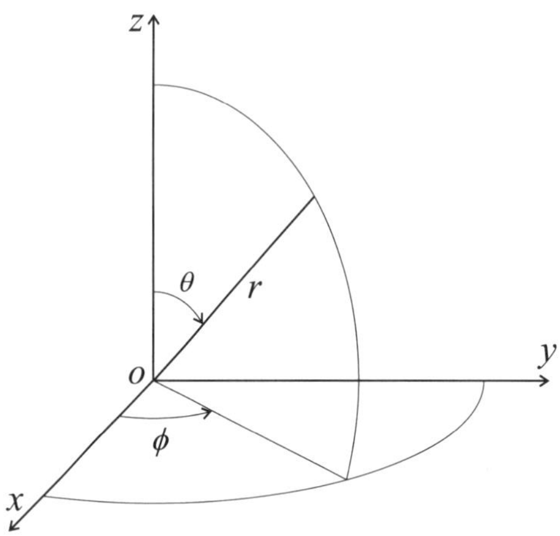

Figure 3.3 Spherical coordinates.

where the Laplacian operator ∇^2 becomes

$$\nabla^2 = \frac{\partial^2}{\partial r^2} + \frac{2}{r}\frac{\partial}{\partial r} + \frac{1}{r^2 \sin\theta}\frac{\partial}{\partial \theta}\left(\sin\theta \frac{\partial}{\partial \theta}\right) + \frac{1}{r^2 \sin^2\theta}\frac{\partial^2}{\partial \phi^2} \tag{3.14}$$

Thus, we find that the heat conduction equation has the same form in various coordinate systems, but the expression for the Laplacian operator is different for each coordinate system.

Example 3.1 Determine the transient heat conduction equation in the cylindrical coordinate system.

Solution The amount of heat which remains in the element of volume due to the heat flow along the r direction is

$$
\begin{aligned}
(dQ_r)_{in} - (dQ_r)_{out} &= q_r r\, d\theta\, dz\, dt - (q_r + dq_r)(r + dr)\, d\theta\, dz\, dt \\
&= -q_r\, dr\, d\theta\, dz\, dt - dq_r r\, d\theta\, dz\, dt - dq_r\, dr\, d\theta\, dz\, dt \\
&\cong -q_r\, dr\, d\theta\, dz\, dt - dq_r r\, d\theta\, dz\, dt \\
&= -\left(\frac{1}{r} q_r + \frac{\partial q_r}{\partial r}\right) r\, dr\, d\theta\, dz\, dt \\
&= \lambda_r \frac{1}{r}\frac{\partial T}{\partial r} r\, dr\, d\theta\, dz\, dt + \frac{\partial}{\partial r}\left(\lambda_r \frac{\partial T}{\partial r}\right) r\, dr\, d\theta\, dz\, dt \qquad \text{(a)}
\end{aligned}
$$

The amount of heat which remains in the element of volume due to the heat flow along the θ direction is

$$
\begin{aligned}
(dQ_\theta)_{in} - (dQ_\theta)_{out} &= q_\theta\, dr\, dz\, dt - (q_\theta + dq_\theta)\, dr\, dz\, dt \\
&= -\frac{1}{r}\frac{\partial q_\theta}{\partial \theta} r\, dr\, d\theta\, dz\, dt \\
&= \frac{1}{r^2}\frac{\partial}{\partial \theta}\left(\lambda_\theta \frac{\partial T}{\partial \theta}\right) r\, dr\, d\theta\, dz\, dt \qquad \text{(b)}
\end{aligned}
$$

The amount of heat which remains in the element of volume due to the heat flow along the z direction is

$$
\begin{aligned}
(dQ_z)_{in} - (dQ_z)_{out} &= q_z r\, dr\, d\theta\, dt - (q_z + dq_z) r\, dr\, d\theta\, dt \\
&= -dq_z r\, dr\, d\theta\, dt \\
&= -\frac{\partial q_z}{\partial z} r\, dr\, d\theta\, dz\, dt \\
&= \frac{\partial}{\partial z}\left(\lambda_z \frac{\partial T}{\partial z}\right) r\, dr\, d\theta\, dz\, dt \qquad \text{(c)}
\end{aligned}
$$

The amount of heat dQ_1 which remains in the element of volume is

$$
dQ_1 = \left[\frac{\partial}{\partial r}\left(\lambda_r \frac{\partial T}{\partial r}\right) + \lambda_r \frac{1}{r}\frac{\partial T}{\partial r} + \frac{1}{r^2}\frac{\partial}{\partial \theta}\left(\lambda_\theta \frac{\partial T}{\partial \theta}\right) + \frac{\partial}{\partial z}\left(\lambda_z \frac{\partial T}{\partial z}\right)\right] r\, dr\, d\theta\, dz\, dt \qquad \text{(d)}
$$

The amount of heat dQ_2 is due to an internal heat generation Q in unit volume in unit time

$$
dQ_2 = Q r\, dr\, d\theta\, dz\, dt \qquad \text{(e)}
$$

The amount of heat dQ_3 is due to the temperature rise dT in unit volume

$$
dQ_3 = c\rho\, dT\, r\, dr\, d\theta\, dz \qquad \text{(f)}
$$

The energy balance gives

$$c\rho\frac{\partial T}{\partial t} = \frac{\partial}{\partial r}\left(\lambda_r\frac{\partial T}{\partial r}\right) + \lambda_r\frac{1}{r}\frac{\partial T}{\partial r} + \frac{1}{r^2}\frac{\partial}{\partial \theta}\left(\lambda_\theta\frac{\partial T}{\partial \theta}\right) + \frac{\partial}{\partial z}\left(\lambda_z\frac{\partial T}{\partial z}\right) + Q \quad \text{(g)}$$

When the material is homogeneous and isotropic, Eq. (g) reduces to Eq. (3.11).

Example 3.2 Determine the transient heat conduction equation in the spherical coordinate system.

Solution The amount of heat which remains in the element of volume due to the heat flow along the r direction is

$$\begin{aligned}(dQ_r)_{in} - (dQ_r)_{out} &= q_r r^2 \sin\theta\, d\theta\, d\phi\, dt \\ &\quad - \left(q_r + \frac{\partial q_r}{\partial r}dr\right)(r+dr)^2 \sin\theta\, d\theta\, d\phi\, dt \\ &= -\left[2q_r r\, dr + q_r(dr)^2 + \frac{\partial q_r}{\partial r}r^2\, dr \right. \\ &\quad \left. + 2\frac{\partial q_r}{\partial r}r(dr)^2 + \frac{\partial q_r}{\partial r}(dr)^3\right]\sin\theta\, d\theta\, d\phi\, dt \\ &\cong -\left(\frac{\partial q_r}{\partial r} + \frac{2}{r}q_r\right)r^2 \sin\theta\, dr\, d\theta\, d\phi\, dt \\ &= \left[\frac{\partial}{\partial r}\left(\lambda_r\frac{\partial T}{\partial r}\right) + \lambda_r\frac{2}{r}\frac{\partial T}{\partial r}\right]dV\, dt \qquad \text{(a)}\end{aligned}$$

The amount of heat which remains in the element of volume due to the heat flow along the θ direction is

$$\begin{aligned}(dQ_\theta)_{in} - (dQ_\theta)_{out} &= q_\theta r \sin\theta\, dr\, d\theta\, d\phi\, dt \\ &\quad - \left(q_\theta + \frac{\partial q_\theta}{\partial \theta}d\theta\right) r \sin(\theta + d\theta)\, dr\, d\phi\, dt \\ &= q_\theta r \sin\theta\, dr\, d\theta\, d\phi\, dt \\ &\quad - \left(q_\theta + \frac{\partial q_\theta}{\partial \theta}d\theta\right) r(\sin\theta\cos d\theta + \cos\theta \sin d\theta)\, dr\, d\phi\, dt \\ &\cong -\left(\frac{1}{r}\frac{\partial q_\theta}{\partial \theta} + \frac{\cos\theta}{r\sin\theta}q_\theta\right) r^2 \sin\theta\, dr\, d\theta\, d\phi\, dt \\ &= \left[\frac{1}{r^2}\frac{\partial}{\partial \theta}\left(\lambda_\theta\frac{\partial T}{\partial \theta}\right) + \frac{\cos\theta}{r^2\sin\theta}\lambda_\theta\frac{\partial T}{\partial \theta}\right]dV\, dt \qquad \text{(b)}\end{aligned}$$

The amount of heat which remains in the element of volume due to the heat flow along the ϕ direction is

$$
\begin{aligned}
(dQ_\phi)_{in} - (dQ_\phi)_{out} &= q_\phi r\, dr\, d\theta\, dt - \left(q_\phi + \frac{\partial q_\phi}{\partial \phi}\, d\phi\right) r\, dr\, d\theta\, dt \\
&= -\frac{\partial q_\phi}{\partial \phi} r\, dr\, d\theta\, d\phi\, dt \\
&= -\frac{1}{r \sin\theta}\frac{\partial q_\phi}{\partial \phi}\, dV\, dt \\
&= \frac{1}{r^2 \sin^2\theta}\frac{\partial}{\partial \phi}\left(\lambda_\phi \frac{\partial T}{\partial \phi}\right) dV\, dt
\end{aligned} \tag{c}
$$

The amount of heat dQ_1 is

$$
\begin{aligned}
dQ_1 = \Bigg[&\frac{\partial}{\partial r}\left(\lambda_r \frac{\partial T}{\partial r}\right) + \lambda_r \frac{2}{r}\frac{\partial T}{\partial r} + \frac{1}{r^2 \sin\theta}\frac{\partial}{\partial \theta}\left(\lambda_\theta \sin\theta \frac{\partial T}{\partial \theta}\right) \\
&+ \frac{1}{r^2 \sin^2\theta}\frac{\partial}{\partial \phi}\left(\lambda_\phi \frac{\partial T}{\partial \phi}\right)\Bigg] dV\, dt
\end{aligned} \tag{d}
$$

The amount of heat dQ_2 is due to an internal heat generation Q in unit volume in unit time

$$
dQ_2 = Q\, dV\, dt \tag{e}
$$

The amount of heat dQ_3 is due to the temperature rise dT in unit volume

$$
dQ_3 = c\rho\, dT\, dV \tag{f}
$$

The energy balance gives

$$
\begin{aligned}
c\rho \frac{\partial T}{\partial t} = &\frac{\partial}{\partial r}\left(\lambda_r \frac{\partial T}{\partial r}\right) + \lambda_r \frac{2}{r}\frac{\partial T}{\partial r} + \frac{1}{r^2 \sin\theta}\frac{\partial}{\partial \theta}\left(\lambda_\theta \sin\theta \frac{\partial T}{\partial \theta}\right) \\
&+ \frac{1}{r^2 \sin^2\theta}\frac{\partial}{\partial \phi}\left(\lambda_\phi \frac{\partial T}{\partial \phi}\right) + Q
\end{aligned} \tag{g}
$$

When the material is homogeneous and isotropic, Eq. (g) reduces to Eq. (3.13).

3.2 Boundary and Initial Conditions

The equation governing the temperature field in a homogeneous isotropic solid is given by Eq. (3.5). It is necessary to consider the boundary and initial conditions in order to obtain the temperature field in a solid.

3.2.1 Boundary Conditions

We list some surface conditions to be prescribed for the heat transfer and for the temperature on the boundary surface.

[1] Prescribed surface temperature
The surface temperature is prescribed as a function of position and time:

$$T(P, t) = \Theta(P, t) \tag{3.15}$$

where P is a point on the boundary surface and $\Theta(P, t)$ is a prescribed function of the position and time.

[2] Adiabatic surface
When there is no heat flux through the boundary surface, the boundary condition is expressed via Eq. (3.1):

$$\frac{\partial T(P, t)}{\partial n} = 0 \tag{3.16}$$

where $\partial/\partial n$ denotes differentiation along the external normal direction n.

[3] Prescribed heat flux through the boundary surface
When the heat flux is prescribed on the boundary surface, the boundary condition follows Eq. (3.1):

$$-\lambda \frac{\partial T(P, t)}{\partial n} = q(P, t) \tag{3.17}$$

where $q(P, t)$ denotes the heat flux on the boundary surface.

[4] Heat transfer by convection
In many cases, the surface of the solid is in contact with fluid or gas. When heat transfer between the boundary surface of the solid and the surrounding medium occurs by convection, as shown in Fig. 3.4, the heat Q_c transferred is

$$Q_c = h_c(T - \Theta)\, ds\, dt \tag{3.18}$$

where h_c denotes the heat transfer coefficient with dimension $[\mathrm{W/(m^2 \cdot K)}]$, ds is a surface area element, and Θ is the temperature of the surrounding medium which is a

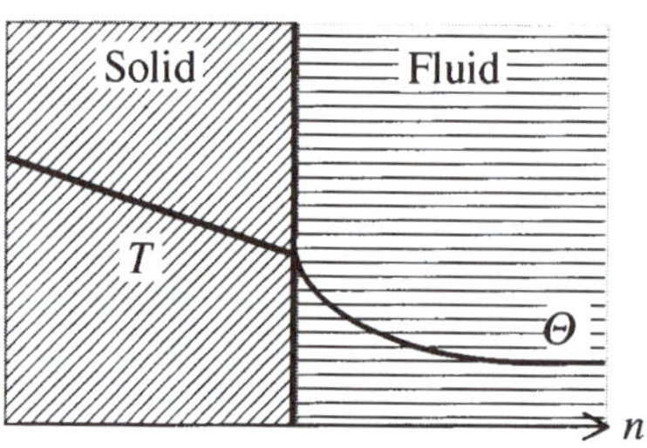

Figure 3.4 Convection.

given function of position and time. The amount of heat which flows through the surface of the solid due to conduction is given by the Fourier law (3.1)

$$-\lambda \frac{\partial T}{\partial n}\, ds\, dt \tag{a}$$

If there is heat generation q_b per unit area per unit time on the boundary surface, it may be expressed by

$$q_b\, ds\, dt \tag{b}$$

The sum of the amount of heat due to heat conduction and the amount of heat due to surface heat generation should be equal to the amount of heat Q_c due to convection on the boundary surface

$$-\lambda \frac{\partial T}{\partial n}\, ds\, dt + q_b\, ds\, dt = h_c(T - \Theta)\, ds\, dt \tag{c}$$

or, after dividing by $ds\, dt$,

$$-\lambda \frac{\partial T}{\partial n} + q_b = h_c(T - \Theta) \tag{3.19}$$

[5] Heat transfer by radiation

A body at absolute temperature T loses heat by radiation. The amount of heat Q_r lost by radiation is

$$Q_r = \sigma F(T^4 - \Theta^4)\, ds\, dt \tag{3.20}$$

where $\sigma = 5.67 \times 10^{-8}\ \mathrm{W/(m^2 \cdot K^4)}$ is the Stefan–Boltzmann constant, F is the emissivity of the surface, and Θ is the absolute temperature of the surrounding medium. Thus, the boundary condition due to heat transfer by radiation is given by

$$-\lambda \frac{\partial T}{\partial n} + q_b = \sigma F(T^4 - \Theta^4) \qquad \text{on the boundary surface} \tag{3.21}$$

This equation is not a linear function of temperature, so this condition offers some difficulty in obtaining the temperature field in the solid from the Fourier heat conduction equation. If the temperature difference between T and Θ is not large, it may be approximated to a linear form

$$\begin{aligned} Q_r &= \sigma F(T^4 - \Theta^4)\, ds\, dt \\ &= \sigma F(T - \Theta)(T + \Theta)(T^2 + \Theta^2)\, ds\, dt \\ &\cong \sigma F(T - \Theta)(2\Theta)(2\Theta^2)\, ds\, dt \\ &= 4\sigma F\Theta^3(T - \Theta)\, ds\, dt \end{aligned} \tag{d}$$

or

$$Q_r = h_r(T - \Theta)\, ds\, dt \tag{3.22}$$

where

$$h_r = 4\sigma F \Theta^3 \tag{3.23}$$

The boundary condition due to heat transfer by radiation is now

$$-\lambda \frac{\partial T}{\partial n} + q_b = h_r(T - \Theta) \tag{3.24}$$

Equation (3.24) has the same form as Eq. (3.19) which represents heat transfer by convection.

[6] Heat transfer by convection and radiation
In many cases, heat transfer on the boundary occurs both by convection and radiation. The boundary condition in this case is given by

$$-\lambda \frac{\partial T}{\partial n} + q_b = h_c(T - \Theta) + \sigma F(T^4 - \Theta^4) \tag{3.25}$$

If the temperature difference between T and Θ is not large, Eq. (3.25) may be approximated by use of a compound coefficient h, as follows:

$$-\lambda \frac{\partial T}{\partial n} + q_b = h(T - \Theta) \tag{3.26}$$

where

$$h = h_c + h_r \tag{e}$$

with h_r given by Eq. (3.23).

[7] Contact between two solids
It very often happens that the surface of a solid is in contact with the surface of another solid. If the surfaces of two solids are in perfect thermal contact, the temperature on the contact surface and the heat flow through the contact surface are the same for both solids

$$T_1 = T_2 \tag{3.27}$$

and

$$\lambda_1 \frac{\partial T_1}{\partial n} = \lambda_2 \frac{\partial T_2}{\partial n} \tag{3.28}$$

where subscripts 1 and 2, respectively, refer to solid 1 and 2 and n is the common normal direction on the contact surface.

The seven boundary conditions discussed may be present simultaneously on different portions of the boundary.

3.2.2 Initial Condition

For solution of the transient heat conduction equation (3.5′), it is necessary to specify an initial condition which expresses the temperature distribution in the solid at initial time,

$$T = \Phi(P) \tag{3.29}$$

where $\Phi(P)$ is the initial temperature distribution and P is a point in the solid. On the other hand, if the temperature has reached a steady state, it is not necessary to specify an initial condition.

3.3 One-Dimensional Heat Conduction Problems in Cartesian Coordinates

We now present some examples of one-dimensional temperature distributions in a solid to demonstrate how to solve the Fourier heat conduction equation (3.5).

3.3.1 One-Dimensional Heat Conduction Equation

When we consider one-dimensional heat conduction problems of a homogeneous isotropic solid, Eq. (3.5) simplifies to the form

$$c\rho \frac{\partial T}{\partial t} = \lambda \frac{\partial^2 T}{\partial x^2} + Q \tag{3.30}$$

If there is no heat generation, as is usually the case, Eq. (3.30) reduces to

$$\frac{\partial T}{\partial t} = \kappa \frac{\partial^2 T}{\partial x^2} \tag{3.31}$$

where $\kappa = \lambda / c\rho$. Furthermore, if the temperature distribution does not depend on time, Eq. (3.31) reduces to

$$\frac{d^2 T}{dx^2} = 0 \tag{3.32}$$

Example 3.3 Determine a one-dimensional heat conduction equation for a non-homogeneous isotropic body.

Solution Equation (3.2) reduces to

$$c\rho \frac{\partial T}{\partial t} = \frac{\partial}{\partial x}\left[\lambda(x)\frac{\partial T}{\partial x}\right] + Q \tag{a}$$

$$c\rho \frac{\partial T}{\partial t} = \frac{\partial}{\partial x}\left[\lambda(x)\frac{\partial T}{\partial x}\right] \quad \text{if no heat generation} \tag{b}$$

$$\frac{d}{dx}\left[\lambda(x)\frac{dT}{dx}\right] = 0 \quad \text{for a steady state with no heat generation} \tag{c}$$

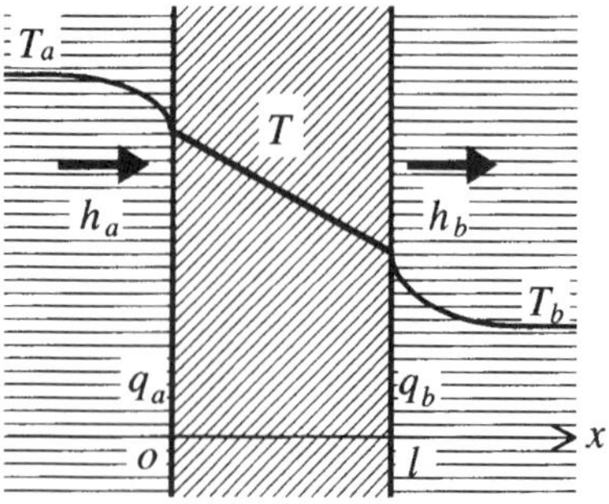

Figure 3.5 Strip.

3.3.2 One-Dimensional Temperature in the Steady State

We consider a strip shown in Fig. 3.5. Both boundary surfaces of width l are in contact with different media. If there is heat generation on both surfaces, the temperature problem is

Governing equation:

$$\frac{d^2T}{dx^2} = 0 \tag{a}$$

Boundary conditions:

$$\lambda \frac{dT}{dx} + q_a = h_a(T - T_a) \qquad \text{on } x = 0 \tag{b}$$

$$-\lambda \frac{dT}{dx} + q_b = h_b(T - T_b) \qquad \text{on } x = l \tag{c}$$

where T_a and T_b are the temperatures of the surrounding media, q_a and q_b are surface heat generations per unit area per unit time, h_a and h_b are the heat transfer coefficients, and subscripts a and b denote boundaries at $x = 0$ and $x = l$, respectively. A general solution of Eq. (a) is

$$T = Ax + B \tag{d}$$

The coefficients A and B can be determined from the boundary conditions (b) and (c):

$$\begin{aligned} A &= \frac{h_a h_b[(T_b - T_a) + (q_b/h_b - q_a/h_a)]}{\lambda(h_a + h_b) + h_a h_b l} \\ B &= T_a + \frac{q_a}{h_a} + \frac{\lambda h_b[(T_b - T_a) + (q_b/h_b - q_a/h_a)]}{\lambda(h_a + h_b) + h_a h_b l} \end{aligned} \tag{e}$$

Thus, after substituting A and B into (d) the temperature is written as

$$T = T_a + (T_b - T_a)\frac{h_b(h_a x + \lambda)}{\lambda(h_a + h_b) + h_a h_b l} + \frac{q_a}{h_a} + \left(\frac{q_b}{h_b} - \frac{q_a}{h_a}\right)\frac{h_b(h_a x + \lambda)}{\lambda(h_a + h_b) + h_a h_b l} \tag{3.33}$$

The steady temperature in the strip with various other boundary conditions is given as follows:

(1) No surface heat generation

$$T = T_a + (T_b - T_a)\frac{h_b(h_a x + \lambda)}{\lambda(h_a + h_b) + h_a h_b l} \tag{3.34}$$

(2) No surface heat generation and prescribed surface temperatures T_a and T_b

$$T = T_a + (T_b - T_a)\frac{x}{l} \tag{3.35}$$

(3) Constant heat flux q_b flows into the strip at the right surface ($x = l$) and the surface temperature T_a at the left surface ($x = 0$) is prescribed

$$T = T_a + \frac{q_b}{\lambda}x \tag{3.36}$$

(4) Constant heat flux q_a flows into the strip at the left surface ($x = 0$) and the surface temperature T_b at the right surface ($x = l$) is prescribed

$$T = T_b + \frac{q_a}{\lambda}(l - x) \tag{3.37}$$

Example 3.4 Determine a one-dimensional steady temperature in a non-homogeneous isotropic strip when boundary surface temperatures on $x = 0$ and $x = l$ are prescribed at T_a and T_b, respectively.

Solution The steady heat conduction equation in the one-dimensional non-homogeneous isotropic strip is

$$\frac{d}{dx}\left[\lambda(x)\frac{dT}{dx}\right] = 0 \tag{a}$$

The boundary conditions are

$$T = T_a \qquad \text{on } x = 0 \tag{b}$$

$$T = T_b \qquad \text{on } x = l \tag{c}$$

A general solution of Eq. (a) is

$$T(x) = C_0 + C_1 \int_0^x \frac{1}{\lambda(x)} dx \tag{d}$$

From Eqs. (b), (c), and (d) the temperature is

$$T(x) = T_a + (T_b - T_a) \frac{\displaystyle\int_0^x \frac{1}{\lambda(x)} dx}{\displaystyle\int_0^l \frac{1}{\lambda(x)} dx} \tag{e}$$

3.3.3 One-Dimensional Transient Temperature; Separation of Variables

There are many solution techniques to analyze the heat conduction equation. Two techniques will be discussed here. One is separation of variables and the other is the Laplace transform technique. We will first consider one-dimensional, transient heat flow in a strip using the separation of variables method.

When the initial temperature in the strip, which is bounded by the surfaces $x = 0$ and $x = l$, is $T = T_i(x)$, and the temperatures on the surfaces $x = 0$ and $x = l$ are kept at a constant and zero, respectively, the differential equation to be solved is

Governing equation:

$$\frac{\partial T}{\partial t} = \kappa \frac{\partial^2 T}{\partial x^2} \tag{a}$$

Boundary conditions:

$$T = T_a \qquad \text{on } x = 0 \tag{b}$$

$$T = 0 \qquad \text{on } x = l \tag{c}$$

Initial condition:

$$T = T_i(x) \qquad \text{at } t = 0 \tag{d}$$

We assume that the temperature can be expressed by a product of two functions, one is a function of position only, and the other is a function of time only:

$$T(x, t) = f(x)g(t) \tag{e}$$

Substituting Eq. (e) into Eq. (a), we obtain

$$\frac{1}{\kappa g(t)} \frac{dg(t)}{dt} = \frac{1}{f(x)} \frac{d^2 f(x)}{dx^2} \tag{f}$$

The left side of Eq. (f) depends only on time and the right side depends only on position, so that Eq. (f) should be a constant for both sides to be equal to each other. Taking the constant as $-s^2$, Eq. (f) leads to two separate equations:

$$\frac{dg(t)}{dt} + \kappa s^2 g(t) = 0 \tag{g}$$

$$\frac{d^2 f(x)}{dx^2} + s^2 f(x) = 0 \tag{h}$$

The general solutions of above equations are

$$g(t) = E, \quad f(x) = Fx + G \qquad \text{for } s = 0 \tag{i}$$

$$g(t) = He^{-\kappa s^2 t}, \quad f(x) = I \sin sx + J \cos sx \qquad \text{for } s \neq 0 \tag{j}$$

where E, F, G, H, I and J are constants. These solutions must exist for arbitrary values of s. Therefore, using solutions (i) and (j) we may write a general solution of the form (e):

$$T(x, t) = (A \sin sx + B \cos sx)e^{-\kappa s^2 t} + Cx + D \tag{k}$$

where

$$A = IH, \qquad B = JH, \qquad C = EF, \qquad D = EG \tag{l}$$

Substitution of Eq. (k) into the boundary conditions (b) and (c) gives the following relations between the coefficients:

$$Be^{-\kappa s^2 t} + D = T_a \tag{m}$$

$$(A \sin sl + B \cos sl)e^{-\kappa s^2 t} + Cl + D = 0 \tag{n}$$

Equations (m) and (n) can be satisfied, provided that:

$$D = T_a, \qquad B = 0, \qquad C = -\frac{T_a}{l} \tag{o}$$

and

$$\sin sl = 0 \tag{3.38}$$

where s is any root of Eq. (3.38). The roots of Eq. (3.38) which are called eigenvalues are given by

$$s_n = \frac{n\pi}{l} \qquad (n = 1, 2, 3, \ldots) \tag{3.39}$$

and the number of such eigenvalues is infinite. Then, the temperature $T(x,t)$ can be developed into an infinite series

$$T(x,t) = T_a\left(1-\frac{x}{l}\right) + \sum_{n=1}^{\infty} A_n \sin s_n x\, e^{-\kappa s_n^2 t} \tag{3.40}$$

where A_n are unknown coefficients. The coefficients A_n in Eq. (3.40) can be determined using the initial condition (d):

$$\sum_{n=1}^{\infty} A_n \sin s_n x = T_i(x) - T_a\left(1-\frac{x}{l}\right) \tag{p}$$

Multiplying both sides of Eq. (p) by $\sin s_m x$, integrating it from zero to l, and using the integrals

$$\begin{aligned}
\int_0^l \sin s_n x \sin s_m x\, dx &= -\frac{\sin(s_n+s_m)l}{2(s_n+s_m)} + \frac{\sin(s_n-s_m)l}{2(s_n-s_m)} = 0 \qquad \text{for } m \neq n \\
\int_0^l \sin^2 s_n x\, dx &= -\frac{\sin 2s_n l}{4s_n} + \frac{l}{2} = \frac{l}{2} \qquad \text{for } m = n
\end{aligned} \tag{q}$$

we obtain the coefficients A_n:

$$A_n = \frac{2}{l}\int_0^l \left[T_i(x) - T_a\left(1-\frac{x}{l}\right)\right] \sin s_n x\, dx \tag{3.41}$$

Thus, the temperature can be expressed as

$$\begin{aligned}
T(x,t) &= T_a\left(1-\frac{x}{l}\right) \\
&+ \frac{2}{l}\sum_{n=1}^{\infty}\left\{\int_0^l \left[T_i(x) - T_a\left(1-\frac{x}{l}\right)\right] \sin n\pi\frac{x}{l}\, dx\right\} \sin n\pi\frac{x}{l}\, e^{-\kappa(n\pi/l)^2 t}
\end{aligned} \tag{3.42}$$

or, in dimensionless form

$$\begin{aligned}
T(X,F) &= T_a(1-X) \\
&+ 2\sum_{n=1}^{\infty}\left\{\int_0^1 [T_i(X) - T_a(1-X)] \sin n\pi X\, dX\right\} \sin n\pi X\, e^{-(n\pi)^2 F}
\end{aligned} \tag{3.43}$$

where the dimensionless quantities F and X are

$$F = \frac{\kappa t}{l^2}, \qquad X = \frac{x}{l} \tag{3.44}$$

and F is called the Fourier's number.

Next, we consider the same problem with different boundary conditions, namely, with heat transfer boundary conditions

$$\lambda \frac{\partial T}{\partial x} = h_a(T - T_a) \qquad \text{on } x = 0 \tag{r}$$

$$-\lambda \frac{\partial T}{\partial x} = h_b(T - T_b) \qquad \text{on } x = l \tag{s}$$

The temperature T which satisfies the governing equation (a) can be expressed using the separation of variables in the form

$$T(x,t) = A_0 x + B_0 + \sum_{n=1}^{\infty} (A_n \sin s_n x + B_n \cos s_n x) e^{-\kappa s_n^2 t} \tag{t}$$

Substitution of the above equation into the boundary conditions (r) and (s) gives the following relations between the coefficients:

$$\lambda A_0 - h_a B_0 = -h_a T_a, \qquad (\lambda + h_b l) A_0 + h_b B_0 = h_b T_b \tag{u}$$

$$\begin{gathered} \lambda s_n A_n - h_a B_n = 0 \\ (h_b \sin s_n l + \lambda s_n \cos s_n l) A_n + (h_b \cos s_n l - \lambda s_n \sin s_n l) B_n = 0 \end{gathered} \tag{v}$$

The coefficients A_0 and B_0 are obtained from Eq. (u) in the form

$$A_0 = \frac{(T_b - T_a) h_a h_b}{\lambda(h_a + h_b) + h_a h_b l}, \qquad B_0 = T_a + \frac{(T_b - T_a)\lambda h_b}{\lambda(h_a + h_b) + h_a h_b l} \tag{w}$$

and from the first of Eqs. (v), we obtain $A_n / B_n = h_a / \lambda s_n$. Substitution of this relation into the second relation in Eqs. (v) gives

$$\left\{ \left(h_a h_b - \lambda^2 s_n^2 \right) \sin s_n l + \lambda s_n (h_a + h_b) \cos s_n l \right\} B_n = 0 \tag{x}$$

In order to obtain a non-trivial solution of Eq. (x), s_n must be any positive root of the transcendental equation

$$\tan sl = \frac{\lambda s (h_a + h_b)}{\lambda^2 s^2 - h_a h_b} \tag{3.45}$$

The roots of this equation are called eigenvalues and are given by the intersections of the plots of the periodic function $\tan sl$ with period π / l and non-periodic function $\lambda s (h_a + h_b)/(\lambda^2 s^2 - h_a h_b)$, and the number of such intersections is infinite. The eigenvalues are denoted here by s_n $(n = 1, 2, \ldots)$. Then, the temperature $T(x,t)$ can be developed in an infinite series

$$T(x,t) = A_0 x + B_0 + \sum_{n=1}^{\infty} A_n \left(\sin s_n x + \frac{\lambda s_n}{h_a} \cos s_n x \right) e^{-\kappa s_n^2 t} \tag{3.46}$$

where $A_n \quad (n = 1, 2, \dots)$ are unknown coefficients. The coefficients A_n in Eq. (3.46) are determined from the initial condition (d)

$$\sum_{n=1}^{\infty} A_n \left(\sin s_n x + \frac{\lambda s_n}{h_a} \cos s_n x \right) = T_i(x) - (A_0 x + B_0) \tag{y}$$

Multiplying both sides of Eq. (y) by $(\sin s_m x + (\lambda s_m / h_a) \cos s_m x)$, and integrating from zero to l, the coefficients A_n can be obtained

$$A_n = \frac{2h_a^2(\lambda^2 s_n^2 + h_b^2) \int_0^l [T_i(x) - (A_0 x + B_0)] \left(\sin s_n x + \dfrac{\lambda s_n}{h_a} \cos s_n x \right) dx}{l(\lambda^2 s_n^2 + h_a^2)(\lambda^2 s_n^2 + h_b^2) + \lambda(h_a + h_b)(\lambda^2 s_n^2 + h_a h_b)} \tag{z}$$

Thus, the temperature is

$$\begin{aligned} T(x,t) &= T_a + (T_b - T_a)\frac{h_b(h_a x + \lambda)}{\lambda(h_a + h_b) + h_a h_b l} \\ &\quad + 2\sum_{n=1}^{\infty} \frac{(\lambda^2 s_n^2 + h_b^2)(h_a \sin s_n x + \lambda s_n \cos s_n x) e^{-\kappa s_n^2 t}}{l(\lambda^2 s_n^2 + h_a^2)(\lambda^2 s_n^2 + h_b^2) + \lambda(h_a + h_b)(\lambda^2 s_n^2 + h_a h_b)} \\ &\quad \times \int_0^l \left\{ T_i(x) - \left[T_a + (T_b - T_a)\frac{h_b(h_a x + \lambda)}{\lambda(h_a + h_b) + h_a h_b l} \right] \right\} \\ &\quad \times (h_a \sin s_n x + \lambda s_n \cos s_n x)\, dx \end{aligned} \tag{3.47}$$

or, in dimensionless form

$$\begin{aligned} T(X,F) &= T_a + (T_b - T_a)\frac{B_{ib}(B_{ia} X + 1)}{B_{ia} + B_{ib} + B_{ia} B_{ib}} \\ &\quad + 2\sum_{n=1}^{\infty} \frac{(\gamma_n^2 + B_{ib}^2)(B_{ia} \sin \gamma_n X + \gamma_n \cos \gamma_n X) e^{-\gamma_n^2 F}}{(\gamma_n^2 + B_{ia}^2)(\gamma_n^2 + B_{ib}^2) + (B_{ia} + B_{ib})(\gamma_n^2 + B_{ia} B_{ib})} \\ &\quad \times \int_0^1 \left\{ T_i(X) - \left[T_a + (T_b - T_a)\frac{B_{ib}(B_{ia} X + 1)}{B_{ia} + B_{ib} + B_{ia} B_{ib}} \right] \right\} \\ &\quad \times (B_{ia} \sin \gamma_n X + \gamma_n \cos \gamma_n X)\, dX \end{aligned} \tag{3.48}$$

where the dimensionless quantities B_{ia} and B_{ib}, which are called the Biot's numbers, at the lower and upper surface, respectively, are denoted by

$$B_{ia} = \frac{h_a l}{\lambda}, \qquad B_{ib} = \frac{h_b l}{\lambda} \tag{3.49}$$

and γ_n is the nth positive root of the transcendental equation

$$\tan\gamma = \frac{\gamma(B_{ia} + B_{ib})}{\gamma^2 - B_{ia}B_{ib}} \tag{3.50}$$

Next, we will determine the one-dimensional temperature in the strip with the initial temperature $T_i(x)$ subjected to specified boundary conditions.

[1] Prescribed surface temperatures $T = T_a$ at $x = 0$, and $T = T_b$ at $x = l$

$$T(X, F) = T_a + (T_b - T_a)X + 2\sum_{n=1}^{\infty}\left\{\int_0^1 [T_i(X) - T_a - (T_b - T_a)X]\sin n\pi X\,dX\right\} \times \sin n\pi X e^{-(n\pi)^2 F} \tag{3.51}$$

[2] Prescribed surface temperature $T = T_a$ at $x = 0$ and adiabatic condition $\partial T/\partial x = 0$ at $x = l$

$$T(X, F) = T_a + 2\sum_{n=1}^{\infty}\left\{\int_0^1 [T_i(X) - T_a]\sin\left(n - \tfrac{1}{2}\right)\pi X\,dX\right\} \times \sin\left(n - \tfrac{1}{2}\right)\pi X e^{-[(2n-1)\frac{\pi}{2}]^2 F} \tag{3.52}$$

[3] Prescribed surface temperature $T = T_a$ at $x = 0$ and constant heat flux $-\lambda(\partial T/\partial x) = q_b$ at $x = l$

$$T(X, F) = T_a - \frac{q_b l}{\lambda}X + 2\sum_{n=1}^{\infty}\left\{\int_0^1 \left[T_i(X) - \left(T_a - \frac{q_b l}{\lambda}X\right)\right]\sin\left(n - \tfrac{1}{2}\right)\pi X\,dX\right\} \times \sin\left(n - \tfrac{1}{2}\right)\pi X e^{-[(2n-1)\frac{\pi}{2}]^2 F} \tag{3.53}$$

[4] Heat transfer boundary conditions at $x = 0$ and at $x = l$. The heat transfer coefficients at $x = 0$ and at $x = l$ are the same.

$$T(X, F) = T_a + (T_b - T_a)\frac{B_i X + 1}{B_i + 2} + 2\sum_{n=1}^{\infty}\frac{(B_i\sin\gamma_n X + \gamma_n\cos\gamma_n X)e^{-\gamma_n^2 F}}{\gamma_n^2 + B_i^2 + 2B_i} \times \int_0^1 \left\{T_i(X) - \left[T_a + (T_b - T_a)\frac{B_i X + 1}{B_i + 2}\right]\right\} \times (B_i\sin\gamma_n X + \gamma_n\cos\gamma_n X)\,dX \tag{3.54}$$

where $B_i = hl/\lambda$ and γ_n is the nth root of the equation

$$\tan\gamma = \frac{2\gamma B_i}{\gamma^2 - B_i^2} \tag{3.55}$$

Example 3.5 When the initial temperature in Eq. (3.48) equals zero, determine the temperature distribution.

Solution Calculate the following integrals

$$\begin{aligned}
&\int_0^1 (B_{ia}\sin\gamma_n X + \gamma_n\cos\gamma_n X)dX \\
&= \left[-\frac{B_{ia}}{\gamma_n}\cos\gamma_n X + \sin\gamma_n X\right]_0^1 = \frac{1}{\gamma_n}[B_{ia} - \cos\gamma_n(B_{ia} - \gamma_n\tan\gamma_n)] \\
&= \frac{1}{\gamma_n}\left[B_{ia} + \frac{B_{ib}\left(\gamma_n^2 + B_{ia}^2\right)}{\gamma_n^2 - B_{ia}B_{ib}}\cos\gamma_n\right]
\end{aligned} \tag{3.56}$$

$$\begin{aligned}
&\int_0^1 X(B_{ia}\sin\gamma_n X + \gamma_n\cos\gamma_n X)\,dX \\
&= \frac{1}{\gamma_n^2}\{-\gamma_n + \cos\gamma_n[\gamma_n(1 - B_{ia}) + (\gamma_n^2 + B_{ia})\tan\gamma_n]\} \\
&= \frac{1}{\gamma_n}\left[-1 + \frac{\left(\gamma_n^2 + B_{ia}^2\right)(1 + B_{ib})}{\gamma_n^2 - B_{ia}B_{ib}}\cos\gamma_n\right]
\end{aligned} \tag{3.57}$$

Substitution of above equations into Eq. (3.48) yields

$$\begin{aligned}
T(X,F) = {} & T_a + (T_b - T_a)\frac{B_{ib}(B_{ia}X + 1)}{B_{ia} + B_{ib} + B_{ia}B_{ib}} \\
& - 2\sum_{n=1}^{\infty}\frac{\left(\gamma_n^2 + B_{ib}^2\right)\left[T_a B_{ia}\left(\gamma_n^2 - B_{ia}B_{ib}\right) + T_b B_{ib}\left(\gamma_n^2 + B_{ia}^2\right)\cos\gamma_n\right]}{\gamma_n\left(\gamma_n^2 - B_{ia}B_{ib}\right)\left[\left(\gamma_n^2 + B_{ia}^2\right)\left(\gamma_n^2 + B_{ib}^2\right) + (B_{ia} + B_{ib})\left(\gamma_n^2 + B_{ia}B_{ib}\right)\right]} \\
& \times (B_{ia}\sin\gamma_n X + \gamma_n\cos\gamma_n X)e^{-\gamma_n^2 F}
\end{aligned} \tag{3.58}$$

3.3.4 One-Dimensional Transient Temperature; Laplace Transform

In this subsection, the solution of the one-dimensional transient heat conduction equation is considered by using the Laplace transform. If, for example, a function $K(p, t)$ is a known function of two variables t and p, and the integral

$$f_I(p) = \int_0^\infty f(t)K(p,t)\,dt \tag{3.59}$$

is convergent, then a function $f_I(p)$ of variable p is defined, and this function is called the integral transform of the function $f(t)$ with the kernel $K(p,t)$. If we select the kernel to be

$$K(p,t) = e^{-pt} \tag{3.60}$$

the integral

$$L\{f(t)\} \equiv \bar{f}(p) = \int_0^\infty f(t)e^{-pt}\,dt \tag{3.61}$$

is called the Laplace transform of the function $f(t)$ with respect to the variable t. The variable p is complex and its real part is positive and large enough to make the integral (3.61) convergent. A few examples of Laplace transforms are calculated:

$$L\{1\} = \int_0^\infty e^{-pt}\,dt = \left[-\frac{e^{-pt}}{p}\right]_0^\infty = \frac{1}{p} \tag{a}$$

$$L\{t\} = \int_0^\infty te^{-pt}\,dt = \left[-t\frac{e^{-pt}}{p}\right]_0^\infty + \frac{1}{p}\int_0^\infty e^{-pt}\,dt = \frac{1}{p^2} \tag{b}$$

$$L\{e^{-at}f(t)\} = \int_0^\infty f(t)e^{-(p+a)t}\,dt = \bar{f}(p+a) \tag{c}$$

$$\begin{aligned} L\left\{\frac{df(t)}{dt}\right\} &= \int_0^\infty e^{-pt}\frac{df(t)}{dt}dt = \left[f(t)e^{-pt}\right]_0^\infty + \int_0^\infty pf(t)e^{-pt}\,dt \\ &= p\bar{f}(p) - f(0) \end{aligned} \tag{d}$$

The determination of $f(t)$ from $\bar{f}(p)$ is called the inverse Laplace transform. The inverse Laplace transform is performed by the use of the inversion theorem:

$$f(t) = L^{-1}\left[\bar{f}(p)\right] = \frac{1}{2\pi i}\int_{\gamma-i\infty}^{\gamma+i\infty} \bar{f}(p)e^{pt}\,dp \tag{3.62}$$

where $i = \sqrt{-1}$ and γ is taken to be large enough so that all the singularities of $\bar{f}(p)$ lie to the left of the line of integration.

The theorems and formulas for the Laplace transform which are often used in heat conduction problems are given in Table 3.1.*

---

*A more extensive table of Laplace transforms related to heat conduction problems appears in: H. S. Carslaw and J. C. Jaeger, Conduction of Heat in Solids, Clarendon Press, Oxford (1959), pp. 494–496. For inversion of some Laplace transforms of exponential type consult: R. B. Hetnarski, An Algorithm for Generating Some Inverse Laplace Transforms of Exponential Form, J. Appl. Math. Phys., vol. 26 (1975), pp. 249–253. Extensive tables of transforms may be found in: (a) A. Erdélyi (Editor), Bateman Manuscript Project, Tables of Integral Transforms, vol. 1, McGraw-Hill, New York (1954); (b) V. A. Ditkin and A. P. Prudnikov, Reference Book on Operational Calculus (in Russian), Vizshaya Shkola, Moscow (1965).

Table 3.1 Table of Laplace transforms

$f(t)$	$\bar{f}(p)$
1	$1/p$
t	$1/p^2$
t^n	$n!/p^{n+1}$ $(n = 1, 2, \ldots)$
e^{-at}	$1/(p+a)$
e^{at}	$1/(p-a)$
$\sin at$	$a/(p^2+a^2)$
$\cos at$	$p/(p^2+a^2)$
$\sinh at$	$a/(p^2-a^2)$
$\cosh at$	$p/(p^2-a^2)$
$\operatorname{erfc}\dfrac{x}{2\sqrt{\kappa t}}$	$\exp(-x\sqrt{p/\kappa})/p$
$\sqrt{\dfrac{\kappa}{\pi t}}\exp(-x^2/4\kappa t)$	$\exp(-x\sqrt{p/\kappa})/\sqrt{p/\kappa}$
$f_1(t) + f_2(t)$	$\bar{f}_1(p) + \bar{f}_2(p)$
$e^{-at} f(t)$	$\bar{f}(p+a)$
$e^{at} f(t)$	$\bar{f}(p-a)$
$f(at), a > 0$	$\dfrac{1}{a}\bar{f}\left(\dfrac{p}{a}\right)$
$tf(t)$	$-\dfrac{d\bar{f}(p)}{dp}$
$(-t)^n f(t)$	$\dfrac{d^n \bar{f}(p)}{dp^n}$
$\dfrac{df(t)}{dt}$	$p\bar{f}(p) - f(0)$
$\dfrac{d^2 f(t)}{dt^2}$	$p^2\bar{f}(p) - pf(0) - \dfrac{df(0)}{dt}$
$\dfrac{d^n f(t)}{dt^n}$	$p^n\bar{f}(p) - \sum_{k=0}^{n-1} p^{n-k-1}\dfrac{d^k f(0)}{dt^k}$
$f_1(t) * f_2(t) = \int_0^t f_1(s) f_2(t-s)\,ds = \int_0^t f_1(t-s) f_2(s)\,ds$	$\bar{f}_1(p)\cdot\bar{f}_2(p)$
$\int_0^t f(s)\,ds$	$\bar{f}(p)/p$

As an example of the use of the Laplace transform method, we consider the temperature in a semi-infinite body $x \geq 0$ with zero initial temperature and prescribed surface temperature T_a. The temperature field is determined through:

Governing equation:

$$\frac{\partial T}{\partial t} = \kappa \frac{\partial^2 T}{\partial x^2} \tag{e}$$

Boundary condition:

$$T = T_a \qquad \text{on } x = 0 \tag{f}$$

Initial condition:

$$T = 0 \qquad \text{at } t = 0 \tag{g}$$

Applying the Laplace transform with respect to t to Eq. (e) and using the initial condition (g), we obtain

Governing equation:

$$\frac{d^2\overline{T}}{dx^2} - q^2\overline{T} = 0 \tag{e$'$}$$

Boundary condition:

$$\overline{T} = \frac{T_a}{p} \qquad \text{on } x = 0 \tag{f$'$}$$

where

$$q = \sqrt{\frac{p}{\kappa}} \tag{h}$$

and its real part is positive. The Laplace transform reduces the partial differential equation (e) to the ordinary differential equation (e′) which automatically contains the initial condition. A general solution to Eq. (e′) is

$$\overline{T} = Ae^{qx} + Be^{-qx} \tag{i}$$

where A and B are constants. The temperature is finite as $x \to \infty$. Therefore, the constant A must be equal to zero and the temperature $\overline{T}$ which satisfies Eq. (f′) is

$$\overline{T} = \frac{T_a}{p}e^{-qx} \tag{j}$$

It follows from Table 3.1 that the temperature is

$$T = T_a \text{erfc}\left(\frac{x}{2\sqrt{\kappa t}}\right) \tag{3.63}$$

where

$$\text{erfc}\left(\frac{x}{2\sqrt{\kappa t}}\right) = \frac{2}{\sqrt{\pi}}\int_{x/2\sqrt{\kappa t}}^{\infty} e^{-\xi^2}\, d\xi \tag{3.64}$$

is called the complementary error function.

Next, we consider the temperature in a semi-infinite body with zero initial temperature and with prescribed heat flux q_a at $x = 0$. The differential equation to be solved is Eq. (e), the initial condition (g), and the boundary condition is

$$-\lambda \frac{\partial T}{\partial x} = q_a \qquad \text{on } x = 0 \tag{k}$$

Performing the Laplace transform on Eq. (k), we have the transformed boundary condition

$$-\lambda \frac{d\overline{T}}{dx} = \frac{q_a}{p} \qquad \text{on } x = 0 \tag{k$'$}$$

The temperature $\overline{T}$ which satisfies Eqs. (i) and (k′) is

$$\overline{T} = \frac{q_a}{\lambda} \frac{1}{p} \frac{e^{-qx}}{q} \tag{l}$$

The inverse Laplace transform of the function (l) can be performed by the theorem in Table 3.1:

$$L^{-1}\left[\bar{f}_1(p) \cdot \bar{f}_2(p)\right] = \int_0^t f_1(s) f_2(t-s)\, ds = \int_0^t f_1(t-s) f_2(s)\, ds \tag{3.65}$$

This theorem is called the convolution theorem, the superposition theorem, or Duhamel's theorem. The inversions of $1/p$ and e^{-qx}/q are, respectively

$$L^{-1}\left[\frac{1}{p}\right] = 1, \qquad L^{-1}\left[\frac{e^{-qx}}{q}\right] = \sqrt{\frac{\kappa}{\pi t}} e^{-x^2/4\kappa t} \tag{m}$$

Using now the convolution theorem, we obtain the temperature in the form

$$\begin{aligned} T &= \frac{q_a}{\lambda} \int_0^t \sqrt{\frac{\kappa}{\pi s}} e^{-x^2/4\kappa s}\, ds \\ &= \frac{q_a}{\lambda} \left[2\sqrt{\frac{\kappa t}{\pi}} e^{-x^2/4\kappa t} - x \operatorname{erfc} \frac{x}{2\sqrt{\kappa t}} \right] \end{aligned} \tag{3.66}$$

In the above examples we were fortunate because the inverse Laplace transform of $\overline{T}$ appears in Table 3.1. If the inverse Laplace transform of $\overline{T}$ does not appear in this table and in other more extensive tables, it is necessary to determine T from $\overline{T}$ by the use of the inversion theorem for the Laplace transform.

We consider a strip which is at zero initial temperature and whose surfaces are kept at constant temperature T_a at $x = 0$ and T_b at $x = l$. The differential equation to be solved is

$$\frac{\partial T}{\partial t} = \kappa \frac{\partial^2 T}{\partial x^2} \tag{n}$$

with boundary conditions

$$T = T_a \qquad \text{on } x = 0 \tag{o}$$

$$T = T_b \qquad \text{on } x = l \tag{p}$$

and the initial condition

$$T = 0 \qquad \text{at } t = 0 \tag{q}$$

Applying the Laplace transform with respect to t to above equations, we obtain

Governing equation:

$$\frac{d^2\overline{T}}{dx^2} - q^2\overline{T} = 0 \tag{n$'$}$$

Boundary conditions:

$$\overline{T} = \frac{T_a}{p} \qquad \text{on } x = 0 \tag{o$'$}$$

$$\overline{T} = \frac{T_b}{p} \qquad \text{on } x = l \tag{p$'$}$$

where $q = \sqrt{p/\kappa}$. A general solution of Eq. (n$'$) is

$$\overline{T} = Ae^{qx} + Be^{-qx} \tag{r}$$

The determination of unknown constants A and B from boundary conditions (o$'$) and (p$'$) gives the temperature $\overline{T}$:

$$\overline{T} = \frac{T_a \sinh q(l-x) + T_b \sinh qx}{p \sinh ql} \tag{s}$$

The inverse Laplace transform of $\overline{T}$ cannot be found in Table 3.1 and in other more extensive tables, so this temperature must be evaluated by the use of the inversion theorem for the Laplace transform (3.62)

$$T(x,t) = \frac{1}{2\pi i}\int_{\gamma - i\infty}^{\gamma + i\infty} e^{pt}\,\frac{T_a \sinh q(l-x) + T_b \sinh qx}{p \sinh ql}\,dp \tag{3.67}$$

If $\overline{T}(p)$ is a single-valued function of the complex variable p in the inner region of the contour given in Fig. 3.6, and it tends to zero as $R \to \infty$, the inversion theorem reduces to the complex contour integral (see Appendix A):

$$T(x,t) = \frac{1}{2\pi i}\int_{\gamma - i\infty}^{\gamma + i\infty} e^{pt}\overline{T}(p)\,dp = \frac{1}{2\pi i}\lim_{R\to\infty}\oint e^{pt}\overline{T}(p)\,dp \tag{3.68}$$

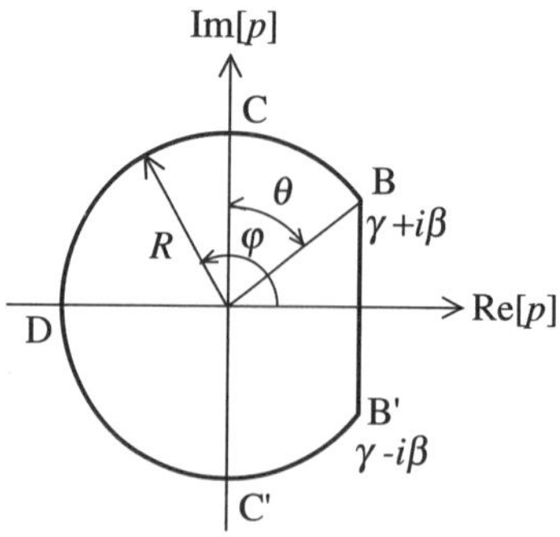

Figure 3.6 Contour.

Let us consider whether Eq. (s) is a single-valued function of p. Expanding the hyperbolic sine functions in Eq. (s) in a Taylor series, $\overline{T}(p)$ reduces to

$$\overline{T}(p)=\frac{T_a \sinh q(l-x)+T_b \sinh qx}{p\sinh ql}$$

$$=\frac{1}{p}\frac{\left(T_a\left[\frac{(l-x)}{1!}+\frac{p(l-x)^3}{3!\kappa}+\frac{p^2(l-x)^5}{5!\kappa^2}+\cdots\right]+T_b\left[\frac{x}{1!}+\frac{px^3}{3!\kappa}+\frac{p^2x^5}{5!\kappa^2}+\cdots\right]\right)}{\frac{l}{1!}+\frac{pl^3}{3!\kappa}+\frac{p^2l^5}{5!\kappa^2}+\cdots} \tag{t}$$

From Eq. (t), $\overline{T}(p)$ turns out to be a single-valued function of p and to tend to zero as $|p| = R \to \infty$. Therefore, the inversion theorem in this case reduces to the complex contour integral (3.68). The complex contour integral (3.68) is equal to the sum of the residues at the poles of its integrand by Cauchy's residue theorem. The integrand $e^{pt}\overline{T}(p)$ has a pole at $p = 0$ and poles at $p = -\kappa(n\pi/l)^2$, $(n = 1, 2, \ldots)$, and the residues at each pole are

$$T_a\left(1-\frac{x}{l}\right)+T_b\frac{x}{l} \qquad \text{for a pole at } p=0 \tag{3.69}$$

$$\frac{2}{\pi}\frac{(-1)^n}{n}\left[T_a \sin n\pi\left(1-\frac{x}{l}\right)+T_b \sin n\pi\frac{x}{l}\right]e^{-\kappa n^2\pi^2 t/l^2}$$
$$\text{for a pole at } p=-\kappa(n\pi/l)^2, \quad (n=1,2,\ldots) \tag{3.70}$$

The temperature is now calculated to be

$$T(x,t)=T_a\left(1-\frac{x}{l}\right)+T_b\frac{x}{l}$$
$$+\frac{2}{\pi}\sum_{n=1}^{\infty}\frac{(-1)^n}{n}\left[T_a \sin n\pi\left(1-\frac{x}{l}\right)+T_b \sin n\pi\frac{x}{l}\right]e^{-\kappa n^2\pi^2 t/l^2} \tag{3.71}$$

and its dimensionless form with $X = x/l$ and $F = \kappa t/l^2$ is

$$T(X, F) = T_a(1 - X) + T_b X + \frac{2}{\pi}\sum_{n=1}^{\infty}\frac{(-1)^n}{n}\left[T_a \sin n\pi(1 - X) + T_b \sin n\pi X\right] e^{-n^2\pi^2 F} \tag{3.72}$$

As another example, we consider a strip which is at the initial temperature $T_i(x)$ and its surfaces are kept at constant temperatures T_a at $x = 0$ and T_b at $x = l$. This problem is the same as the previous problem except for the initial temperature. The Laplace transform of the governing equation is

$$\frac{d^2\overline{T}}{dx^2} - q^2\overline{T} = -\frac{1}{\kappa}T_i(x) \tag{u}$$

where as before $q = \sqrt{p/\kappa}$. We obtain the temperature by the use of the method of variation of parameters (see Problem 3.5)

$$\overline{T} = Ae^{qx} + Be^{-qx} - \frac{1}{\kappa q}\int_0^x T_i(\eta)\sinh q(x - \eta)\, d\eta \tag{v}$$

The determination of unknown constants A and B from boundary conditions (o′) and (p′) gives the temperature $\overline{T}$:

$$\overline{T} = \frac{T_a \sinh q(l - x) + T_b \sinh qx}{p \sinh ql} + \frac{\sinh qx}{\kappa q \sinh ql}\int_0^l T_i(\eta)\sinh q(l - \eta)\, d\eta - \frac{1}{\kappa q}\int_0^x T_i(\eta)\sinh q(x - \eta)\, d\eta \tag{w}$$

This temperature $\overline{T}$ cannot be found in Table 3.1 of Laplace transforms, hence it must be evaluated by the use of the inversion theorem (3.62). The first term in Eq. (w) is equal to the expression in Eq. (s), so the inverse Laplace transform of this term is given by Eq. (3.71). Now consider the third term, which is a single-valued function and has no singular points in the inner region with the contour given in Fig. 3.6. The integral of the third term along the contour is zero by the Cauchy's integral theorem.

The second term is a single-valued function and has singular points of the first order in the inner region with the contour given in Fig. 3.6. The inverse Laplace transform of the second term reduces to the complex integral along the contour and this integral is equal to the sum of the residues at the pole of the first order at points $p = -\kappa n^2\pi^2/l^2$ $(n = 1, 2, 3, \ldots)$ of its integrand

$$L^{-1}\left[\frac{\sinh qx}{\kappa q \sinh ql}\int_0^l T_i(\eta)\sinh q(l - \eta)\, d\eta\right] = \frac{2}{l}\sum_{n=1}^{\infty}(-1)^{n+1}\sin n\pi\frac{x}{l}\int_0^l T_i(\eta)\sin n\pi\left(1 - \frac{\eta}{l}\right) d\eta\, e^{-\kappa n^2\pi^2 t/l^2} \tag{x}$$

Combining the inverse Laplace transforms, we receive the temperature

$$\begin{aligned} T(x,t) &= T_a\left(1-\frac{x}{l}\right)+T_b\frac{x}{l} \\ &+\frac{2}{\pi}\sum_{n=1}^{\infty}\frac{(-1)^n}{n}\left[T_a\sin n\pi\left(1-\frac{x}{l}\right)+T_b\sin n\pi\frac{x}{l}\right]e^{-\kappa n^2\pi^2 t/l^2} \\ &+\frac{2}{l}\sum_{n=1}^{\infty}(-1)^{n+1}\sin n\pi\frac{x}{l}\int_0^l T_i(\eta)\sin n\pi\left(1-\frac{\eta}{l}\right)d\eta\, e^{-\kappa n^2\pi^2 t/l^2} \end{aligned} \tag{3.73}$$

In the dimensionless form with $X = x/l$ and $F = \kappa t/l^2$ the temperature is expressed by

$$\begin{aligned} T(X,F) &= T_a(1-X)+T_bX \\ &+\frac{2}{\pi}\sum_{n=1}^{\infty}\frac{(-1)^n}{n}\left[T_a\sin n\pi(1-X)+T_b\sin n\pi X\right]e^{-n^2\pi^2F} \\ &+2\sum_{n=1}^{\infty}(-1)^{n+1}\sin n\pi X\int_0^1 T_i(\xi)\sin n\pi(1-\xi)\,d\xi\, e^{-n^2\pi^2F} \end{aligned} \tag{3.74}$$

Example 3.6 Determine a one-dimensional steady temperature in a plate from Eq. (3.74) when an initial temperature is a linear function of a position.

Solution Integral in Eq. (3.74) is

$$\begin{aligned} &\int_0^1 T_i(\xi)\sin n\pi(1-\xi)\,d\xi \\ &\quad= \int_0^1 (C_0+C_1\xi)\sin n\pi(1-\xi)\,d\xi \\ &\quad= \frac{1}{n\pi}(C_0+C_1\xi)\cos n\pi(1-\xi)\Big|_0^1 - \int_0^1\frac{C_1}{n\pi}\cos n\pi(1-\xi)\,d\xi \\ &\quad= \frac{1}{n\pi}\left[C_0+C_1-(-1)^nC_0\right]+\frac{C_1}{n^2\pi^2}\sin n\pi(1-\xi)\Big|_0^1 \\ &\quad= \frac{1}{n\pi}\left\{\left[1-(-1)^n\right]C_0+C_1\right\} \end{aligned}$$

The temperature is

$$\begin{aligned} T(X,F) &= T_a(1-X)+T_bX+\frac{2}{\pi}\sum_{n=1}^{\infty}\frac{(-1)^n}{n} \\ &\quad\times\Big(T_a\sin n\pi(1-X)+\left\{T_b+\left[(-1)^n-1\right]C_0-C_1\right\}\sin n\pi X\Big)e^{-n^2\pi^2F} \end{aligned} \tag{a}$$

Example 3.7 Determine a one-dimensional steady temperature of each layer in a two-layered plate with thickness l_i $(i = 1, 2)$ when on the boundary surfaces are the heat transfer boundary conditions (3.19).

Solution The temperature of each layer is

$$T_i = A_i + B_i x \qquad (i = 1, 2) \tag{a}$$

The boundary condition at each boundary is

$$\lambda_1 \frac{dT_1}{dx} = h_a(T_1 - T_a) \qquad \text{on } x = 0 \tag{b}$$

$$T_1 = T_2, \quad \lambda_1 \frac{dT_1}{dx} = \lambda_2 \frac{dT_2}{dx} \qquad \text{on } x = l_1 \tag{c}$$

$$-\lambda_2 \frac{dT_2}{dx} = h_b(T_2 - T_b) \qquad \text{on } x = l_1 + l_2 \tag{d}$$

We get from Eqs. (a) to (d)

$$A_1 = T_a + \frac{T_b - T_a}{D} h_b \lambda_1 \lambda_2, \qquad B_1 = \frac{T_b - T_a}{D} h_a h_b \lambda_2$$

$$A_2 = T_a + \frac{T_b - T_a}{D} h_b \left[h_a(\lambda_2 - \lambda_1) l_1 + \lambda_1 \lambda_2\right],$$

$$B_2 = \frac{T_b - T_a}{D} h_a h_b \lambda_1$$

where

$$D = h_a h_b (l_1 \lambda_2 + l_2 \lambda_1) + \lambda_1 \lambda_2 (h_a + h_b)$$

Then, the temperature of each layer is

$$T_1 = T_a + \frac{T_b - T_a}{D} h_b \lambda_2 (\lambda_1 + h_a x) \tag{e}$$

$$T_2 = T_a + \frac{T_b - T_a}{D} h_b \left[h_a(\lambda_2 - \lambda_1) l_1 + \lambda_1 \lambda_2 + \lambda_1 h_a x\right] \tag{f}$$

3.4 One-Dimensional Heat Conduction Problems in Cylindrical Coordinates

3.4.1 Heat Conduction Equation in Cylindrical Coordinates

The heat conduction equation for a homogeneous, isotropic cylinder is expressed by Eq. (3.11) in cylindrical coordinates shown in Fig. 3.2 as

$$c\rho \frac{\partial T}{\partial t} = \lambda \left(\frac{\partial^2 T}{\partial r^2} + \frac{1}{r} \frac{\partial T}{\partial r} + \frac{1}{r^2} \frac{\partial^2 T}{\partial \theta^2} + \frac{\partial^2 T}{\partial z^2} \right) + Q \tag{3.75}$$

If the temperature in the cylinder does not vary along the axis z, the heat conduction equation (3.75) reduces to

$$c\rho \frac{\partial T}{\partial t} = \lambda \left(\frac{\partial^2 T}{\partial r^2} + \frac{1}{r}\frac{\partial T}{\partial r} + \frac{1}{r^2}\frac{\partial^2 T}{\partial \theta^2} \right) + Q \tag{3.76}$$

When the temperature in the cylinder is independent of the angle θ, the heat conduction equation (3.75) becomes

$$c\rho \frac{\partial T}{\partial t} = \lambda \left(\frac{\partial^2 T}{\partial r^2} + \frac{1}{r}\frac{\partial T}{\partial r} + \frac{\partial^2 T}{\partial z^2} \right) + Q \tag{3.77}$$

If the temperature in the cylinder varies neither along axis θ nor along axis z, the heat conduction equation (3.75) reduces to

$$c\rho \frac{\partial T}{\partial t} = \lambda \left(\frac{\partial^2 T}{\partial r^2} + \frac{1}{r}\frac{\partial T}{\partial r} \right) + Q \tag{3.78}$$

If, in addition, there is no heat generation, the equation simplifies to

$$\frac{\partial T}{\partial t} = \kappa \left(\frac{\partial^2 T}{\partial r^2} + \frac{1}{r}\frac{\partial T}{\partial r} \right) \tag{3.79}$$

Furthermore, in the steady state case, Eq. (3.79) reduces to

$$\frac{d^2 T}{dr^2} + \frac{1}{r}\frac{dT}{dr} = 0 \tag{3.80}$$

3.4.2 One-Dimensional Temperature in the Steady State

When both boundary surfaces of a long hollow cylinder with inner radius a and outer radius b are surrounded by media with different temperatures, the one-dimensional temperature equation in the hollow cylinder without internal heat generation is

Governing equation:

$$\frac{d^2 T}{dr^2} + \frac{1}{r}\frac{dT}{dr} = 0 \tag{a}$$

Boundary conditions:

$$\lambda \frac{dT}{dr} = h_a(T - T_a) \qquad \text{on } r = a \tag{b}$$

$$-\lambda \frac{dT}{dr} = h_b(T - T_b) \qquad \text{on } r = b \tag{c}$$

where T_a and T_b are the temperatures of the surrounding media, h_a and h_b are the heat transfer coefficients, and subscripts a and b correspond to surfaces $r = a$ and $r = b$, respectively. The general solution of Eq. (a) is

$$T = A + B \ln r \tag{d}$$

The coefficients A and B can be determined from the boundary conditions (b) and (c).

$$A = T_a - (T_b - T_a)\left[\frac{\ln a - \dfrac{\lambda}{h_a a}}{\ln \dfrac{b}{a} + \dfrac{\lambda}{h_a a} + \dfrac{\lambda}{h_b b}}\right], \qquad B = \left(\frac{T_b - T_a}{\ln \dfrac{b}{a} + \dfrac{\lambda}{h_a a} + \dfrac{\lambda}{h_b b}}\right) \tag{e}$$

Thus the temperature is

$$T = T_a + (T_b - T_a)\left[\frac{\ln \dfrac{r}{a} + \dfrac{\lambda}{h_a a}}{\ln \dfrac{b}{a} + \dfrac{\lambda}{h_a a} + \dfrac{\lambda}{h_b b}}\right] \tag{3.81}$$

There are some special cases of Eq. (3.81).

[1] Prescribed surface temperatures T_a and T_b at the inner and outer surfaces, respectively:

$$T = T_a + (T_b - T_a)\left[\frac{\ln \dfrac{r}{a}}{\ln \dfrac{b}{a}}\right] \tag{3.82}$$

[2] Prescribed surface temperature T_a at the inner surface and heat transfer at the surface between the outer surface and the surrounding medium with temperature T_b:

$$T = T_a + (T_b - T_a)\left[\frac{\ln \dfrac{r}{a}}{\ln \dfrac{b}{a} + \dfrac{\lambda}{h_b b}}\right] \tag{3.83}$$

[3] Prescribed surface temperature T_b at the outer surface and heat transfer at the surface between the inner surface and the surrounding medium with temperature T_a:

$$T = T_a + (T_b - T_a)\left[\frac{\ln \dfrac{r}{a} + \dfrac{\lambda}{h_a a}}{\ln \dfrac{b}{a} + \dfrac{\lambda}{h_a a}}\right] \tag{3.84}$$

[4] Constant heat flux $q_a (=\lambda\,(dT/dr))$ at the inner surface $r = a$ and heat transfer between the outer surface and the surrounding medium with temperature T_b:

$$T = T_b + \frac{q_a a}{\lambda}\left(\ln\frac{r}{b} - \frac{\lambda}{h_b b}\right) \tag{3.85}$$

[5] Constant heat flux $q_b (= -\lambda\,(dT/dr))$ at the outer surface $r = b$ and heat transfer between the inner surface and the surrounding medium with temperature T_a:

$$T = T_a - \frac{q_b b}{\lambda}\left(\ln\frac{r}{a} + \frac{\lambda}{h_a a}\right) \tag{3.86}$$

Example 3.8 Determine a one-dimensional steady temperature in a two-layered hollow cylinder with an inner radius a, a bonded radius c and an outer radius b when the temperatures on the boundary surfaces are prescribed at T_a and T_b, respectively.

Solution The temperature of each layer is

$$T_i = A_i + B_i \ln r \qquad (i = 1, 2) \tag{a}$$

The boundary condition at each boundary is

$$T_1 = T_a \qquad \text{on } r = a \tag{b}$$

$$T_1 = T_2, \quad \lambda_1\frac{dT_1}{dr} = \lambda_2\frac{dT_2}{dr} \qquad \text{on } r = c \tag{c}$$

$$T_2 = T_b \qquad \text{on } r = b \tag{d}$$

We get from Eqs. (a) to (d)

$$A_1 = T_a - \frac{T_b - T_a}{\lambda_1 \ln\dfrac{b}{c} + \lambda_2 \ln\dfrac{c}{a}}\lambda_2 \ln a, \qquad B_1 = \frac{T_b - T_a}{\lambda_1 \ln\dfrac{b}{c} + \lambda_2 \ln\dfrac{c}{a}}\lambda_2$$

$$A_2 = T_a + \frac{T_b - T_a}{\lambda_1 \ln\dfrac{b}{c} + \lambda_2 \ln\dfrac{c}{a}}\left(\lambda_2 \ln\frac{c}{a} - \lambda_1 \ln c\right), \qquad B_2 = \frac{T_b - T_a}{\lambda_1 \ln\dfrac{b}{c} + \lambda_2 \ln\dfrac{c}{a}}\lambda_1$$

Then, the temperature of each layer is

$$T_1 = T_a + \frac{T_b - T_a}{\lambda_1 \ln\dfrac{b}{c} + \lambda_2 \ln\dfrac{c}{a}}\lambda_2 \ln\frac{r}{a} \tag{e}$$

$$T_2 = T_a + \frac{T_b - T_a}{\lambda_1 \ln\dfrac{b}{c} + \lambda_2 \ln\dfrac{c}{a}}\left(\lambda_2 \ln\frac{c}{a} + \lambda_1 \ln\frac{r}{c}\right) \tag{f}$$

3.4.3 One-Dimensional Transient Temperature in a Solid Cylinder

For a solid cylinder of radius a with initial temperature $T_i(r)$ and surface temperature T_a, the equations to be solved are

Governing equation:

$$\frac{\partial T}{\partial t} = \kappa\left(\frac{\partial^2 T}{\partial r^2} + \frac{1}{r}\frac{\partial T}{\partial r}\right) \tag{a}$$

Boundary condition:

$$T = T_a \qquad \text{on } r = a \tag{b}$$

Initial condition:

$$T = T_i(r) \qquad \text{at } t = 0 \tag{c}$$

We use the method of the separation of variables. If the temperature can be expressed by

$$T(r,t) = f(r)g(t) \tag{d}$$

equation (a) reduces to

$$\frac{1}{\kappa g(t)}\frac{dg(t)}{dt} = \frac{1}{f(r)}\left[\frac{d^2 f(r)}{dr^2} + \frac{1}{r}\frac{df(r)}{dr}\right] \equiv -s^2 \tag{e}$$

Equation (e) leads to two separate equations

$$\frac{dg(t)}{dt} + s^2\kappa g(t) = 0 \tag{f}$$

$$\frac{d^2 f(r)}{dr^2} + \frac{1}{r}\frac{df(r)}{dr} + s^2 f(r) = 0 \tag{3.87}$$

Equation (3.87) is the Bessel's differential equation of order zero† and it has two particular solutions

$$J_0(sr) \quad \text{and} \quad Y_0(sr) \tag{3.88}$$

where $J_0(sr)$ is the Bessel function of the first kind of order zero, and $Y_0(sr)$ is the Bessel function of the second kind of order zero. For more information, see Appendix B.

†G. N. Watson, Theory of Bessel Functions (2nd ed.), Cambridge University Press, Cambridge (1944).

The general solutions of Eqs. (f) and (3.87) are

$$g(t) = E, \quad f(r) = F + G \ln r \qquad \text{for } s = 0 \tag{g}$$

$$g(t) = He^{-\kappa s^2 t}, \quad f(r) = IJ_0(sr) + JY_0(sr) \qquad \text{for } s \neq 0 \tag{h}$$

These solutions must exist for arbitrary values of s. Therefore, using the solutions (g) and (h), we write a general solution of the form (d):

$$T(r,t) = A + B \ln r + [CJ_0(sr) + DY_0(sr)]e^{-\kappa s^2 t} \tag{i}$$

where

$$A = EF, \qquad B = EG, \qquad C = HI, \qquad D = HJ \tag{j}$$

As the Bessel function $Y_0(sr)$ and $\ln r$ are infinite at $r = 0$, B and D must be equal to zero, and the temperature for this problem is

$$T(r,t) = A + CJ_0(sr)e^{-\kappa s^2 t} \tag{k}$$

Substitution of Eq. (k) into the boundary condition (b) gives the following relations:

$$A = T_a, \quad J_0(sa) = 0 \tag{l}$$

The equation $J_0(sa) = 0$ has an infinite number of real positive roots s_n $(n = 1, 2, 3, \ldots)$. Thus, the temperature $T(r,t)$ can be developed into an infinite series called the Bessel series

$$T(r,t) = T_a + \sum_{n=1}^{\infty} A_n J_0(s_n r)e^{-\kappa s_n^2 t} \tag{3.89}$$

where A_n are unknown coefficients. The coefficients A_n in Eq. (3.89) can be determined from the initial condition (c)

$$\sum_{n=1}^{\infty} A_n J_0(s_n r) = T_i(r) - T_a \tag{m}$$

Multiplying both sides of Eq. (m) by $rJ_0(s_m r)$, integrating from zero to a, and using the integration (see Eqs. (B.31) and (B.32) in Appendix B, we find

$$\int_0^a J_0(s_m r)J_0(s_n r)r\,dr$$

$$= \frac{a}{s_m^2 - s_n^2}[s_m J_1(s_m a)J_0(s_n a) - s_n J_1(s_n a)J_0(s_m a)] = 0, \qquad m \neq n \tag{n}$$

$$\int_0^a J_0(s_m r)J_0(s_n r)r\,dr = \frac{1}{2}a^2 J_1^2(s_n a), \qquad m = n$$

where the coefficients A_n are

$$A_n = \frac{2}{a^2 J_1^2(s_n a)} \int_0^a [T_i(r) - T_a] J_0(s_n r) r\, dr \tag{o}$$

Finally, the temperature can be expressed as

$$T(r,t) = T_a - \frac{2}{a^2} \sum_{n=1}^{\infty} \left\{ \frac{1}{J_1^2(s_n a)} \int_0^a [T_a - T_i(r)] J_0(s_n r) r\, dr \right\} J_0(s_n r) e^{-\kappa s_n^2 t} \tag{3.90}$$

and its dimensionless form with dimensionless quantities $F = \kappa t/a^2$ and $\rho = r/a$ is

$$T(\rho,F) = T_a - 2 \sum_{n=1}^{\infty} \left\{ \frac{1}{J_1^2(p_n)} \int_0^1 [T_a - T_i(\rho)] J_0(p_n \rho) \rho\, d\rho \right\} J_0(p_n \rho) e^{-p_n^2 F} \tag{3.91}$$

where $p_n = s_n a$.

Temperatures with various boundary conditions are listed as follows:

[1] Prescribed surface temperature T_a at the boundary surface $r = a$ and constant initial temperature T_i:

$$T(r,t) = T_a - 2(T_a - T_i) \sum_{n=1}^{\infty} \frac{J_0(s_n r)}{s_n a J_1(s_n a)} e^{-\kappa s_n^2 t} \tag{3.92}$$

where s_n are the roots of the equation $J_0(s_n a) = 0$.

[2] Heat transfer between the surface and the surrounding medium with temperature T_a and initial temperature $T_i(r)$:

$$T(r,t) = T_a - \frac{2}{a^2} \sum_{n=1}^{\infty} \frac{\lambda^2 s_n^2 J_0(s_n r)}{\left(\lambda^2 s_n^2 + h_a^2\right) J_0^2(s_n a)} \times \int_0^a [T_a - T_i(r)] J_0(s_n r) r\, dr e^{-\kappa s_n^2 t} \tag{3.93}$$

where s_n are the roots of the equation $\lambda s_n J_1(s_n a) - h_a J_0(s_n a) = 0$.

[3] Heat transfer between the surface and the surrounding medium with temperature T_a and constant initial temperature T_i:

$$T(r,t) = T_a - 2(T_a - T_i) \sum_{n=1}^{\infty} \frac{\lambda h_a J_0(s_r)}{a\left(\lambda^2 s_n^2 + h_a^2\right) J_0(s_n a)} e^{-\kappa s_n^2 t} \tag{3.94}$$

where s_n are the roots of the equation $\lambda s_n J_1(s_n a) - h_a J_0(s_n a) = 0$.

3.4.4 One-Dimensional Transient Temperature in a Hollow Cylinder

Consider a long hollow cylinder with inner radius a and outer radius b. The initial temperature in the cylinder is $T_i(r)$ and heat is exchanged between the inner and outer surfaces and surrounding media at uniform temperatures T_a and T_b, respectively. The problem to be solved consists of

Governing equation:

$$\frac{\partial T}{\partial t} = \kappa \left(\frac{\partial^2 T}{\partial r^2} + \frac{1}{r}\frac{\partial T}{\partial r} \right) \tag{a}$$

Boundary conditions:

$$\lambda \frac{\partial T}{\partial r} = h_a (T - T_a) \qquad \text{on } r = a \tag{b}$$

$$-\lambda \frac{\partial T}{\partial r} = h_b (T - T_b) \qquad \text{on } r = b \tag{c}$$

Initial condition:

$$T = T_i(r) \qquad \text{at } t = 0 \tag{d}$$

Applying the Laplace transform with respect to t to the above equations, we obtain

Governing equation:

$$\frac{d^2\overline{T}}{dr^2} + \frac{1}{r}\frac{d\overline{T}}{dr} - \frac{p}{\kappa}\overline{T} = -\frac{1}{\kappa}T_i(r) \tag{a$'$}$$

Boundary conditions:

$$\lambda \frac{d\overline{T}}{dr} = h_a \left(\overline{T} - \frac{T_a}{p} \right) \qquad \text{on } r = a \tag{b$'$}$$

$$-\lambda \frac{d\overline{T}}{dr} = h_b \left(\overline{T} - \frac{T_b}{p} \right) \qquad \text{on } r = b \tag{c$'$}$$

We find the temperature by the use of the method of variation of parameters

$$\overline{T} = AI_0(qr) + BK_0(qr) - \frac{1}{\kappa}\int_a^r T_i(\eta)\eta[I_0(qr)K_0(q\eta) - I_0(q\eta)K_0(qr)]\,d\eta \tag{e}$$

where $q = \sqrt{p/\kappa}$. The determination of the unknown constants A and B from the boundary conditions (b$'$) and (c$'$) gives the temperature $\overline{T}$:

$$\overline{T} = \frac{1}{pD}[h_a T_a\{[\lambda q K_1(qb) - h_b K_0(qb)]I_0(qr) + [\lambda q I_1(qb) + h_b I_0(qb)]K_0(qr)\}$$
$$+ h_b T_b\{[\lambda q K_1(qa) + h_a K_0(qa)]I_0(qr) + [\lambda q I_1(qa) - h_a I_0(qa)]K_0(qr)\}]$$

$$+\frac{1}{D}\{[\lambda q K_1(qa)+h_a K_0(qa)]I_0(qr)+[\lambda q I_1(qa)-h_a I_0(qa)]K_0(qr)\}$$
$$\times\frac{1}{\kappa}\int_a^b T_i(\eta)\eta\{[\lambda q K_1(qb)-h_b K_0(qb)]I_0(q\eta)$$
$$+[\lambda q I_1(qb)+h_b I_0(qb)]K_0(q\eta)\}\,d\eta$$
$$-\frac{1}{\kappa}\int_a^r T_i(\eta)\eta[I_0(qr)K_0(q\eta)-I_0(q\eta)K_0(qr)]\,d\eta \tag{3.95}$$

where

$$D=[\lambda q I_1(qb)+h_b I_0(qb)][\lambda q K_1(qa)+h_a K_0(qa)]$$
$$-[\lambda q I_1(qa)-h_a I_0(qa)][\lambda q K_1(qb)-h_b K_0(qb)] \tag{f}$$

The inverse Laplace transform of Eq. (3.95) reduces to calculation of the sum of the residues at the poles in the inner region with the contour given in Fig. 3.6. Equation (3.95) has poles at $p=0$ and $p=-\kappa s_n^2$, and s_n are the positive roots of the equation:

$$[h_a J_0(sa)+\lambda s J_1(sa)]\,[h_b Y_0(sb)-\lambda s Y_1(sb)]$$
$$-[h_b J_0(sb)-\lambda s J_1(sb)]\,[h_a Y_0(sa)+\lambda s Y_1(sa)]=0 \tag{g}$$

The residue at $p=0$ is

$$T_a+(T_b-T_a)\left[\frac{\ln\frac{r}{a}+\frac{\lambda}{h_a a}}{\ln\frac{b}{a}+\frac{\lambda}{h_a a}+\frac{\lambda}{h_b b}}\right] \tag{h}$$

The residues at $p=-\kappa s_n^2$ are

$$-\pi\sum_{n=1}^{\infty}\frac{(T_a h_a-T_b G_n h_b)f(s_n,r)}{h_a^2+\lambda^2 s_n^2-G_n^2\left(h_b^2+\lambda^2 s_n^2\right)}e^{-\kappa s_n^2 t}$$
$$-\frac{\pi^2}{2}\sum_{n=1}^{\infty}\frac{s_n^2 f(s_n,r)e^{-\kappa s_n^2 t}}{h_a^2+\lambda^2 s_n^2-G_n^2\left(h_b^2+\lambda^2 s_n^2\right)}\int_a^b T_i(\eta)f(s_n,\eta)\eta\,d\eta \tag{i}$$

where $f(s_n,r)$ and G_n are given by

$$f(s_n,r)=[h_a Y_0(s_n a)+\lambda s_n Y_1(s_n a)]J_0(s_n r)$$
$$-[h_a J_0(s_n a)+\lambda s_n J_1(s_n a)]Y_0(s_n r) \tag{j}$$
$$G_n=\frac{h_a J_0(s_n a)+\lambda s_n J_1(s_n a)}{h_b J_0(s_n b)-\lambda s_n J_1(s_n b)}=\frac{h_a Y_0(s_n a)+\lambda s_n Y_1(s_n a)}{h_b Y_0(s_n b)-\lambda s_n Y_1(s_n b)} \tag{k}$$

The temperature, therefore, is

$$T(r,t) = T_a + (T_b - T_a)\left[\frac{\ln\dfrac{r}{a} + \dfrac{\lambda}{h_a a}}{\ln\dfrac{b}{a} + \dfrac{\lambda}{h_a a} + \dfrac{\lambda}{h_b b}}\right]$$

$$- \pi \sum_{n=1}^{\infty} \frac{(T_a h_a - T_b G_n h_b) f(s_n, r)}{h_a^2 + \lambda^2 s_n^2 - G_n^2\left(h_b^2 + \lambda^2 s_n^2\right)} e^{-\kappa s_n^2 t}$$

$$- \frac{\pi^2}{2} \sum_{n=1}^{\infty} \frac{s_n^2 f(s_n, r) e^{-\kappa s_n^2 t}}{h_a^2 + \lambda^2 s_n^2 - G_n^2\left(h_b^2 + \lambda^2 s_n^2\right)} \int_a^b T_i(\eta) f(s_n, \eta)\eta \, d\eta \qquad (3.96)$$

or

$$T(r,t) = T_a + (T_b - T_a)\left[\frac{\ln\dfrac{r}{a} + \dfrac{\lambda}{h_a a}}{\ln\dfrac{b}{a} + \dfrac{\lambda}{h_a a} + \dfrac{\lambda}{h_b b}}\right]$$

$$- \frac{\pi^2}{2} \sum_{n=1}^{\infty} \frac{s_n^2 f(s_n, r) e^{-\kappa s_n^2 t}}{h_a^2 + \lambda^2 s_n^2 - G_n^2\left(h_b^2 + \lambda^2 s_n^2\right)}$$

$$\times \int_a^b \left[T_i(\eta) - T_a - (T_b - T_a)\left(\frac{\ln\dfrac{\eta}{a} + \dfrac{\lambda}{h_a a}}{\ln\dfrac{b}{a} + \dfrac{\lambda}{h_a a} + \dfrac{\lambda}{h_b b}}\right)\right] f(s_n, \eta)\eta \, d\eta \qquad (3.96')$$

where s_n are the positive roots of Eq. (g).

There are some special cases of Eq. (3.96).

[1] Prescribed surface temperatures T_a and T_b at surfaces $r = a$ and $r = b$, respectively:

$$T = T_a + (T_b - T_a)\left[\frac{\ln\dfrac{r}{a}}{\ln\dfrac{b}{a}}\right]$$

$$- \pi \sum_{n=1}^{\infty} \frac{T_a J_0(s_n b) - T_b J_0(s_n a)}{J_0^2(s_n b) - J_0^2(s_n a)} J_0(s_n b) f(s_n, r) e^{-\kappa s_n^2 t}$$

$$- \frac{\pi^2}{2} \sum_{n=1}^{\infty} \frac{s_n^2 J_0^2(s_n b) f(s_n, r)}{J_0^2(s_n b) - J_0^2(s_n a)} \int_a^b T_i(\eta) f(s_n, \eta)\eta \, d\eta \cdot e^{-\kappa s_n^2 t} \qquad (3.97)$$

where

$$f(s_n, r) = Y_0(s_n a) J_0(s_n r) - J_0(s_n a) Y_0(s_n r) \tag{l}$$

and s_n are the positive roots of the equation $f(s, b) = 0$.

[2] Prescribed surface temperature T_a at the inner surface $r = a$ and heat transfer between the outer surface and the surrounding medium with temperature T_b:

$$\begin{aligned} T = T_a + (T_b - T_a) &\left[\frac{\ln \dfrac{r}{a}}{\ln \dfrac{b}{a} + \dfrac{\lambda}{h_b b}} \right] \\ &- \pi \sum_{n=1}^{\infty} \frac{(T_a - T_b G_n h_b) f(s_n, r)}{1 - G_n^2 \left(h_b^2 + \lambda^2 s_n^2\right)} e^{-\kappa s_n^2 t} \\ &- \frac{\pi^2}{2} \sum_{n=1}^{\infty} \frac{s_n^2 f(s_n, r) e^{-\kappa s_n^2 t}}{1 - G_n^2 \left(h_b^2 + \lambda^2 s_n^2\right)} \int_a^b T_i(\eta) f(s_n, \eta) \eta \, d\eta \end{aligned} \tag{3.98}$$

where $f(s_n, r)$ and G_n are given by

$$f(s_n, r) = Y_0(s_n a) J_0(s_n r) - J_0(s_n a) Y_0(s_n r) \tag{m}$$

$$G_n = \frac{J_0(s_n a)}{h_b J_0(s_n b) - \lambda s_n J_1(s_n b)} = \frac{Y_0(s_n a)}{h_b Y_0(s_n b) - \lambda s_n Y_1(s_n b)} \tag{n}$$

and s_n are positive roots of the equation

$$[h_b Y_0(sb) - \lambda s Y_1(sb)] J_0(sa) - [h_b J_0(sb) - \lambda s J_1(sb)] Y_0(sa) = 0 \tag{o}$$

[3] Prescribed surface temperature T_b at the outer surface $r = b$ and heat transfer between the inner surface and the surrounding medium with temperature T_a:

$$\begin{aligned} T = T_a + (T_b - T_a) &\left[\frac{\ln \dfrac{r}{a} + \dfrac{\lambda}{h_a a}}{\ln \dfrac{b}{a} + \dfrac{\lambda}{h_a a}} \right] - \pi \sum_{n=1}^{\infty} \frac{(T_a h_a - T_b G_n) f(s_n, r)}{h_a^2 + \lambda^2 s_n^2 - G_n^2} e^{-\kappa s_n^2 t} \\ &- \frac{\pi^2}{2} \sum_{n=1}^{\infty} \frac{s_n^2 f(s_n, r) e^{-\kappa s_n^2 t}}{h_a^2 + \lambda^2 s_n^2 - G_n^2} \int_a^b T_i(\eta) f(s_n, \eta) \eta \, d\eta \end{aligned} \tag{3.99}$$

where $f(s_n, r)$ and G_n are given by

$$\begin{aligned} f(s_n, r) = &[h_a Y_0(s_n a) + \lambda s_n Y_1(s_n a)] J_0(s_n r) \\ &- [h_a J_0(s_n a) + \lambda s_n J_1(s_n a)] Y_0(s_n r) \end{aligned} \tag{p}$$

$$G_n = \frac{h_a J_0(s_n a) + \lambda s_n J_1(s_n a)}{J_0(s_n b)} = \frac{h_a Y_0(s_n a) + \lambda s_n Y_1(s_n a)}{Y_0(s_n b)} \tag{q}$$

and s_n are the positive roots of the equation $f(s, b) = 0$.

[4] Constant heat flux $q_a (=\lambda\, (\partial T/\partial r))$ at the inner surface $r = a$ and heat transfer between the outer surface and the surrounding medium with temperature T_b:

$$T = T_b + \frac{q_a a}{\lambda}\left(\ln\frac{r}{b} - \frac{\lambda}{h_b b}\right) + \pi \sum_{n=1}^{\infty} \frac{(q_a + T_b G_n h_b \lambda s_n) f(s_n, r)}{\lambda s_n \left[1 - G_n^2 \left(h_b^2 + \lambda^2 s_n^2\right)\right]} e^{-\kappa s_n^2 t}$$
$$- \frac{\pi^2}{2} \sum_{n=1}^{\infty} \frac{s_n^2 f(s_n, r) e^{-\kappa s_n^2 t}}{1 - G_n^2 \left(h_b^2 + \lambda^2 s_n^2\right)} \int_a^b T_i(\eta) f(s_n, \eta) \eta \, d\eta \tag{3.100}$$

where $f(s_n, r)$ and G_n are given by

$$f(s_n, r) = Y_1(s_n a) J_0(s_n r) - J_1(s_n a) Y_0(s_n r) \tag{r}$$

$$G_n = \frac{J_1(s_n a)}{h_b J_0(s_n b) - \lambda s_n J_1(s_n b)} = \frac{Y_1(s_n a)}{h_b Y_0(s_n b) - \lambda s_n Y_1(s_n b)} \tag{s}$$

and s_n are the positive roots of the equation

$$[h_b Y_0(sb) - \lambda s Y_1(sb)]\, J_1(sa) - [h_b J_0(sb) - \lambda s J_1(sb)]\, Y_1(sa) = 0 \tag{t}$$

[5] Constant heat flux $q_b (= -\lambda\, (\partial T/\partial r))$ at the outer surface $r = b$ and heat transfer between the inner surface and the surrounding medium with temperature T_a:

$$T = T_a - \frac{q_b b}{\lambda}\left(\ln\frac{r}{a} + \frac{\lambda}{h_a a}\right) - \pi \sum_{n=1}^{\infty} \frac{(T_a h_a \lambda s_n - q_b G_n) f(s_n, r)}{\lambda s_n \left(h_a^2 + \lambda^2 s_n^2 - G_n^2\right)} e^{-\kappa s_n^2 t}$$
$$- \frac{\pi^2}{2} \sum_{n=1}^{\infty} \frac{s_n^2 f(s_n, r) e^{-\kappa s_n^2 t}}{h_a^2 + \lambda^2 s_n^2 - G_n^2} \int_a^b T_i(\eta) f(s_n, \eta) \eta \, d\eta \tag{3.101}$$

where $f(s_n, r)$ and G_n are given by

$$f(s_n, r) = [h_a Y_0(s_n a) + \lambda s_n Y_1(s_n a)] J_0(s_n r) - [h_a J_0(s_n a) + \lambda s_n J_1(s_n a)] Y_0(s_n r) \tag{u}$$

$$G_n = \frac{h_a J_0(s_n a) + \lambda s_n J_1(s_n a)}{J_1(s_n b)} = \frac{h_a Y_0(s_n a) + \lambda s_n Y_1(s_n a)}{Y_1(s_n b)} \tag{v}$$

and s_n are the positive roots of the equation

$$[h_a Y_0(sa) + \lambda s Y_1(sa)] J_1(sb) - [h_a J_0(sa) + \lambda s J_1(sa)] Y_1(sb) = 0 \tag{w}$$

Example 3.9 Determine the temperature in a hollow cylinder when the initial temperature T_i in Eq. (3.97) is constant.

Solution We calculate the integral:

$$\begin{aligned}\int_a^b f(s_n, r) r\,dr &= \frac{1}{s_n}\left[rY_0(s_n a)J_1(s_n r) - rJ_0(s_n a)Y_1(s_n r)\right]_a^b \\ &= \frac{b}{s_n}\left[Y_0(s_n a)J_1(s_n b) - J_0(s_n a)Y_1(s_n b)\right] \\ &\quad - \frac{a}{s_n}\left[Y_0(s_n a)J_1(s_n a) - J_0(s_n a)Y_1(s_n a)\right]\end{aligned} \tag{a}$$

Taking $f(s_n, b) = 0$ and Eq. (B.12) in Appendix B into consideration, Eq. (a) reduces to

$$\begin{aligned}\int_a^b f(s_n, r) r\,dr &= \frac{b}{s_n}\left[Y_0(s_n a)J_1(s_n b) - J_0(s_n a)Y_1(s_n b)\right] - \frac{a}{s_n}\frac{2}{\pi s_n a} \\ &= \frac{bJ_0(s_n a)}{s_n J_0(s_n b)}\left[Y_0(s_n b)J_1(s_n b) - J_0(s_n b)Y_1(s_n b)\right] - \frac{2}{\pi s_n^2} \\ &= \frac{2}{\pi s_n^2}\frac{J_0(s_n a) - J_0(s_n b)}{J_0(s_n b)}\end{aligned} \tag{b}$$

Substitution of Eq. (b) into Eq. (3.97) gives

$$\begin{aligned}T &= T_a + (T_b - T_a)\left[\frac{\ln \frac{r}{a}}{\ln \frac{b}{a}}\right] \\ &\quad - \pi \sum_{n=1}^{\infty} \frac{(T_a - T_i)J_0(s_n b) - (T_b - T_i)J_0(s_n a)}{J_0^2(s_n b) - J_0^2(s_n a)} J_0(s_n b) f(s_n, r) e^{-\kappa s_n^2 t}\end{aligned} \tag{c}$$

3.5 One-Dimensional Heat Conduction Problems in Spherical Coordinates

3.5.1 Heat Conduction Equation in Spherical Coordinates

The heat conduction equation of a homogeneous, isotropic sphere is expressed by Eq. (3.13) in spherical coordinates shown in Fig. 3.3 as

$$c\rho\frac{\partial T}{\partial t} = \lambda\left[\frac{\partial^2 T}{\partial r^2} + \frac{2}{r}\frac{\partial T}{\partial r} + \frac{1}{r^2 \sin\theta}\frac{\partial}{\partial\theta}\left(\sin\theta\frac{\partial T}{\partial\theta}\right) + \frac{1}{r^2\sin^2\theta}\frac{\partial^2 T}{\partial\phi^2}\right] + Q \tag{3.102}$$

When the temperature in the sphere is independent of the coordinate ϕ, the heat conduction equation (3.102) reduces to

$$c\rho\frac{\partial T}{\partial t} = \lambda\left[\frac{\partial^2 T}{\partial r^2} + \frac{2}{r}\frac{\partial T}{\partial r} + \frac{1}{r^2 \sin\theta}\frac{\partial}{\partial\theta}\left(\sin\theta\frac{\partial T}{\partial\theta}\right)\right] + Q \tag{3.103}$$

If the temperature in the sphere varies only along the r axis, the heat conduction equation (3.102) reduces to

$$c\rho \frac{\partial T}{\partial t} = \lambda \left(\frac{\partial^2 T}{\partial r^2} + \frac{2}{r} \frac{\partial T}{\partial r} \right) + Q \tag{3.104}$$

Moreover, if there is no internal heat generation, the equation becomes

$$\frac{\partial T}{\partial t} = \kappa \left(\frac{\partial^2 T}{\partial r^2} + \frac{2}{r} \frac{\partial T}{\partial r} \right) \tag{3.105}$$

Furthermore, Eq. (3.105) reduces to the steady state equation

$$\frac{d^2 T}{dr^2} + \frac{2}{r} \frac{dT}{dr} = 0 \tag{3.106}$$

Introducing $\Theta = rT$, Eqs. (3.104), (3.105), and (3.106) are written in a new variable Θ

$$c\rho \frac{\partial \Theta}{\partial t} = \lambda \frac{\partial^2 \Theta}{\partial r^2} + rQ \tag{3.107}$$

$$\frac{\partial \Theta}{\partial t} = \kappa \frac{\partial^2 \Theta}{\partial r^2} \tag{3.108}$$

$$\frac{d^2 \Theta}{dr^2} = 0 \tag{3.109}$$

We note that the heat conduction equations (3.107), (3.108), and (3.109) in the new variable Θ are of the same forms as Eqs. (3.30), (3.31) and (3.32) in Cartesian coordinates.

3.5.2 One-Dimensional Temperature in the Steady State

When both boundary surfaces of a hollow sphere with inner radius a and outer radius b are surrounded by media with temperatures, T_a and T_b, respectively, the equations for the temperature in the sphere without internal heat generation are

Governing equation:

$$\frac{d^2 T}{dr^2} + \frac{2}{r} \frac{dT}{dr} = 0 \tag{a}$$

Boundary conditions:

$$\lambda \frac{dT}{dr} = h_a (T - T_a) \qquad \text{on } r = a \tag{b}$$

$$-\lambda \frac{dT}{dr} = h_b (T - T_b) \qquad \text{on } r = b \tag{c}$$

where T_a and T_b are the temperatures of surrounding media, h_a and h_b are the heat transfer coefficients, and subscripts a and b mean boundary surfaces on $r = a$ and $r = b$. The general solution of Eq. (a) is

$$T = A + \frac{B}{r} \tag{d}$$

The coefficients A and B can be determined from the boundary conditions (b) and (c)

$$A = T_a + (T_b - T_a)\left[\frac{1 + \dfrac{\lambda}{h_a a}}{1 - \dfrac{a}{b} + \dfrac{\lambda}{h_a a} + \dfrac{a}{b}\dfrac{\lambda}{h_b b}}\right], \qquad B = -\frac{a(T_b - T_a)}{1 - \dfrac{a}{b} + \dfrac{\lambda}{h_a a} + \dfrac{a}{b}\dfrac{\lambda}{h_b b}} \tag{e}$$

Thus

$$T = T_a + (T_b - T_a)\left[\frac{1 + \dfrac{\lambda}{h_a a} - \dfrac{a}{r}}{1 - \dfrac{a}{b} + \dfrac{\lambda}{h_a a} + \dfrac{a}{b}\dfrac{\lambda}{h_b b}}\right] \tag{3.110}$$

There are some special cases of Eq. (3.110) which we shall now discuss.

[1] Prescribed surface temperatures T_a and T_b at both surfaces at $r = a$ and $r = b$, respectively:

$$T = T_a + (T_b - T_a)\left[\frac{1 - \dfrac{a}{r}}{1 - \dfrac{a}{b}}\right] \tag{3.111}$$

[2] Prescribed surface temperature T_a at the inner surface $r = a$ and the heat transfer between the outer surface and the surrounding medium with temperature T_b:

$$T = T_a + (T_b - T_a)\left[\frac{1 - \dfrac{a}{r}}{1 - \dfrac{a}{b} + \dfrac{a}{b}\dfrac{\lambda}{h_b b}}\right] \tag{3.112}$$

[3] Prescribed surface temperature T_b at the outer surface $r = b$ and the heat transfer between the inner surface and the surrounding medium with temperature T_a:

$$T = T_a + (T_b - T_a)\left[\frac{1 + \dfrac{\lambda}{h_a a} - \dfrac{a}{r}}{1 - \dfrac{a}{b} + \dfrac{\lambda}{h_a a}}\right] \tag{3.113}$$

[4] Constant heat flux $q_a(=\lambda\,(dT/dr))$ at the inner surface $r = a$ and heat transfer between the outer surface and the surrounding medium with temperature T_b:

$$T = T_b - \frac{q_a a}{\lambda}\frac{a}{b}\left(\frac{b}{r} - 1 + \frac{\lambda}{h_b b}\right) \tag{3.114}$$

[5] Constant heat flux $q_b(= -\lambda\,(dT/dr))$ at the outer surface $r = b$ and heat transfer between the inner surface and the surrounding medium with temperature T_a:

$$T = T_a - \frac{q_b b}{\lambda}\frac{b}{a}\left(1 - \frac{a}{r} + \frac{\lambda}{h_a a}\right) \tag{3.115}$$

Example 3.10 Determine the one-dimensional steady temperature in a two-layered hollow sphere with an inner radius a, a bonded radius c and an outer radius b, when at the boundary surfaces are heat transfer boundary conditions.

Solution The temperature of each layer is from Eq. (d) in Subsection 3.5.2

$$T_i = A_i + \frac{B_i}{r} \qquad (i = 1, 2) \tag{a}$$

The boundary condition at each boundary is

$$\lambda_1 \frac{dT_1}{dr} = h_a(T_1 - T_a) \qquad \text{on } r = a \tag{b}$$

$$T_1 = T_2, \quad \lambda_1 \frac{dT_1}{dr} = \lambda_2 \frac{dT_2}{dr} \qquad \text{on } r = c \tag{c}$$

$$-\lambda_2 \frac{dT_2}{dr} = h_b(T_2 - T_b) \qquad \text{on } r = b \tag{d}$$

We get from Eqs. (a) to (d)

$$A_1 = T_a - \frac{T_b - T_a}{D} c b^2 h_b \lambda_2^2 (h_a a + \lambda_1), \qquad B_1 = \frac{T_b - T_a}{D} a^2 b^2 c h_a h_b \lambda_2^2$$

$$A_2 = T_a + \frac{T_b - T_a}{D} b^2 h_b \lambda_2 \left[a^2 h_a(\lambda_2 - \lambda_1) - c\lambda_2(h_a a + \lambda_1)\right]$$

$$B_2 = \frac{T_b - T_a}{D} a^2 b^2 c h_a h_b \lambda_1 \lambda_2$$

where

$$D = a^2 b^2 h_a h_b \lambda_2(\lambda_2 - \lambda_1) - c b^2 h_b \lambda_2^2 (h_a a + \lambda_1) + c a^2 h_a \lambda_1 \lambda_2 (h_b b - \lambda_2)$$

Then, the temperature of each layer is

$$T_1 = T_a - \frac{T_b - T_a}{D} cb^2 h_b \lambda_2^2 \left[\lambda_1 + h_a a \left(1 - \frac{a}{r}\right)\right] \tag{e}$$

$$T_2 = T_a + \frac{T_b - T_a}{D} b^2 h_b \lambda_2 \left[a^2 h_a(\lambda_2 - \lambda_1) - c\lambda_2(h_a a + \lambda_1) + ach_a\lambda_1 \frac{a}{r}\right] \tag{f}$$

3.5.3 One-Dimensional Transient Temperature in a Solid Sphere

Consider a solid sphere of radius a with initial temperature $T_i(r)$. When there is heat transfer between the boundary surface and a surrounding medium with temperature T_a, the equations to be solved are

Governing equation:

$$\frac{\partial T}{\partial t} = \kappa \left(\frac{\partial^2 T}{\partial r^2} + \frac{2}{r}\frac{\partial T}{\partial r}\right) \tag{a}$$

Boundary condition:

$$-\lambda \frac{\partial T}{\partial r} = h_a (T - T_a) \qquad \text{on } r = a \tag{b}$$

Initial condition:

$$T = T_i(r) \qquad \text{at } t = 0 \tag{c}$$

Applying the Laplace transform with respect to t to above equations, we obtain

Governing equation:

$$\frac{d^2\overline{T}}{dr^2} + \frac{2}{r}\frac{d\overline{T}}{dr} - \frac{p}{\kappa}\overline{T} = -\frac{1}{\kappa} T_i(r) \tag{a$'$}$$

Boundary condition:

$$-\lambda \frac{d\overline{T}}{dr} = h_a \left(\overline{T} - \frac{T_a}{p}\right) \qquad \text{on } r = a \tag{b$'$}$$

We obtain the temperature with the method of variation of parameters

$$\overline{T} = A\frac{1}{r}\sinh qr + B\frac{1}{r}\cosh qr - \frac{1}{\kappa q r}\int_0^r T_i(\eta)\eta \sinh q(r - \eta)\, d\eta \tag{d}$$

where $q = \sqrt{p/\kappa}$. The temperature must be bounded at $r = 0$, hence Eq. (d) reduces to

$$\overline{T} = A\frac{1}{r}\sinh qr - \frac{1}{\kappa q r}\int_0^r T_i(\eta)\eta \sinh q(r - \eta)\, d\eta \tag{e}$$

Determination of the unknown constant A from boundary condition (b′) gives the temperature $\overline{T}$:

$$
\begin{aligned}
\overline{T} = {} & \frac{\sinh qr}{r}\left[\frac{1}{\lambda qa\cosh qa + (h_a a - \lambda)\sinh qa}\right] \\
& \times \left\{ T_a \frac{h_a a^2}{p} + \frac{1}{\kappa q} \right. \\
& \left. \times \int_0^a T_i(\eta)\eta[\lambda qa\cosh q(a-\eta) + (h_a a - \lambda)\sinh q(a-\eta)]\,d\eta \right\} \\
& - \frac{1}{\kappa q r}\int_0^r T_i(\eta)\eta\sinh q(r-\eta)\,d\eta
\end{aligned} \tag{f}
$$

Performing the inverse Laplace transform on Eq. (f) gives the temperature in the form

$$
\begin{aligned}
T(r,t) = T_a + 2\frac{a}{r}\sum_{n=1}^{\infty} & \left[\frac{\lambda}{(\lambda^2 s_n^2 a^2 + h_a^2 a^2 - \lambda h_a a)\sin s_n a}\right]\sin s_n r \\
& \times \left[-T_a h_a a + \left(\frac{(\lambda - h_a a)^2 + \lambda^2 s_n^2 a^2}{\lambda a^2}\right)\sin s_n a \right. \\
& \left. \times \int_0^a T_i(\eta)\eta\sin s_n\eta\,d\eta \right] e^{-\kappa s_n^2 t}
\end{aligned} \tag{3.116}
$$

or

$$
\begin{aligned}
T(r,t) = T_a + 2\sum_{n=1}^{\infty} & \frac{(\lambda - h_a a)^2 + \lambda^2 s_n^2 a^2}{\lambda^2 s_n^2 a^2 + h_a^2 a^2 - \lambda h_a a}\,\frac{\sin s_n r}{ar} e^{-\kappa s_n^2 t} \\
& \times \int_0^a [T_i(\eta) - T_a]\eta\sin s_n\eta\,d\eta
\end{aligned} \tag{3.116′}
$$

where s_n are the positive roots of

$$
\tan sa = \frac{\lambda sa}{\lambda - h_a a} \tag{3.117}
$$

The temperatures with various boundary conditions are listed as follows:

[1] Prescribed surface temperature T_a at $r = a$ and constant initial temperature T_i:

$$
T = T_a + 2(T_a - T_i)\sum_{n=1}^{\infty}(-1)^n \frac{\sin n\pi \dfrac{r}{a}}{n\pi \dfrac{r}{a}} e^{-\kappa n^2\pi^2 t/a^2} \tag{3.118}
$$

[2] Prescribed surface temperature T_a at $r = a$ and initial temperature $T_i(r)$:

$$T = T_a + 2\sum_{n=1}^{\infty} \frac{\sin n\pi \dfrac{r}{a}}{ar} e^{-\kappa n^2\pi^2 t/a^2} \int_0^a [T_i(\eta) - T_a]\,\eta \sin n\pi \frac{\eta}{a}\, d\eta \tag{3.119}$$

[3] Heat transfer between the boundary surface and the surrounding medium with the temperature T_a and constant initial temperature T_i:

$$T = T_a - 2(T_a - T_i)\sum_{n=1}^{\infty} \frac{\lambda h_a a^2 \sin s_n r}{r\left(\lambda^2 s_n^2 a^2 + h_a^2 a^2 - \lambda h_a a\right)\sin s_n a} e^{-\kappa s_n^2 t} \tag{3.120}$$

where h_a denotes the heat transfer coefficient and the s_n are positive roots of Eq. (3.117).

3.5.4 One-Dimensional Transient Temperature in a Hollow Sphere

Consider a hollow sphere with inner radius a and outer radius b. The initial temperature in the sphere is $T_i(r)$ and heat is exchanged at the inner and outer surfaces with surrounding media at uniform temperatures T_a and T_b at the inner and outer surfaces, respectively. The problem to be solved consists of

Governing equation:

$$\frac{\partial T}{\partial t} = \kappa\left(\frac{\partial^2 T}{\partial r^2} + \frac{2}{r}\frac{\partial T}{\partial r}\right) \tag{a}$$

Boundary conditions:

$$\lambda \frac{\partial T}{\partial r} = h_a(T - T_a) \qquad \text{on } r = a \tag{b}$$

$$-\lambda \frac{\partial T}{\partial r} = h_b(T - T_b) \qquad \text{on } r = b \tag{c}$$

Initial condition:

$$T = T_i(r) \qquad \text{at } t = 0 \tag{d}$$

We use the method of separation of variables. If the temperature can be expressed by

$$T(r, t) = f(r)g(t) \tag{e}$$

and substituting Eq. (e) into Eq. (a), we get two separate equations

$$\frac{dg(t)}{dt} + \kappa s^2 g(t) = 0 \tag{f}$$

$$\frac{d^2 f(r)}{dr^2} + \frac{2}{r}\frac{df(r)}{dr} + s^2 f(r) = 0 \tag{g}$$

where s is a constant. The general solutions of Eqs. (f) and (g) are

$$g(t) = A, \quad f(r) = B + \frac{C}{r} \qquad \text{for } s = 0 \tag{h}$$

$$g(t) = De^{-\kappa s^2 t}, \quad f(r) = E\frac{\sin sr}{r} + F\frac{\cos sr}{r} \qquad \text{for } s \neq 0 \tag{i}$$

These solutions must exist for arbitrary values of s, so that the temperature T can be expressed, by rewriting the coefficients, in the form

$$T(r,t) = A_0 + B_0\frac{1}{r} + \sum_{n=1}^{\infty}\left(A_n\frac{\sin s_n r}{r} + B_n\frac{\cos s_n r}{r}\right)e^{-\kappa s_n^2 t} \tag{3.121}$$

Substitution of Eq. (3.121) into the boundary conditions (b) and (c) gives the coefficients A_0 and B_0, and B_n

$$A_0 = T_a + (T_b - T_a)\frac{1 + \dfrac{\lambda}{h_a a}}{1 - \dfrac{a}{b} + \dfrac{\lambda}{h_a a} + \dfrac{a}{b}\dfrac{\lambda}{h_b b}}$$

$$B_0 = -\frac{a(T_b - T_a)}{1 - \dfrac{a}{b} + \dfrac{\lambda}{h_a a} + \dfrac{a}{b}\dfrac{\lambda}{h_b b}} \tag{j}$$

$$B_n = -\frac{(h_a a + \lambda)\sin s_n a - \lambda s_n a \cos s_n a}{(h_a a + \lambda)\cos s_n a + \lambda s_n a \sin s_n a}A_n \qquad (n \neq 0)$$

and s_n must be any positive root of the transcendental equation

$$\begin{aligned}&\left[(h_a a + \lambda)(h_b b - \lambda) - \lambda^2 s^2 ab\right]\sin s(b-a)\\ &+\lambda s\left[b(h_a a + \lambda) + a(h_b b - \lambda)\right]\cos s(b-a) = 0\end{aligned} \tag{k}$$

Then, the temperature is given by the following equation:

$$T(r,t) = T_s(r) + \sum_{n=1}^{\infty}\frac{C_n}{r}\left[(h_a a + \lambda)\sin s_n(r-a) + \lambda s_n a \cos s_n(r-a)\right]e^{-\kappa s_n^2 t} \tag{l}$$

where $T_s(r)$ means the steady temperature given by

$$T_s(r) = T_a + (T_b - T_a)\left[\frac{1 + \dfrac{\lambda}{h_a a} - \dfrac{a}{r}}{1 - \dfrac{a}{b} + \dfrac{\lambda}{h_a a} + \dfrac{a}{b}\dfrac{\lambda}{h_b b}}\right] \tag{m}$$

and

$$C_n = \frac{A_n}{(h_a a + \lambda)\cos s_n a + \lambda s_n a \sin s_n a} \tag{n}$$

The coefficients C_n in Eq. (l) can be determined by the initial condition (d)

$$\sum_{n=1}^{\infty} \frac{C_n}{r}\left[(h_a a + \lambda)\sin s_n(r-a) + \lambda s_n a \cos s_n(r-a)\right] = T_i(r) - T_s(r) \tag{o}$$

Multiplying both sides of Eq. (o) by $r[(h_a a + \lambda)\sin s_m(r-a) + \lambda s_m a \cos s_m(r-a)]$, integrating from a to b, and using the relations:

$$\begin{aligned}
&\int_a^b [(h_a a + \lambda)\sin s_n(r-a) + \lambda s_n a \cos s_n(r-a)] \\
&\quad \times [(h_a a + \lambda)\sin s_m(r-a) + \lambda s_m a \cos s_m(r-a)]\, dr = 0 \qquad \text{for } s_m \neq s_n \\
&\int_a^b [(h_a a + \lambda)\sin s_n(r-a) + \lambda s_n a \cos s_n(r-a)]^2\, dr \\
&\quad = \frac{1}{2}(b-a)\left[(h_a a + \lambda)^2 + \lambda^2 s_n^2 a^2\right] \\
&\qquad + \frac{\lambda\,[b(h_a a + \lambda) + a(h_b b - \lambda)]\left[(h_a a + \lambda)(h_b b - \lambda) + \lambda^2 s_n^2 ab\right]}{2\left[(h_b b - \lambda)^2 + \lambda^2 s_n^2 b^2\right]} \qquad \text{for } s_m = s_n
\end{aligned} \tag{p}$$

the coefficients C_n can be obtained

$$C_n = \frac{\left(\begin{array}{l} 2\left[(h_b b - \lambda)^2 + \lambda^2 s_n^2 b^2\right] \displaystyle\int_a^b [T_i(\eta) - T_s(\eta)]\eta \\ \quad \times [(h_a a + \lambda)\sin s_n(\eta - a) + \lambda s_n a \cos s_n(\eta - a)]\, d\eta \end{array}\right)}{\left(\begin{array}{l} (b-a)\left[(h_a a + \lambda)^2 + \lambda^2 s_n^2 a^2\right]\left[(h_b b - \lambda)^2 + \lambda^2 s_n^2 b^2\right] \\ \quad + \lambda[b(h_a a + \lambda) + a(h_b b - \lambda)]\left[(h_a a + \lambda)(h_b b - \lambda) + \lambda^2 s_n^2 ab\right] \end{array}\right)} \tag{q}$$

Therefore, the temperature is

$$T(r,t) = T_s(r) + \sum_{n=1}^{\infty} \frac{C_n}{r}\left[(h_a a + \lambda)\sin s_n(r-a) + \lambda s_n a \cos s_n(r-a)\right] e^{-\kappa s_n^2 t} \tag{3.122}$$

where the steady temperature $T_s(r)$ and coefficients C_n are given by Eqs. (m) and (q), respectively, and s_n are the roots of the transcendental equation (k).

There are some special cases of Eq. (3.122).

[1] Prescribed surface temperatures T_a and T_b at both surfaces at $r = a$ and $r = b$, respectively:

$$T = T_a + (T_b - T_a)\left[\frac{1 - \dfrac{a}{r}}{1 - \dfrac{a}{b}}\right] + \frac{2}{(b-a)r}\sum_{n=1}^{\infty} \sin n\pi\left(\frac{r-a}{b-a}\right) e^{-\kappa n^2\pi^2 t/(b-a)^2} \times \int_a^b \left[T_i(\eta) - T_a - (T_b - T_a)\left(\frac{1 - \dfrac{a}{\eta}}{1 - \dfrac{a}{b}}\right)\right] \eta \sin n\pi\left(\frac{\eta - a}{b-a}\right) d\eta \tag{3.123}$$

[2] Prescribed surface temperature T_a at the inner surface $r = a$ and heat transfer between the outer surface and the surrounding medium with temperature T_b:

$$T = T_a + (T_b - T_a)\left[\frac{1 - \dfrac{a}{r}}{1 - \dfrac{a}{b} + \dfrac{a}{b}\dfrac{\lambda}{h_b b}}\right] + \frac{2}{r}\sum_{n=1}^{\infty} \sin s_n(r-a) e^{-\kappa s_n^2 t} \times \frac{\lambda^2 s_n^2 b^2 + (h_b b - \lambda)^2}{(b-a)\left[\lambda^2 s_n^2 b^2 + (h_b b - \lambda)^2\right] + \lambda b(h_b b - \lambda)} \times \int_a^b \left[T_i(\eta) - T_a - \frac{(T_b - T_a)\left(1 - \dfrac{a}{\eta}\right)}{1 - \dfrac{a}{b} + \dfrac{a}{b}\dfrac{\lambda}{h_b b}}\right] \eta \sin s_n(\eta - a)\, d\eta \tag{3.124}$$

where s_n are the positive roots of the equation

$$(h_b b - \lambda)\sin s(b-a) + \lambda s b \cos s(b-a) = 0 \tag{r}$$

[3] Prescribed surface temperature T_b at the outer surface $r = b$ and heat transfer between the inner surface and the surrounding medium with temperature T_a:

$$T = T_a + (T_b - T_a)\left[\frac{1 + \dfrac{\lambda}{h_a a} - \dfrac{a}{r}}{1 - \dfrac{a}{b} + \dfrac{\lambda}{h_a a}}\right] + \frac{2}{r}\sum_{n=1}^{\infty} \frac{(h_a a + \lambda)\sin s_n(r-a) + \lambda s_n a \cos s_n(r-a)}{(b-a)\left[\lambda^2 s_n^2 a^2 + (h_a a + \lambda)^2\right] + \lambda a(h_a a + \lambda)} e^{-\kappa s_n^2 t}$$

$$\times \int_a^b \left[T_i(\eta) - T_a - (T_b - T_a) \left(\frac{1 + \dfrac{\lambda}{h_a a} - \dfrac{a}{\eta}}{1 - \dfrac{a}{b} + \dfrac{\lambda}{h_a a}} \right) \right]$$

$$\times \eta[(h_a a + \lambda) \sin s_n(\eta - a) + \lambda s_n a \cos s_n(\eta - a)]\, d\eta \tag{3.125}$$

where s_n are the positive roots of the equation

$$(h_a a + \lambda) \sin s(b - a) + \lambda s a \cos s(b - a) = 0 \tag{s}$$

[**4**] Constant heat flux $q_a (= \lambda\, (\partial T / \partial r))$ at the inner surface $r = a$ and heat transfer between the outer surface and the surrounding medium with temperature T_b:

$$T = T_b - \frac{q_a a}{\lambda} \frac{a}{b} \left(\frac{b}{r} - 1 + \frac{\lambda}{h_b b} \right)$$

$$+ \frac{2}{r} \sum_{n=1}^{\infty} \frac{[\lambda^2 s_n^2 b^2 + (h_b b - \lambda)^2][\sin s_n(r - a) + s_n a \cos s_n(r - a)]}{\left(\begin{array}{c} (b - a)\big(1 + s_n^2 a^2\big)\big[\lambda^2 s_n^2 b^2 + (h_b b - \lambda)^2\big] \\ + [b\lambda + a(h_b b - \lambda)]\big[\lambda s_n^2 ab + (h_b b - \lambda)\big] \end{array} \right)}$$

$$\times e^{-\kappa s_n^2 t} \int_a^b \left[T_i(\eta) - T_b + \frac{q_a a}{\lambda} \frac{a}{b} \left(\frac{b}{\eta} - 1 + \frac{\lambda}{h_b b} \right) \right]$$

$$\times \eta[\sin s_n(\eta - a) + s_n a \cos s_n(\eta - a)]\, d\eta \tag{3.126}$$

where s_n are the roots of the equation

$$\big(h_b b - \lambda - \lambda s^2 ab\big) \sin s(b - a) + s\, [a(h_b b - \lambda) + b\lambda] \cos s(b - a) = 0 \tag{t}$$

[**5**] Constant heat flux $q_b (= -\lambda\, (\partial T / \partial r))$ at the outer surface $r = b$ and heat transfer between the inner surface and the surrounding medium with temperature T_a:

$$T = T_a - \frac{q_b b}{\lambda} \frac{b}{a} \left(1 - \frac{a}{r} + \frac{\lambda}{h_a a} \right)$$

$$+ \frac{2}{r} \sum_{n=1}^{\infty} \frac{\big(s_n^2 b^2 + 1\big)[(h_a a + \lambda) \sin s_n(r - a) + \lambda s_n a \cos s_n(r - a)]}{\left(\begin{array}{c} (b - a)\big(1 + s_n^2 b^2\big)\big[\lambda^2 s_n^2 a^2 + (h_a a + \lambda)^2\big] \\ + [b(h_a a + \lambda) - a\lambda]\big[\lambda s_n^2 ab - (h_a a + \lambda)\big] \end{array} \right)}$$

$$\times e^{-\kappa s_n^2 t} \int_a^b \left[T_i(\eta) - T_a + \frac{q_b b}{\lambda} \frac{b}{a} \left(1 - \frac{a}{\eta} + \frac{\lambda}{h_a a} \right) \right]$$

$$\times \eta[(h_a a + \lambda) \sin s_n(\eta - a) + \lambda s_n a \cos s_n(\eta - a)]\, d\eta \tag{3.127}$$

where s_n are the roots of the equation

$$\big(h_a a + \lambda + \lambda s^2 ab\big) \sin s(b - a) - s\, [b(h_a a + \lambda) - a\lambda] \cos s(b - a) = 0 \tag{u}$$

Problems

3.1 When a surface of a strip with a width l is kept at constant temperature T_a, and the other surface is cooled by heat transfer between the surface and the surrounding medium with temperature T_b, find the steady temperature in the strip.

Answer

$$T = T_a + (T_b - T_a)\left[\frac{h_b x}{\lambda + h_b l}\right]$$

3.2 Prove Eq. (z) in Subsection 3.3.3.

3.3 Prove Eqs. (3.51)–(3.54).

3.4 When the boundary conditions of a strip are $T = T_a$ on $x = 0$ and $T = T_b$ on $x = l$, find the transient temperature in the strip with zero initial temperature by separation of variables.

Answer

$$T(x,t) = T_a + (T_b - T_a)\frac{x}{l} + \frac{2}{\pi}\sum_{n=1}^{\infty}\frac{1}{n}[(-1)^n T_b - T_a]\sin n\pi\frac{x}{l}e^{-\kappa(n\pi/l)^2 t}$$

3.5 Derive Eq. (v) in Subsection 3.3.4.

Hint The complementary solution of Eq. (u) in Subsection 3.3.4 is

$$\overline{T}_c = Ae^{qx} + Be^{-qx} \tag{a}$$

The particular solution can be expressed by the above equation with A and B as functions of a variable x.

$$\overline{T}_p = A(x)e^{qx} + B(x)e^{-qx} \tag{b}$$

Differentiation of Eq. (b) with respect to x gives

$$\frac{d\overline{T}_p}{dx} = \frac{dA(x)}{dx}e^{qx} + \frac{dB(x)}{dx}e^{-qx} + qA(x)e^{qx} - qB(x)e^{-qx} \tag{c}$$

We assume

$$\frac{dA(x)}{dx}e^{qx} + \frac{dB(x)}{dx}e^{-qx} = 0 \tag{d}$$

then, Eq. (c) reduces to

$$\frac{d\overline{T}_p}{dx} = qA(x)e^{qx} - qB(x)e^{-qx} \tag{e}$$

Substitution of Eq. (b) and of the derivative with respect to x of Eq. (e) into Eq. (u) gives

$$\frac{dA(x)}{dx} q e^{qx} - \frac{dB(x)}{dx} q e^{-qx} = -\frac{1}{\kappa} T_i(x) \tag{f}$$

From Eqs. (d) and (f), we get

$$\frac{dA(x)}{dx} = -\frac{T_i}{2q\kappa} e^{-qx}, \qquad \frac{dB(x)}{dx} = \frac{T_i}{2q\kappa} e^{qx} \tag{g}$$

From Eq. (g), we obtain

$$A(x) = -\frac{1}{2qk} \int_0^x T_i(\eta) e^{-q\eta}\, d\eta, \qquad B(x) = \frac{1}{2qk} \int_0^x T_i(\eta) e^{q\eta}\, d\eta \tag{h}$$

Substituting Eq. (h) into Eq. (b) leads to

$$\begin{aligned} \overline{T}_p &= A(x) e^{qx} + B(x) e^{-qx} \\ &= -\frac{1}{2qk} \int_0^x T_i(\eta) \left(e^{q(x-\eta)} - e^{-q(x-\eta)}\right) d\eta \\ &= -\frac{1}{qk} \int_0^x T_i(\eta) \sinh q(x-\eta)\, d\eta \end{aligned} \tag{i}$$

3.6 When there is no thermal contact resistance at the contact surface ($r = c$), find the steady temperature in the composite hollow cylinder with the boundary conditions of heat transfer by convection and radiation at the inner surface ($r = a$) and outer surface ($r = b$), where the thermal conductivities of inner and outer cylinders are λ_1 and λ_2, respectively.

Answer

$$T_1 = T_a + (T_b - T_a) \frac{\lambda_2 h_b b}{D} \left(h_a a \ln \frac{r}{a} + \lambda_1 \right) \qquad \text{for } a \le r \le c$$

$$T_2 = T_a + (T_b - T_a) \frac{\lambda_2 h_b b}{D} \left[h_a a \left(\ln \frac{c}{a} + \frac{\lambda_1}{\lambda_2} \ln \frac{r}{c} \right) + \lambda_1 \right] \qquad \text{for } c \le r \le b$$

where

$$D = \lambda_1 \lambda_2 (h_a a + h_b b) + h_a a h_b b \left(\lambda_1 \ln \frac{b}{c} + \lambda_2 \ln \frac{c}{a} \right)$$

3.7 Derive Eq. (3.93).

3.8 Derive Eqs. (3.97)–(3.101).

3.9 Determine the temperature (3.96) using the method of separation of variables.

3.10 Derive Eqs. (3.111)–(3.115).

3.11 Derive Eqs. (3.123)–(3.127).

3.12 A solid cylinder of radius a is at initial temperature $T_i(r)$ and heat transfer occurs between the surface and the surrounding medium. When the temperature T_a of the surrounding medium changes with time t, find the transient temperature in the solid cylinder.

Answer

$$T = \frac{2\kappa\lambda h_a}{a}\int_0^t T_a(\tau)\sum_{n=1}^{\infty}\frac{s_n^2 J_0(s_n r)}{(\lambda^2 s_n^2 + h_a^2) J_0(s_n a)} e^{-\kappa s_n^2(t-\tau)}\, d\tau$$

$$+ \frac{2\lambda^2}{a^2}\sum_{n=1}^{\infty}\frac{s_n^2 J_0(s_n r)}{(\lambda^2 s_n^2 + h_a^2) J_0^2(s_n a)} e^{-\kappa s_n^2 t}\int_0^a T_i(\eta)\eta J_0(s_n\eta)\, d\eta$$

where s_n are the positive roots of the equation:

$$\lambda s J_1(sa) - h_a J_0(sa) = 0.$$

Chapter 4

BASIC EQUATIONS OF THERMOELASTICITY

The theory of linearized thermoelasticity for three-dimensional bodies is formulated. Stress, strain, and force equilibrium relations along with compatibility conditions are derived in Cartesian coordinates. The connection between the equations of linearized thermoelasticity and the Navier's governing equations, which are a set of vector partial differential equations, is established. The alternate formulation in terms of stresses leads to the Beltrami–Michell compatibility equations. Boundary conditions are derived for: (1) prescribed surface stresses; (2) prescribed surface displacements; (3) prescribed stresses on part of the boundary surface, and prescribed displacements on the remainder of the boundary surface. A detailed analysis of the solution of the Navier's equations is carried out wherein the Goodier's thermoelastic displacement potential is used in conjunction with generalized Boussinesq harmonic functions. General solutions for the potential function are derived in the Cartesian coordinate system. Next, the equations of equilibrium and the formulas for strain in cylindrical and spherical coordinate systems are discussed and, in this context, the Navier's governing equations and associated solutions are derived. Lastly, the basic equations for multiply-connected bodies are treated. Since multi-valuedness of rotations and displacements may be produced in a multiply-connect region, even if stresses and strains are single-valued, the appropriate conditions for rotations and displacements are discussed.

4.1 Stress in a Cartesian Coordinate System

4.1.1 Equations of Equilibrium

We consider an elastic body with an arbitrary shape subjected to temperature change and external forces. The internal forces will be produced in the body by the action of

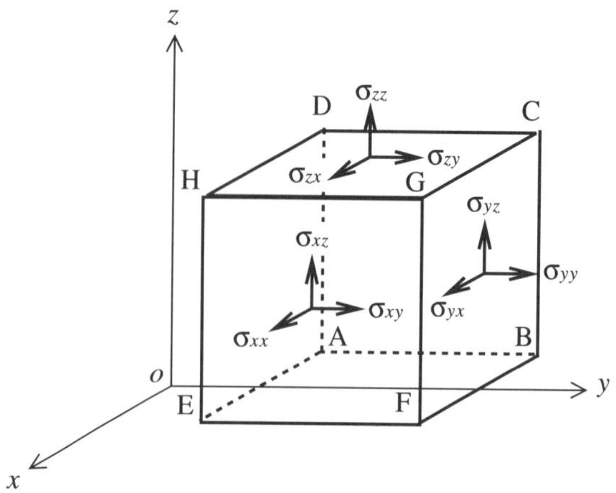

Figure 4.1 A rectangular parallelepiped element.

the temperature change and external forces. A rectangular parallelepiped element at a point in the body, such as in Fig. 4.1, is considered. We take Cartesian coordinate axes x, y, z parallel to the sides of the rectangular parallelepiped element. The components of stress acting on the plane EFGH (x plane) are presented by σ_{xx}, σ_{xy}, and σ_{xz}, in which the first subscript letter denotes a plane acted on by the stress, and the second indicates the direction of the component of the stress. If the two subscript letters are the same, the stress denotes normal stress, but if not, the stress is a shearing stress. The plane is called the x plane, when the outward-drawn normal direction of the plane coincides with the direction of the x axis of coordinate system: the y and z planes are similarly defined. The plane is called positive, when the outer-drawn normal direction of the plane coincides with the positive direction of the axis of coordinate system, and negative when the outer-drawn normal direction of the plane coincides with the negative direction of the axis. The stress is taken positive when the direction of the component of stress acting on the positive (or negative) plane coincides with the positive (or negative) direction of the axis, and negative when the direction of the component of stress acting on the positive (or negative) plane is negative (or positive). Therefore, the stresses as shown in Fig. 4.1 are positive. There are nine components of stress acting on the planes of the rectangular parallelepiped element, that is

the components of stress on x plane:	$\sigma_{xx}, \sigma_{xy}, \sigma_{xz}$
the components of stress on y plane:	$\sigma_{yx}, \sigma_{yy}, \sigma_{yz}$
the components of stress on z plane:	$\sigma_{zx}, \sigma_{zy}, \sigma_{zz}$

We consider the equilibrium of the surface forces acting on each plane and the body force acting at the center of the rectangular element as shown in Fig. 4.2. When the derivation of the components of stress is taken into account due to small changes

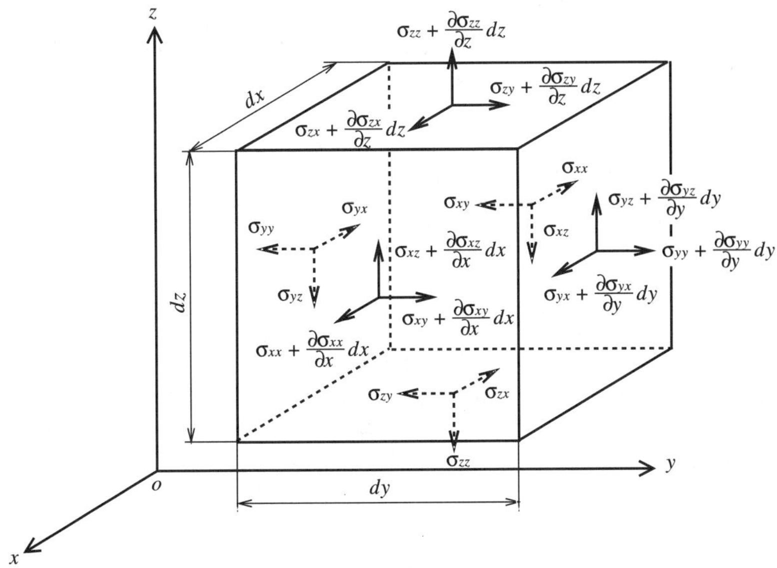

Figure 4.2 Equilibrium of forces on a rectangular parallelepiped element.

dx, dy, dz, the components of stress acting on each plane are

$$\begin{array}{llll}
\sigma_{xx} + \dfrac{\partial \sigma_{xx}}{\partial x}\,dx, & \sigma_{xy} + \dfrac{\partial \sigma_{xy}}{\partial x}\,dx, & \sigma_{xz} + \dfrac{\partial \sigma_{xz}}{\partial x}\,dx & \text{on positive } x \text{ plane} \\
\sigma_{yx} + \dfrac{\partial \sigma_{yx}}{\partial y}\,dy, & \sigma_{yy} + \dfrac{\partial \sigma_{yy}}{\partial y}\,dy, & \sigma_{yz} + \dfrac{\partial \sigma_{yz}}{\partial y}\,dy & \text{on positive } y \text{ plane} \\
\sigma_{zx} + \dfrac{\partial \sigma_{zx}}{\partial z}\,dz, & \sigma_{zy} + \dfrac{\partial \sigma_{zy}}{\partial z}\,dz, & \sigma_{zz} + \dfrac{\partial \sigma_{zz}}{\partial z}\,dz & \text{on positive } z \text{ plane} \\
\sigma_{xx}, & \sigma_{xy}, & \sigma_{xz} & \text{on negative } x \text{ plane} \\
\sigma_{yx}, & \sigma_{yy}, & \sigma_{yz} & \text{on negative } y \text{ plane} \\
\sigma_{zx}, & \sigma_{zy}, & \sigma_{zz} & \text{on negative } z \text{ plane}
\end{array}$$

We denote the components of body force per unit volume acting at the center of the rectangular element by F_x, F_y, F_z. The equilibrium of the forces in the x direction acting on the element is

$$\begin{aligned}
&\left(\sigma_{xx} + \frac{\partial \sigma_{xx}}{\partial x}\,dx\right) dy\,dz - \sigma_{xx}\,dy\,dz \\
&\quad + \left(\sigma_{yx} + \frac{\partial \sigma_{yx}}{\partial y}\,dy\right) dz\,dx - \sigma_{yx}\,dz\,dx \\
&\quad + \left(\sigma_{zx} + \frac{\partial \sigma_{zx}}{\partial z}\,dz\right) dx\,dy - \sigma_{zx}\,dx\,dy + F_x\,dx\,dy\,dz = 0
\end{aligned} \tag{a}$$

The two other equations for equilibrium of the forces in the y and z directions can be obtained in the same way. After simplification, we get from these equations

$$\begin{aligned}
\frac{\partial\sigma_{xx}}{\partial x} + \frac{\partial\sigma_{yx}}{\partial y} + \frac{\partial\sigma_{zx}}{\partial z} + F_x = 0 \\
\frac{\partial\sigma_{xy}}{\partial x} + \frac{\partial\sigma_{yy}}{\partial y} + \frac{\partial\sigma_{zy}}{\partial z} + F_y = 0 \\
\frac{\partial\sigma_{xz}}{\partial x} + \frac{\partial\sigma_{yz}}{\partial y} + \frac{\partial\sigma_{zz}}{\partial z} + F_z = 0
\end{aligned} \tag{4.1}$$

Equations (4.1) are called the differential equations of equilibrium and must be satisfied at all points in the body. When a new notation (x_1, x_2, x_3) is used instead of (x, y, z) and σ_{ij} denotes the component of stress in the direction of the x_j axis on the x_i plane, the equations of equilibrium (4.1) are expressed in a more simple form:

$$\sum_{j=1}^{3} \sigma_{ji,j} + F_i = 0 \qquad (i = 1, 2, 3) \tag{4.1$'$}$$

where the comma denotes partial differentiation with respect to a following variable. Using a convention that if a certain index is used twice it means the summation with respect to that index for all admissible values of the index, the summation symbol can be omitted. Thus, Eqs. (4.1′) can be written in the index notation:

$$\sigma_{ji,j} + F_i = 0 \qquad (i, j = 1, 2, 3) \tag{4.1$''$}$$

Moreover, we will assume that subscripts in Roman alphabet letters, such as i and j, will take values of 1, 2, and 3 only. Thus, from now on we will not specify these values.

We consider the equilibrium of the moments by the forces acting on the rectangular parallelepiped element with respect to the center of the element. The equilibrium of moments about a line through the center of the element and parallel to the x axis is

$$\begin{aligned}
&\left(\sigma_{yz} + \frac{\partial\sigma_{yz}}{\partial y}\,dy\right) dx\,dz\frac{dy}{2} + \sigma_{yz}\,dx\,dz\frac{dy}{2} \\
&- \left(\sigma_{zy} + \frac{\partial\sigma_{zy}}{\partial z}\,dz\right) dy\,dx\frac{dz}{2} - \sigma_{zy}\,dy\,dx\frac{dz}{2} = 0
\end{aligned} \tag{b}$$

The two other equations for equilibrium of the moments can be obtained in the same way. After omitting higher infinitesimal terms and dividing by $dx\,dy\,dz$, we obtain

$$\sigma_{xy} = \sigma_{yx}, \qquad \sigma_{yz} = \sigma_{zy}, \qquad \sigma_{zx} = \sigma_{xz} \tag{4.2}$$

or in index notation

$$\sigma_{ij} = \sigma_{ji} \tag{4.2'}$$

Equations (4.2′) are more general than (4.2) because they include normal stresses as well. The shearing stresses acting on two perpendicular surfaces and with the direction perpendicular to the line of intersection of the planes are equal, and these shearing stresses are called complementary shearing stresses. Hence the components of stress are symmetric, and there are six distinct components.

When a point in the body appears at the surface of the body, a rectangular parallelepiped element cannot be made. In such case, we can make a small tetrahedral element (Fig. 4.3) with an oblique surface as a boundary surface instead of the rectangular parallelepiped element. If N is the external normal to the small oblique surface ABC, the direction cosines l, m, n are defined by

$$\cos(N, x) = l, \qquad \cos(N, y) = m, \qquad \cos(N, z) = n \tag{4.3}$$

If dA is the area of the oblique surface ABC, the areas of the three other surfaces are obtained by projecting the small area of the oblique surface on the x, y, and z planes

$$l\,dA, \qquad m\,dA, \qquad n\,dA \tag{4.4}$$

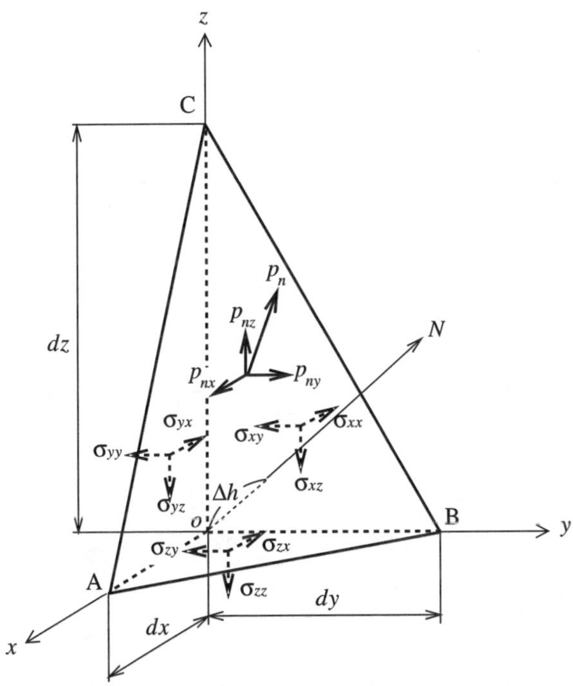

Figure 4.3 Equilibrium of forces on a small tetrahedron element.

Denoting the three components of stress, parallel to the coordinate axes, acting on the oblique surface ABC by p_{nx}, p_{ny}, p_{nz}, the equilibrium of the forces in the x direction acting on the element is given by

$$p_{nx}\, dA - \sigma_{xx}\, l\, dA - \sigma_{yx}\, m\, dA - \sigma_{zx}\, n\, dA + F_x \frac{dAdh}{3} = 0 \qquad \text{(c)}$$

where dh denotes a distance from the point O to the oblique surface ABC. The two other equations for equilibrium of forces in the y and z directions can be obtained in the same way. After dividing by dA, and proceeding with the infinitesimal volume of the element to zero, we obtain

$$\begin{aligned} \sigma_{xx} l + \sigma_{yx} m + \sigma_{zx} n &= p_{nx} \\ \sigma_{xy} l + \sigma_{yy} m + \sigma_{zy} n &= p_{ny} \\ \sigma_{xz} l + \sigma_{yz} m + \sigma_{zz} n &= p_{nz} \end{aligned} \qquad (4.5)$$

or in the index notation

$$\sigma_{ji} n_j = p_{ni} \qquad (4.5')$$

where n_j denote the direction cosines between the external normal of the oblique surface ABC and each axis. Equations (4.5) are called Cauchy's fundamental relations and present the relation between the traction acting on the boundary surface and the stresses on the internal rectangular surfaces, so Eqs. (4.5) can be used as the boundary conditions for stress.

4.1.2 Coordinate Transformation of Stress

The stress produced at a point in the body can be expressed by six components with respect to the given coordinate system. As the coordinate system can be arbitrarily selected, there are many expressions for stress components corresponding to the coordinate systems. We must consider how to express the six components of stress referred to one coordinate system in terms of those referred to another system. Let the components of stress referred to an old Cartesian coordinate system (x, y, z) be denoted by $\sigma_{xx}, \sigma_{xy}, \ldots$, the components referred to a new coordinate system (x', y', z') be denoted by $\sigma_{x'x'}, \sigma_{x'y'}, \ldots$, and the direction cosines between the axes of the old Cartesian coordinate system and the new be given in Table 4.1.

Table 4.1 The direction cosines between two coordinate systems

	x	y	z
x'	l_1	m_1	n_1
y'	l_2	m_2	n_2
z'	l_3	m_3	n_3

When the oblique surface ABC of a small tetrahedral element in Fig. 4.3 is selected as the x' plane, the components, in the direction of x, y, z axes, of traction on the x' plane are given from Eqs. (4.5) by

$$\begin{aligned} p_{x'x} &= \sigma_{xx} l_1 + \sigma_{yx} m_1 + \sigma_{zx} n_1 \\ p_{x'y} &= \sigma_{xy} l_1 + \sigma_{yy} m_1 + \sigma_{zy} n_1 \\ p_{x'z} &= \sigma_{xz} l_1 + \sigma_{yz} m_1 + \sigma_{zz} n_1 \end{aligned} \tag{a}$$

Since the components along the x', y', z' axes of these tractions give the components of stress in the x', y', z' axes on the x' plane, we obtain

$$\begin{aligned} \sigma_{x'x'} &= p_{x'x} l_1 + p_{x'y} m_1 + p_{x'z} n_1 \\ \sigma_{x'y'} &= p_{x'x} l_2 + p_{x'y} m_2 + p_{x'z} n_2 \\ \sigma_{x'z'} &= p_{x'x} l_3 + p_{x'y} m_3 + p_{x'z} n_3 \end{aligned} \tag{b}$$

Substitution of Eqs. (a) into (b) yields the components of stress on the x' plane referred to the new coordinate system (x', y', z') in terms of the components referred to the old coordinate system (x, y, z). The two other equations for the components of stress on the y' and z' planes can be obtained in the same manner

$$\begin{aligned} \sigma_{x'x'} &= \sigma_{xx} l_1^2 + \sigma_{yy} m_1^2 + \sigma_{zz} n_1^2 + 2(\sigma_{xy} l_1 m_1 + \sigma_{yz} m_1 n_1 + \sigma_{zx} n_1 l_1) \\ \sigma_{y'y'} &= \sigma_{xx} l_2^2 + \sigma_{yy} m_2^2 + \sigma_{zz} n_2^2 + 2(\sigma_{xy} l_2 m_2 + \sigma_{yz} m_2 n_2 + \sigma_{zx} n_2 l_2) \\ \sigma_{z'z'} &= \sigma_{xx} l_3^2 + \sigma_{yy} m_3^2 + \sigma_{zz} n_3^2 + 2(\sigma_{xy} l_3 m_3 + \sigma_{yz} m_3 n_3 + \sigma_{zx} n_3 l_3) \\ \sigma_{x'y'} &= \sigma_{xx} l_1 l_2 + \sigma_{yy} m_1 m_2 + \sigma_{zz} n_1 n_2 \\ &\quad + \sigma_{xy}(l_1 m_2 + l_2 m_1) + \sigma_{yz}(m_1 n_2 + m_2 n_1) + \sigma_{zx}(n_1 l_2 + n_2 l_1) \\ \sigma_{y'z'} &= \sigma_{xx} l_2 l_3 + \sigma_{yy} m_2 m_3 + \sigma_{zz} n_2 n_3 \\ &\quad + \sigma_{xy}(l_2 m_3 + l_3 m_2) + \sigma_{yz}(m_2 n_3 + m_3 n_2) + \sigma_{zx}(n_2 l_3 + n_3 l_2) \\ \sigma_{z'x'} &= \sigma_{xx} l_3 l_1 + \sigma_{yy} m_3 m_1 + \sigma_{zz} n_3 n_1 \\ &\quad + \sigma_{xy}(l_3 m_1 + l_1 m_3) + \sigma_{yz}(m_3 n_1 + m_1 n_3) + \sigma_{zx}(n_3 l_1 + n_1 l_3) \end{aligned} \tag{4.6}$$

These equations are called the formulae for coordinate transformation of stress components. Let us rename (x', y', z') by (x_1', x_2', x_3'), and (x, y, z) by (x_1, x_2, x_3), and let $l_{i'k}$ denote the direction cosines between the axis $x_{i'}$ of the new Cartesian coordinate system (x_1', x_2', x_3') and the axis x_k of the old (x_1, x_2, x_3). Equations (4.6) can be expressed in a more simple form:

$$\sigma_{i'j'} = l_{i'k} l_{j'l} \sigma_{kl} \qquad (i', j' = 1, 2, 3) \tag{4.6'}$$

4.2 Strain in a Cartesian Coordinate System

4.2.1 Components of Strain

We consider an elastic body of arbitrary shape subjected to temperature change and external forces. When a point A in the elastic body undergoes a small displacement **u**, consider the deformation of a rectangular parallelepiped element $dx\,dy\,dz$ at the point A in the body, such as in Fig. 4.4. When a small element AB parallel to the y axis becomes an element A*B* after small deformation, the length of the component in the y direction of the element A*B* is

$$(\mathrm{A^*B^*})_y = dy + u_y(\mathrm{B}) - u_y(\mathrm{A}) = dy + \frac{\partial u_y(\mathrm{A})}{\partial y}\,dy \tag{a}$$

The normal strain at the point A in the y direction is

$$\epsilon_{yy} = \frac{(\mathrm{A^*B^*})_y - \mathrm{AB}}{\mathrm{AB}} = \frac{\partial u_y}{\partial y} \tag{b}$$

The normal strains at the point A in the x and z directions are given in the same manner. Thus, the normal strains at point A are

$$\epsilon_{xx} = \frac{\partial u_x}{\partial x}, \qquad \epsilon_{yy} = \frac{\partial u_y}{\partial y}, \qquad \epsilon_{zz} = \frac{\partial u_z}{\partial z} \tag{4.7}$$

Let us define the shearing strain γ_{yz} as the distortion of the angle between the elements AB and AD. Since the shearing strain γ_{yz} is expressed by the sum of the small

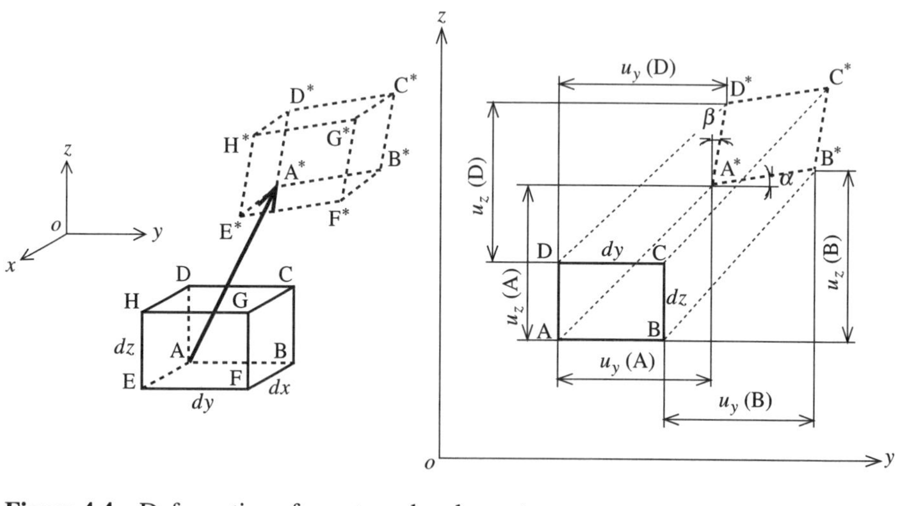

Figure 4.4 Deformation of a rectangular element.

angle between the element A*B* and the y axis, and the small angle between the element A*D* and the z axis, the shearing strain γ_{yz} is given by

$$\gamma_{yz}=\alpha+\beta\cong\tan\alpha+\tan\beta=\frac{u_z(\mathrm{B})-u_z(\mathrm{A})}{(\mathrm{A^*B^*})_y}+\frac{u_y(\mathrm{D})-u_y(\mathrm{A})}{(\mathrm{A^*D^*})_z}=\frac{\partial u_z}{\partial y}+\frac{\partial u_y}{\partial z} \tag{c}$$

The shearing strains γ_{xy} and γ_{zx} at point A are given in the same manner. Thus, the shearing strains at point A are

$$\gamma_{xy}=\frac{\partial u_y}{\partial x}+\frac{\partial u_x}{\partial y},\qquad \gamma_{yz}=\frac{\partial u_z}{\partial y}+\frac{\partial u_y}{\partial z},\qquad \gamma_{zx}=\frac{\partial u_x}{\partial z}+\frac{\partial u_z}{\partial x} \tag{4.8}$$

If we define new shearing strains ϵ_{xy}, ϵ_{yz}, and ϵ_{zx} as half of the shearing strains γ_{xy}, γ_{yz}, and γ_{zx}:

$$\epsilon_{xy}=\frac{\gamma_{xy}}{2},\qquad \epsilon_{yz}=\frac{\gamma_{yz}}{2},\qquad \epsilon_{zx}=\frac{\gamma_{zx}}{2} \tag{4.9}$$

the strains are

$$\begin{aligned}
&\epsilon_{xx}=\frac{\partial u_x}{\partial x},\qquad \epsilon_{yy}=\frac{\partial u_y}{\partial y},\qquad \epsilon_{zz}=\frac{\partial u_z}{\partial z}\\
&\epsilon_{xy}=\frac{1}{2}\left(\frac{\partial u_y}{\partial x}+\frac{\partial u_x}{\partial y}\right),\qquad \epsilon_{yz}=\frac{1}{2}\left(\frac{\partial u_z}{\partial y}+\frac{\partial u_y}{\partial z}\right)\\
&\epsilon_{zx}=\frac{1}{2}\left(\frac{\partial u_x}{\partial z}+\frac{\partial u_z}{\partial x}\right)
\end{aligned} \tag{4.10}$$

When we use new notations (u_1, u_2, u_3) instead of (u_x, u_y, u_z) for displacements and $(\epsilon_{11}, \epsilon_{22}, \epsilon_{33}, \epsilon_{12}, \epsilon_{23}, \epsilon_{31})$ instead of $(\epsilon_{xx}, \epsilon_{yy}, \epsilon_{zz}, \epsilon_{xy}, \epsilon_{yz}, \epsilon_{zx})$ for strains, Eqs. (4.10) are expressed in a more simple form:

$$\epsilon_{ij}=\tfrac{1}{2}(u_{i,j}+u_{j,i})\qquad (i,j=1,2,3) \tag{4.10$'$}$$

It is clear from Eq. (4.10′) that the components of strain are symmetric:

$$\epsilon_{ij}=\epsilon_{ji}\qquad (i,j=1,2,3) \tag{4.11}$$

Hence there are six distinct components of strain.

4.2.2 Compatibility Conditions

When a body is subjected to elastic deformation, the three components of the displacement can be arbitrarily determined as a function of x, y, z at each point. The six components of strain at each point are defined by differentiation of the three components of displacement with respect to the position coordinates. On the other hand, if the six components of strain are independently defined as functions of x, y, z, then we have six equations to determine the three unknown displacement components. This system of partial differential equations does not, in general, give a solution for the three unknown components of displacement unless the six components of strain relate to each other. Let us consider the relations among the six components of strain in this case.

Differentiating the first of Eqs. (4.10) twice with respect to y, and the second twice with respect to x, and adding them, we get

$$\frac{\partial^2 \epsilon_{xx}}{\partial y^2} + \frac{\partial^2 \epsilon_{yy}}{\partial x^2} = \frac{\partial^3 u_x}{\partial y^2 \partial x} + \frac{\partial^3 u_y}{\partial x^2 \partial y} = 2\frac{\partial^2 \epsilon_{xy}}{\partial x \partial y}$$

We can obtain five additional equations in the same way. Then, the six equations are

$$\begin{aligned}
&\frac{\partial^2 \epsilon_{xx}}{\partial y^2} + \frac{\partial^2 \epsilon_{yy}}{\partial x^2} = 2\frac{\partial^2 \epsilon_{xy}}{\partial x \partial y}, \qquad \frac{\partial^2 \epsilon_{xx}}{\partial y \partial z} = \frac{\partial}{\partial x}\left(-\frac{\partial \epsilon_{yz}}{\partial x} + \frac{\partial \epsilon_{zx}}{\partial y} + \frac{\partial \epsilon_{xy}}{\partial z}\right)\\
&\frac{\partial^2 \epsilon_{yy}}{\partial z^2} + \frac{\partial^2 \epsilon_{zz}}{\partial y^2} = 2\frac{\partial^2 \epsilon_{yz}}{\partial y \partial z}, \qquad \frac{\partial^2 \epsilon_{yy}}{\partial z \partial x} = \frac{\partial}{\partial y}\left(\frac{\partial \epsilon_{yz}}{\partial x} - \frac{\partial \epsilon_{zx}}{\partial y} + \frac{\partial \epsilon_{xy}}{\partial z}\right)\\
&\frac{\partial^2 \epsilon_{zz}}{\partial x^2} + \frac{\partial^2 \epsilon_{xx}}{\partial z^2} = 2\frac{\partial^2 \epsilon_{zx}}{\partial z \partial x}, \qquad \frac{\partial^2 \epsilon_{zz}}{\partial x \partial y} = \frac{\partial}{\partial z}\left(\frac{\partial \epsilon_{yz}}{\partial x} + \frac{\partial \epsilon_{zx}}{\partial y} - \frac{\partial \epsilon_{xy}}{\partial z}\right)
\end{aligned} \tag{4.12}$$

or

$$\epsilon_{ij,kl} + \epsilon_{kl,ij} - \epsilon_{ik,jl} - \epsilon_{jl,ik} = 0 \tag{4.12$'$}$$

These equations are called the Saint-Venant compatibility equations. The components of strain must satisfy these equations if the solution of the components of displacement under the known strain is to exist.

4.2.3 Coordinate Transformation of Strain

As stated earlier, the strain produced at a point in the body due to elastic deformation can be expressed by six components with respect to the given coordinate system. As the coordinate system can be arbitrarily selected, there are many expressions for the strain components corresponding to the coordinate systems. Let us consider how to express the six components of strain, referred to any coordinate system, in terms of those referred to another system. Let the components of displacement and strain referred to the coordinate system (x_1, x_2, x_3) be denoted by (u_1, u_2, u_3) and $(\epsilon_{11}, \epsilon_{22}, \epsilon_{33}, \epsilon_{12}, \epsilon_{23}, \epsilon_{31})$, respectively, and the components referred to a new coordinate system $(x_{1'}, x_{2'}, x_{3'})$, be denoted by $(u_{1'}, u_{2'}, u_{3'})$ and $(\epsilon_{1'1'}, \epsilon_{2'2'}, \epsilon_{3'3'}, \epsilon_{1'2'}, \epsilon_{2'3'}, \epsilon_{3'1'})$, respectively. Let $l_{i'k}$ denote the direction cosine between the $x_{i'}$ axis of the new Cartesian coordinate system $(x_{1'}, x_{2'}, x_{3'})$ and the x_k axis of the old (x_1, x_2, x_3). The relationships between the components of coordinates in the new coordinate system and the old, and between the components of the displacement in each coordinate system are

$$\begin{aligned} x_{i'} &= l_{i'j} x_j, \quad x_i = l_{j'i} x_{j'} \\ u_{i'} &= l_{i'j} u_j, \quad u_i = l_{j'i} u_{j'} \end{aligned} \qquad (i, i' = 1, 2, 3) \tag{4.13}$$

Using the strain–displacement relations (4.10′), the coordinate transformation of strain components is

$$
\begin{aligned}
\epsilon_{i'j'} &= \tfrac{1}{2}(u_{i',j'} + u_{j',i'}) = \tfrac{1}{2}\left(\frac{\partial u_{i'}}{\partial x_k}\frac{\partial x_k}{\partial x_{j'}} + \frac{\partial u_{j'}}{\partial x_k}\frac{\partial x_k}{\partial x_{i'}}\right) \\
&= \tfrac{1}{2}(l_{j'k}l_{i'l}u_{l,k} + l_{i'k}l_{j'l}u_{l,k}) = \tfrac{1}{2}l_{i'k}l_{j'l}(u_{k,l} + u_{l,k}) \\
&= l_{i'k}l_{j'l}\epsilon_{kl} \qquad (i', j' = 1, 2, 3)
\end{aligned}
\tag{4.14}
$$

The coordinate transformation (4.14) of strain components agrees with the coordinate transformation (4.6′) of stress components. Thus, the coordinate transformation (4.14) of strain components follows those for stress (Eqs. (4.6)).

4.3 Governing Equations of Thermoelasticity

4.3.1 Constitutive Equations

We consider constitutive equations in the elastic body under mechanical and thermal loads. The constitutive equations for a homogeneous, isotropic body are

$$
\begin{aligned}
\epsilon_{xx} &= \frac{1}{E}[\sigma_{xx} - \nu(\sigma_{yy} + \sigma_{zz})] + \alpha\tau = \frac{1}{2G}\left(\sigma_{xx} - \frac{\nu}{1+\nu}\Theta\right) + \alpha\tau \\
\epsilon_{yy} &= \frac{1}{E}[\sigma_{yy} - \nu(\sigma_{zz} + \sigma_{xx})] + \alpha\tau = \frac{1}{2G}\left(\sigma_{yy} - \frac{\nu}{1+\nu}\Theta\right) + \alpha\tau \\
\epsilon_{zz} &= \frac{1}{E}[\sigma_{zz} - \nu(\sigma_{xx} + \sigma_{yy})] + \alpha\tau = \frac{1}{2G}\left(\sigma_{zz} - \frac{\nu}{1+\nu}\Theta\right) + \alpha\tau \\
\epsilon_{xy} &= \frac{\sigma_{xy}}{2G}, \quad \epsilon_{yz} = \frac{\sigma_{yz}}{2G}, \quad \epsilon_{zx} = \frac{\sigma_{zx}}{2G}
\end{aligned}
\tag{4.15}
$$

where τ is the temperature change from the reference temperature (where the reference temperature is the temperature of the body in the unstrained state):

$$
\tau = T - T_0 \tag{4.16}
$$

and E is the Young's modulus, G is the shear modulus, ν is the Poisson's ratio and α is the coefficient of linear thermal expansion. The temperature change τ should not be confused with shearing stress τ in Eqs. (1.9) and (1.15). These moduli (E, G, ν) are called the material constants. Also in Eqs. (4.15),

$$
\Theta = \sigma_{xx} + \sigma_{yy} + \sigma_{zz} \tag{4.17}
$$

These material properties are in general dependent on the temperature. When the temperature change is not so large, the influence of the temperature change on their values is small in most engineering problems, so that the material properties are assumed to be constant in this book unless otherwise stated. With the Kronecker's symbol defined as

$$\delta_{ij} = \begin{cases} 1 & \text{for } i = j \\ 0 & \text{for } i \neq j \end{cases} \tag{a}$$

Eqs. (4.15) can be written in an alternative form:

$$\begin{aligned} \epsilon_{ij} &= \frac{1}{2G}\left(\sigma_{ij} - \frac{\nu}{1+\nu}\Theta\delta_{ij}\right) + \alpha\tau\delta_{ij} \\ &= \frac{1}{2G}\left(\sigma_{ij} - \frac{\nu}{1+\nu}\sigma_{kk}\delta_{ij}\right) + \alpha\tau\delta_{ij} \qquad (i, j = 1, 2, 3) \end{aligned} \tag{4.15$'$}$$

Equations (4.15) and (4.15′) are called the generalized Hooke's law or the Duhamel–Neumann relations.

Equations (4.15) can be solved for the components of stress and give

$$\begin{aligned} \sigma_{xx} &= 2\mu\epsilon_{xx} + \lambda e - \beta\tau, & \sigma_{xy} &= 2\mu\epsilon_{xy} \\ \sigma_{yy} &= 2\mu\epsilon_{yy} + \lambda e - \beta\tau, & \sigma_{yz} &= 2\mu\epsilon_{yz} \\ \sigma_{zz} &= 2\mu\epsilon_{zz} + \lambda e - \beta\tau, & \sigma_{zx} &= 2\mu\epsilon_{zx} \end{aligned} \tag{4.18}$$

or in index notation

$$\sigma_{ij} = 2\mu\epsilon_{ij} + (\lambda e - \beta\tau)\delta_{ij} \qquad (i, j = 1, 2, 3) \tag{4.18$'$}$$

where e is the dilatation

$$e = \epsilon_{kk} = \epsilon_{xx} + \epsilon_{yy} + \epsilon_{zz} \tag{4.19}$$

λ and μ are the Lamé elastic constants, and β is the thermoelastic constant. The relationship between the elastic constants $(E, G, \nu, \lambda, \mu)$ and the thermoelastic constant β is given by

$$\begin{aligned} 2G &= \frac{E}{1+\nu}, & \lambda &= \frac{\nu E}{(1+\nu)(1-2\nu)} = \frac{2\nu G}{1-2\nu} \\ \mu &= G, & \beta &= \frac{\alpha E}{1-2\nu} = \alpha(3\lambda + 2\mu) \end{aligned} \tag{4.20}$$

The relation between the dilatation e and the sum of the normal stresses Θ is from Eqs. (4.18)

$$e = \frac{\Theta}{3K_v} + 3\alpha\tau \tag{4.21}$$

where the bulk modulus K_v is

$$K_v = \frac{E}{3(1-2\nu)} \tag{4.22}$$

4.3.2 Displacement Equations

The equilibrium equations (4.1) may be expressed in terms of the components of displacement by use of the generalized Hooke's law (4.18) and the definition of strain (4.10)

$$
\begin{aligned}
&\mu\nabla^2 u_x + (\lambda+\mu)\frac{\partial e}{\partial x} - \beta\frac{\partial \tau}{\partial x} + F_x = 0\\
&\mu\nabla^2 u_y + (\lambda+\mu)\frac{\partial e}{\partial y} - \beta\frac{\partial \tau}{\partial y} + F_y = 0\\
&\mu\nabla^2 u_z + (\lambda+\mu)\frac{\partial e}{\partial z} - \beta\frac{\partial \tau}{\partial z} + F_z = 0
\end{aligned}
\tag{4.23}
$$

or in index notation

$$\mu\nabla^2 u_i + (\lambda+\mu)u_{k,ki} - \beta\tau_{,i} + F_i = 0 \qquad (i = 1, 2, 3) \tag{4.23$'$}$$

where ∇^2 is the Laplacian operator defined as:

$$\nabla^2 = \frac{\partial^2}{\partial x^2} + \frac{\partial^2}{\partial y^2} + \frac{\partial^2}{\partial z^2} \tag{a}$$

The equilibrium equations (4.23) expressed in terms of the components of displacement are called the displacement equations of thermoelasticity or Navier's equations of thermoelasticity. Thermoelastic problems are generally reduced to solve these displacement equations under the boundary conditions. Once the solution for the components of the displacement are obtained from Eqs. (4.23), the components of the strain and stress can be determined from Eqs. (4.10) and (4.18), respectively. Equations (4.23) may be expressed in an alternative form

$$
\begin{aligned}
&(\lambda+2\mu)\frac{\partial e}{\partial x} - 2\mu\left(\frac{\partial \omega_z}{\partial y} - \frac{\partial \omega_y}{\partial z}\right) - \beta\frac{\partial \tau}{\partial x} + F_x = 0\\
&(\lambda+2\mu)\frac{\partial e}{\partial y} - 2\mu\left(\frac{\partial \omega_x}{\partial z} - \frac{\partial \omega_z}{\partial x}\right) - \beta\frac{\partial \tau}{\partial y} + F_y = 0\\
&(\lambda+2\mu)\frac{\partial e}{\partial z} - 2\mu\left(\frac{\partial \omega_y}{\partial x} - \frac{\partial \omega_x}{\partial y}\right) - \beta\frac{\partial \tau}{\partial z} + F_z = 0
\end{aligned}
\tag{4.23$''$}
$$

or in index notation

$$(\lambda+2\mu)u_{k,ki} - 2\mu\gamma_{ijk}\omega_{k,j} - \beta\tau_{,i} + F_i = 0 \tag{4.23$'''$}$$

where

$$\omega_x = \frac{1}{2}\left(\frac{\partial u_z}{\partial y} - \frac{\partial u_y}{\partial z}\right), \qquad \omega_y = \frac{1}{2}\left(\frac{\partial u_x}{\partial z} - \frac{\partial u_z}{\partial x}\right), \qquad \omega_z = \frac{1}{2}\left(\frac{\partial u_y}{\partial x} - \frac{\partial u_x}{\partial y}\right) \tag{4.24}$$

and γ_{ijk} is the alternating tensor, also known as permutation symbol, and is defined by

$$\gamma_{ijk} = \begin{cases} 1 & \text{if } ijk \text{ represents an even permutation of 123} \\ 0 & \text{if any two of the } ijk \text{ indices are equal} \\ -1 & \text{if } ijk \text{ represents an odd permutation of 123} \end{cases} \tag{4.25}$$

The displacement equations (4.23) and (4.23″) may be expressed by use of the displacement vector $\boldsymbol{u}$ in the following three equivalent forms:

$$\mu\nabla^2\boldsymbol{u} + (\lambda+\mu)\,\text{grad div}\,\boldsymbol{u} - \beta\,\text{grad}\,\tau + \boldsymbol{F} = 0 \tag{4.23a}$$

$$(\lambda+2\mu)\,\text{grad div}\,\boldsymbol{u} - \mu\,\text{curl curl}\,\boldsymbol{u} - \beta\,\text{grad}\,\tau + \boldsymbol{F} = 0 \tag{4.23b}$$

$$(\lambda+2\mu)\nabla^2\boldsymbol{u} + (\lambda+\mu)\,\text{curl curl}\,\boldsymbol{u} - \beta\,\text{grad}\,\tau + \boldsymbol{F} = 0 \tag{4.23c}$$

Example 4.1 Derive Eqs. (4.23).

Solution Substitution of Eqs. (4.18) into the first of Eqs. (4.1) gives

$$\begin{aligned}
&\frac{\partial\sigma_{xx}}{\partial x} + \frac{\partial\sigma_{yx}}{\partial y} + \frac{\partial\sigma_{zx}}{\partial z} + F_x \\
&= \frac{\partial}{\partial x}(2\mu\epsilon_{xx} + \lambda e - \beta\tau) + 2\mu\frac{\partial\epsilon_{yx}}{\partial y} + 2\mu\frac{\partial\epsilon_{zx}}{\partial z} + F_x \\
&= \mu\left(2\frac{\partial^2 u_x}{\partial x^2} + \frac{\partial^2 u_x}{\partial y^2} + \frac{\partial^2 u_y}{\partial x\partial y} + \frac{\partial^2 u_x}{\partial z^2} + \frac{\partial^2 u_z}{\partial x\partial z}\right) + \lambda\frac{\partial e}{\partial x} - \beta\frac{\partial\tau}{\partial x} + F_x \\
&= \mu\left[\nabla^2 u_x + \frac{\partial}{\partial x}\left(\frac{\partial u_x}{\partial x} + \frac{\partial u_y}{\partial y} + \frac{\partial u_z}{\partial z}\right)\right] + \lambda\frac{\partial e}{\partial x} - \beta\frac{\partial\tau}{\partial x} + F_x \\
&= \mu\nabla^2 u_x + (\lambda+\mu)\frac{\partial e}{\partial x} - \beta\frac{\partial\tau}{\partial x} + F_x = 0
\end{aligned}$$

The second and third equation in Eqs. (4.23) may be obtained in the same way.

4.3.3 Compatibility Equations in Terms of Stress

Let us express the compatibility equations (4.12) in terms of the components of stress. Eliminating the components of strain in Eq. (4.12) by use of Eqs. (4.15) and (4.1), we have

$$\begin{aligned}
&\nabla^2\sigma_{xx} + \frac{1}{1+\nu}\frac{\partial^2\Theta}{\partial x^2} + \alpha E\left(\frac{1}{1-\nu}\nabla^2\tau + \frac{1}{1+\nu}\frac{\partial^2\tau}{\partial x^2}\right) \\
&= -\left[\frac{\nu}{1-\nu}\left(\frac{\partial F_x}{\partial x} + \frac{\partial F_y}{\partial y} + \frac{\partial F_z}{\partial z}\right) + 2\frac{\partial F_x}{\partial x}\right]
\end{aligned}$$

$$\nabla^2\sigma_{yy} + \frac{1}{1+\nu}\frac{\partial^2\Theta}{\partial y^2} + \alpha E\left(\frac{1}{1-\nu}\nabla^2\tau + \frac{1}{1+\nu}\frac{\partial^2\tau}{\partial y^2}\right)$$
$$= -\left[\frac{\nu}{1-\nu}\left(\frac{\partial F_x}{\partial x} + \frac{\partial F_y}{\partial y} + \frac{\partial F_z}{\partial z}\right) + 2\frac{\partial F_y}{\partial y}\right]$$

$$\nabla^2\sigma_{zz} + \frac{1}{1+\nu}\frac{\partial^2\Theta}{\partial z^2} + \alpha E\left(\frac{1}{1-\nu}\nabla^2\tau + \frac{1}{1+\nu}\frac{\partial^2\tau}{\partial z^2}\right) \tag{4.26}$$
$$= -\left[\frac{\nu}{1-\nu}\left(\frac{\partial F_x}{\partial x} + \frac{\partial F_y}{\partial y} + \frac{\partial F_z}{\partial z}\right) + 2\frac{\partial F_z}{\partial z}\right]$$

$$\nabla^2\sigma_{xy} + \frac{1}{1+\nu}\frac{\partial^2\Theta}{\partial x\partial y} + \frac{\alpha E}{1+\nu}\frac{\partial^2\tau}{\partial x\partial y} = -\left(\frac{\partial F_x}{\partial y} + \frac{\partial F_y}{\partial x}\right)$$

$$\nabla^2\sigma_{yz} + \frac{1}{1+\nu}\frac{\partial^2\Theta}{\partial y\partial z} + \frac{\alpha E}{1+\nu}\frac{\partial^2\tau}{\partial y\partial z} = -\left(\frac{\partial F_y}{\partial z} + \frac{\partial F_z}{\partial y}\right)$$

$$\nabla^2\sigma_{zx} + \frac{1}{1+\nu}\frac{\partial^2\Theta}{\partial z\partial x} + \frac{\alpha E}{1+\nu}\frac{\partial^2\tau}{\partial z\partial x} = -\left(\frac{\partial F_z}{\partial x} + \frac{\partial F_x}{\partial z}\right)$$

or in index notation

$$\nabla^2\sigma_{ij} + \frac{1}{1+\nu}\sigma_{kk,ij} + \alpha E\left(\frac{1}{1-\nu}\nabla^2\tau\delta_{ij} + \frac{1}{1+\nu}\tau_{,ij}\right)$$
$$= -\left(\frac{\nu}{1-\nu}F_{k,k}\delta_{ij} + F_{i,j} + F_{j,i}\right) \tag{4.26$'$}$$

These equations are called the compatibility equations in terms of the stress for thermoelasticity, or the Beltrami–Michell compatibility equations for thermoelasticity.

4.3.4 Boundary Conditions

Let us consider boundary conditions which prescribe the displacements and the external forces at the boundary surface. There are two types of governing equations needed for solution of the thermoelasticity problems, one type is Navier's equations (4.23) and the other is the equilibrium equations (4.1) and Beltrami–Michell compatibility equations (4.26). When we use the governing equations to solve thermoelasticity problems, it is necessary to consider the boundary conditions. There are three kinds of boundary conditions.

[1] First boundary-value problem

When the stress is prescribed over the entire boundary surface, the problem is called the first boundary-value problem. This boundary condition can be expressed from

Eqs. (4.5) as

$$
\begin{aligned}
\sigma_{xx}l + \sigma_{yx}m + \sigma_{zx}n &= p_{nx} \\
\sigma_{xy}l + \sigma_{yy}m + \sigma_{zy}n &= p_{ny} \\
\sigma_{xz}l + \sigma_{yz}m + \sigma_{zz}n &= p_{nz}
\end{aligned}
\tag{4.27}
$$

or in index notation

$$\sigma_{ji}n_j = p_{ni} \qquad (i = 1, 2, 3) \tag{4.27$'$}$$

where p_{ni} denotes the prescribed surface traction, and l, m, n, and n_j denote the direction cosines between the external normal and each axis.

[2] Second boundary-value problem

When the displacement is prescribed over the entire boundary surface, the problem is called the second boundary-value problem. This boundary condition can be expressed as

$$u_x = \bar{u}_x, \qquad u_y = \bar{u}_y, \qquad u_z = \bar{u}_z \tag{4.28}$$

or in index notation

$$u_i = \bar{u}_i \qquad (i = 1, 2, 3) \tag{4.28$'$}$$

where displacements with an over bar denote the prescribed boundary displacements.

[3] Mixed boundary-value problem

When the displacement is prescribed over a part of the boundary surface and the stress over the rest of the boundary surface, the problem is referred to as a mixed boundary-value problem.* This boundary condition can be expressed as

$$
\begin{aligned}
\sigma_{ji}n_j &= p_{ni} \qquad (i = 1, 2, 3) \text{ on the part of the boundary} \\
u_i &= \bar{u}_i \qquad (i = 1, 2, 3) \text{ on the rest of the boundary}
\end{aligned}
\tag{4.29}
$$

4.3.5 Body-Force Analogy

Navier's equations (4.23) for thermoelasticity become Navier's equations for isothermal elasticity if the gradient of temperature change is taken as a body force. The generalized Hooke's law (4.18$'$) can be rewritten in the form

$$\sigma_{ij} = \sigma^*_{ij} - \beta\tau\delta_{ij} \qquad (i, j = 1, 2, 3) \tag{4.30}$$

where

$$\sigma^*_{ij} = 2\mu\epsilon_{ij} + \lambda\epsilon_{kk}\delta_{ij} \qquad (i, j, k = 1, 2, 3) \tag{a}$$

*See: R. B. Hetnarski and J. Ignaczak, Mathematical Theory of Elasticity, to be published. The First and Second boundary-value problems are treated there as special cases of the Mixed boundary-value problem.

Equations (a) correspond to Hooke's law for isothermal conditions. Substitution of Eq. (4.30) into Eq. (4.1″) leads to

$$\sigma^*_{ji,j} + F^*_i = 0 \tag{4.31}$$

where

$$F^*_i = F_i - \beta\tau_{,i} \tag{b}$$

Equation (b) may be regarded as the body force which contains the imaginary body forces $-\beta\tau_{,i}$, so that equilibrium equations (4.31) may be considered as the equilibrium equations under isothermal conditions. The boundary conditions (4.27′) may be written as

$$\sigma^*_{ij} n_j = p^*_{ni} \qquad (i = 1, 2, 3) \tag{4.32}$$

where

$$p^*_{ni} = p_{ni} + \beta\tau n_i \qquad (i = 1, 2, 3) \tag{c}$$

This analysis shows that thermal stresses can be obtained by addition of the pressure $\beta\tau\delta_{ij}$ to the isothermal stresses σ^*_{ij} which can be obtained by solving the isothermal governing equations with imaginary body forces $\beta\tau_{,i}$ and imaginary surface tractions $\beta\tau n_i$.

4.4 General Solution of Navier's Equations

4.4.1 Thermoelastic Displacement Potential

Let us consider the solutions of Navier's equations (4.23′) for thermoelasticity. When we assume that body forces are absent, the boundary-value problem may be written in the form

$$\mu\nabla^2 u_i + (\lambda + \mu) u_{k,ki} = \beta\tau_{,i} \qquad \text{in the body } D \tag{4.33}$$

$$\sigma_{ji} n_j = p_{ni} \qquad \text{on the part of the boundary } B_1 \tag{4.34}$$

$$u_i = \bar{u}_i \qquad \text{on the rest of the boundary } B_2 \tag{4.35}$$

The general solution of Eq. (4.33) for the components of displacement may be expressed as the sum of a complementary solution u_i^c and a particular solution u_i^p

$$u_i = u_i^c + u_i^p \tag{4.36}$$

where the complementary solution u_i^c and the particular solution u_i^p are governed by the following linear differential equations

$$\mu\nabla^2 u_i^c + (\lambda + \mu) u_{k,ki}^c = 0 \tag{4.37}$$

$$\mu\nabla^2 u_i^p + (\lambda+\mu)u_{k,ki}^p = \beta\tau_{,i} \tag{4.38}$$

We focus our attention on the particular solution u_i^p of the innonhomogeneous differential equation (4.38). The particular solution u_i^p may be expressed in terms of a scalar potential function as follows:

$$u_i^p = \Phi_{,i} \tag{4.39}$$

where the Φ is called the Goodier's thermoelastic displacement potential. Substituting Eq. (4.39) into Eq. (4.38), we get

$$[(\lambda+2\mu)\nabla^2\Phi - \beta\tau]_{,i} = 0$$

The above equation will be satisfied, if Φ satisfies the equation:

$$\nabla^2\Phi = K\tau \tag{4.40}$$

where

$$K = \frac{\beta}{\lambda+2\mu} = \frac{1+\nu}{1-\nu}\alpha \tag{4.41}$$

Equation (4.40) is the governing equation for the thermoelastic displacement potential Φ. Substituting Eq. (4.39) into the generalized Hooke's law (4.18′), the thermal stresses in terms of the thermoelastic displacement potential Φ as a particular solution of Eq. (4.38) are

$$\sigma_{ij}^p = 2\mu\Phi_{,ij} + (\lambda\Phi_{,kk} - \beta\tau)\delta_{ij} \tag{4.42}$$

or

$$\sigma_{ij}^p = 2\mu(\Phi_{,ij} - \nabla^2\Phi\delta_{ij}) = 2\mu(\Phi_{,ij} - K\tau\delta_{ij}) \tag{4.42$'$}$$

It should be noted that the thermal stresses (4.42) obtained from the thermoelastic displacement potential Φ in general do not satisfy the boundary conditions (4.34) and (4.35).

Next, we determine the particular solution of Eq. (4.40). One solution of Eq. (4.40) follows from potential theory† as follows:

$$\begin{aligned}\Phi(P) &= -\frac{K}{4\pi}\int_V \frac{\tau(Q)}{r(P,Q)}\,dV \\ &= -\frac{K}{4\pi}\iiint_V \frac{\tau(x_1', x_2', x_3')\,dx_1'\,dx_2'\,dx_3'}{\sqrt{(x_1-x_1')^2+(x_2-x_2')^2+(x_3-x_3')^2}}\end{aligned} \tag{4.43}$$

†See: Oliver Dimon Kellog, Foundation of Potential Theory, Dover Publication, New York (1953), p. 219.

where $r(P, Q)$ denotes the distance from the point $P(x_1, x_2, x_3)$ to the point $Q(x'_1, x'_2, x'_3)$ in the region of the body. When the temperature change is given, the thermoelastic displacement potential may be determined by integrating Eq. (4.43).

It is, however, difficult to perform the integration (4.43) in many cases. If transient thermoelasticity is considered, then the heat conduction equation is given by Eq. (3.8):

$$\frac{\partial \tau}{\partial t} = \kappa \nabla^2 \tau \tag{4.44}$$

The thermoelastic displacement potential is determined in the following manner. Differentiating Eq. (4.40) with respect to time t, and substituting it into Eq. (4.44), we get

$$\nabla^2 \left(\frac{\partial \Phi}{\partial t} - \kappa K \tau \right) = 0$$

A solution of this equation can be obtained from

$$\Phi = \kappa K \int_{t_r}^{t} \tau \, dt + \Phi_r + (t - t_r)\Phi_0 \tag{4.45}$$

where t_r denotes the reference time, and Φ_r and Φ_0 denote solutions of the following Poisson and harmonic equations, respectively:

$$\nabla^2 \Phi_r = K \tau_r \tag{4.46}$$

$$\nabla^2 \Phi_0 = 0 \tag{4.47}$$

The function Φ_0 is usually selected in such a way that a thermoelastic displacement potential Φ takes a finite value at an arbitrary time if the thermal stresses will be finite at an arbitrary time. The function Φ_r is a thermoelastic displacement potential with temperature change τ_r at a reference time (usually, at the initial time or final time). The function Φ_r can be determined by Eq. (4.43). However, the temperature change at the reference time is uniform or zero in the body in many cases, so the determination of Φ_r is easy in such cases:

$$\begin{aligned} \Phi_r &= 0 && \text{for } \tau_r(x_1, x_2, x_3) = 0 \\ \Phi_r &= \frac{KC}{6} r^2 = \frac{KC}{6}\left(x_1^2 + x_2^2 + x_3^2\right) && \text{for } \tau_r(x_1, x_2, x_3) = C \end{aligned} \tag{4.48}$$

where C denotes a constant reference temperature. If the temperature change at the reference time is zero, the thermoelastic displacement potential Φ is given by

$$\Phi = \kappa K \int_{t_r}^{t} \tau \, dt + (t - t_r)\Phi_0 \tag{4.49}$$

Moreover, if the first term on the right-hand side in Eq. (4.49) is finite at any time, the thermoelastic displacement potential Φ reduces to

$$\Phi = \kappa K \int_{t_r}^{t} \tau \, dt \tag{4.50}$$

Example 4.2 When a three-dimensional body with zero initial temperature and without internal heat generation is heated on its surface and the temperature in the body is finite, the temperature in the body may be frequently expressed by

$$T = T_s(x, y, z) + \sum_{n=1}^{\infty} e^{-\kappa s_n^2 t} f_n(x, y, z) \tag{a}$$

Find the thermoelastic displacement potential Φ due to temperature (a).

Solution Selecting the reference temperature as the zero initial temperature, the temperature change is

$$\tau = T_s(x, y, z) + \sum_{n=1}^{\infty} e^{-\kappa s_n^2 t} f_n(x, y, z) \tag{b}$$

Since the initial temperature change is zero, the function Φ_r becomes zero. Making the initial time equal zero, the first term on right-hand side in Eq. (4.49) is

$$\begin{aligned} \kappa K \int_{t_r}^{t} \tau \, dt &= \kappa K \int_{0}^{t} \left[T_s(x, y, z) + \sum_{n=1}^{\infty} e^{-\kappa s_n^2 t} f_n(x, y, z) \right] dt \\ &= \kappa K t T_s(x, y, z) - K \sum_{n=1}^{\infty} \frac{1}{s_n^2} e^{-\kappa s_n^2 t} f_n(x, y, z) \end{aligned} \tag{c}$$

Expression (c) becomes infinite as $t \to \infty$, and the steady temperature $T_s(x, y, z)$ is a harmonic function, so the function Φ_0 is selected as $\Phi_0 = -\kappa K T_s(x, y, z)$. Therefore, the thermoelastic displacement potential Φ becomes

$$\Phi = -K \sum_{n=1}^{\infty} \frac{1}{s_n^2} e^{-\kappa s_n^2 t} f_n(x, y, z) \tag{d}$$

4.4.2 Displacement Potentials

In Subsection 4.4.1, the particular solution of Navier's equations (4.23′) for thermoelasticity was discussed. Let us consider the complementary solutions for Navier's equations (4.23′). The differential equation for the complementary solutions given by Eq. (4.37) can be expressed in the vector form as follows:

$$\mu \nabla^2 \boldsymbol{u}^c + (\lambda + \mu) \,\text{grad div}\, \boldsymbol{u}^c = 0 \tag{4.51}$$

The displacement vector $\boldsymbol{u}^c$ may generally be expressed as the sum of the displacement vector $\boldsymbol{v}$ without dilatation, and the displacement vector $\boldsymbol{w}$ without rotation:

$$\boldsymbol{u}^c = \boldsymbol{v} + \boldsymbol{w} \tag{4.52}$$

where

$$\text{div}\, \boldsymbol{v} = 0, \qquad \text{curl}\, \boldsymbol{w} = 0 \tag{a}$$

The displacement vectors $\boldsymbol{v}$ and $\boldsymbol{w}$ may be expressed as

$$\boldsymbol{v} = \text{curl}\, \boldsymbol{s}, \qquad \boldsymbol{w} = \text{grad}\, \varphi \tag{b}$$

Substitution of Eqs. (4.52) and (b) into Eq. (4.51) yields

$$\begin{aligned} \mu \nabla^2 \boldsymbol{u}^c + (\lambda + \mu)\, \text{grad div}\, \boldsymbol{u}^c &= \mu \nabla^2 (\text{grad}\, \varphi + \text{curl}\, \boldsymbol{s}) + (\lambda + \mu) \text{grad}\, \nabla^2 \varphi \\ &= \mu \nabla^2 \left(\frac{\lambda + 2\mu}{\mu} \text{grad}\, \varphi + \text{curl}\, \boldsymbol{s} \right) = 0 \end{aligned} \tag{c}$$

A solution of Eq. (c) may be expressed as

$$\frac{\lambda + 2\mu}{\mu} \text{grad}\, \varphi + \text{curl}\, \boldsymbol{s} = \boldsymbol{B} \tag{d}$$

where components B_i of vector $\boldsymbol{B}$ are harmonic functions:

$$\nabla^2 B_i = 0 \qquad (i = 1, 2, 3) \tag{4.53}$$

Applying the divergence operator to Eq. (d), we get

$$\frac{\lambda + 2\mu}{\mu} \nabla^2 \varphi = \text{div}\, \boldsymbol{B} \tag{e}$$

A solution of Eq. (e) is

$$\varphi = \frac{\mu}{2(\lambda + 2\mu)} (\boldsymbol{r} \cdot \boldsymbol{B} + A) \tag{f}$$

where A is a harmonic function. Then, the displacement vector $\boldsymbol{u}^c$ is expressed as

$$\boldsymbol{u}^c = \text{curl}\, \boldsymbol{s} + \text{grad}\, \varphi = \boldsymbol{B} - \frac{1}{4(1 - \nu)} \text{grad}\, (\boldsymbol{r} \cdot \boldsymbol{B} + A) \tag{4.54}$$

The displacement vector $\boldsymbol{u}^c$ can be expressed in terms of the four Papkovich–Neuber harmonic functions A and $\boldsymbol{B} = (B_1, B_2, B_3)$.

When we rewrite the four harmonic functions A and $\boldsymbol{B} = (B_1, B_2, B_3)$ in the form

$$\boldsymbol{B} = -4(1 - \nu) \boldsymbol{\Psi} = -4(1 - \nu)(\Psi_1, \Psi_2, \Psi_3), \qquad A = -4(1 - \nu) \Psi_0 \tag{g}$$

the displacement vector $\boldsymbol{u}^c$ can be expressed in terms of the four Boussinesq–Papkovich harmonic functions $\boldsymbol{\Psi} = (\Psi_1, \Psi_2, \Psi_3)$ and Ψ_0 as

$$\boldsymbol{u}^c = \text{grad}\,(\boldsymbol{r} \cdot \boldsymbol{\Psi} + \Psi_0) - 4(1-\nu)\boldsymbol{\Psi} \tag{4.55}$$

or in index notation

$$u_i^c = (x_k \Psi_k + \Psi_0)_{,i} - 4(1-\nu)\Psi_i \qquad (i = 1, 2, 3) \tag{4.55$'$}$$

where Ψ_0 and Ψ_i are harmonic functions

$$\nabla^2 \Psi_0 = 0, \qquad \nabla^2 \Psi_i = 0 \tag{4.56}$$

The displacement vector $\boldsymbol{u}^c$ expressed by the four harmonic functions $(\Psi_0, \Psi_1, \Psi_2, \Psi_3)$ may be expressed by three other harmonic functions.

Expressing the four functions by

$$\Psi_0 = 4(1-\nu)\varphi_0 - x_k \varphi_{0,k}, \qquad \Psi_i = \Phi_i + \varphi_{0,i} \tag{h}$$

and substituting Eqs. (h) into Eq. (4.55), we get

$$u_i^c = -(3 - 4\nu)\Phi_i + x_k \Phi_{k,i} \tag{4.57}$$

where these three functions must satisfy

$$\nabla^2 \Phi_i = 0 \tag{4.58}$$

When we rewrite the four harmonic functions Ψ_0 and $\boldsymbol{\Psi}$ in the form

$$\boldsymbol{\Psi} = \boldsymbol{\psi} - \frac{1}{2(1-\nu)} \text{curl}\, \boldsymbol{\vartheta}, \qquad \Psi_0 = \varphi + \frac{1}{2(1-\nu)} \boldsymbol{r} \cdot \text{curl}\, \boldsymbol{\vartheta} \tag{i}$$

the displacement vector $\boldsymbol{u}^c$ can be expressed in terms of the generalized Boussinesq harmonic functions as

$$\boldsymbol{u}^c = \text{grad}\, \varphi + 2\, \text{curl}\, \boldsymbol{\vartheta} + \text{grad}\, (\boldsymbol{r} \cdot \boldsymbol{\psi}) - 4(1-\nu)\boldsymbol{\psi} \tag{4.59}$$

where $\boldsymbol{\vartheta} = (\vartheta_1, \vartheta_2, \vartheta_3)$ and $\boldsymbol{\psi} = (\psi_1, \psi_2, \psi_3)$, and these functions must satisfy

$$\nabla^2 \varphi = 0, \qquad \nabla^2 \boldsymbol{\vartheta} = 0, \qquad \nabla^2 \boldsymbol{\psi} = 0 \tag{4.60}$$

If we select these functions as

$$\vartheta_1 = \vartheta_2 = \psi_1 = \psi_2 = 0, \qquad \vartheta_3 = \vartheta, \qquad \psi_3 = \psi \tag{j}$$

the displacement vector $\boldsymbol{u}^c$ can be expressed in terms of the Boussinesq harmonic functions φ, ϑ, and ψ as

$$\boldsymbol{u}^c = \text{grad}\, \varphi + 2\, \text{curl}\, [0, 0, \vartheta] + \text{grad}\, (z\psi) - 4(1-\nu)[0, 0, \psi] \tag{4.61}$$

Thus, the Boussinesq functions must satisfy the relations

$$\nabla^2\varphi = 0, \qquad \nabla^2\vartheta = 0, \qquad \nabla^2\psi = 0 \tag{4.62}$$

It is clear from this discussion that the complementary solutions $\boldsymbol{u}^c$ of Eq. (4.37) may be expressed in terms of harmonic functions. Hence, the general solutions of the Navier's equations (4.33) for thermoelasticity can be expressed in terms of Goodier's thermoelastic displacement potential (4.43) or (4.45) and the displacement potential functions. A typical example of the general solution of Navier's equations (4.33) is as follows:

$$u_i = \Phi_{,i} + (x_k\Psi_k + \Psi_0)_{,i} - 4(1-\nu)\Psi_i \qquad (i = 1, 2, 3) \tag{4.63}$$

where

$$\nabla^2\Phi = K\tau, \qquad \nabla^2\Psi_0 = 0, \qquad \nabla^2\Psi_i = 0 \tag{4.64}$$

The other solution is

$$\boldsymbol{u} = \text{grad}\,\Phi + \text{grad}\,\varphi + 2\,\text{curl}\,[0, 0, \vartheta] + \text{grad}\,(z\psi) - 4(1-\nu)[0, 0, \psi] \tag{4.65}$$

where

$$\nabla^2\Phi = K\tau, \qquad \nabla^2\varphi = 0, \qquad \nabla^2\vartheta = 0, \qquad \nabla^2\psi = 0 \tag{4.66}$$

Example 4.3 Express the strain and the stress by Eq. (4.65).

Solution Equation (4.65) may be rewritten as

$$\begin{aligned}
u_x &= \frac{\partial\Phi}{\partial x} + \frac{\partial\varphi}{\partial x} + 2\frac{\partial\vartheta}{\partial y} + z\frac{\partial\psi}{\partial x} \\
u_y &= \frac{\partial\Phi}{\partial y} + \frac{\partial\varphi}{\partial y} - 2\frac{\partial\vartheta}{\partial x} + z\frac{\partial\psi}{\partial y} \\
u_z &= \frac{\partial\Phi}{\partial z} + \frac{\partial\varphi}{\partial z} + z\frac{\partial\psi}{\partial z} - (3-4\nu)\psi
\end{aligned} \tag{a}$$

The displacement (a) gives the strain:

$$\begin{aligned}
\epsilon_{xx} &= \frac{\partial^2\Phi}{\partial x^2} + \frac{\partial^2\varphi}{\partial x^2} + 2\frac{\partial^2\vartheta}{\partial x\partial y} + z\frac{\partial^2\psi}{\partial x^2} \\
\epsilon_{yy} &= \frac{\partial^2\Phi}{\partial y^2} + \frac{\partial^2\varphi}{\partial y^2} - 2\frac{\partial^2\vartheta}{\partial x\partial y} + z\frac{\partial^2\psi}{\partial y^2} \\
\epsilon_{zz} &= \frac{\partial^2\Phi}{\partial z^2} + \frac{\partial^2\varphi}{\partial z^2} + z\frac{\partial^2\psi}{\partial z^2} - 2(1-2\nu)\frac{\partial\psi}{\partial z} \\
\epsilon_{xy} &= \frac{\partial^2\Phi}{\partial x\partial y} + \frac{\partial^2\varphi}{\partial x\partial y} + \frac{\partial^2\vartheta}{\partial y^2} - \frac{\partial^2\vartheta}{\partial x^2} + z\frac{\partial^2\psi}{\partial x\partial y} \\
\epsilon_{yz} &= \frac{\partial^2\Phi}{\partial y\partial z} + \frac{\partial^2\varphi}{\partial y\partial z} - \frac{\partial^2\vartheta}{\partial x\partial z} + z\frac{\partial^2\psi}{\partial y\partial z} - (1-2\nu)\frac{\partial\psi}{\partial y}
\end{aligned}$$

$$\epsilon_{zx} = \frac{\partial^2 \Phi}{\partial z \partial x} + \frac{\partial^2 \varphi}{\partial z \partial x} + \frac{\partial^2 \vartheta}{\partial y \partial z} + z\frac{\partial^2 \psi}{\partial z \partial x} - (1-2\nu)\frac{\partial \psi}{\partial x}$$

$$e = \nabla^2 \Phi - 2(1-2\nu)\frac{\partial \psi}{\partial z} = K\tau - 2(1-2\nu)\frac{\partial \psi}{\partial z} \tag{b}$$

Substitution of Eqs. (b) into Eqs. (4.18) yields

$$\begin{aligned}
\sigma_{xx} &= 2\mu\left(\frac{\partial^2 \Phi}{\partial x^2} - K\tau + \frac{\partial^2 \varphi}{\partial x^2} + 2\frac{\partial^2 \vartheta}{\partial x \partial y} + z\frac{\partial^2 \psi}{\partial x^2} - 2\nu\frac{\partial \psi}{\partial z}\right)\\
\sigma_{yy} &= 2\mu\left(\frac{\partial^2 \Phi}{\partial y^2} - K\tau + \frac{\partial^2 \varphi}{\partial y^2} - 2\frac{\partial^2 \vartheta}{\partial x \partial y} + z\frac{\partial^2 \psi}{\partial y^2} - 2\nu\frac{\partial \psi}{\partial z}\right)\\
\sigma_{zz} &= 2\mu\left[\frac{\partial^2 \Phi}{\partial z^2} - K\tau + \frac{\partial^2 \varphi}{\partial z^2} + z\frac{\partial^2 \psi}{\partial z^2} - 2(1-\nu)\frac{\partial \psi}{\partial z}\right]\\
\sigma_{xy} &= 2\mu\left(\frac{\partial^2 \Phi}{\partial x \partial y} + \frac{\partial^2 \varphi}{\partial x \partial y} + \frac{\partial^2 \vartheta}{\partial y^2} - \frac{\partial^2 \vartheta}{\partial x^2} + z\frac{\partial^2 \psi}{\partial x \partial y}\right)\\
\sigma_{yz} &= 2\mu\left[\frac{\partial^2 \Phi}{\partial y \partial z} + \frac{\partial^2 \varphi}{\partial y \partial z} - \frac{\partial^2 \vartheta}{\partial x \partial z} + z\frac{\partial^2 \psi}{\partial y \partial z} - (1-2\nu)\frac{\partial \psi}{\partial y}\right]\\
\sigma_{zx} &= 2\mu\left[\frac{\partial^2 \Phi}{\partial z \partial x} + \frac{\partial^2 \varphi}{\partial z \partial x} + \frac{\partial^2 \vartheta}{\partial y \partial z} + z\frac{\partial^2 \psi}{\partial z \partial x} - (1-2\nu)\frac{\partial \psi}{\partial x}\right]
\end{aligned} \tag{c}$$

where following relations are used:

$$\lambda K\tau - \beta\tau = \left(\lambda\frac{\beta}{\lambda+2\mu} - \beta\right)\tau = -2\mu\frac{\beta}{\lambda+2\mu}\tau = -2\mu K\tau$$

$$-2(1-2\nu)\lambda\frac{\partial \psi}{\partial z} = -2(1-2\nu)\frac{2\mu\nu}{1-2\nu}\frac{\partial \psi}{\partial z} = -2\nu 2\mu\frac{\partial \psi}{\partial z}$$

4.4.3 General Solutions for the Potential Function

It is found from Subsections 4.4.1 and 4.4.2 that thermoelastic problems can be reduced to a Poisson equation for the thermoelastic displacement potential, and Laplace equations for the displacement potential functions. The general solution of the Poisson equation is given by Eq. (4.43) or Eq. (4.45). Therefore, we here consider the general solution of the Laplace equation

$$\nabla^2 \varphi = \left(\frac{\partial^2}{\partial x^2} + \frac{\partial^2}{\partial y^2} + \frac{\partial^2}{\partial z^2}\right)\varphi = 0 \tag{4.67}$$

using separation of variables.

We assume that the harmonic function $\varphi(x, y, z)$ can be expressed by the product of three unknown functions, each of only one variable

$$\varphi(x, y, z) = f(x)g(y)h(z) \tag{a}$$

Substituting Eq. (a) into Eq. (4.67), we get

$$\frac{1}{f(x)}\frac{d^2 f(x)}{dx^2} + \frac{1}{g(y)}\frac{d^2 g(y)}{dy^2} + \frac{1}{h(z)}\frac{d^2 h(z)}{dz^2} = 0 \tag{b}$$

Equation (b) may be satisfied if the functions are selected as:

$$\frac{d^2 f(x)}{dx^2} + a^2 f(x) = 0 \tag{c}$$

$$\frac{d^2 g(y)}{dy^2} + b^2 g(y) = 0 \tag{d}$$

$$\frac{d^2 h(z)}{dz^2} - (a^2 + b^2) h(z) = 0 \tag{e}$$

The general solutions of equations (c), (d), and (e) are

$$f(x) = \begin{pmatrix} 1 \\ x \end{pmatrix} \quad \text{for } a = 0, \qquad f(x) = \begin{pmatrix} \sin ax \\ \cos ax \end{pmatrix} \quad \text{for } a \neq 0 \tag{f}$$

$$g(y) = \begin{pmatrix} 1 \\ y \end{pmatrix} \quad \text{for } b = 0, \qquad g(y) = \begin{pmatrix} \sin by \\ \cos by \end{pmatrix} \quad \text{for } b \neq 0 \tag{g}$$

$$h(z) = \begin{pmatrix} 1 \\ z \end{pmatrix} \quad \text{for} \quad a^2 + b^2 = 0$$

$$h(z) = \begin{pmatrix} \sinh \sqrt{a^2+b^2}\, z \\ \cosh \sqrt{a^2+b^2}\, z \end{pmatrix} \quad \text{or} \quad \begin{pmatrix} \exp\left(\sqrt{a^2+b^2}\, z\right) \\ \exp\left(-\sqrt{a^2+b^2}\, z\right) \end{pmatrix} \quad \text{for } a^2 + b^2 \neq 0 \tag{h}$$

Then the general solutions of harmonic function may be expressed by the product of Eqs. (f), (g), and (h):

$$\begin{gathered} \begin{pmatrix} 1 \\ x \end{pmatrix}\begin{pmatrix} 1 \\ y \end{pmatrix}\begin{pmatrix} 1 \\ z \end{pmatrix}, \quad \begin{pmatrix} 1 \\ x \end{pmatrix}\begin{pmatrix} \sin by \\ \cos by \end{pmatrix}\begin{pmatrix} \sinh bz \\ \cosh bz \end{pmatrix} \\ \begin{pmatrix} \sin ax \\ \cos ax \end{pmatrix}\begin{pmatrix} 1 \\ y \end{pmatrix}\begin{pmatrix} \sinh az \\ \cosh az \end{pmatrix} \\ \begin{pmatrix} \sin ax \\ \cos ax \end{pmatrix}\begin{pmatrix} \sin by \\ \cos by \end{pmatrix}\begin{pmatrix} \sinh \sqrt{a^2+b^2}\, z \\ \cosh \sqrt{a^2+b^2}\, z \end{pmatrix} \end{gathered} \tag{4.68}$$

or

$$\begin{gathered} \begin{pmatrix} 1 \\ x \end{pmatrix}\begin{pmatrix} 1 \\ y \end{pmatrix}\begin{pmatrix} 1 \\ z \end{pmatrix}, \quad \begin{pmatrix} 1 \\ x \end{pmatrix}\begin{pmatrix} \sin by \\ \cos by \end{pmatrix}\begin{pmatrix} \exp(bz) \\ \exp(-bz) \end{pmatrix} \\ \begin{pmatrix} \sin ax \\ \cos ax \end{pmatrix}\begin{pmatrix} 1 \\ y \end{pmatrix}\begin{pmatrix} \exp(az) \\ \exp(-az) \end{pmatrix} \\ \begin{pmatrix} \sin ax \\ \cos ax \end{pmatrix}\begin{pmatrix} \sin by \\ \cos by \end{pmatrix}\begin{pmatrix} \exp\left(\sqrt{a^2+b^2}\, z\right) \\ \exp\left(-\sqrt{a^2+b^2}\, z\right) \end{pmatrix} \end{gathered} \tag{4.68$'$}$$

4.5 Cylindrical Coordinate System

4.5.1 Stress in a Cylindrical Coordinate System

We consider the equilibrium of forces acting on a body in the cylindrical coordinate system (r, θ, z) shown in Fig. 4.5. When F_r, F_θ, F_z denote the components of the body force per unit volume acting on the cylindrical element, the equilibrium equation of the forces in the r direction acting on the element is

$$\begin{aligned}
&\left(\sigma_{rr} + \frac{\partial \sigma_{rr}}{\partial r}\,dr\right)(r+dr)\,d\theta\,dz - \sigma_{rr} r\,d\theta\,dz \\
&\quad + \left(\sigma_{zr} + \frac{\partial \sigma_{zr}}{\partial z}\,dz\right)\left[\pi(r+dr)^2 - \pi r^2\right]\frac{d\theta}{2\pi} - \sigma_{zr}\left[\pi(r+dr)^2 - \pi r^2\right]\frac{d\theta}{2\pi} \\
&\quad + \left(\sigma_{\theta r} + \frac{\partial \sigma_{\theta r}}{\partial \theta}\,d\theta\right)dr\,dz\cos\frac{d\theta}{2} - \sigma_{\theta r}\,dr\,dz\cos\frac{d\theta}{2} \\
&\quad - \left(\sigma_{\theta\theta} + \frac{\partial \sigma_{\theta\theta}}{\partial \theta}\,d\theta\right)dr\,dz\sin\frac{d\theta}{2} - \sigma_{\theta\theta}\,dr\,dz\sin\frac{d\theta}{2} \\
&\quad + F_r\left[\pi(r+dr)^2 - \pi r^2\right]\frac{d\theta}{2\pi}\,dz = 0 \qquad \text{(a)}
\end{aligned}$$

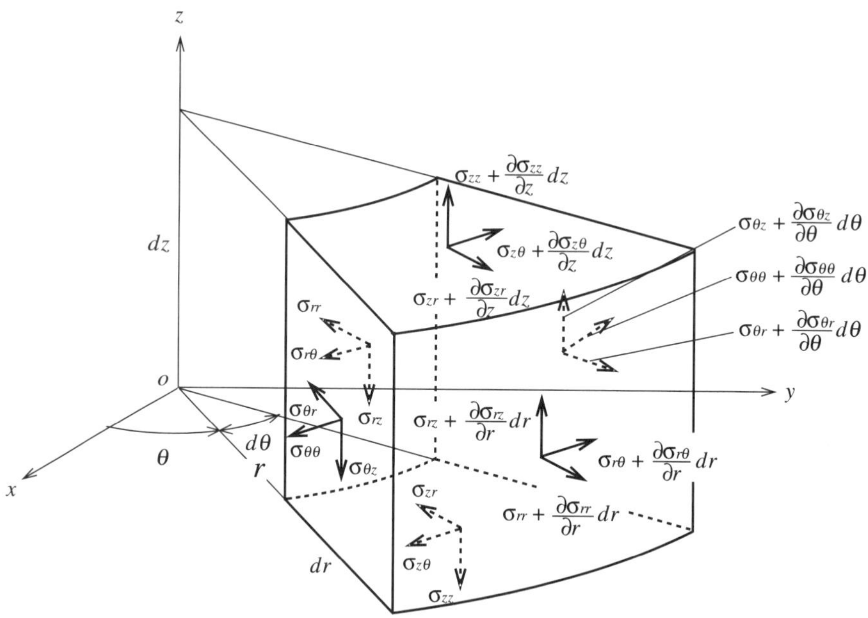

Figure 4.5 Equilibrium of forces on a cylindrical element.

After dividing Eq. (a) by $r\,dr\,d\theta\,dz$ and omitting higher infinitesimal terms, we obtain

$$\frac{\partial \sigma_{rr}}{\partial r} + \frac{1}{r}\frac{\partial \sigma_{\theta r}}{\partial \theta} + \frac{\sigma_{zr}}{\partial z} + \frac{\sigma_{rr} - \sigma_{\theta\theta}}{r} + F_r = 0 \tag{b}$$

The two other equations for equilibrium of forces in the θ and z directions can be obtained in the same manner. Thus, the equations of equilibrium for the cylindrical coordinate system are

$$\begin{aligned}
\frac{\partial \sigma_{rr}}{\partial r} + \frac{1}{r}\frac{\partial \sigma_{\theta r}}{\partial \theta} + \frac{\partial \sigma_{zr}}{\partial z} + \frac{\sigma_{rr} - \sigma_{\theta\theta}}{r} + F_r = 0 \\
\frac{\partial \sigma_{r\theta}}{\partial r} + \frac{1}{r}\frac{\partial \sigma_{\theta\theta}}{\partial \theta} + \frac{\partial \sigma_{z\theta}}{\partial z} + 2\frac{\sigma_{r\theta}}{r} + F_\theta = 0 \\
\frac{\partial \sigma_{rz}}{\partial r} + \frac{1}{r}\frac{\partial \sigma_{\theta z}}{\partial \theta} + \frac{\partial \sigma_{zz}}{\partial z} + \frac{\sigma_{rz}}{r} + F_z = 0
\end{aligned} \tag{4.69}$$

Next, we consider the coordinate transformation of stress between cylindrical and Cartesian coordinate systems. The direction cosines given in Table 4.1 are

$$l_1 = m_2 = \cos\theta, \quad m_1 = -l_2 = \sin\theta, \quad n_3 = 1, \quad n_1 = n_2 = l_3 = m_3 = 0 \tag{c}$$

Substituting Eqs. (c) into Eqs. (4.6), we obtain the components of stress in the cylindrical coordinate system expressed in terms of those of the Cartesian coordinate system:

$$\begin{aligned}
\sigma_{rr} &= \sigma_{xx}\cos^2\theta + \sigma_{yy}\sin^2\theta + \sigma_{xy}\sin 2\theta \\
\sigma_{\theta\theta} &= \sigma_{xx}\sin^2\theta + \sigma_{yy}\cos^2\theta - \sigma_{xy}\sin 2\theta \\
\sigma_{zz} &= \sigma_{zz} \\
\sigma_{r\theta} &= -\tfrac{1}{2}(\sigma_{xx} - \sigma_{yy})\sin 2\theta + \sigma_{xy}\cos 2\theta \\
\sigma_{\theta z} &= \sigma_{yz}\cos\theta - \sigma_{zx}\sin\theta \\
\sigma_{zr} &= \sigma_{zx}\cos\theta + \sigma_{yz}\sin\theta
\end{aligned} \tag{4.70}$$

4.5.2 Strain in a Cylindrical Coordinate System

We consider the components of strain in the cylindrical coordinate system. Following Subsection 4.2.1, we define the components of strain and dilatation e in the cylindrical coordinate system in terms of the components of displacement

$$\begin{aligned}
&\epsilon_{rr} = \frac{\partial u_r}{\partial r}, \qquad \epsilon_{\theta\theta} = \frac{u_r}{r} + \frac{1}{r}\frac{\partial u_\theta}{\partial \theta}, \qquad \epsilon_{zz} = \frac{\partial u_z}{\partial z} \\
&\epsilon_{r\theta} = \frac{1}{2}\left(\frac{1}{r}\frac{\partial u_r}{\partial \theta} + \frac{\partial u_\theta}{\partial r} - \frac{u_\theta}{r}\right), \qquad \epsilon_{\theta z} = \frac{1}{2}\left(\frac{\partial u_\theta}{\partial z} + \frac{1}{r}\frac{\partial u_z}{\partial \theta}\right) \\
&\epsilon_{zr} = \frac{1}{2}\left(\frac{\partial u_r}{\partial z} + \frac{\partial u_z}{\partial r}\right), \qquad e = \frac{\partial u_r}{\partial r} + \frac{u_r}{r} + \frac{1}{r}\frac{\partial u_\theta}{\partial \theta} + \frac{\partial u_z}{\partial z}
\end{aligned} \tag{4.71}$$

where u_r, u_θ, u_z are the components of displacement in the r, θ, z directions, respectively.

The compatibility conditions of strain in the cylindrical coordinate system can be obtained by eliminating the components of the displacement in Eqs. (4.71). We obtain

$$\begin{aligned}
2\frac{\partial^2(r\epsilon_{r\theta})}{\partial r\partial\theta} &= \frac{\partial^2\epsilon_{rr}}{\partial\theta^2} + r\frac{\partial^2(r\epsilon_{\theta\theta})}{\partial r^2} - r\frac{\partial\epsilon_{rr}}{\partial r} \\
2\frac{\partial^2\epsilon_{zr}}{\partial r\partial z} &= \frac{\partial^2\epsilon_{zz}}{\partial r^2} + \frac{\partial^2\epsilon_{rr}}{\partial z^2} \\
2\frac{\partial^2(r\epsilon_{\theta z})}{\partial\theta\partial z} &= r^2\frac{\partial^2\epsilon_{\theta\theta}}{\partial z^2} - 2r\frac{\partial\epsilon_{zr}}{\partial z} + r\frac{\partial\epsilon_{zz}}{\partial r} + \frac{\partial^2\epsilon_{zz}}{\partial\theta^2} \\
\frac{\partial}{\partial z}\left(-\frac{\partial\epsilon_{r\theta}}{\partial z} + \frac{1}{r}\frac{\partial\epsilon_{zr}}{\partial\theta} + \frac{\partial\epsilon_{\theta z}}{\partial r}\right) &= \frac{\partial^2}{\partial r\partial\theta}\left(\frac{\epsilon_{zz}}{r}\right) + \frac{1}{r}\frac{\partial\epsilon_{\theta z}}{\partial z} \\
\frac{\partial}{\partial\theta}\left(\frac{\partial\epsilon_{r\theta}}{\partial z} - \frac{1}{r}\frac{\partial\epsilon_{zr}}{\partial\theta} + \frac{\partial\epsilon_{\theta z}}{\partial r}\right) &= \frac{\partial^2(r\epsilon_{\theta\theta})}{\partial r\partial z} - \frac{\partial\epsilon_{rr}}{\partial z} - \frac{1}{r}\frac{\partial\epsilon_{\theta z}}{\partial\theta} \\
\frac{\partial}{\partial r}\left(\frac{\partial\epsilon_{r\theta}}{\partial z} + \frac{1}{r}\frac{\partial\epsilon_{zr}}{\partial\theta} - \frac{\partial\epsilon_{\theta z}}{\partial r}\right) &= \frac{1}{r}\frac{\partial^2\epsilon_{rr}}{\partial\theta\partial z} - \frac{2}{r}\frac{\partial\epsilon_{r\theta}}{\partial z} + \frac{\partial}{\partial r}\left(\frac{\epsilon_{\theta z}}{r}\right)
\end{aligned} \tag{4.72}$$

Next, we consider the coordinate transformation of strain between a cylindrical coordinate system and a Cartesian coordinate system. The components (r, θ, z) expressed in terms of components (x, y, z) are

$$x = r\cos\theta, \quad y = r\sin\theta, \quad z = z, \quad r = \sqrt{x^2+y^2}, \quad \theta = \tan^{-1}\frac{y}{x} \tag{4.73}$$

The relations between the components of displacement in the two coordinate systems are

$$u_x = u_r\cos\theta - u_\theta\sin\theta, \qquad u_y = u_r\sin\theta + u_\theta\cos\theta, \qquad u_z = u_z \tag{4.74}$$

Substitution of the direction cosines between the two coordinate systems into Eq. (4.14) gives the relations between the components of strain in the two coordinate systems:

$$\begin{aligned}
\epsilon_{rr} &= \epsilon_{xx}\cos^2\theta + \epsilon_{yy}\sin^2\theta + \epsilon_{xy}\sin 2\theta \\
\epsilon_{\theta\theta} &= \epsilon_{xx}\sin^2\theta + \epsilon_{yy}\cos^2\theta - \epsilon_{xy}\sin 2\theta \\
\epsilon_{zz} &= \epsilon_{zz} \\
\epsilon_{r\theta} &= -\tfrac{1}{2}(\epsilon_{xx} - \epsilon_{yy})\sin 2\theta + \epsilon_{xy}\cos 2\theta \\
\epsilon_{\theta z} &= \epsilon_{yz}\cos\theta - \epsilon_{zx}\sin\theta \\
\epsilon_{zr} &= \epsilon_{zx}\cos\theta + \epsilon_{yz}\sin\theta
\end{aligned} \tag{4.75}$$

4.5.3 Generalized Hooke's Law in a Cylindrical Coordinate System

The constitutive equations, or the generalized Hooke's law, for a homogeneous, isotropic body in a cylindrical coordinate system are

$$\begin{aligned}
\epsilon_{rr} &= \frac{1}{E}[\sigma_{rr} - \nu(\sigma_{\theta\theta} + \sigma_{zz})] + \alpha\tau = \frac{1}{2G}\left(\sigma_{rr} - \frac{\nu}{1+\nu}\Theta\right) + \alpha\tau \\
\epsilon_{\theta\theta} &= \frac{1}{E}[\sigma_{\theta\theta} - \nu(\sigma_{zz} + \sigma_{rr})] + \alpha\tau = \frac{1}{2G}\left(\sigma_{\theta\theta} - \frac{\nu}{1+\nu}\Theta\right) + \alpha\tau \\
\epsilon_{zz} &= \frac{1}{E}[\sigma_{zz} - \nu(\sigma_{rr} + \sigma_{\theta\theta})] + \alpha\tau = \frac{1}{2G}\left(\sigma_{zz} - \frac{\nu}{1+\nu}\Theta\right) + \alpha\tau \\
\epsilon_{r\theta} &= \frac{\sigma_{r\theta}}{2G}, \qquad \epsilon_{\theta z} = \frac{\sigma_{\theta z}}{2G}, \qquad \epsilon_{zr} = \frac{\sigma_{zr}}{2G}
\end{aligned} \tag{4.76}$$

where $\Theta = \sigma_{rr} + \sigma_{\theta\theta} + \sigma_{zz}$. The alternative form is

$$\begin{aligned}
\sigma_{rr} &= 2\mu\epsilon_{rr} + \lambda e - \beta\tau, & \sigma_{r\theta} &= 2\mu\epsilon_{r\theta} \\
\sigma_{\theta\theta} &= 2\mu\epsilon_{\theta\theta} + \lambda e - \beta\tau, & \sigma_{\theta z} &= 2\mu\epsilon_{\theta z} \\
\sigma_{zz} &= 2\mu\epsilon_{zz} + \lambda e - \beta\tau, & \sigma_{zr} &= 2\mu\epsilon_{zr}
\end{aligned} \tag{4.77}$$

where $e = \epsilon_{rr} + \epsilon_{\theta\theta} + \epsilon_{zz}$.

4.5.4 Navier's Equations of Thermoelasticity in a Cylindrical Coordinate System

Navier's equations (4.23′) for thermoelasticity can be expressed in a cylindrical coordinate system as

$$\begin{aligned}
(\lambda + 2\mu)\frac{\partial e}{\partial r} - 2\mu\left(\frac{1}{r}\frac{\partial \omega_z}{\partial \theta} - \frac{\partial \omega_\theta}{\partial z}\right) - \beta\frac{\partial \tau}{\partial r} + F_r = 0 \\
(\lambda + 2\mu)\frac{1}{r}\frac{\partial e}{\partial \theta} - 2\mu\left(\frac{\partial \omega_r}{\partial z} - \frac{\partial \omega_z}{\partial r}\right) - \beta\frac{1}{r}\frac{\partial \tau}{\partial \theta} + F_\theta = 0 \\
(\lambda + 2\mu)\frac{\partial e}{\partial z} - \frac{2\mu}{r}\left[\frac{\partial (r\omega_\theta)}{\partial r} - \frac{\partial \omega_r}{\partial \theta}\right] - \beta\frac{\partial \tau}{\partial z} + F_z = 0
\end{aligned} \tag{4.78}$$

where

$$\begin{aligned}
\omega_r &= \frac{1}{2}\left(\frac{1}{r}\frac{\partial u_z}{\partial \theta} - \frac{\partial u_\theta}{\partial z}\right), \qquad \omega_\theta = \frac{1}{2}\left(\frac{\partial u_r}{\partial z} - \frac{\partial u_z}{\partial r}\right) \\
\omega_z &= \frac{1}{2r}\left[\frac{\partial (ru_\theta)}{\partial r} - \frac{\partial u_r}{\partial \theta}\right]
\end{aligned} \tag{4.79}$$

The solution of Navier's equations (4.78) without the body force in the cylindrical coordinate system can be expressed, for example, by the Boussinesq harmonic functions:

$$
\begin{aligned}
u_r &= \frac{\partial \Phi}{\partial r} + \frac{\partial \varphi}{\partial r} + \frac{2}{r}\frac{\partial \vartheta}{\partial \theta} + z\frac{\partial \psi}{\partial r} \\
u_\theta &= \frac{1}{r}\frac{\partial \Phi}{\partial \theta} + \frac{1}{r}\frac{\partial \varphi}{\partial \theta} - 2\frac{\partial \vartheta}{\partial r} + \frac{z}{r}\frac{\partial \psi}{\partial \theta} \\
u_z &= \frac{\partial \Phi}{\partial z} + \frac{\partial \varphi}{\partial z} + z\frac{\partial \psi}{\partial z} - (3-4\nu)\psi
\end{aligned}
\tag{4.80}
$$

where the four functions Φ, φ, ϑ, and ψ must satisfy

$$
\nabla^2 \Phi = K\tau, \qquad \nabla^2 \varphi = 0, \qquad \nabla^2 \vartheta = 0, \qquad \nabla^2 \psi = 0 \tag{4.81}
$$

and

$$
\nabla^2 = \frac{\partial^2}{\partial r^2} + \frac{1}{r}\frac{\partial}{\partial r} + \frac{1}{r^2}\frac{\partial^2}{\partial \theta^2} + \frac{\partial^2}{\partial z^2}.
$$

The general solutions of the Laplace equation in the cylindrical coordinate system are given by

$$
\begin{aligned}
&\begin{pmatrix} 1 \\ \ln r \end{pmatrix}\begin{pmatrix} 1 \\ \theta \end{pmatrix}\begin{pmatrix} 1 \\ z \end{pmatrix}, &&\begin{pmatrix} r^b \\ r^{-b} \end{pmatrix}\begin{pmatrix} \sin b\theta \\ \cos b\theta \end{pmatrix}\begin{pmatrix} 1 \\ z \end{pmatrix} \\
&\begin{pmatrix} J_0(ar) \\ Y_0(ar) \end{pmatrix}\begin{pmatrix} 1 \\ \theta \end{pmatrix}\begin{pmatrix} e^{az} \\ e^{-az} \end{pmatrix}, &&\begin{pmatrix} J_b(ar) \\ Y_b(ar) \end{pmatrix}\begin{pmatrix} \sin b\theta \\ \cos b\theta \end{pmatrix}\begin{pmatrix} e^{az} \\ e^{-az} \end{pmatrix} \\
&\begin{pmatrix} I_0(ar) \\ K_0(ar) \end{pmatrix}\begin{pmatrix} 1 \\ \theta \end{pmatrix}\begin{pmatrix} \sin az \\ \cos az \end{pmatrix}, &&\begin{pmatrix} I_b(ar) \\ K_b(ar) \end{pmatrix}\begin{pmatrix} \sin b\theta \\ \cos b\theta \end{pmatrix}\begin{pmatrix} \sin az \\ \cos az \end{pmatrix}
\end{aligned}
\tag{4.82}
$$

where a and b are constants, $J_b(ar)$ and $Y_b(ar)$ are the Bessel functions, $I_b(ar)$ and $K_b(ar)$ are the modified Bessel functions, $\binom{\sinh az}{\cosh az}$ can be used instead of $\binom{e^{az}}{e^{-az}}$. The third and forth terms in Eqs. (4.82) can mainly be used for the semi-infinite body and the thick plate, and the fifth and sixth terms can mainly be used for the cylinder.

Example 4.4 Express the strain and the stress from Eqs. (4.80).

Solution The displacement (4.80) gives the strain:

$$
\begin{aligned}
\epsilon_{rr} &= \frac{\partial^2 \Phi}{\partial r^2} + \frac{\partial^2 \varphi}{\partial r^2} + \frac{2}{r}\frac{\partial^2 \vartheta}{\partial r \partial \theta} - \frac{2}{r^2}\frac{\partial \vartheta}{\partial \theta} + z\frac{\partial^2 \psi}{\partial r^2} \\
\epsilon_{\theta\theta} &= \frac{1}{r}\frac{\partial \Phi}{\partial r} + \frac{1}{r^2}\frac{\partial^2 \Phi}{\partial \theta^2} + \frac{1}{r}\frac{\partial \varphi}{\partial r} + \frac{1}{r^2}\frac{\partial^2 \varphi}{\partial \theta^2} + \frac{2}{r^2}\frac{\partial \vartheta}{\partial \theta} - \frac{2}{r}\frac{\partial^2 \vartheta}{\partial r \partial \theta} + \frac{z}{r}\frac{\partial \psi}{\partial r} + \frac{z}{r^2}\frac{\partial^2 \psi}{\partial \theta^2} \\
\epsilon_{zz} &= \frac{\partial^2 \Phi}{\partial z^2} + \frac{\partial^2 \varphi}{\partial z^2} + z\frac{\partial^2 \psi}{\partial z^2} - 2(1-2\nu)\frac{\partial \psi}{\partial z}
\end{aligned}
$$

$$\epsilon_{r\theta} = \frac{1}{r}\frac{\partial^2 \Phi}{\partial r \partial \theta} - \frac{1}{r^2}\frac{\partial \Phi}{\partial \theta} + \frac{1}{r}\frac{\partial^2 \varphi}{\partial r \partial \theta} - \frac{1}{r^2}\frac{\partial \varphi}{\partial \theta} - \frac{\partial^2 \vartheta}{\partial r^2} + \frac{1}{r}\frac{\partial \vartheta}{\partial r} + \frac{1}{r^2}\frac{\partial^2 \vartheta}{\partial \theta^2}$$

$$+ \frac{z}{r}\frac{\partial^2 \psi}{\partial r \partial \theta} - \frac{z}{r^2}\frac{\partial \psi}{\partial \theta} \tag{a}$$

$$\epsilon_{\theta z} = \frac{1}{r}\frac{\partial^2 \Phi}{\partial z \partial \theta} + \frac{1}{r}\frac{\partial^2 \varphi}{\partial z \partial \theta} - \frac{\partial^2 \vartheta}{\partial r \partial z} + \frac{z}{r}\frac{\partial^2 \psi}{\partial z \partial \theta} - (1-2\nu)\frac{1}{r}\frac{\partial \psi}{\partial \theta}$$

$$\epsilon_{zr} = \frac{\partial^2 \Phi}{\partial r \partial z} + \frac{\partial^2 \varphi}{\partial r \partial z} + \frac{1}{r}\frac{\partial^2 \vartheta}{\partial \theta \partial z} + z\frac{\partial^2 \psi}{\partial r \partial z} - (1-2\nu)\frac{\partial \psi}{\partial r}$$

$$e = \nabla^2 \Phi - 2(1-2\nu)\frac{\partial \psi}{\partial z} = K\tau - 2(1-2\nu)\frac{\partial \psi}{\partial z}$$

Substitution of Eqs. (a) into Eqs. (4.18) yields

$$\sigma_{rr} = 2\mu\left(\frac{\partial^2 \Phi}{\partial r^2} - K\tau + \frac{\partial^2 \varphi}{\partial r^2} + \frac{2}{r}\frac{\partial^2 \vartheta}{\partial r \partial \theta} - \frac{2}{r^2}\frac{\partial \vartheta}{\partial \theta} + z\frac{\partial^2 \psi}{\partial r^2} - 2\nu\frac{\partial \psi}{\partial z}\right)$$

$$\sigma_{\theta\theta} = 2\mu\left(\frac{1}{r}\frac{\partial \Phi}{\partial r} + \frac{1}{r^2}\frac{\partial^2 \Phi}{\partial \theta^2} - K\tau + \frac{1}{r}\frac{\partial \varphi}{\partial r} + \frac{1}{r^2}\frac{\partial^2 \varphi}{\partial \theta^2} + \frac{2}{r^2}\frac{\partial \vartheta}{\partial \theta} - \frac{2}{r}\frac{\partial^2 \vartheta}{\partial r \partial \theta}\right.$$

$$\left. + \frac{z}{r}\frac{\partial \psi}{\partial r} + \frac{z}{r^2}\frac{\partial^2 \psi}{\partial \theta^2} - 2\nu\frac{\partial \psi}{\partial z}\right)$$

$$\sigma_{zz} = 2\mu\left[\frac{\partial^2 \Phi}{\partial z^2} - K\tau + \frac{\partial^2 \varphi}{\partial z^2} + z\frac{\partial^2 \psi}{\partial z^2} - 2(1-\nu)\frac{\partial \psi}{\partial z}\right]$$

$$\sigma_{r\theta} = 2\mu\left(\frac{1}{r}\frac{\partial^2 \Phi}{\partial r \partial \theta} - \frac{1}{r^2}\frac{\partial \Phi}{\partial \theta} + \frac{1}{r}\frac{\partial^2 \varphi}{\partial r \partial \theta} - \frac{1}{r^2}\frac{\partial \varphi}{\partial \theta} - \frac{\partial^2 \vartheta}{\partial r^2} + \frac{1}{r}\frac{\partial \vartheta}{\partial r} + \frac{1}{r^2}\frac{\partial^2 \vartheta}{\partial \theta^2}\right.$$

$$\left. + \frac{z}{r}\frac{\partial^2 \psi}{\partial r \partial \theta} - \frac{z}{r^2}\frac{\partial \psi}{\partial \theta}\right)$$

$$\sigma_{\theta z} = 2\mu\left[\frac{1}{r}\frac{\partial^2 \Phi}{\partial z \partial \theta} + \frac{1}{r}\frac{\partial^2 \varphi}{\partial z \partial \theta} - \frac{\partial^2 \vartheta}{\partial r \partial z} + \frac{z}{r}\frac{\partial^2 \psi}{\partial z \partial \theta} - (1-2\nu)\frac{1}{r}\frac{\partial \psi}{\partial \theta}\right]$$

$$\sigma_{zr} = 2\mu\left[\frac{\partial^2 \Phi}{\partial r \partial z} + \frac{\partial^2 \varphi}{\partial r \partial z} + \frac{1}{r}\frac{\partial^2 \vartheta}{\partial \theta \partial z} + z\frac{\partial^2 \psi}{\partial r \partial z} - (1-2\nu)\frac{\partial \psi}{\partial r}\right] \tag{b}$$

where following relations are used:

$$\lambda K\tau - \beta\tau = \left(\lambda\frac{\beta}{\lambda + 2\mu} - \beta\right)\tau = -2\mu\frac{\beta}{\lambda + 2\mu}\tau = -2\mu K\tau$$

$$-2(1-2\nu)\lambda\frac{\partial \psi}{\partial z} = -2(1-2\nu)\frac{2\mu\nu}{1-2\nu}\frac{\partial \psi}{\partial z} = -2\nu 2\mu\frac{\partial \psi}{\partial z}$$

4.6 Spherical Coordinate System

4.6.1 Stress in a Spherical Coordinate System

We consider the equilibrium of forces acting on a body in the spherical coordinate system (r, θ, ϕ) shown in Fig. 4.6. When F_r, F_θ, F_ϕ denote the components of the body force per unit volume acting on the spherical element, the equilibrium of forces in terms of r, θ,

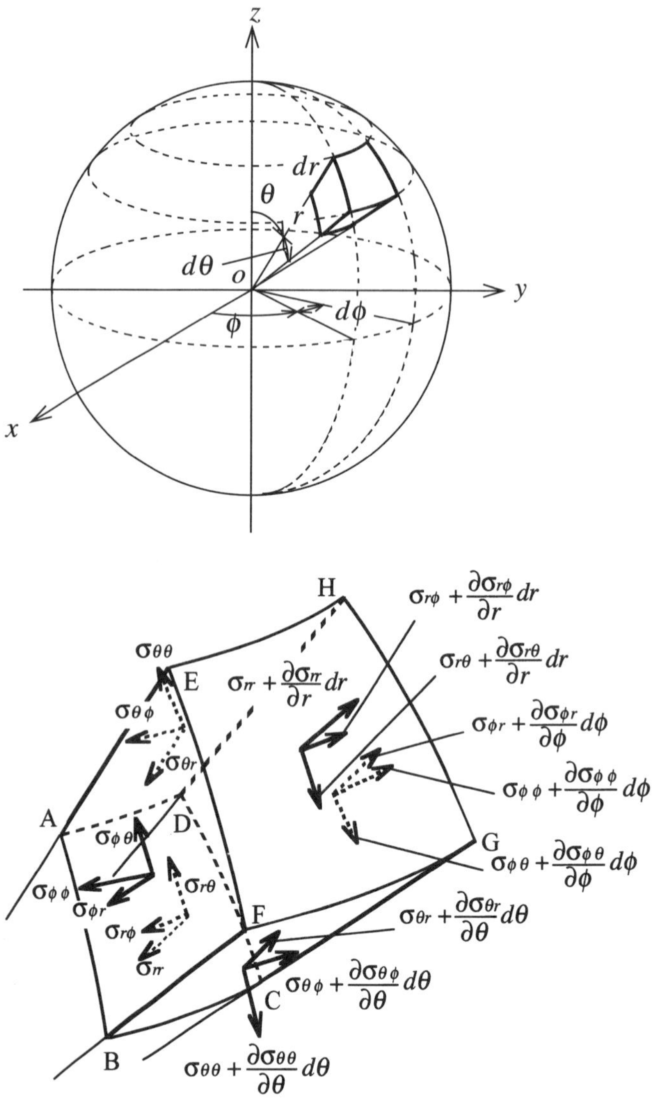

Figure 4.6 Equilibrium of forces on a spherical element.

and φ are given by

$$
\begin{aligned}
&\frac{\partial \sigma_{rr}}{\partial r}+\frac{1}{r}\frac{\partial \sigma_{\theta r}}{\partial \theta}+\frac{1}{r\sin\theta}\frac{\partial \sigma_{\phi r}}{\partial \phi}+\frac{1}{r}(2\sigma_{rr}-\sigma_{\theta\theta}-\sigma_{\phi\phi}+\sigma_{\theta r}\cot\theta)+F_r=0\\
&\frac{\partial \sigma_{r\theta}}{\partial r}+\frac{1}{r}\frac{\partial \sigma_{\theta\theta}}{\partial \theta}+\frac{1}{r\sin\theta}\frac{\partial \sigma_{\phi\theta}}{\partial \phi}+\frac{1}{r}\left[(\sigma_{\theta\theta}-\sigma_{\phi\phi})\cot\theta+3\sigma_{r\theta}\right]+F_\theta=0 \qquad (4.83)\\
&\frac{\partial \sigma_{r\phi}}{\partial r}+\frac{1}{r}\frac{\partial \sigma_{\theta\phi}}{\partial \theta}+\frac{1}{r\sin\theta}\frac{\partial \sigma_{\phi\phi}}{\partial \phi}+\frac{1}{r}(3\sigma_{r\phi}+2\sigma_{\theta\phi}\cot\theta)+F_\phi=0
\end{aligned}
$$

Next, we consider the coordinate transformation of stress between the spherical coordinate system and the Cartesian coordinate system. The direction cosines given in Table 4.1 are

$$
\begin{aligned}
&l_1=\sin\theta\cos\phi, \quad m_1=\sin\theta\sin\phi, \quad n_1=\cos\theta, \quad l_2=\cos\theta\cos\phi\\
&m_2=\cos\theta\sin\phi, \quad n_2=-\sin\theta, \quad l_3=-\sin\phi, \quad m_3=\cos\phi, \quad n_3=0
\end{aligned} \qquad \text{(a)}
$$

Substituting Eqs. (a) into Eqs. (4.6), we obtain the components of stress in the spherical coordinate system expressed in terms of those of the Cartesian coordinate system:

$$
\begin{aligned}
\sigma_{rr} &= \sigma_{xx}\sin^2\theta\cos^2\phi+\sigma_{yy}\sin^2\theta\sin^2\phi+\sigma_{zz}\cos^2\theta\\
&\quad+\sigma_{xy}\sin^2\theta\sin 2\phi+\sigma_{yz}\sin 2\theta\sin\phi+\sigma_{zx}\sin 2\theta\cos\phi\\
\sigma_{\theta\theta} &= \sigma_{xx}\cos^2\theta\cos^2\phi+\sigma_{yy}\cos^2\theta\sin^2\phi+\sigma_{zz}\sin^2\theta\\
&\quad+\sigma_{xy}\cos^2\theta\sin 2\phi-\sigma_{yz}\sin 2\theta\sin\phi-\sigma_{zx}\sin 2\theta\cos\phi\\
\sigma_{\phi\phi} &= \sigma_{xx}\sin^2\phi+\sigma_{yy}\cos^2\phi-\sigma_{xy}\sin 2\phi\\
\sigma_{r\theta} &= \tfrac{1}{2}\sin 2\theta\,(\sigma_{xx}\cos^2\phi+\sigma_{yy}\sin^2\phi-\sigma_{zz}) \qquad (4.84)\\
&\quad+\tfrac{1}{2}\sigma_{xy}\sin 2\theta\sin 2\phi+\sigma_{yz}\cos 2\theta\sin\phi+\sigma_{zx}\cos 2\theta\cos\phi\\
\sigma_{\theta\phi} &= -\tfrac{1}{2}\cos\theta\sin 2\phi\,(\sigma_{xx}-\sigma_{yy})\\
&\quad+\sigma_{xy}\cos\theta\cos 2\phi-\sigma_{yz}\sin\theta\cos\phi+\sigma_{zx}\sin\theta\sin\phi\\
\sigma_{\phi r} &= -\tfrac{1}{2}\sin\theta\sin 2\phi\,(\sigma_{xx}-\sigma_{yy})\\
&\quad+\sigma_{xy}\sin\theta\cos 2\phi+\sigma_{yz}\cos\theta\cos\phi-\sigma_{zx}\cos\theta\sin\phi
\end{aligned}
$$

4.6.2 Strain in a Spherical Coordinate System

We consider the components of strain in a spherical coordinate system. In the same manner as in Subsection 4.2.1, we define the components of strain and dilatation e in terms of the components of displacement. We obtain

$$\begin{aligned}
\epsilon_{rr} &= \frac{\partial u_r}{\partial r}, \qquad \epsilon_{\theta\theta} = \frac{u_r}{r} + \frac{1}{r}\frac{\partial u_\theta}{\partial \theta} \\
\epsilon_{\phi\phi} &= \frac{u_r}{r} + \cot\theta \frac{u_\theta}{r} + \frac{1}{r\sin\theta}\frac{\partial u_\phi}{\partial \phi}, \qquad \epsilon_{r\theta} = \frac{1}{2}\left(\frac{1}{r}\frac{\partial u_r}{\partial \theta} + \frac{\partial u_\theta}{\partial r} - \frac{u_\theta}{r}\right) \\
\epsilon_{\theta\phi} &= \frac{1}{2}\left(\frac{1}{r}\frac{\partial u_\phi}{\partial \theta} - \cot\theta \frac{u_\phi}{r} + \frac{1}{r\sin\theta}\frac{\partial u_\theta}{\partial \phi}\right) \\
\epsilon_{\phi r} &= \frac{1}{2}\left(\frac{1}{r\sin\theta}\frac{\partial u_r}{\partial \phi} + \frac{\partial u_\phi}{\partial r} - \frac{u_\phi}{r}\right) \\
e &= \frac{\partial u_r}{\partial r} + 2\frac{u_r}{r} + \frac{1}{r}\frac{\partial u_\theta}{\partial \theta} + \cot\theta \frac{u_\theta}{r} + \frac{1}{r\sin\theta}\frac{\partial u_\phi}{\partial \phi}
\end{aligned} \tag{4.85}$$

where u_r, u_θ, u_ϕ are the components of displacement in the r, θ, ϕ directions, respectively.

The compatibility conditions of strain in the spherical coordinate system may be obtained by eliminating the displacement components in Eqs. (4.85).

$$\begin{aligned}
&-\frac{\epsilon_{\theta\theta,\phi\phi}}{r^2\sin^2\theta} + \frac{\epsilon_{\theta\theta,\theta}}{r^2\tan\theta} - \frac{\epsilon_{\theta\theta,r}}{r} + \frac{2\epsilon_{r\theta,\theta}}{r^2} + \frac{2\epsilon_{r\theta}}{r^2\tan\theta} - \frac{2\epsilon_{\theta\theta}}{r^2} - \frac{\epsilon_{\phi\phi,\theta\theta}}{r^2} \\
&\quad - \frac{\epsilon_{\phi\phi,r}}{r} - \frac{2\epsilon_{\phi\phi,\theta}}{r^2\tan\theta} + \frac{\epsilon_{\phi r}}{r^2\sin\theta} + \frac{2\cos\theta\,\epsilon_{\theta\phi,\phi}}{r^2\sin^2\theta} + \frac{2\epsilon_{\theta\phi,\theta\phi}}{r^2\sin\theta} + \frac{2\epsilon_{rr}}{r^2} = 0 \\
&\frac{2\epsilon_{\phi r,\phi r}}{r\sin\theta} + \frac{\epsilon_{\phi r,\phi}}{r^2\sin\theta} - \frac{\epsilon_{rr,\phi\phi}}{r^2\sin^2\theta} - \frac{2\epsilon_{\phi\phi,r}}{r} - \epsilon_{\phi\phi,rr} \\
&\quad + \frac{2\epsilon_{r\theta,r}}{r\tan\theta} + \frac{2\epsilon_{r\theta}}{r^2\tan\theta} - \frac{\epsilon_{rr,\theta}}{r^2\tan\theta} + \frac{\epsilon_{rr,r}}{r} = 0 \\
&\frac{2\epsilon_{r\theta,r\theta}}{r} - \frac{2\epsilon_{\theta\theta,r}}{r} - \epsilon_{\theta\theta,rr} - \frac{\epsilon_{rr,r}}{r} - \frac{\epsilon_{rr,\theta\theta}}{r^2} + \frac{2\epsilon_{r\theta,\theta}}{r^2} = 0 \\
&\frac{\epsilon_{rr,\theta\phi}}{r^2\sin\theta} - \frac{\epsilon_{r\theta,\phi r}}{r\sin\theta} - \frac{\epsilon_{r\theta,\phi}}{r^2\sin\theta} + \frac{\epsilon_{\phi r}}{r^2\tan\theta} - \frac{\epsilon_{\phi r,\theta}}{r^2} - \frac{\epsilon_{\phi r,r\theta}}{r} \\
&\quad + \epsilon_{\theta\phi,rr} + \frac{2\epsilon_{\theta\phi,r}}{r} + \frac{\epsilon_{\phi r,r}}{r\tan\theta} - \frac{\cos\theta\,\epsilon_{rr,\phi}}{r^2\sin^2\theta} = 0 \\
&\frac{\epsilon_{\theta\theta,\phi r}}{r\sin\theta} - \frac{\epsilon_{\theta\phi,r\theta}}{r} - \frac{2\epsilon_{\theta\phi,r}}{r\tan\theta} + \frac{\cos\theta\,\epsilon_{r\theta,\phi}}{r^2\sin^2\theta} + \frac{\epsilon_{\phi r,\theta}}{r^2\tan\theta} \\
&\quad - \frac{\epsilon_{r\theta,\theta\phi}}{r^2\sin\theta} + \frac{\epsilon_{\phi r,\theta\theta}}{r^2} - \frac{\cos 2\theta\,\epsilon_{\phi r}}{r^2\sin^2\theta} - \frac{\epsilon_{rr,\phi}}{r^2\sin\theta} = 0 \\
&\frac{\epsilon_{\phi\phi,r\theta}}{r} + \frac{\epsilon_{\phi\phi,r}}{r\tan\theta} - \frac{\epsilon_{\phi r,\theta\phi}}{r^2\sin\theta} + \frac{2\epsilon_{r\theta}}{r^2} - \frac{\epsilon_{rr,\theta}}{r^2} \\
&\quad - \frac{\epsilon_{\theta\phi,\phi r}}{r\sin\theta} + \frac{\epsilon_{r\theta,\phi\phi}}{r^2\sin^2\theta} - \frac{\cos\theta\,\epsilon_{\phi r,\phi}}{r^2\sin^2\theta} - \frac{\epsilon_{\theta\theta,r}}{r\tan\theta} = 0
\end{aligned} \tag{4.86}$$

Next, we consider the coordinate transformation of the strain between the spherical coordinate system and the Cartesian coordinate system. The relation between the components (r, θ, ϕ) and the components (x, y, z) may be expressed by

$$\begin{gathered} x = r\sin\theta\cos\phi, \qquad y = r\sin\theta\sin\phi, \qquad z = r\cos\theta \\ r^2 = x^2 + y^2 + z^2, \qquad \theta = \tan^{-1}\left(\frac{\sqrt{x^2+y^2}}{z}\right), \qquad \phi = \tan^{-1}\left(\frac{y}{x}\right) \end{gathered} \tag{4.87}$$

The relation between the components of displacement in the two coordinate systems is

$$\begin{aligned} u_x &= u_r \sin\theta\cos\phi + u_\theta \cos\theta\cos\phi - u_\phi \sin\phi \\ u_y &= u_r \sin\theta\sin\phi + u_\theta \cos\theta\sin\phi + u_\phi \cos\phi \\ u_z &= u_r \cos\theta - u_\theta \sin\theta \end{aligned} \tag{4.88}$$

Substitution of the direction cosines between the two coordinate systems into Eq. (4.14) gives the relation between the components of strain of the two coordinate systems:

$$\begin{aligned} \epsilon_{rr} &= \epsilon_{xx}\sin^2\theta\cos^2\phi + \epsilon_{yy}\sin^2\theta\sin^2\phi + \epsilon_{zz}\cos^2\theta \\ &\quad + \epsilon_{xy}\sin^2\theta\sin 2\phi + \epsilon_{yz}\sin 2\theta\sin\phi + \epsilon_{zx}\sin 2\theta\cos\phi \\ \epsilon_{\theta\theta} &= \epsilon_{xx}\cos^2\theta\cos^2\phi + \epsilon_{yy}\cos^2\theta\sin^2\phi + \epsilon_{zz}\sin^2\theta \\ &\quad + \epsilon_{xy}\cos^2\theta\sin 2\phi - \epsilon_{yz}\sin 2\theta\sin\phi - \epsilon_{zx}\sin 2\theta\cos\phi \\ \epsilon_{\phi\phi} &= \epsilon_{xx}\sin^2\phi + \epsilon_{yy}\cos^2\phi - \epsilon_{xy}\sin 2\phi \\ \epsilon_{r\theta} &= \tfrac{1}{2}\sin 2\theta\,(\epsilon_{xx}\cos^2\phi + \epsilon_{yy}\sin^2\phi - \epsilon_{zz}) \\ &\quad + \tfrac{1}{2}\epsilon_{xy}\sin 2\theta\sin 2\phi + \epsilon_{yz}\cos 2\theta\sin\phi + \epsilon_{zx}\cos 2\theta\cos\phi \\ \epsilon_{\theta\phi} &= -\tfrac{1}{2}\cos\theta\sin 2\phi\,(\epsilon_{xx} - \epsilon_{yy}) \\ &\quad + \epsilon_{xy}\cos\theta\sin 2\phi - \epsilon_{yz}\sin\theta\cos\phi + \epsilon_{zx}\sin\theta\sin\phi \\ \epsilon_{\phi r} &= -\tfrac{1}{2}\sin\theta\sin 2\phi\,(\epsilon_{xx} - \epsilon_{yy}) \\ &\quad + \epsilon_{xy}\sin\theta\cos 2\phi + \epsilon_{yz}\cos\theta\cos\phi - \epsilon_{zx}\cos\theta\sin\phi \end{aligned} \tag{4.89}$$

4.6.3 Generalized Hooke's Law in a Spherical Coordinate System

The constitutive equations for a homogeneous, isotropic body in a spherical coordinate system are

$$\epsilon_{rr} = \frac{1}{E}[\sigma_{rr} - \nu(\sigma_{\theta\theta} + \sigma_{\phi\phi})] + \alpha\tau = \frac{1}{2G}\left(\sigma_{rr} - \frac{\nu}{1+\nu}\Theta\right) + \alpha\tau$$

$$\epsilon_{\theta\theta} = \frac{1}{E}[\sigma_{\theta\theta} - \nu(\sigma_{\phi\phi} + \sigma_{rr})] + \alpha\tau = \frac{1}{2G}\left(\sigma_{\theta\theta} - \frac{\nu}{1+\nu}\Theta\right) + \alpha\tau$$

$$\epsilon_{\phi\phi} = \frac{1}{E}[\sigma_{\phi\phi} - \nu(\sigma_{rr} + \sigma_{\theta\theta})] + \alpha\tau = \frac{1}{2G}\left(\sigma_{\phi\phi} - \frac{\nu}{1+\nu}\Theta\right) + \alpha\tau \tag{4.90}$$

$$\epsilon_{r\theta} = \frac{\sigma_{r\theta}}{2G}, \qquad \epsilon_{\theta\phi} = \frac{\sigma_{\theta\phi}}{2G}, \qquad \epsilon_{\phi r} = \frac{\sigma_{\phi r}}{2G}$$

where $\Theta = \sigma_{rr} + \sigma_{\theta\theta} + \sigma_{\phi\phi}$. The alternative form is

$$\begin{aligned} \sigma_{rr} &= 2\mu\epsilon_{rr} + \lambda e - \beta\tau, & \sigma_{r\theta} &= 2\mu\epsilon_{r\theta} \\ \sigma_{\theta\theta} &= 2\mu\epsilon_{\theta\theta} + \lambda e - \beta\tau, & \sigma_{\theta\phi} &= 2\mu\epsilon_{\theta\phi} \\ \sigma_{\phi\phi} &= 2\mu\epsilon_{\phi\phi} + \lambda e - \beta\tau, & \sigma_{\phi r} &= 2\mu\epsilon_{\phi r} \end{aligned} \tag{4.91}$$

where $e = \epsilon_{rr} + \epsilon_{\theta\theta} + \epsilon_{\phi\phi}$.

4.6.4 Navier's Equations of Thermoelasticity in a Spherical Coordinate System

Navier's equations (4.23″) of thermoelasticity can be expressed in a spherical coordinate system as

$$\begin{aligned} (\lambda + 2\mu)\frac{\partial e}{\partial r} - \frac{2\mu}{r\sin\theta}\left[\frac{\partial(\omega_\phi \sin\theta)}{\partial\theta} - \frac{\partial\omega_\theta}{\partial\phi}\right] - \beta\frac{\partial\tau}{\partial r} + F_r &= 0 \\ (\lambda + 2\mu)\frac{1}{r}\frac{\partial e}{\partial\theta} - \frac{2\mu}{r\sin\theta}\left[\frac{\partial\omega_r}{\partial\phi} - \sin\theta\frac{\partial(r\omega_\phi)}{\partial r}\right] - \beta\frac{1}{r}\frac{\partial\tau}{\partial\theta} + F_\theta &= 0 \\ (\lambda + 2\mu)\frac{1}{r\sin\theta}\frac{\partial e}{\partial\phi} - \frac{2\mu}{r}\left[\frac{\partial(r\omega_\theta)}{\partial r} - \frac{\partial\omega_r}{\partial\theta}\right] - \beta\frac{1}{r\sin\theta}\frac{\partial\tau}{\partial\phi} + F_\phi &= 0 \end{aligned} \tag{4.92}$$

where

$$\omega_r = \frac{1}{2r\sin\theta}\left[\frac{\partial(u_\phi\sin\theta)}{\partial\theta} - \frac{\partial u_\theta}{\partial\phi}\right], \qquad \omega_\theta = \frac{1}{2r\sin\theta}\left[\frac{\partial u_r}{\partial\phi} - \sin\theta\frac{\partial(ru_\phi)}{\partial r}\right]$$

$$\omega_\phi = \frac{1}{2r}\left[\frac{\partial(ru_\theta)}{\partial r} - \frac{\partial u_r}{\partial\theta}\right] \tag{4.93}$$

The solution of Navier's equations (4.92) without the body force in the spherical coordinate system can be expressed, for example, by the Boussinesq harmonic functions:

$$\begin{aligned} u_r &= \frac{\partial\Phi}{\partial r} + \frac{\partial\varphi}{\partial r} + \frac{2}{r}\frac{\partial\vartheta}{\partial\phi} + r\cos\theta\frac{\partial\psi}{\partial r} - (3-4\nu)\psi\cos\theta \\ u_\theta &= \frac{1}{r}\frac{\partial\Phi}{\partial\theta} + \frac{1}{r}\frac{\partial\varphi}{\partial\theta} + \frac{2}{r\tan\theta}\frac{\partial\vartheta}{\partial\phi} + \cos\theta\frac{\partial\psi}{\partial\theta} + (3-4\nu)\psi\sin\theta \\ u_\phi &= \frac{1}{r\sin\theta}\frac{\partial\Phi}{\partial\phi} + \frac{1}{r\sin\theta}\frac{\partial\varphi}{\partial\phi} - 2\sin\theta\frac{\partial\vartheta}{\partial r} - 2\frac{\cos\theta}{r}\frac{\partial\vartheta}{\partial\theta} + \frac{1}{\tan\theta}\frac{\partial\psi}{\partial\phi} \end{aligned} \tag{4.94}$$

where the four functions must satisfy

$$\nabla^2 \Phi = K\tau, \qquad \nabla^2 \varphi = 0, \qquad \nabla^2 \vartheta = 0, \qquad \nabla^2 \psi = 0 \tag{4.95}$$

in which

$$\nabla^2 = \frac{\partial^2}{\partial r^2} + \frac{2}{r}\frac{\partial}{\partial r} + \frac{1}{r^2}\frac{\partial^2}{\partial \theta^2} + \frac{1}{r^2 \tan\theta}\frac{\partial}{\partial \theta} + \frac{1}{r^2 \sin^2\theta}\frac{\partial^2}{\partial \phi^2} \tag{4.96}$$

The general solutions of the Laplace equation in the spherical coordinate system are given by

$$\begin{gathered}
\begin{pmatrix} r^{\nu} \\ r^{-(\nu+1)} \end{pmatrix} \begin{pmatrix} P_\nu(\cos\theta) \\ Q_\nu(\cos\theta) \end{pmatrix} \begin{pmatrix} 1 \\ \phi \end{pmatrix} \\
\begin{pmatrix} r^{\nu} \\ r^{-(\nu+1)} \end{pmatrix} \begin{pmatrix} P_\nu^{\mu}(\cos\theta) \\ Q_\nu^{\mu}(\cos\theta) \end{pmatrix} \begin{pmatrix} \sin\mu\phi \\ \cos\mu\phi \end{pmatrix}
\end{gathered} \tag{4.97}$$

where μ and ν are constants, $P_\nu(\cos\theta)$ is the Legendre function of the first kind, $Q_\nu(\cos\theta)$ is the Legendre function of the second kind, $P_\nu^{\mu}(\cos\theta)$ is the associated Legendre function of the first kind, and $Q_\nu^{\mu}(\cos\theta)$ is the associated Legendre function of the second kind. For more information, see Appendix C.

4.7 Multiply Connected Bodies

4.7.1 Basic Equations for Multiply Connected Bodies

For the discussion which follows, we need to distinguish between two kinds of regions in space. A region with a property that every simple closed curve[‡] within it can be continuously reduced to a point without having to leave the region at any stage is called simply connected. Examples are the interior of a sphere, the exterior of a sphere, and the space between two concentric spheres. A region that does not have this property is called multiply connected. Examples are a torus and the space between two infinitely long coaxial cylinders.

The distinction between simply connected regions and multiply connected regions applies also for plane regions. An example of a plane simply connected region is the interior of a circle, and an example of a plane multiply connected region is the area between two concentric circles.

We consider a multiply connected body bounded by an exterior surface B_0 and interior surfaces B_i $(i = 1, 2, \ldots, m)$ which is called $(m+1)$-tuply connected body, shown in Fig. 4.7. Let the components of displacement $u_i^0 (= u_i(P_0))$ and the rotations $\omega_{ij}^0 (= \omega_{ij}(P_0))$ be known at a point $P_0 (= P(x_1^0, x_2^0, x_3^0))$ in the multiply connected body.

[‡]We define a simple closed curve, or simple closed contour, as a closed, sectionally smooth curve which does not cross itself.

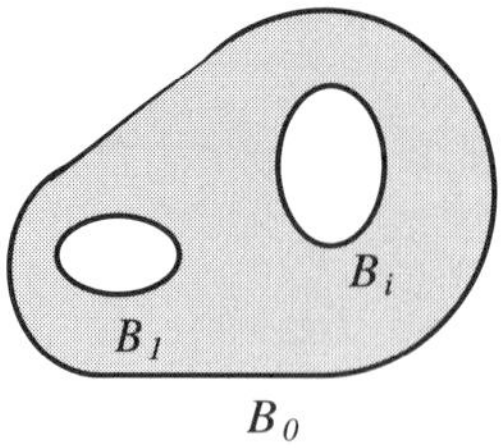

Figure 4.7 A multiply connected body.

The displacements $u_i(P)$ at any other point $P(x_1, x_2, x_3)$ may be expressed in terms of the strains ϵ_{ij} and the rotations ω_{ik}

$$u_i(P) = u_i^0 + \int_{P_0}^{P} du_i = u_i^0 + \int_{P_0}^{P} u_{i,k}\, d\xi_k = u_i^0 + \int_{P_0}^{P} \epsilon_{ik}\, d\xi_k + \int_{P_0}^{P} \omega_{ik}\, d\xi_k \quad \text{(a)}$$

by use of the relation

$$u_{i,k} = \tfrac{1}{2}(u_{i,k} + u_{k,i}) + \tfrac{1}{2}(u_{i,k} - u_{k,i}) = \epsilon_{ik} + \omega_{ik}$$

The last term can be integrated by parts

$$\int_{P_0}^{P} \omega_{ik}\, d\xi_k = \int_{P_0}^{P} \omega_{ik}\, d(\xi_k - x_k) = \left(x_k - x_k^0\right)\omega_{ik}^0 + \int_{P_0}^{P} (x_k - \xi_k)\omega_{ik,j}\, d\xi_j$$

Hence

$$u_i(P) = u_i^0 + \left(x_k - x_k^0\right)\omega_{ik}^0 + \int_{P_0}^{P} [\epsilon_{ij} + (x_k - \xi_k)\omega_{ik,j}]\, d\xi_j \qquad \text{(b)}$$

But the term $\omega_{ik,j}$ in the integrand reduces to

$$\omega_{ik,j} = \tfrac{1}{2}(u_{i,k} - u_{k,i})_{,j} = \epsilon_{ij,k} - \epsilon_{kj,i}$$

Hence

$$u_i(P) = u_i^0 + \left(x_k - x_k^0\right)\omega_{ik}^0 + \int_{P_0}^{P} \left[\epsilon_{ij} + (x_k - \xi_k)(\epsilon_{ij,k} - \epsilon_{kj,i})\right] d\xi_j \qquad (4.98)$$

where the integrand is a known function.

We shall cut the multiply connected region by a sufficient number of interior surfaces B_{ci} $(i = 1, 2, \ldots, m)$ in order to change it to a simply connected body shown in Fig. 4.8. Since the displacements $u_i(P)$ given by Eq. (4.98) must be independent of the path of integration in such a body, the integrand must be an exact differential. The necessary

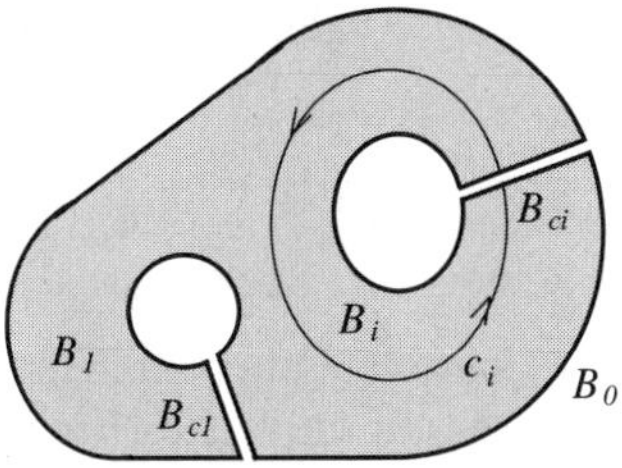

Figure 4.8 A simply connected body.

and sufficient condition for the integrand in Eq. (4.98) to be an exact differential for the simply connected body is

$$[\epsilon_{il} + (x_k - \xi_k)(\epsilon_{il,k} - \epsilon_{kl,i})]_{,j} = [\epsilon_{ij} + (x_k - \xi_k)(\epsilon_{ij,k} - \epsilon_{kj,i})]_{,l} \qquad \text{(c)}$$

Rearranging Eq. (c) gives

$$\epsilon_{il,j} - \delta_{kj}(\epsilon_{il,k} - \epsilon_{kl,i}) - \epsilon_{ij,l} + \delta_{kl}(\epsilon_{ij,k} - \epsilon_{kj,i}) + (x_k - \xi_k)(\epsilon_{il,kj} - \epsilon_{kl,ij} - \epsilon_{ij,kl} + \epsilon_{kj,il}) = 0$$

Thus, we have

$$(x_k - \xi_k)(\epsilon_{il,kj} - \epsilon_{kl,ij} - \epsilon_{ij,kl} + \epsilon_{kj,il}) = 0 \qquad \text{(d)}$$

Since Eq. (d) must be true for an arbitrary choice of $(x_k - \xi_k)$, we get

$$\epsilon_{ij,kl} + \epsilon_{kl,ij} - \epsilon_{jk,li} - \epsilon_{li,jk} = 0 \qquad (i, j, k, l = 1, 2, 3) \qquad (4.99)$$

which is the compatibility equation, see Eq. (4.12′).

Next, we consider the multiply connected region in Fig. 4.7. The rotations $\omega_{ij}(P)$ at any other point $P(x_1, x_2, x_3)$ can be expressed in terms of the strains ϵ_{ij} and the rotations ω_{ik}

$$\omega_{ij}(P) = \omega_{ij}^0 + \int_{P_0}^{P} \omega_{ij,k}\, d\xi_k = \omega_{ij}^0 + \int_{P_0}^{P} (\epsilon_{ik,j} - \epsilon_{kj,i})\, d\xi_k \qquad (4.100)$$

If the rotations and displacements are single-valued for the multiply connected region, the rotations and the displacements at the arriving point of point P which coincides with the starting point must have the same values as the starting point P_0. The integral path may be arbitrarily selected. If we select any closed integral path C_l $(l = 1, 2, \ldots, m)$ to surround only one inner boundary B_l $(l = 1, 2, \ldots, m)$, then the single-valuedness of the rotations and the displacements requires that

$$\oint_{C_l} (\epsilon_{ik,j} - \epsilon_{jk,i})\, d\xi_k = 0 \qquad (l = 1, 2, \ldots, m) \qquad (4.101)$$

$$\oint_{C_l} [\epsilon_{ik} + (x_j^0 - \xi_j)(\epsilon_{ik,j} - \epsilon_{jk,i})]\, d\xi_k = 0 \qquad (4.102)$$

By use of the single-valuedness of the rotations (4.101), the single-valuedness of the displacements (4.102) reduces to

$$\oint_{C_l} [\epsilon_{ik} - \xi_j(\epsilon_{ik,j} - \epsilon_{jk,i})]\, d\xi_k = 0 \tag{4.103}$$

Therefore, Eqs. (4.101) and (4.103) must be satisfied for any closed integral path in the multiply connected region in order to insure the single-valuedness of the rotations and the displacements.

The basic equations of the linear thermoelastic problem in the $(m + 1)$-tuply connected bodies are summarized as follows:

Equilibrium equation:

$$\sigma_{ji,j} + F_i = 0 \tag{4.104}$$

Boundary conditions:

$$\sigma_{ji} n_j = p_{ni} \tag{4.105}$$

Compatibility equation:

$$\epsilon_{ij,kl} + \epsilon_{kl,ij} - \epsilon_{jk,li} - \epsilon_{il,jk} = 0 \tag{4.106}$$

Single-valuedness of rotations:

$$\oint_{C_l} (\epsilon_{ik,j} - \epsilon_{jk,i})\, d\xi_k = 0 \qquad (l = 1, 2, \ldots, m) \tag{4.107}$$

Single-valuedness of displacements:

$$\oint_{C_l} [\epsilon_{ik} - \xi_j(\epsilon_{ik,j} - \epsilon_{jk,i})]\, d\xi_k = 0 \qquad (l = 1, 2, \ldots, m) \tag{4.108}$$

It is seen that the basic equation system of the linear thermoelastic problem for the multiply connected body consists of Eqs. (4.104) to (4.108), while the basic equation system for the simply connected body consists of Eqs. (4.104) to (4.106).

4.7.2 Temperature Distributions with Zero Thermal Stress

We consider temperature distributions with zero thermal stress in a simply or a multiply connected body which is free of surface tractions and body forces. The components of stress are zero by hypothesis, so that the generalized Hooke's law (4.15′) becomes

$$\epsilon_{ij} = \alpha\tau\delta_{ij} \tag{a}$$

Substitution of Eq. (a) into the compatibility equations (4.12) yields

$$\tau_{,ij} = 0 \tag{b}$$

From Eq. (b) it follows that the temperature change must be linear

$$\tau = A_0 + A_1 x + A_2 y + A_3 z \tag{4.109}$$

where A_i $(i = 0, 1, 2, 3)$ are arbitrary constants.

Next, we show that the single-valuedness of rotations (4.107) and the single-valuedness of displacements (4.108) are satisfied when the stress is zero and the temperature change is expressed by a linear function of a position (4.109). The single-valuedness of rotations (4.107) is clearly satisfied by substitution of $\sigma_{ij} = 0$ and $\tau = A_0 + A_1 x + A_2 y + A_3 z$ into Eq. (4.107)

$$\begin{aligned}\oint_{C_l} (\epsilon_{ik,j} - \epsilon_{jk,i})\, d\xi_k &= \alpha \oint_{C_l} (\tau_{,j}\, d\xi_i - \tau_{,i}\, d\xi_j) \\ &= \alpha \left(A_j \oint_{C_l} d\xi_i - A_i \oint_{C_l} d\xi_j \right) = 0\end{aligned} \tag{4.110}$$

By use of the linear distribution of temperature, the single-valuedness of displacements (4.108) becomes

$$\begin{aligned}&\oint_{C_l} [\epsilon_{ik} - \xi_j(\epsilon_{ik,j} - \epsilon_{jk,i})]\, d\xi_k \\ &\quad = \alpha \oint_{C_l} [\delta_{ik}\tau - \xi_j(\delta_{ik}\tau_{,j} - \delta_{jk}\tau_{,i})]\, d\xi_k \\ &\quad = \alpha \left(\oint_{C_l} \tau\, d\xi_i - \oint_{C_l} \xi_j \tau_{,j}\, d\xi_i + \oint_{C_l} \xi_j \tau_{,i}\, d\xi_j \right) \\ &\quad = \alpha \left(\oint_{C_l} (A_0 + A_j \xi_j)\, d\xi_i - A_j \oint_{C_l} \xi_j\, d\xi_i + A_i \oint_{C_l} \xi_j\, d\xi_j \right) \\ &\quad = \alpha A_0 \oint_{C_l} d\xi_i + \alpha A_i \oint_{C_l} \xi_j\, d\xi_j\end{aligned}$$

The last two integrals are obviously zero. Therefore,

$$\oint_{C_l} [\epsilon_{ik} - \xi_j(\epsilon_{ik,j} - \epsilon_{jk,i})]\, d\xi_k = 0 \tag{4.111}$$

The single-valuedness of displacements (4.108) is then satisfied. If a body is free of stress and has displacement components at least three times differentiable throughout, the temperature change in the body must be linear.

Problems

4.1 When a continuous body moves under mechanical and thermal loads, prove the equations of motion are

$$\sigma_{ji,j} + F_i = \rho \ddot{u}_i \qquad (i = 1, 2, 3) \tag{4.1'''}$$

where u_i denotes the components of displacement, ρ is the density and a dot denotes partial differentiation with respect to time.

4.2 Derive the compatibility equations (4.12′).

4.3 Show that the strain components transform by the following equations:

$$\begin{aligned}
\epsilon_{x'x'} &= \epsilon_{xx} l_1^2 + \epsilon_{yy} m_1^2 + \epsilon_{zz} n_1^2 + 2(\epsilon_{xy} l_1 m_1 + \epsilon_{yz} m_1 n_1 + \epsilon_{zx} n_1 l_1)\\
\epsilon_{y'y'} &= \epsilon_{xx} l_2^2 + \epsilon_{yy} m_2^2 + \epsilon_{zz} n_2^2 + 2(\epsilon_{xy} l_2 m_2 + \epsilon_{yz} m_2 n_2 + \epsilon_{zx} n_2 l_2)\\
\epsilon_{z'z'} &= \epsilon_{xx} l_3^2 + \epsilon_{yy} m_3^2 + \epsilon_{zz} n_3^2 + 2(\epsilon_{xy} l_3 m_3 + \epsilon_{yz} m_3 n_3 + \epsilon_{zx} n_3 l_3)\\
\epsilon_{x'y'} &= \epsilon_{xx} l_1 l_2 + \epsilon_{yy} m_1 m_2 + \epsilon_{zz} n_1 n_2 + \epsilon_{xy}(l_1 m_2 + l_2 m_1)\\
&\quad + \epsilon_{yz}(m_1 n_2 + m_2 n_1) + \epsilon_{zx}(n_1 l_2 + n_2 l_1)\\
\epsilon_{y'z'} &= \epsilon_{xx} l_2 l_3 + \epsilon_{yy} m_2 m_3 + \epsilon_{zz} n_2 n_3 + \epsilon_{xy}(l_2 m_3 + l_3 m_2)\\
&\quad + \epsilon_{yz}(m_2 n_3 + m_3 n_2) + \epsilon_{zx}(n_2 l_3 + n_3 l_2)\\
\epsilon_{z'x'} &= \epsilon_{xx} l_3 l_1 + \epsilon_{yy} m_3 m_1 + \epsilon_{zz} n_3 n_1 + \epsilon_{xy}(l_3 m_1 + l_1 m_3)\\
&\quad + \epsilon_{yz}(m_3 n_1 + m_1 n_3) + \epsilon_{zx}(n_3 l_1 + n_1 l_3)
\end{aligned} \tag{4.14'}$$

4.4 Prove Navier's equations of thermoelasticity with motion of the body to be given by

$$\begin{aligned}
\mu \nabla^2 u_i + (\lambda + \mu) u_{k,ki} - \beta \tau_{,i} + F_i &= \rho \ddot{u}_i \qquad (i = 1, 2, 3)\\
(\lambda + 2\mu) u_{k,ki} - 2\mu \gamma_{ijk} \omega_{k,j} - \beta \tau_{,i} + F_i &= \rho \ddot{u}_i \qquad (i = 1, 2, 3)\\
\mu \nabla^2 \boldsymbol{u} + (\lambda + \mu)\, \text{grad div}\, \boldsymbol{u} - \beta\, \text{grad}\, \tau + \boldsymbol{F} &= \rho\, \ddot{\boldsymbol{u}}\\
(\lambda + 2\mu)\, \text{grad div}\, \boldsymbol{u} - \mu\, \text{curl curl}\, \boldsymbol{u} - \beta\, \text{grad}\, \tau + \boldsymbol{F} &= \rho\, \ddot{\boldsymbol{u}}\\
(\lambda + 2\mu) \nabla^2 \boldsymbol{u} + (\lambda + \mu)\, \text{curl curl}\, \boldsymbol{u} - \beta\, \text{grad}\, \tau + \boldsymbol{F} &= \rho\, \ddot{\boldsymbol{u}}
\end{aligned} \tag{4.23''''}$$

4.5 Derive Eqs. (4.26′).

4.6 Derive the equations of equilibrium (4.69) in a cylindrical coordinate system.

4.7 Prove that the general solutions of Laplace equation in a cylindrical coordinate system are given by the Eqs. (4.82).

4.8 Derive the equilibrium Eqs. (4.83) in a spherical coordinate system.

4.9 Prove that the general solutions of the Laplace equation in the spherical coordinate system are given by Eqs. (4.97).

Chapter 5

PLANE THERMOELASTIC PROBLEMS

5.1 Plane Strain and Plane Stress

The basic treatment of plane thermoelastic problems is presented in detail. The special cases of plane strain and plane stress are introduced along with derivations of the associated governing equations. Three techniques are introduced: (a) the thermal stress function method, which is the analogue of the Airy stress function method for isothermal problems; (b) the complex variable method; and (c) the potential method. The thermal stress function technique is a powerful tool for plane problems. The stresses are defined by partial differentiation of the thermal stress function, whereas the displacement components are expressed in terms of the thermal stress function and an additional plane function. The complex representation for both simply-connected and multiply-connected bodies is discussed. Moreover, the method of conformal mapping is presented in order to treat bodies with complex boundary contours. Lastly, the plane counterpart of the 3D Navier's solution is considered.

5.1.1 Introduction

This chapter treats the solutions of two physically special classes of problems known as plane thermoelastic problems, that is, plane strain and plane stress. A plane strain problem arises in a long body under mechanical and thermal loads distributed uniformly along the longitudinal direction and acting perpendicular to that direction. On the other hand, a plane stress problem appears in a plane plate under mechanical and thermal loads that are uniformly distributed over the thickness and are perpendicular to the cross section of the plate. Plane problems constitute an important part of thermoelasticity due to the large number of applications and the simplicity of the mathematical treatment.

5.1.2 Plane Strain

Consider a long body, shown in Fig. 5.1. Let mechanical and thermal loads be uniformly distributed along the longitudinal axis of the long body, and let them act perpendicular to that axis. Then, the long body, with possible exception of end regions, undergoes a displacement consisting of three components: one component is along the axis of the long body, which we take as z axis, being a linear function of z; and two other components are functions of x and y only. Thus, the deformation in the long body is

$$u_x = u_x(x, y), \qquad u_y = u_y(x, y), \qquad u_z = \epsilon_0 z + \epsilon_1 \tag{5.1}$$

These displacements lead to the following strain components from the definitions (4.10) of strain

$$\begin{aligned} &\epsilon_{xx} = \epsilon_{xx}(x, y), \qquad \epsilon_{yy} = \epsilon_{yy}(x, y), \qquad \epsilon_{zz} = \epsilon_0 \text{ (constant)} \\ &\epsilon_{xy} = \epsilon_{xy}(x, y), \qquad \epsilon_{yz} = 0, \qquad \epsilon_{xz} = 0 \end{aligned} \tag{5.2}$$

The body in which strains are defined by Eqs. (5.2) is said to be in a state of plane strain. Since the deformation of the long body in a state of plane strain is independent of the variable z, it is sufficient to consider only the deformation of any one cross section of the long body perpendicular to the z axis. Using the generalized Hooke's law (4.18), the components of stress are given by

$$\begin{aligned} \sigma_{xx} &= (\lambda + 2\mu)\epsilon_{xx} + \lambda(\epsilon_{yy} + \epsilon_0) - \beta\tau \\ \sigma_{yy} &= (\lambda + 2\mu)\epsilon_{yy} + \lambda(\epsilon_{xx} + \epsilon_0) - \beta\tau \\ \sigma_{xy} &= 2\mu\epsilon_{xy} \\ \sigma_{zz} &= (\lambda + 2\mu)\epsilon_0 + \lambda(\epsilon_{xx} + \epsilon_{yy}) - \beta\tau \\ \sigma_{xz} &= \sigma_{yz} = 0 \end{aligned} \tag{5.3}$$

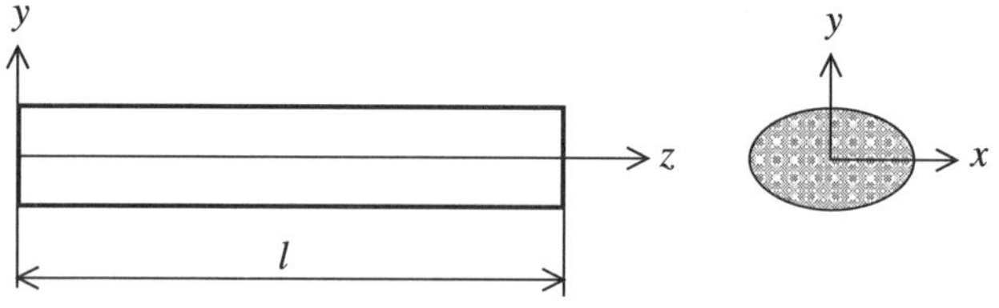

Figure 5.1 A long body for plane strain analysis.

or

$$
\begin{aligned}
\epsilon_{xx} &= \frac{1}{E'}(\sigma_{xx} - \nu'\sigma_{yy}) + \alpha'\tau - \nu\epsilon_0 \\
\epsilon_{yy} &= \frac{1}{E'}(\sigma_{yy} - \nu'\sigma_{xx}) + \alpha'\tau - \nu\epsilon_0 \\
\epsilon_{xy} &= \frac{1}{2G}\sigma_{xy} \\
\epsilon_{zz} &= \frac{1}{E}[\sigma_{zz} - \nu(\sigma_{xx} + \sigma_{yy})] + \alpha\tau = \epsilon_0 \\
\epsilon_{xz} &= \epsilon_{yz} = 0
\end{aligned}
\tag{5.3'}
$$

where

$$
E' = \frac{E}{1-\nu^2}, \qquad \nu' = \frac{\nu}{1-\nu}, \qquad \alpha' = (1+\nu)\alpha \tag{5.4}
$$

The normal stress σ_{zz} acts on the cross sections of the long body, and is necessary to maintain the long body in the state of plane strain. The stress distribution in a state of plane strain is not two-dimensional but three-dimensional. The normal stress σ_{zz} is dependent on the normal stresses σ_{xx} and σ_{yy}. From the fourth equation in Eqs. (5.3′) we have

$$
\sigma_{zz} = \nu(\sigma_{xx} + \sigma_{yy}) - \alpha E\tau + E\epsilon_0 \tag{5.5}
$$

Substituting Eqs. (5.3) into the third equation of equilibrium, Eq. (4.1), and neglecting the body force yields

$$
\frac{\partial \tau}{\partial z} = 0
$$

The temperature change τ must be independent of the z variable,

$$
\tau = \tau(x, y) \tag{5.6}
$$

Thus, the stresses, strains, and temperature change are expressed in terms of the variables x and y.

The equilibrium equations in the absence of body forces reduce to

$$
\frac{\partial \sigma_{xx}}{\partial x} + \frac{\partial \sigma_{yx}}{\partial y} = 0, \qquad \frac{\partial \sigma_{xy}}{\partial x} + \frac{\partial \sigma_{yy}}{\partial y} = 0 \tag{5.7}
$$

The compatibility equations (4.12) reduce to

$$
\frac{\partial^2 \epsilon_{xx}}{\partial y^2} + \frac{\partial^2 \epsilon_{yy}}{\partial x^2} = 2\frac{\partial^2 \epsilon_{xy}}{\partial x \partial y} \tag{5.8}
$$

The substitution of Eqs. (5.3) into Eqs. (5.7) gives Navier's equations for plane strain,

$$\mu\nabla^2 u_x + (\lambda+\mu)\frac{\partial e}{\partial x} - \beta\frac{\partial \tau}{\partial x} = 0, \qquad \mu\nabla^2 u_y + (\lambda+\mu)\frac{\partial e}{\partial y} - \beta\frac{\partial \tau}{\partial y} = 0 \tag{5.9}$$

where

$$\nabla^2 = \frac{\partial^2}{\partial x^2} + \frac{\partial^2}{\partial y^2}, \qquad e = \epsilon_{xx} + \epsilon_{yy} + \epsilon_0$$

The boundary conditions in a state of plane strain are specified along the contour of the long body:

$$\sigma_{xx}l + \sigma_{yx}m = p_{nx}, \qquad \sigma_{xy}l + \sigma_{yy}m = p_{ny} \tag{5.10}$$

The distributions of stress on the end planes of the long body are given in the form:

$$\sigma_{zz} = \nu(\sigma_{xx} + \sigma_{yy}) - \alpha E\tau + E\epsilon_0, \qquad \sigma_{xz} = \sigma_{yz} = 0 \tag{5.11}$$

The normal stress $\sigma_{zz} = \nu(\sigma_{xx} + \sigma_{yy}) - \alpha E\tau + E\epsilon_0$ is necessary to maintain the long body in a state of the plane strain. However, there are many practical problems in which the end planes are free of stress.

Let us consider the case in which the end planes of the long body are free of stress. The desired solution can be obtained by superposing an additional solution onto the solution of the plane problem. The additional solution (σ_{ij}^*) is that of the isothermal problem, and its boundary conditions for the prism must be in the form

$$\left.\begin{aligned} &\sigma_{zz}^* = -\sigma_{zz} = -\nu(\sigma_{xx} + \sigma_{yy}) + \alpha E\tau - E\epsilon_0 \\ &\sigma_{xz}^* = \sigma_{yz}^* = 0 \end{aligned}\right\} \quad \text{on the end planes}$$

$$\sigma_{xx}^* l + \sigma_{yx}^* m = 0, \quad \sigma_{xy}^* l + \sigma_{yy}^* m = 0 \qquad \text{on the lateral surfaces} \tag{5.12}$$

The additional solution represents a three-dimensional problem so that it is difficult to obtain it in exact form. If the length of the long body is sufficiently large compared to the cross sectional dimensions, an approximate solution with sufficient accuracy can be obtained with bending theory. The longitudinal force P and bending moments M_x, M_y produced by the normal stress $\sigma_{zz} = \nu(\sigma_{xx} + \sigma_{yy}) - \alpha E\tau + E\epsilon_0$ on the ends are calculated as

$$\begin{aligned} P &= \iint_A \sigma_{zz}\,dA = \iint_A [\nu(\sigma_{xx} + \sigma_{yy}) - \alpha E\tau + E\epsilon_0]\,dx\,dy \\ M_x &= \iint_A \sigma_{zz} y\,dA = \iint_A [\nu(\sigma_{xx} + \sigma_{yy}) - \alpha E\tau + E\epsilon_0] y\,dx\,dy \\ M_y &= \iint_A \sigma_{zz} x\,dA = \iint_A [\nu(\sigma_{xx} + \sigma_{yy}) - \alpha E\tau + E\epsilon_0] x\,dx\,dy \end{aligned} \tag{a}$$

where M_x and M_y denote the moments with respect to the x and y axes, respectively. If the constant normal strain ϵ_0 is selected as

$$\epsilon_0 = \frac{1}{EA} \iint_A [-\nu(\sigma_{xx} + \sigma_{yy}) + \alpha E \tau]\, dx\, dy \tag{5.13}$$

the longitudinal force P becomes zero on the end surfaces. When the origin of the coordinate system is taken at the centroid of the cross section of the prism, the last integrals on the right-hand side in the second and third of Eqs. (a) $\iint_A E\epsilon_0 y\, dA$, $\iint_A E\epsilon_0 x\, dA$ reduce to zero because of the constant normal strain ϵ_0

$$\iint_A E\epsilon_0 x\, dA = 0, \qquad \iint_A E\epsilon_0 y\, dA = 0 \tag{b}$$

The boundary condition of the additional problem reduces to

$$\begin{aligned} \overline{M}_x &= -M_x = -\iint_A [\nu(\sigma_{xx} + \sigma_{yy}) - \alpha E \tau] y\, dx\, dy \\ \overline{M}_y &= -M_y = -\iint_A [\nu(\sigma_{xx} + \sigma_{yy}) - \alpha E \tau] x\, dx\, dy \end{aligned} \tag{c}$$

By use of the results in Section 2.5, the normal stress σ_{zz}^* of the additional problem produced in the long body due to the bending moments $\overline{M}_x, \overline{M}_y$ is expressed by

$$\sigma_{zz}^* = \left[\frac{I_x \overline{M}_y - I_{xy} \overline{M}_x}{I_x I_y - I_{xy}^2} x \right] + \left[\frac{I_y \overline{M}_x - I_{xy} \overline{M}_y}{I_x I_y - I_{xy}^2} y \right] \tag{5.14}$$

After superposing the solution (5.14) on the solution of the plane strain problem, the solution in the prism with stress free boundaries can be obtained. However, the resulting deformation is in general no longer two-dimensional.

5.1.3 Plane Stress

When a thin plate is subjected to mechanical and thermal loads that are independent of the thickness direction and are parallel to the mid-plane of the plate, the stress distribution in the plate may be approximately described by a state of plane stress. A state of plane stress is said to exist when the components of stress are characterized by

$$\begin{gathered} \sigma_{xx} = \sigma_{xx}(x, y), \qquad \sigma_{yy} = \sigma_{yy}(x, y), \qquad \sigma_{xy} = \sigma_{xy}(x, y) \\ \sigma_{zz} = \sigma_{zx} = \sigma_{zy} = 0 \end{gathered} \tag{5.15}$$

Using the generalized Hooke's law (4.15), the components of strain associated with plane stress are given by

$$\begin{aligned}
\epsilon_{xx} &= \frac{1}{E}(\sigma_{xx} - \nu\sigma_{yy}) + \alpha\tau \\
\epsilon_{yy} &= \frac{1}{E}(\sigma_{yy} - \nu\sigma_{xx}) + \alpha\tau \\
\epsilon_{xy} &= \frac{1}{2G}\sigma_{xy} \\
\epsilon_{zz} &= -\frac{\nu}{E}(\sigma_{xx} + \sigma_{yy}) + \alpha\tau \\
\epsilon_{zx} &= \epsilon_{yz} = 0
\end{aligned} \tag{5.16}$$

or

$$\begin{aligned}
\sigma_{xx} &= (\lambda' + 2\mu)\epsilon_{xx} + \lambda'\epsilon_{yy} - \beta'\tau \\
\sigma_{yy} &= (\lambda' + 2\mu)\epsilon_{yy} + \lambda'\epsilon_{xx} - \beta'\tau \\
\sigma_{xy} &= 2\mu\epsilon_{xy} \\
\sigma_{zz} &= (\lambda + 2\mu)\epsilon_{zz} + \lambda(\epsilon_{xx} + \epsilon_{yy}) - \beta\tau = 0 \\
\sigma_{zx} &= \sigma_{yz} = 0
\end{aligned} \tag{5.16$'$}$$

where

$$\lambda' = \frac{2\mu\lambda}{\lambda + 2\mu}, \qquad \beta' = \frac{2\mu\beta}{\lambda + 2\mu}$$

It should be noted that the normal strain ϵ_{zz} is not zero in general.

The equilibrium equations in the absence of body forces reduce to

$$\frac{\partial \sigma_{xx}}{\partial x} + \frac{\partial \sigma_{yx}}{\partial y} = 0, \qquad \frac{\partial \sigma_{xy}}{\partial x} + \frac{\partial \sigma_{yy}}{\partial y} = 0 \tag{5.17}$$

The compatibility equations (4.12) reduce to

$$\frac{\partial^2 \epsilon_{xx}}{\partial y^2} + \frac{\partial^2 \epsilon_{yy}}{\partial x^2} = 2\frac{\partial^2 \epsilon_{xy}}{\partial x \partial y} \tag{5.18}$$

Substitution of Eqs. (5.16′) into Eqs. (5.17) gives Navier's equations for the plane stress

$$\begin{aligned}
\mu\nabla^2 u_x + (\lambda' + \mu)\frac{\partial e}{\partial x} - \beta'\frac{\partial \tau}{\partial x} &= 0, \\
\mu\nabla^2 u_y + (\lambda' + \mu)\frac{\partial e}{\partial y} - \beta'\frac{\partial \tau}{\partial y} &= 0
\end{aligned} \tag{5.19}$$

where

$$e = \epsilon_{xx} + \epsilon_{yy}$$

The boundary conditions for a state of plane stress are specified in the form

$$\sigma_{xx}l + \sigma_{yx}m = p_{nx}, \qquad \sigma_{xy}l + \sigma_{yy}m = p_{ny} \tag{5.20}$$

Let us assume that the thin plate, shown in Fig. 5.2, has thickness $2h$ which is small compared with other dimensions, the surfaces $z = \pm h$ are free of stress, and the plane $z = 0$ is the middle plane of the plate. Moreover, we shall assume that the temperature change is symmetric with respect to the middle plane and the variation of the temperature change along the z axis is small. We consider the third equation of the equilibrium equations (4.1)

$$\frac{\partial \sigma_{xz}}{\partial x} + \frac{\partial \sigma_{yz}}{\partial y} + \frac{\partial \sigma_{zz}}{\partial z} = 0 \tag{a}$$

Since the stresses on the surfaces are zero

$$\sigma_{zx} = 0, \quad \sigma_{zy} = 0, \quad \sigma_{zz} = 0 \qquad \text{at } z = \pm h \tag{b}$$

Equation (a) reduces to

$$\frac{\partial \sigma_{zz}}{\partial z} = 0 \qquad \text{at } z = \pm h \tag{c}$$

The normal stress σ_{zz} and its derivative with respect to z are zero on the surfaces $z = \pm h$. This fact suggests that the variation of the normal stress σ_{zz} along the z axis is small if the thickness of the plate is small. Therefore, we can assume that the average value of

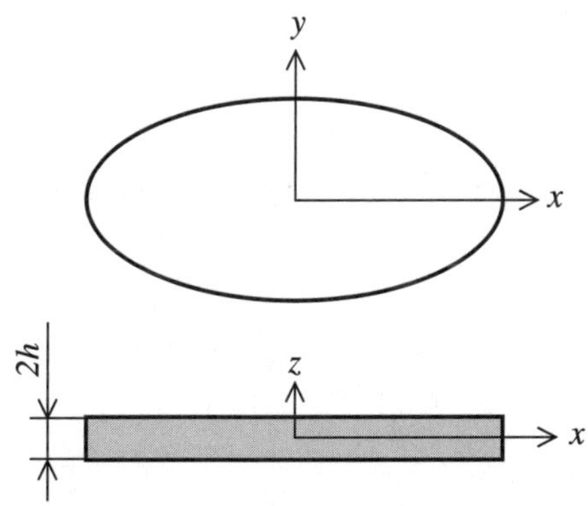

Figure 5.2 A body (thin plate) for plane stress analysis.

the normal stress may be neglected:

$$\bar{\sigma}_{zz} = \frac{1}{2h}\int_{-h}^{h} \sigma_{zz}(x, y, z)\, dz \approx 0 \tag{5.21}$$

where the upper bar indicates the average value:

$$\bar{\sigma}_{\alpha\beta}(x, y) = \frac{1}{2h}\int_{-h}^{h} \sigma_{\alpha\beta}(x, y, z)\, dz \qquad (\alpha, \beta = x, y) \tag{d}$$

The average values of the temperature change, displacement, and stress are independent of the z variable. These physical quantities may satisfy the same governing equations of plane stress. Let us integrate the first and second equilibrium equations (4.1) with respect to the z variable from $-h$ to h in order to prove the matter. Thus,

$$\begin{aligned} \frac{1}{2h}\int_{-h}^{h}\left(\frac{\partial\sigma_{xx}}{\partial x} + \frac{\partial\sigma_{yx}}{\partial y} + \frac{\partial\sigma_{zx}}{\partial z}\right) dz = 0 \\ \frac{1}{2h}\int_{-h}^{h}\left(\frac{\partial\sigma_{xy}}{\partial x} + \frac{\partial\sigma_{yy}}{\partial y} + \frac{\partial\sigma_{zy}}{\partial z}\right) dz = 0 \end{aligned} \tag{e}$$

The third term in the integrands in Eqs. (e) is zero:

$$\int_{-h}^{h}\left(\frac{\partial\sigma_{zx}}{\partial z}, \frac{\partial\sigma_{zy}}{\partial z}\right) dz = [\sigma_{zx}(x, y, z), \sigma_{zy}(x, y, z)]_{-h}^{h} = 0 \tag{f}$$

so that Eqs. (e) give the equilibrium equations for the average stresses

$$\frac{\partial\bar{\sigma}_{xx}}{\partial x} + \frac{\partial\bar{\sigma}_{yx}}{\partial y} = 0, \qquad \frac{\partial\bar{\sigma}_{xy}}{\partial x} + \frac{\partial\bar{\sigma}_{yy}}{\partial y} = 0 \tag{5.22}$$

The stress–strain relations for the average values of the temperature change, displacement, and stress become

$$\begin{aligned} \bar{\epsilon}_{xx} &= \frac{1}{E}(\bar{\sigma}_{xx} - \nu\bar{\sigma}_{yy}) + \alpha\bar{\tau} \\ \bar{\epsilon}_{yy} &= \frac{1}{E}(\bar{\sigma}_{yy} - \nu\bar{\sigma}_{xx}) + \alpha\bar{\tau} \\ \bar{\epsilon}_{xy} &= \frac{1}{2G}\bar{\sigma}_{xy} \end{aligned} \tag{5.23}$$

or

$$\begin{aligned} \bar{\sigma}_{xx} &= (\lambda' + 2\mu)\bar{\epsilon}_{xx} + \lambda'\bar{\epsilon}_{yy} - \beta'\bar{\tau} \\ \bar{\sigma}_{yy} &= (\lambda' + 2\mu)\bar{\epsilon}_{yy} + \lambda'\bar{\epsilon}_{xx} - \beta'\bar{\tau} \\ \bar{\sigma}_{xy} &= 2\mu\bar{\epsilon}_{xy} \end{aligned} \tag{5.23$'$}$$

The compatibility equation is given by

$$\frac{\partial^2 \bar{\epsilon}_{xx}}{\partial y^2} + \frac{\partial^2 \bar{\epsilon}_{yy}}{\partial x^2} = 2\frac{\partial^2 \bar{\epsilon}_{xy}}{\partial x \partial y} \tag{5.24}$$

The Navier's equations are

$$\begin{aligned} \mu\nabla^2 \bar{u}_x + (\lambda' + \mu)\frac{\partial \bar{e}}{\partial x} - \beta'\frac{\partial \bar{\tau}}{\partial x} &= 0, \\ \mu\nabla^2 \bar{u}_y + (\lambda' + \mu)\frac{\partial \bar{e}}{\partial y} - \beta'\frac{\partial \bar{\tau}}{\partial y} &= 0 \end{aligned} \tag{5.25}$$

where $\bar{e} = \bar{\epsilon}_{xx} + \bar{\epsilon}_{yy}$. The problem governed by the system of Eqs. (5.22) to (5.25) is called generalized plane stress*. Generalized plane stress gives useful information on the deformation of the thin plate.

5.1.4 Governing Equations of Plane Problems

When the governing equations of both plane strain with constant axial strain and plane stress are carefully examined, it is noticed that both governing equations are formally the same. If we takc

$$\begin{aligned}
E^* &= \begin{cases} E' = \dfrac{E}{1-\nu^2} & \text{for plane strain} \\ E & \text{for plane stress} \end{cases} \\
\nu^* &= \begin{cases} \nu' = \dfrac{\nu}{1-\nu} & \text{for plane strain} \\ \nu & \text{for plane stress} \end{cases} \\
\alpha^* &= \begin{cases} \alpha' = (1+\nu)\alpha & \text{for plane strain} \\ \alpha & \text{for plane stress} \end{cases} \\
\lambda^* &= \begin{cases} \lambda & \text{for plane strain} \\ \lambda' = \dfrac{2\mu\lambda}{\lambda + 2\mu} & \text{for plane stress} \end{cases} \\
\beta^* &= \begin{cases} \beta & \text{for plane strain} \\ \beta' = \dfrac{2\mu\beta}{\lambda + 2\mu} & \text{for plane stress} \end{cases} \\
c^* &= \begin{cases} \nu\epsilon_0 & \text{for plane strain} \\ 0 & \text{for plane stress} \end{cases}
\end{aligned} \tag{5.26}$$

*See the detailed discussion in B. A. Boley and J. H. Weiner, Theory of Thermal Stresses, John Wiley & Sons, New York (1960), p. 116.

the unified systems of the governing equations for both cases are as follows:

The generalized Hooke's law is

$$\begin{aligned}\epsilon_{xx} &= \frac{1}{E^*}\left(\sigma_{xx} - \nu^*\sigma_{yy}\right) + \alpha^*\tau - c^* \\ \epsilon_{yy} &= \frac{1}{E^*}\left(\sigma_{yy} - \nu^*\sigma_{xx}\right) + \alpha^*\tau - c^* \\ \epsilon_{xy} &= \frac{1}{2G}\sigma_{xy}\end{aligned} \tag{5.27}$$

or

$$\begin{aligned}\sigma_{xx} &= (\lambda^* + 2\mu)\epsilon_{xx} + \lambda^*\epsilon_{yy} - \beta^*\tau \\ \sigma_{yy} &= (\lambda^* + 2\mu)\epsilon_{yy} + \lambda^*\epsilon_{xx} - \beta^*\tau \\ \sigma_{xy} &= 2\mu\epsilon_{xy}\end{aligned} \tag{5.27$'$}$$

The equilibrium equations in the absence of body forces are

$$\frac{\partial \sigma_{xx}}{\partial x} + \frac{\partial \sigma_{yx}}{\partial y} = 0, \qquad \frac{\partial \sigma_{xy}}{\partial x} + \frac{\partial \sigma_{yy}}{\partial y} = 0 \tag{5.28}$$

The compatibility equation is

$$\frac{\partial^2 \epsilon_{xx}}{\partial y^2} + \frac{\partial^2 \epsilon_{yy}}{\partial x^2} = 2\frac{\partial^2 \epsilon_{xy}}{\partial x \partial y} \tag{5.29}$$

The Navier's equations are

$$\begin{aligned}\mu\nabla^2 u_x + (\lambda^* + \mu)\frac{\partial e}{\partial x} - \beta^*\frac{\partial \tau}{\partial x} &= 0 \\ \mu\nabla^2 u_y + (\lambda^* + \mu)\frac{\partial e}{\partial y} - \beta^*\frac{\partial \tau}{\partial y} &= 0\end{aligned} \tag{5.30}$$

where $e = \epsilon_{xx} + \epsilon_{yy} + c^*$. The boundary conditions are

$$\sigma_{xx}l + \sigma_{yx}m = p_{nx}, \qquad \sigma_{xy}l + \sigma_{yy}m = p_{ny} \tag{5.31}$$

It is, therefore, found that both plane strain and plane stress problems must satisfy the unified system of governing equations (5.27) to (5.31).

Example 5.1 Express the compatibility equation (5.29) by the stress components.

Solution Substitution of Eqs. (5.27) into Eqs. (5.28) gives

$$\frac{1}{E^*}\frac{\partial^2\sigma_{xx}}{\partial y^2} - \frac{\nu^*}{E^*}\frac{\partial^2\sigma_{yy}}{\partial y^2} + \alpha^*\frac{\partial^2\tau}{\partial y^2} + \frac{1}{E^*}\frac{\partial^2\sigma_{yy}}{\partial x^2} - \frac{\nu^*}{E^*}\frac{\partial^2\sigma_{xx}}{\partial x^2} + \alpha^*\frac{\partial^2\tau}{\partial x^2} = 2\frac{1+\nu^*}{E^*}\frac{\partial^2\sigma_{xy}}{\partial x\partial y} \tag{a}$$

Simplification of Eq. (a) reduces it to

$$\frac{\partial^2\sigma_{xx}}{\partial y^2} - 2\frac{\partial^2\sigma_{xy}}{\partial x\partial y} + \frac{\partial^2\sigma_{yy}}{\partial x^2} - \nu^*\left(\frac{\partial^2\sigma_{xx}}{\partial x^2} + 2\frac{\partial^2\sigma_{xy}}{\partial x\partial y} + \frac{\partial^2\sigma_{yy}}{\partial y^2}\right) = -\alpha^* E^*\nabla^2\tau \tag{b}$$

5.2 Thermal Stress Function

5.2.1 Thermal Stress Function

There are two main methods used to solve plane thermoelastic problems. The first method pertains to solution of the differential equations of equilibrium (5.28) together with the compatibility equation (5.29), the boundary conditions (5.31) and the generalized Hooke's law (5.27). The second pertains to solution of Navier's equations (5.30) together with the boundary conditions (5.31) and the generalized Hooke's law (5.27).

The first method consists of a new function called the thermal stress function. Let us introduce a thermal stress function χ related to the components of stress as follows:

$$\sigma_{xx} = \frac{\partial^2\chi}{\partial y^2}, \qquad \sigma_{yy} = \frac{\partial^2\chi}{\partial x^2}, \qquad \sigma_{xy} = -\frac{\partial^2\chi}{\partial x\partial y} \tag{5.32}$$

It is easily found that the equilibrium equations (5.28) are automatically satisfied by introduction of the thermal stress function χ. Since the solution of the problem should satisfy the compatibility equation (5.27), substitution of Eqs. (5.32) into Eq. (5.29), and using Eqs. (5.27), yields

$$\nabla^4\chi = -\alpha^* E^*\nabla^2\tau \tag{5.33}$$

or

$$\nabla^4\chi = -k\nabla^2\tau \tag{5.33$'$}$$

where

$$\nabla^4 = \nabla^2\nabla^2 = \left(\frac{\partial^2}{\partial x^2} + \frac{\partial^2}{\partial y^2}\right)\left(\frac{\partial^2}{\partial x^2} + \frac{\partial^2}{\partial y^2}\right) = \frac{\partial^4}{\partial x^4} + 2\frac{\partial^4}{\partial x^2\partial y^2} + \frac{\partial^4}{\partial y^4} \tag{5.34}$$

$$k = \begin{cases} \dfrac{\alpha E}{1-\nu} & \text{for plane strain} \\ \alpha E & \text{for plane stress} \end{cases} \tag{5.35}$$

The differential equation (5.33) or (5.33′) is called the fundamental equation of the thermal stress function χ for plane thermoelastic problems.

The general solution of (5.33′) for the thermal stress function χ may be expressed as the sum of the complementary solution χ_c and the particular solution χ_p:

$$\chi = \chi_c + \chi_p \tag{5.36}$$

where the complementary solution χ_c and the particular solution χ_p are governed by a linear homogeneous differential equation and a linear non-homogeneous differential equation, respectively:

$$\nabla^4 \chi_c = 0, \qquad \nabla^2 \chi_p = -k\tau \tag{5.37}$$

One solution of the second equation in Eqs. (5.37) may be written according to the potential theory as follows:

$$\chi_p(x, y) = -\frac{k}{2\pi} \iint_A \tau(x', y') \ln \sqrt{(x - x')^2 + (y - y')^2}\, dx'\, dy' \tag{5.38}$$

The formula (5.38) can be used for both transient temperature and steady temperature. If the temperature satisfies the heat conduction equation with no heat generation, we can obtain the particular solution χ_p in the same manner as the thermoelastic displacement potential discussed in Subsection 4.4.1

$$\chi_p = -k\kappa \int_{t_r}^{t} \tau(x, y, t')\, dt' + \chi_{pr} + (t - t_r)\chi_{p0} \tag{5.39}$$

where t_r denotes the reference time, and χ_{pr} and χ_{p0} denote solutions of the following Poisson and Laplace equations, respectively:

$$\nabla^2 \chi_{pr} = -k\tau_r, \qquad \nabla^2 \chi_{p0} = 0 \tag{5.40}$$

The function χ_{p0} is usually selected in such a way that a function χ_p is to have a finite value at an arbitrary time if the thermal stresses are to be finite. The function χ_{pr} is the thermal stress function with temperature change τ_r at reference time (usually, at the initial time or final time).

5.2.2 Displacements by Thermal Stress Function

Let us express the components of displacement by the thermal stress function. The generalized Hooke's law is given by Eqs. (5.27). Rearrangement of Eqs. (5.27) yields

$$\begin{aligned}\frac{\partial u_x}{\partial x} &= \epsilon_{xx} = \frac{1}{E^*}\left(\sigma_{xx} - \nu^*\sigma_{yy}\right) + \alpha^*\tau - c^* \\ &= \frac{1}{2G}\left[-\sigma_{yy} + \frac{1}{1+\nu^*}\left(\sigma_{xx} + \sigma_{yy} + \alpha^* E^* \tau\right)\right] - c^*\end{aligned}$$

$$\frac{\partial u_y}{\partial y} = \epsilon_{yy} = \frac{1}{E^*}\left(\sigma_{yy} - \nu^* \sigma_{xx}\right) + \alpha^* \tau - c^* \tag{a}$$

$$= \frac{1}{2G}\left[-\sigma_{xx} + \frac{1}{1+\nu^*}\left(\sigma_{xx} + \sigma_{yy} + \alpha^* E^* \tau\right)\right] - c^*$$

$$\frac{1}{2}\left(\frac{\partial u_x}{\partial y} + \frac{\partial u_y}{\partial x}\right) = \epsilon_{xy} = \frac{1}{2G}\sigma_{xy} = -\frac{1}{2G}\frac{\partial^2 \chi}{\partial x \partial y}$$

By use of the relation $\sigma_{xx} + \sigma_{yy} = \nabla^2 \chi$, the governing equation (5.33) reduces to

$$\nabla^2\left(\sigma_{xx} + \sigma_{yy} + \alpha^* E^* \tau\right) = 0 \tag{b}$$

Therefore, the function $(\sigma_{xx} + \sigma_{yy} + \alpha^* E^* \tau)$ is a plane harmonic function. Thus, we can express the function $(\sigma_{xx} + \sigma_{yy} + \alpha^* E^* \tau)$ by use of a function ψ in the form

$$\sigma_{xx} + \sigma_{yy} + \alpha^* E^* \tau = \nabla^2 \chi + \alpha^* E^* \tau \equiv \frac{\partial^2 \psi}{\partial x \partial y} \tag{5.41}$$

where

$$\frac{\partial^2}{\partial x \partial y}\nabla^2 \psi = 0 \tag{5.42}$$

Integrating Eqs. (a) by use of Eqs. (5.32) and (5.41), we get

$$u_x = \frac{1}{2G}\left[-\frac{\partial \chi}{\partial x} + \frac{1}{1+\nu^*}\frac{\partial \psi}{\partial y}\right] - c^* x + f_1(y)$$

$$u_y = \frac{1}{2G}\left[-\frac{\partial \chi}{\partial y} + \frac{1}{1+\nu^*}\frac{\partial \psi}{\partial x}\right] - c^* y + f_2(x) \tag{c}$$

where functions $f_1(y)$ and $f_2(x)$ are arbitrary. In order to determine the functions $f_1(y)$ and $f_2(x)$, we substitute of Eqs. (c) into the third equation in Eqs. (a) to obtain

$$\frac{df_1(y)}{dy} + \frac{df_2(x)}{dx} = 0 \tag{d}$$

Therefore, the functions $f_1(y)$ and $f_2(x)$ are expressed in the form

$$f_1(y) = -\omega_0 y + u_{x0}, \qquad f_2(x) = \omega_0 x + u_{y0} \tag{5.43}$$

where ω_0, u_{x0}, and u_{y0} are arbitrary constants. The deformation due to Eqs. (5.43) means rigid displacement and rigid rotation. When the rigid deformation may be neglected, the components of displacement can be expressed in the form

$$u_x = \frac{1}{2G}\left[-\frac{\partial \chi}{\partial x} + \frac{1}{1+\nu^*}\frac{\partial \psi}{\partial y}\right] - c^* x$$
$$u_y = \frac{1}{2G}\left[-\frac{\partial \chi}{\partial y} + \frac{1}{1+\nu^*}\frac{\partial \psi}{\partial x}\right] - c^* y \tag{5.44}$$

where the function ψ satisfies Eq. (5.42).

5.2.3 Boundary Conditions in Terms of Thermal Stress Function

Let us express the boundary conditions in terms of the thermal stress function. The boundary conditions are

$$\sigma_{xx} l + \sigma_{yx} m = p_{nx}, \qquad \sigma_{xy} l + \sigma_{yy} m = p_{ny} \tag{5.45}$$

From Fig. 5.3, the direction cosines can be expressed in the form

$$l = \cos(n, x) = \cos(s, y) = \frac{dy}{ds} = \frac{dx}{dn}$$
$$m = \cos(n, y) = -\cos(s, x) = -\frac{dx}{ds} = \frac{dy}{dn} \tag{5.46}$$

Substituting Eqs. (5.32) and (5.46) into Eqs. (5.45), we get

$$\frac{d}{ds}\left(\frac{\partial \chi}{\partial y}\right) = p_{nx}, \qquad -\frac{d}{ds}\left(\frac{\partial \chi}{\partial x}\right) = p_{ny} \tag{a}$$

Integration of these equations along the boundary contour from some fixed point A to a variable point P gives

$$\frac{\partial \chi(P)}{\partial x} = -\int_A^P p_{ny}\, ds + C_1, \qquad \frac{\partial \chi(P)}{\partial y} = \int_A^P p_{nx}\, ds + C_2 \tag{b}$$

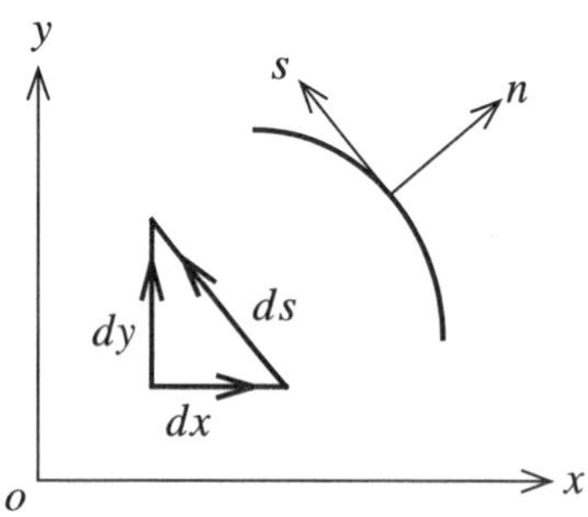

Figure 5.3 Boundary.

From Eqs. (b), we can get $\chi(P)$ and $(\partial\chi(P)/\partial n')$ on the contour:

$$\begin{aligned}\chi(P) &= \int_A^P d\chi + C_3' = \int_A^P \left[\left(\frac{\partial\chi}{\partial x}\right) dx + \left(\frac{\partial\chi}{\partial y}\right) dy\right] + C_3' \\ &= -\iint_A^P p_{ny}\, ds\, dx + \iint_A^P p_{nx}\, ds\, dy + C_1 x + C_2 y + C_3 \\ \frac{\partial\chi(P)}{\partial n'} &= -\cos(n', x)\int_A^P p_{ny}\, ds + \cos(n', y)\int_A^P p_{nx}\, ds \\ &\quad + C_1\cos(n', x) + C_2\cos(n', y)\end{aligned} \tag{5.47}$$

where n' denotes some direction which does not coincide with the direction of the contour, and C_1, C_2, and C_3 are arbitrary integration constants. The boundary conditions of pure thermal stress problems with no external force can be written in the form

$$\begin{aligned}\chi(P) &= C_1 x + C_2 y + C_3 \\ \frac{\partial\chi(P)}{\partial n'} &= C_1\cos(n', x) + C_2\cos(n', y)\end{aligned} \tag{5.48}$$

The arbitrary integration constants C_1, C_2, and C_3 can be taken zero for simply connected regions.

On the other hand, for a multiply connected body whose boundary consists of $m+1$ simple closed contours[†] L_i $(i = 0, 1, \ldots, m)$, Eqs. (5.48) can be rewritten as:

$$\left.\begin{aligned}\chi(P_i) &= C_{1i}x + C_{2i}y + C_{3i} \\ \frac{\partial\chi(P_i)}{\partial n'} &= C_{1i}\cos(n', x) + C_{2i}\cos(n', y)\end{aligned}\right\} \quad \text{on } L_i\ (i = 0, 1, \ldots, m) \tag{5.49}$$

where P_i is an arbitrary point on the i-th boundary contour L_i $(i = 0, 1, \ldots, m)$, C_{1i}, C_{2i}, and C_{3i} are the integration constants on the boundary contour L_i $(i = 0, 1, \ldots, m)$, and the integration constants on only one contour can be zero. When thermoelastic problems in the $(m+1)$-tuply connected body are treated, the unknown constants C_{1i}, C_{2i}, and C_{3i} on L_i $(i = 1, \ldots, m)$ are added, so that the number of unknown constants is $3m$ more than the number of equation of boundary conditions. The lacking $3m$ equations will be discussed in Subsection 5.2.4.

5.2.4 Multiply Connected Bodies

When the thermoelastic problem in a multiply connected body is treated, the conditions of single-valuedness of the displacement and the rotation must be taken into account

---

[†] See the definition in the footnote on page 173.

as mentioned in Subsection 4.7.1. The conditions of single-valuedness of the displacement and the rotation for the plane problem can be expressed from Eqs. (4.103) and (4.101):

$$\oint_{L_i} \left[\epsilon_{\alpha\gamma} - x_\beta(\epsilon_{\alpha\gamma,\beta} - \epsilon_{\beta\gamma,\alpha})\right] dx_\gamma = 0 \qquad (i = 1, \ldots, m) \tag{5.50}$$

$$\oint_{L_i} (\epsilon_{1\gamma,2} - \epsilon_{2\gamma,1})\, dx_\gamma = 0 \qquad (i = 1, \ldots, m) \tag{5.51}$$

where Greek indices mean 1 and 2.

Let us consider a doubly connected body in Fig. 5.4 (a), and cut the body along the surface B in order to make a simply connected body in Fig. 5.4 (b), A being a point on B. If a discontinuity of rotation is produced at a point Q on the surface B, it is defined as:

$$\Delta\omega = \omega^+(Q) - \omega^-(Q) \tag{5.52}$$

where $\omega^+(Q)$ is the limit of $\omega(P)$ as P approaches Q from one side, and $\omega^-(Q)$ is the limit of $\omega(P)$ as P approaches Q from the other side. The discontinuity of rotation is called the dislocation of rotation. In the same way, we can define the dislocation of displacements:

$$\Delta u_\alpha = u_\alpha^+(Q) - u_\alpha^-(Q) \qquad (\alpha = 1, 2) \tag{5.53}$$

Then, the dislocations of rotation and displacements can be written as:

$$\Delta\omega = \oint_L d\omega = -\oint_L d\omega_{12} = -\oint_L (\epsilon_{1\gamma,2} - \epsilon_{2\gamma,1})\, dx_\gamma \tag{5.54}$$

$$\Delta u_\alpha = \oint_L du_\alpha = \oint_L \left[\epsilon_{\alpha\gamma} + \left(x_\beta^Q - x_\beta\right)(\epsilon_{\alpha\gamma,\beta} - \epsilon_{\beta\gamma,\alpha})\right] dx_\gamma \tag{5.55}$$

where L is any closed contour containing a hole, and $\alpha = 1, 2$.

Next, let us express Eqs. (5.54) and (5.55) by the thermal stress function. From the generalized Hooke's law (5.27), the dislocation of rotation (5.54)

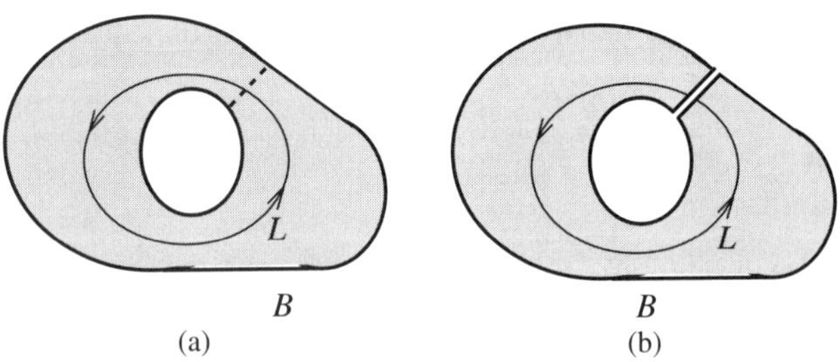

Figure 5.4 A doubly connected body.

reduces to

$$\Delta\omega = -\oint_L \frac{1}{E^*}\left\{\left[\sigma_{11,2} - \nu^*\sigma_{22,2} - (1+\nu^*)\sigma_{12,1} + \alpha^* E^* \tau_{,2}\right] dx_1 \right.$$
$$\left. - \left[\sigma_{22,1} - \nu^*\sigma_{11,1} - (1+\nu^*)\sigma_{12,2} + \alpha^* E^* \tau_{,1}\right] dx_2\right\}$$

By use of the thermal stress function (5.32), it follows that

$$\Delta\omega = -\frac{1}{E^*}\oint_L \left[\left(\nabla^2\chi + \alpha^* E^* \tau\right)_{,2} dx_1 - \left(\nabla^2\chi + \alpha^* E^* \tau\right)_{,1} dx_2\right]$$

From Eqs. (5.46), the dislocation of rotation becomes

$$\Delta\omega = \frac{1}{E^*}\oint_L \frac{\partial}{\partial n}\left(\nabla^2\chi + \alpha^* E^* \tau\right) ds \tag{5.56}$$

From the generalized Hooke's law (5.27) and thermal stress function (5.32), the dislocations of displacement (5.55) reduce to

$$\Delta u_\alpha = -\frac{1+\nu^*}{E^*}\oint_L d(\chi_{,\alpha}) + \frac{1}{E^*}\oint_L d\left[x_\alpha\left(\nabla^2\chi + \alpha^* E^* \tau\right)\right]$$
$$- \frac{1}{E^*}\oint_L \left[x_\alpha d\left(\nabla^2\chi + \alpha^* E^* \tau\right)\right] + \frac{1}{E^*}\oint_L \left(x_\beta^Q - x_\beta\right)$$
$$\times \left[\left(\nabla^2\chi + \alpha^* E^* \tau\right)_{,\beta} dx_\alpha - \left(\nabla^2\chi + \alpha^* E^* \tau\right)_{,\alpha} dx_\beta\right]$$

When $\alpha = 1$, it follows that

$$\Delta u_1 = -\frac{1+\nu^*}{E^*}\oint_L d(\chi_{,1}) + \frac{1}{E^*}\oint_L d\left[x_1\left(\nabla^2\chi + \alpha^* E^* \tau\right)\right]$$
$$- \frac{1}{E^*}\oint_L \left[x_1 d\left(\nabla^2\chi + \alpha^* E^* \tau\right)\right] + \frac{1}{E^*}\oint_L \left(x_2^Q - x_2\right)$$
$$\times \left[\left(\nabla^2\chi + \alpha^* E^* \tau\right)_{,2} dx_1 - \left(\nabla^2\chi + \alpha^* E^* \tau\right)_{,1} dx_2\right]$$

From Eqs. (5.46), the dislocation of displacement Δu_1 becomes

$$\Delta u_1 = -\frac{1+\nu^*}{E^*}\oint_L d(\chi_{,1}) + \frac{1}{E^*}\oint_L d\left[x_1\left(\nabla^2\chi + \alpha^* E^* \tau\right)\right] - x_2^Q \Delta\omega$$
$$- \frac{1}{E^*}\oint_L \left(x_1\frac{\partial}{\partial s} - x_2\frac{\partial}{\partial n}\right)\left(\nabla^2\chi + \alpha^* E^* \tau\right) ds \tag{5.57}$$

In the same way, the dislocation of displacement Δu_2 becomes

$$\Delta u_2 = -\frac{1+\nu^*}{E^*}\oint_L d(\chi_{,2}) + \frac{1}{E^*}\oint_L d\left[x_2\left(\nabla^2\chi + \alpha^* E^* \tau\right)\right] + x_1^Q \Delta\omega$$
$$- \frac{1}{E^*}\oint_L \left(x_1\frac{\partial}{\partial n} + x_2\frac{\partial}{\partial s}\right)\left(\nabla^2\chi + \alpha^* E^* \tau\right) ds \tag{5.58}$$

These equations give the dislocations of rotation and displacements in the double connected body. Applying these results to $(m+1)$-tuply connected body, the dislocations of rotation and displacements with respect to closed integral paths L_i, each of which surrounds only one single hole, are given by

$$(\Delta\omega)_{L_i} = \frac{1}{E^*}\oint_{L_i} \frac{\partial}{\partial n}\left(\nabla^2\chi + \alpha^* E^* \tau\right) ds \quad (i = 1, \ldots, m) \tag{5.59}$$

$$\begin{aligned}(\Delta u_1)_{L_i} &= -\frac{1+\nu^*}{E^*}\oint_{L_i} d(\chi_{,1}) + \frac{1}{E^*}\oint_{L_i} d\left[x_1\left(\nabla^2\chi + \alpha^* E^* \tau\right)\right] - \left(x_2^Q \Delta\omega\right)_{L_i} \\ &\quad - \frac{1}{E^*}\oint_{L_i}\left(x_1\frac{\partial}{\partial s} - x_2\frac{\partial}{\partial n}\right)\left(\nabla^2\chi + \alpha^* E^* \tau\right) ds \quad (i = 1, \ldots, m)\end{aligned} \tag{5.60}$$

$$\begin{aligned}(\Delta u_2)_{L_i} &= -\frac{1+\nu^*}{E^*}\oint_{L_i} d(\chi_{,2}) + \frac{1}{E^*}\oint_{L_i} d\left[x_2\left(\nabla^2\chi + \alpha^* E^* \tau\right)\right] + \left(x_1^Q \Delta\omega\right)_{L_i} \\ &\quad - \frac{1}{E^*}\oint_{L_i}\left(x_1\frac{\partial}{\partial n} + x_2\frac{\partial}{\partial s}\right)\left(\nabla^2\chi + \alpha^* E^* \tau\right) ds \quad (i = 1, \ldots, m)\end{aligned} \tag{5.61}$$

The conditions of single-valuedness of rotation and displacements mean that the dislocations of rotation and displacements are zero. Therefore, the conditions of single-valuedness of rotation and displacements in $(m+1)$-tuply connected body with traction free surfaces become

$$\oint_{L_i} \frac{\partial}{\partial n}\left(\nabla^2\chi + \alpha^* E^* \tau\right) ds = 0 \qquad (i = 1, \ldots, m) \tag{5.62}$$

$$\oint_{L_i}\left(x_1\frac{\partial}{\partial s} - x_2\frac{\partial}{\partial n}\right)\left(\nabla^2\chi + \alpha^* E^* \tau\right) ds = 0 \qquad (i = 1, \ldots, m) \tag{5.63}$$

$$\oint_{L_i}\left(x_1\frac{\partial}{\partial n} + x_2\frac{\partial}{\partial s}\right)\left(\nabla^2\chi + \alpha^* E^* \tau\right) ds = 0 \qquad (i = 1, \ldots, m) \tag{5.64}$$

When the thermoelastic problem in a multiply connected body is treated, conditions (5.62), (5.63), and (5.64) of single-valuedness of rotation and displacements must be taken into account.

5.2.5 Temperature Distribution with Zero Thermal Stress

We consider a temperature distribution with zero thermal stress for a plane problem in a simply or a multiply connected body free of all surface tractions and body forces. The components of stress are zero by hypothesis, so that the generalized Hooke's law (5.27) for a plane problem becomes

$$\epsilon_{\alpha\beta} = (\alpha^*\tau - c^*)\delta_{\alpha\beta} \qquad (\alpha, \beta = 1, 2) \tag{a}$$

Substitution of Eq. (a) into the compatibility equation (5.29) yields

$$\nabla^2 \tau = \left(\frac{\partial^2}{\partial x^2} + \frac{\partial^2}{\partial y^2} \right) \tau = 0 \tag{5.65}$$

Equation (5.65) means that the temperature distribution with zero thermal stress must be a plane harmonic function. In other words, the temperature field with zero thermal stress is governed by steady heat conduction equation with no internal heat generation.

The single-valuedness of rotation (5.62) and the single-valuedness of displacements (5.63) and (5.64) must be satisfied in a multiply connected body. Substitution of Eq. (a) into the single-valuedness of rotation (5.62) and the single-valuedness of displacements (5.63) and (5.64) leads to

$$\oint_{L_i} \frac{\partial \tau}{\partial n} ds = 0 \qquad (i = 1, \ldots, m) \tag{5.66}$$

$$\oint_{L_i} \left(x \frac{\partial \tau}{\partial s} - y \frac{\partial \tau}{\partial n} \right) ds = 0 \qquad (i = 1, \ldots, m) \tag{5.67}$$

$$\oint_{L_i} \left(x \frac{\partial \tau}{\partial n} + y \frac{\partial \tau}{\partial s} \right) ds = 0 \qquad (i = 1, \ldots, m) \tag{5.68}$$

Differentiation of the rotation $\omega = \frac{1}{2}((\partial u_y/\partial x) - (\partial u_x/\partial y))$ with respect to x and y, respectively, gives (taking Eq. (a) into consideration)

$$\begin{aligned} \frac{\partial \omega}{\partial x} &= \frac{1}{2} \frac{\partial}{\partial x} \left(\frac{\partial u_y}{\partial x} - \frac{\partial u_x}{\partial y} \right) = -\frac{\partial^2 u_x}{\partial x \partial y} = -\frac{\partial \epsilon_{xx}}{\partial y} = -\frac{\partial(\alpha^* \tau)}{\partial y} \\ \frac{\partial \omega}{\partial y} &= \frac{1}{2} \frac{\partial}{\partial y} \left(\frac{\partial u_y}{\partial x} - \frac{\partial u_x}{\partial y} \right) = \frac{\partial^2 u_y}{\partial x \partial y} = \frac{\partial \epsilon_{yy}}{\partial x} = \frac{\partial(\alpha^* \tau)}{\partial x} \end{aligned} \tag{b}$$

Functions $\alpha^* \tau$ and ω satisfy the Cauchy–Riemann conditions[‡], so that the rotation ω is a conjugate harmonic function to the harmonic function $\alpha^* \tau$. We can define a complex analytical function $P(z)$ with single-valuedness:

$$P(z) = \tau + i \frac{\omega}{\alpha^*} \tag{c}$$

where $z = x + iy$ is a complex variable and $i^2 = -1$. The zero condition of the dislocation of displacements is expressed by use of the complex function:

$$\Delta u_x + i \Delta u_y = 0 \tag{d}$$

[‡]See: Subsection 5.3.1.

From Eqs. (5.60) and (5.61) it follows that

$$
\begin{aligned}
(\Delta u_x + i\Delta u_y)_{L_i} &= -\alpha^* \oint_{L_i} \left[(x+iy)\frac{\partial \tau}{\partial s} + i(x+iy)\frac{\partial \tau}{\partial n} \right] ds \\
&= -\alpha^* \oint_{L_i} z \left(\frac{\partial \tau}{\partial s} + i\frac{\partial \tau}{\partial n} \right) ds \\
&= -\alpha^* \oint_{L_i} z \left[d\tau + i\left(\frac{\partial \tau}{\partial x} dy - \frac{\partial \tau}{\partial y} dx \right) \right] \\
&= -\alpha^* \oint_{L_i} z \left[\left(\frac{\partial \tau}{\partial x} dx + \frac{\partial \tau}{\partial y} dy \right) + i\left(\frac{\partial \tau}{\partial x} dy - \frac{\partial \tau}{\partial y} dx \right) \right] \\
&= -\alpha^* \oint_{L_i} z \left(\frac{\partial P(z)}{\partial x} dx + \frac{\partial P(z)}{\partial y} dy \right) \\
&= -\alpha^* \oint_{L_i} z\, dP(z) = \alpha^* \oint_{L_i} P(z)\, dz = 0
\end{aligned} \qquad \text{(e)}
$$

The conditions of the single-valuedness of displacements (5.67) and (5.68) can be rewritten in the alternative form:

$$
\oint_{L_i} P(z)\, dz = 0 \qquad (5.69)
$$

Therefore, conditions of temperature with zero thermal stress in the plane problem are summarized as

$$
\nabla^2 \tau = 0, \quad \oint_{L_i} \frac{\partial \tau}{\partial n} ds = 0, \quad \oint_{L_i} P(z)\, dz = 0 \qquad (i = 1, \ldots, m) \qquad (5.70)
$$

Even under the condition of zero thermal stress, the axial stress may not vanish in the case of plane strain:

$$
\sigma_{zz} = -\alpha E \tau + E\epsilon_0 \qquad (5.71)
$$

This axial stress is necessary to maintain the plane strain condition.

5.2.6 General Solution of Thermal Stress Function

The general solutions of Eq. (5.33) for the thermal stress function χ may be expressed as the sum of a complementary solution χ_c and a particular solution χ_p

$$
\chi = \chi_c + \chi_p \qquad (5.72)
$$

where the complementary solution χ_c and the particular solution χ_p are governed by a linear homogeneous differential equation and a linear non-homogeneous differential equation, respectively,

$$
\nabla^4 \chi_c = 0, \qquad \nabla^2 \chi_p = -\alpha^* E^* \tau \qquad (5.73)
$$

One solution of the second equation in Eqs. (5.73) may be written according to the potential theory as follows:

$$\chi_p(x, y) = -\frac{\alpha^* E^*}{2\pi} \iint_A \tau(x', y') \ln \sqrt{(x - x')^2 + (y - y')^2}\, dx'\, dy' \tag{5.74}$$

Let us derive solutions of a biharmonic function of the first equation in Eqs. (5.73). The first equation in Eqs. (5.73) corresponds to two equations:

$$\nabla^2 \varphi(x, y) = 0, \qquad \nabla^2 \chi_c(x, y) = \varphi(x, y) \tag{5.75}$$

By use of separation of variables, the harmonic function $\varphi(x, y)$ can be expressed by

$$\begin{pmatrix} 1 \\ x \end{pmatrix} \begin{pmatrix} 1 \\ y \end{pmatrix}, \quad \begin{pmatrix} \sin ax \\ \cos ax \end{pmatrix} \begin{pmatrix} \sinh ay \\ \cosh ay \end{pmatrix}, \quad \begin{pmatrix} \sinh ax \\ \cosh ax \end{pmatrix} \begin{pmatrix} \sin ay \\ \cos ay \end{pmatrix} \tag{5.76}$$

When the function $\varphi(x, y)$ is harmonic, it is easily found that functions $\varphi(x, y)$, $x\varphi(x, y)$, $y\varphi(x, y)$, and $(x^2 + y^2)\varphi(x, y)$ satisfy the biharmonic differential equation. Therefore, multiplying x, y, and $(x^2 + y^2)$ by Eqs. (5.76), the biharmonic function can be expressed in the form of separation of variables

$$\begin{gathered} \begin{pmatrix} 1 \\ x \\ y \\ x^2 + y^2 \end{pmatrix} \begin{pmatrix} 1 \\ x \end{pmatrix} \begin{pmatrix} 1 \\ y \end{pmatrix} \\ \begin{pmatrix} 1 \\ x \\ y \\ x^2 + y^2 \end{pmatrix} \begin{pmatrix} \sin ax \\ \cos ax \end{pmatrix} \begin{pmatrix} \sinh ay \\ \cosh ay \end{pmatrix} \\ \begin{pmatrix} 1 \\ x \\ y \\ x^2 + y^2 \end{pmatrix} \begin{pmatrix} \sinh ax \\ \cosh ax \end{pmatrix} \begin{pmatrix} \sin ay \\ \cos ay \end{pmatrix} \end{gathered} \tag{5.77}$$

Next, let us show the harmonic function and the biharmonic function in the form of a polynomial. By use of a conjugate harmonic function $q(x, y)$ to a harmonic function $p(x, y)$, we can define a complex analytical function $P(z)$

$$P(z) = p(x, y) + iq(x, y)$$

When the complex analytical function $P(z)$ can be expressed in the form of a polynomial:

$$P(z) = z^n = (x + iy)^n \qquad (n\text{: integer from } -\infty \text{ to } \infty)$$

the harmonic functions are given by

$$p(x, y) = \mathrm{Re}[P(z)], \qquad q(x, y) = \mathrm{Im}[P(z)] \tag{5.78}$$

where Re and Im denote the real part and the imaginary part, respectively. Therefore, the harmonic functions are

$$\begin{gathered} 1, \quad \begin{pmatrix} 1 \\ r^{-2} \end{pmatrix}\begin{pmatrix} x \\ y \end{pmatrix}, \quad \begin{pmatrix} 1 \\ r^{-4} \end{pmatrix}\begin{pmatrix} x^2 - y^2 \\ xy \end{pmatrix} \\ \begin{pmatrix} 1 \\ r^{-6} \end{pmatrix}\begin{pmatrix} x(x^2 - 3y^2) \\ y(y^2 - 3x^2) \end{pmatrix}, \quad \begin{pmatrix} 1 \\ r^{-8} \end{pmatrix}\begin{pmatrix} x^4 - 6x^2y^2 + y^4 \\ 4xy(x^2 - y^2) \end{pmatrix}, \ldots \end{gathered} \tag{5.79}$$

where $r^2 = x^2 + y^2$. Thus, the biharmonic functions can be expressed as

$$\begin{gathered} \begin{pmatrix} 1 \\ x \\ y \\ x^2 + y^2 \end{pmatrix}, \quad \begin{pmatrix} 1 \\ x \\ y \\ x^2 + y^2 \end{pmatrix}\begin{pmatrix} 1 \\ r^{-2} \end{pmatrix}\begin{pmatrix} x \\ y \end{pmatrix} \\ \begin{pmatrix} 1 \\ x \\ y \\ x^2 + y^2 \end{pmatrix}\begin{pmatrix} 1 \\ r^{-4} \end{pmatrix}\begin{pmatrix} x^2 - y^2 \\ xy \end{pmatrix} \\ \begin{pmatrix} 1 \\ x \\ y \\ x^2 + y^2 \end{pmatrix}\begin{pmatrix} 1 \\ r^{-6} \end{pmatrix}\begin{pmatrix} x(x^2 - 3y^2) \\ y(y^2 - 3x^2) \end{pmatrix} \\ \begin{pmatrix} 1 \\ x \\ y \\ x^2 + y^2 \end{pmatrix}\begin{pmatrix} 1 \\ r^{-8} \end{pmatrix}\begin{pmatrix} x^4 - 6x^2y^2 + y^4 \\ 4xy(x^2 - y^2) \end{pmatrix}, \ldots \end{gathered} \tag{5.80}$$

Next, let us show the general solutions of the harmonic equation and biharmonic equation in the polar coordinate system (r, θ). Referring to the general solution (4.82) of the harmonic equation in the cylindrical coordinate system (r, θ, z), the harmonic functions take forms

$$\begin{pmatrix} 1 \\ \ln r \end{pmatrix}\begin{pmatrix} 1 \\ \theta \end{pmatrix}, \quad \begin{pmatrix} r^n \\ r^{-n} \end{pmatrix}\begin{pmatrix} \cos n\theta \\ \sin n\theta \end{pmatrix} \tag{5.81}$$

Next, multiplying $(1, r\cos\theta, r\sin\theta, r^2)$ by Eqs. (5.81), the biharmonic functions are obtained in the forms:

$$\begin{pmatrix} 1 \\ r^2 \\ \ln r \\ r^2 \ln r \end{pmatrix} \begin{pmatrix} 1 \\ \theta \end{pmatrix}, \quad \begin{pmatrix} r \\ \ln r \end{pmatrix} \begin{pmatrix} \theta\cos\theta \\ \theta\sin\theta \end{pmatrix}$$
$$\begin{pmatrix} r \\ r^{-1} \\ r^3 \\ r\ln r \end{pmatrix} \begin{pmatrix} \cos\theta \\ \sin\theta \end{pmatrix}, \quad \begin{pmatrix} r^n \\ r^{-n} \\ r^{n+2} \\ r^{-n+2} \end{pmatrix} \begin{pmatrix} \cos n\theta \\ \sin n\theta \end{pmatrix} \tag{5.82}$$

Example 5.2 Evaluate the steady thermal stress in a strip with width l when both surfaces on the strip are symmetrically heated along y axis by the temperature $T_0 f(y)$.

Solution The origin of the Cartesian coordinate system is placed at the middle point of the strip. The boundary conditions are

$$T = T_0 f(y) \quad \text{on } x = \pm\frac{l}{2}, \qquad T \to \text{finite} \quad \text{at } y \to \infty \tag{a}$$

where the boundary temperature is expressed by

$$T_0 f(y) = \int_0^\infty f^*(p) \cos py\, dp \tag{b}$$

in which

$$f^*(p) = \frac{2T_0}{\pi} \int_0^\infty f(y) \cos py\, dy \tag{b$'$}$$

Taking Eqs. (5.76) into consideration, the two-dimensional steady temperature can be expressed by

$$T = \int_0^\infty A(p) \cosh px \cos py\, dp \tag{c}$$

The boundary conditions give

$$A(p) = \frac{f^*(p)}{\cosh \dfrac{pl}{2}} \tag{d}$$

The temperature is

$$T = \int_0^\infty \frac{f^*(p)}{\cosh \dfrac{pl}{2}} \cosh px \cos py\, dp \tag{e}$$

The thermal stress function may have the form which satisfies Eqs. (5.73)

$$\chi_c = \int_0^\infty (B \cosh px + Cx \sinh px) \cos py \, dp \tag{f}$$

$$\chi_p = -\frac{\alpha^* E^*}{2} \int_0^\infty \frac{f^*(p)}{p \cosh \frac{pl}{2}} x \sinh px \cos py \, dp \tag{g}$$

Then, we get

$$\chi = \chi_c + \chi_p = \int_0^\infty (B \cosh px + Dx \sinh px) \cos py \, dp \tag{h}$$

where

$$D = C - \frac{\alpha^* E^*}{2} \frac{f^*(p)}{p \cosh \frac{pl}{2}}$$

The thermal stresses are

$$\begin{aligned} \sigma_{xx} &= \frac{\partial^2 \chi}{\partial y^2} = -\int_0^\infty p^2 (B \cosh px + Dx \sinh px) \cos py \, dp \\ \sigma_{yy} &= \frac{\partial^2 \chi}{\partial x^2} = \int_0^\infty p[pB \cosh px + D(2 \cosh px + px \sinh px)] \cos py \, dp \\ \sigma_{xy} &= -\frac{\partial^2 \chi}{\partial x \partial y} = \int_0^\infty p[pB \sinh px + D(\sinh px + px \cosh px)] \sin py \, dp \end{aligned} \tag{i}$$

The boundary conditions

$$\sigma_{xx} = \sigma_{xy} = 0 \qquad \text{on } x = \pm \frac{l}{2} \tag{j}$$

determine unknown coefficients

$$B = D = 0 \tag{k}$$

Thermal stresses do not appear in the strip.

Example 5.3 Derive the transient thermal stress in a strip when the temperature in the strip is given by Eq. (3.71).

Solution The temperature given by Eq. (3.71) is

$$\begin{aligned} T(x,t) &= T_a \left(1 - \frac{x}{l}\right) + T_b \frac{x}{l} \\ &+ \frac{2}{\pi} \sum_{n=1}^\infty \frac{(-1)^n}{n} \left[T_a \sin n\pi \left(1 - \frac{x}{l}\right) + T_b \sin n\pi \frac{x}{l} \right] e^{-\kappa n^2 \pi^2 t / l^2} \end{aligned} \tag{a}$$

Since the stress function satisfies Eq. (5.33):

$$\nabla^4 \chi = -\alpha^* E^* \nabla^2 \tau \tag{b}$$

the general solution is

$$\chi = A_0 + A_1 x + A_2 x^2 + A_3 x^3 + \frac{2\alpha^* E^* l^2}{\pi^3} \sum_{n=1}^{\infty} \frac{(-1)^n}{n^3} \left[T_a \sin n\pi \left(1 - \frac{x}{l}\right) + T_b \sin n\pi \frac{x}{l} \right] e^{-\kappa n^2 \pi^2 t / l^2} \tag{c}$$

The thermal stresses are

$$\begin{aligned} \sigma_{yy} &= \frac{\partial^2 \chi}{\partial x^2} = 2A_2 + 6A_3 x \\ &\quad - \frac{2\alpha^* E^*}{\pi} \sum_{n=1}^{\infty} \frac{(-1)^n}{n} \left[T_a \sin n\pi \left(1 - \frac{x}{l}\right) + T_b \sin n\pi \frac{x}{l} \right] e^{-\kappa n^2 \pi^2 t / l^2} \\ \sigma_{xx} &= \frac{\partial^2 \chi}{\partial y^2} = 0 \\ \sigma_{xy} &= -\frac{\partial^2 \chi}{\partial x \partial y} = 0 \end{aligned} \tag{d}$$

The condition $\lim_{\substack{y \to \infty \\ t \to \infty}} \sigma_{yy} = 0$ and the boundary conditions:

$$\sigma_{xx} = 0, \quad \sigma_{xy} = 0 \qquad \text{on } x = 0, l \tag{e}$$

determine unknown coefficients as

$$A_2 = 0, \qquad A_3 = 0 \tag{f}$$

Then, the thermal stress is

$$\sigma_{yy} = -\frac{2\alpha^* E^*}{\pi} \sum_{n=1}^{\infty} \frac{(-1)^n}{n} \left[T_a \sin n\pi \left(1 - \frac{x}{l}\right) + T_b \sin n\pi \frac{x}{l} \right] e^{-\kappa n^2 \pi^2 t / l^2} \tag{g}$$

5.3 Complex Variable Method

5.3.1 Complex Presentation of Thermal Stress Function

The general solutions of Eq. (5.33) for the thermal stress function χ may be expressed as the sum of the complementary solution χ_c and the particular solution χ_p which are governed by Eqs. (5.73). The particular solution χ_p of the second equation of Eqs. (5.73) is given by Eq. (5.74). It will be shown that the biharmonic function χ_c can be represented by two complex functions of the complex variable $z = x + iy$. Using the notation

$$\nabla^2 \chi_c(x, y) \equiv P(x, y) \tag{5.83}$$

the function P is seen to be harmonic, because χ_c is the biharmonic function. Introducing a function $Q(x, y)$ to be the harmonic function, which is conjugate to the harmonic function P, we can define a complex function $f(z)$ to be regular in the body

$$f(z) = P(x, y) + iQ(x, y) \tag{5.84}$$

Furthermore, introducing a new complex function related to the complex function $f(z)$:

$$\varphi(z) = p(x, y) + iq(x, y) = \frac{1}{4}\int f(z)\,dz \tag{5.85}$$

the functions $p(x, y)$ and $q(x, y)$ are harmonic, and satisfy the Cauchy–Riemann conditions[§]:

$$\frac{\partial p}{\partial x} = \frac{\partial q}{\partial y}, \quad \frac{\partial q}{\partial x} = -\frac{\partial p}{\partial y} \tag{5.86}$$

Then,

$$\frac{d\varphi(z)}{dz} \equiv \frac{\partial p}{\partial x} + i\frac{\partial q}{\partial x} = \frac{\partial q}{\partial y} - i\frac{\partial p}{\partial y} = \frac{1}{4}f(z) = \frac{1}{4}(P + iQ)$$

Next, let us introduce a new function $p_1 = \chi_c - px - qy$. It is easily found that the function $p_1 = \chi_c - px - qy$ is harmonic

$$\begin{aligned}\nabla^2 p_1 &= \nabla^2(\chi_c - px - qy)\\ &= \nabla^2\chi_c - 2\left(\frac{\partial p}{\partial x} + \frac{\partial q}{\partial y}\right) - x\nabla^2 p - y\nabla^2 q = \nabla^2\chi_c - P = 0\end{aligned}$$

The biharmonic function χ_c can be represented by three harmonic functions

$$\chi_c = px + qy + p_1 \tag{5.87}$$

where

$$\nabla^2 p = 0, \qquad \nabla^2 q = 0, \qquad \nabla^2 p_1 = 0$$

Introducing a complex function $\psi_1(z)$ with real part p_1, the biharmonic function χ_c can be written in terms of two complex functions

$$\chi_c = \text{Re}\,[\bar{z}\varphi(z) + \psi_1(z)] \tag{5.88}$$

[§] See: N. I. Muskhelishvili, Some Basic Problems of the Mathematical Theory of Elasticity, Noordhoff, Groningen (1963), p. 108.

or

$$\chi_c = \frac{1}{2}\left[\bar{z}\varphi(z) + z\overline{\varphi(z)} + \psi_1(z) + \overline{\psi_1(z)}\right] \tag{5.89}$$

where the upper bar, such as in $\overline{\varphi(z)}$, denotes its conjugate complex function:

$$\bar{z} = x - iy, \qquad \overline{\varphi(z)} = p - iq \tag{5.90}$$

Hence, the thermal stress function χ can be represented by two complex functions and the particular solution χ_p

$$\chi = \chi_c + \chi_p = \frac{1}{2}\left[\bar{z}\varphi(z) + z\overline{\varphi(z)} + \psi_1(z) + \overline{\psi_1(z)}\right] + \chi_p \tag{5.91}$$

5.3.2 Complex Representation of Stress and Displacement

It is found by use of the thermal stress function χ defined by Eqs. (5.32) that

$$\sigma_{xx} + \sigma_{yy} = \nabla^2\chi = \nabla^2\chi_c - \alpha^* E^* \tau \tag{a}$$

$$\begin{aligned}\sigma_{yy} - \sigma_{xx} - 2i\sigma_{xy} &= \frac{\partial^2\chi}{\partial x^2} - \frac{\partial^2\chi}{\partial y^2} + 2i\frac{\partial^2\chi}{\partial x\partial y}\\ &= \left(\frac{\partial}{\partial x} + i\frac{\partial}{\partial y}\right)^2\chi_c + \left(\frac{\partial}{\partial x} + i\frac{\partial}{\partial y}\right)^2\chi_p\end{aligned} \tag{b}$$

It is easily found from Eq. (5.89)

$$\begin{aligned}\frac{\partial\chi_c}{\partial x} &= \frac{1}{2}\left[\varphi(z) + \bar{z}\varphi'(z) + \overline{\varphi(z)} + z\overline{\varphi'(z)} + \psi_1'(z) + \overline{\psi_1'(z)}\right]\\ \frac{\partial\chi_c}{\partial y} &= \frac{i}{2}\left[-\varphi(z) + \bar{z}\varphi'(z) + \overline{\varphi(z)} - z\overline{\varphi'(z)} + \psi_1'(z) - \overline{\psi_1'(z)}\right]\end{aligned}$$

Hence

$$\left(\frac{\partial}{\partial x} + i\frac{\partial}{\partial y}\right)\chi_c = \left[\varphi(z) + z\overline{\varphi'(z)} + \overline{\psi(z)}\right] \tag{5.92}$$

where $\psi(z) \equiv \psi_1'(z)$. By use of the relation

$$\left(\frac{\partial}{\partial x} + i\frac{\partial}{\partial y}\right)^2\chi_c = 2\left[z\overline{\varphi''} + \overline{\psi'(z)}\right] \tag{c}$$

Equation (a) and conjugate complex value of Eq. (b) become

$$\begin{aligned}\sigma_{xx} + \sigma_{yy} &= 4\mathrm{Re}\left[\varphi'(z)\right] - \alpha^* E^* \tau\\ \sigma_{yy} - \sigma_{xx} + 2i\sigma_{xy} &= 2\left[\bar{z}\varphi''(z) + \psi'(z)\right] + \left(\frac{\partial}{\partial x} - i\frac{\partial}{\partial y}\right)^2\chi_p\end{aligned} \tag{5.93}$$

Equations (5.93) are called the general solution of plane thermal stresses by complex functions. The complex functions $\varphi(z)$ and $\psi(z)$ are called the complex stress functions.

Let us express the components of displacement by complex functions. The generalized Hooke's law (5.27) yields

$$
\begin{aligned}
\frac{\partial u_x}{\partial x} &= \epsilon_{xx} = \frac{1}{E^*}\left(\sigma_{xx} - \nu^*\sigma_{yy}\right) + \alpha^*\tau - c^* \\
&= \frac{1}{2G}\left[-\sigma_{yy} + \frac{1}{1+\nu^*}\left(\sigma_{xx} + \sigma_{yy} + \alpha^* E^*\tau\right)\right] - c^* \\
&= \frac{1}{2G}\left(-\frac{\partial^2 \chi_c}{\partial x^2} + \frac{4}{1+\nu^*}\frac{\partial p}{\partial x} - \frac{\partial^2 \chi_p}{\partial x^2}\right) - c^* \qquad \text{(d)}\\
\frac{\partial u_y}{\partial y} &= \epsilon_{yy} = \frac{1}{E^*}\left(\sigma_{yy} - \nu^*\sigma_{xx}\right) + \alpha^*\tau - c^* \\
&= \frac{1}{2G}\left[-\sigma_{xx} + \frac{1}{1+\nu^*}\left(\sigma_{xx} + \sigma_{yy} + \alpha^* E^*\tau\right)\right] - c^* \\
&= \frac{1}{2G}\left(-\frac{\partial^2 \chi_c}{\partial y^2} + \frac{4}{1+\nu^*}\frac{\partial q}{\partial y} - \frac{\partial^2 \chi_p}{\partial y^2}\right) - c^* \qquad \text{(e)}
\end{aligned}
$$

$$\frac{1}{2}\left(\frac{\partial u_x}{\partial y} + \frac{\partial u_y}{\partial x}\right) = \epsilon_{xy} = \frac{1}{2G}\sigma_{xy} = -\frac{1}{2G}\frac{\partial^2 \chi}{\partial x \partial y} \qquad \text{(f)}$$

Integration of Eq. (d) with respect to x and of Eq. (e) with respect to y yields

$$u_x = \frac{1}{2G}\left(-\frac{\partial \chi_c}{\partial x} + \frac{4p}{1+\nu^*} - \frac{\partial \chi_p}{\partial x}\right) - c^* x + f_1(y) \qquad \text{(g)}$$

$$u_y = \frac{1}{2G}\left(-\frac{\partial \chi_c}{\partial y} + \frac{4q}{1+\nu^*} - \frac{\partial \chi_p}{\partial y}\right) - c^* y + f_2(x) \qquad \text{(h)}$$

where functions $f_1(y)$ and $f_2(x)$ are arbitrary. In order to determine the functions $f_1(y)$ and $f_2(x)$, substitution of these equations into Eq. (f) gives

$$\frac{df_1(y)}{dy} + \frac{df_2(x)}{dx} = 0 \qquad \text{(i)}$$

Therefore, the functions $f_1(y)$ and $f_2(x)$ can be expressed in the form

$$f_1(y) = -\omega_0 y + u_{x0}, \qquad f_2(x) = \omega_0 x + u_{y0} \qquad (5.94)$$

where ω_0, u_{x0}, and u_{y0} are arbitrary constants. The deformation due to Eqs. (5.94) means rigid displacements and rigid rotation. When the rigid deformation may be neglected, the components of displacement can be expressed in the form

$$u_x + iu_y = \frac{1}{2G}\left[-\left(\frac{\partial \chi_c}{\partial x} + i\frac{\partial \chi_c}{\partial y}\right) + \frac{4}{1+\nu^*}(p+iq) - \left(\frac{\partial \chi_p}{\partial x} + i\frac{\partial \chi_p}{\partial y}\right)\right] - c^*(x+iy)$$

Hence

$$u_x + iu_y = \frac{1}{2G}\left[\frac{3-\nu^*}{1+\nu^*}\varphi(z) - z\overline{\varphi'(z)} - \overline{\psi(z)} - \left(\frac{\partial \chi_p}{\partial x} + i\frac{\partial \chi_p}{\partial y}\right)\right] - c^*(x+iy) \tag{5.95}$$

Next, let us consider the complex expression of the boundary conditions. It is easily found from Eqs. (a) in Subsection 5.2.3 that

$$(p_{nx} + ip_{ny})\, ds = -id\left(\frac{\partial \chi}{\partial x} + i\frac{\partial \chi}{\partial y}\right) \tag{j}$$

Integrating Eq. (j) along the boundary contour from a fixed point A to a variable point P on the boundary contour, and using Eqs. (5.36) and (5.92), we get

$$\varphi(z) + z\overline{\varphi'(z)} + \overline{\psi(z)} = i\int_A^P (p_{nx} + ip_{ny})\, ds - \left(\frac{\partial \chi_p}{\partial x} + i\frac{\partial \chi_p}{\partial y}\right) + C \tag{5.96}$$

where C is an integration constant. Equation (5.96) is called the boundary condition expressed by complex functions. As the function $\int_A^P (p_{nx} + ip_{ny})\, ds$ is a complex representation of the resultant force acting from the fixed point A to the variable point P on the boundary contour, rewriting the function by $(P_x + iP_y)$, Eq. (5.96) becomes

$$\varphi(z) + z\overline{\varphi'(z)} + \overline{\psi(z)} = i(P_x + iP_y) - \left(\frac{\partial \chi_p}{\partial x} + i\frac{\partial \chi_p}{\partial y}\right) + C \tag{5.96$'$}$$

where (P_x, P_y) denote the components of the resultant force acting from the fixed point A to the variable point P on the boundary contour. In the pure thermoelastic problem without traction, the boundary condition (5.96) reduces to

$$\varphi(z) + z\overline{\varphi'(z)} + \overline{\psi(z)} = -\left(\frac{\partial \chi_p}{\partial x} + i\frac{\partial \chi_p}{\partial y}\right) + C \tag{5.97}$$

Next, we consider the resultant moment about the origin of coordinate systems. The moment M is expressed as

$$M = \int_A^P (p_{ny}x - p_{nx}y)\, ds \tag{5.98}$$

Substituting Eq. (a) in Subsection 5.2.3 into Eq. (5.98), and integrating by parts gives

$$M = -\int_A^P \left[xd\left(\frac{\partial \chi}{\partial x}\right) + yd\left(\frac{\partial \chi}{\partial y}\right)\right] = -\left[x\frac{\partial \chi}{\partial x} + y\frac{\partial \chi}{\partial y} - \chi\right]_A^P$$

Using Eq. (5.91):

$$\chi = \chi_c + \chi_p = \frac{1}{2}\left[\bar{z}\varphi(z) + z\overline{\varphi(z)} + \psi_1(z) + \overline{\psi_1(z)}\,\right] + \chi_p$$

and the relation:

$$x\frac{\partial \chi_c}{\partial x} + y\frac{\partial \chi_c}{\partial y} = \mathrm{Re}\left[z\left(\frac{\partial \chi_c}{\partial x} - i\frac{\partial \chi_c}{\partial y}\right)\right]$$

the resultant moment M becomes

$$M = \mathrm{Re}\left[\psi_1(z) - z\psi_1'(z) - z\bar{z}\varphi'(z)\right]_A^P - \left[x\frac{\partial \chi_p}{\partial x} + y\frac{\partial \chi_p}{\partial y} - \chi_p\right]_A^P \tag{5.99}$$

5.3.3 General Formulae for Multiply Connected Bodies

Let us consider the general representation of complex stress functions for the multiply connected body whose boundary consists of $m+1$ simple closed contours L_i $(i = 0, 1, \ldots, m)$ in Fig. 5.5. We assume that the components of stress and displacement are to be single-valued functions. However, this fact does not mean that the complex stress functions are single-valued functions. As the components of stress and the temperature change are single-valued, the first equation of Eqs. (5.93)

$$\sigma_{xx} + \sigma_{yy} = 4\mathrm{Re}\left[\varphi'(z)\right] - \alpha^* E^* \tau$$

shows that the real part of $\varphi'(z)$ is single-valued. This fact means that the imaginary part of $\varphi'(z)$ may be multi-valued. For example, the imaginary part of a function $\ln(z - z_k)$ is multi-valued and undergoes an increase 2π around one circuit containing a point z_k, though the real part of the function is single-valued. If an increase of this imaginary part

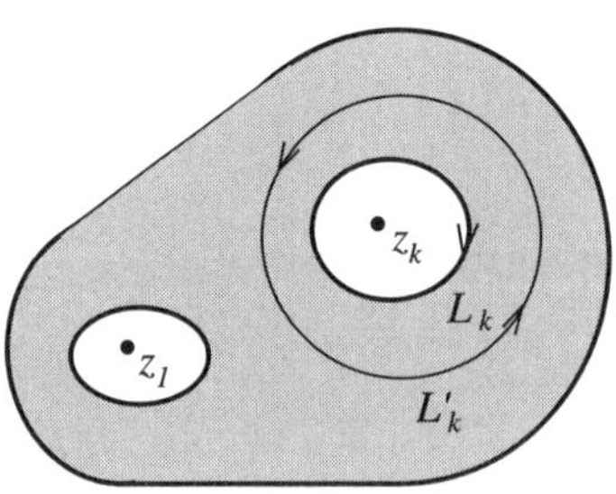

Figure 5.5 A multiply connected body.

is $2\pi A_k$ for one circuit around some contour L'_k containing one of the boundary contours L_k, the function $\varphi'(z)$ can be expressed as:

$$\varphi'(z) = \sum_{k=1}^{m} A_k \ln(z - z_k) + \varphi_1^*(z) \tag{5.100}$$

where z_k denote fixed points which can be arbitrarily chosen inside the contours, L_k, A_k are real constants, and the function $\varphi_1^*(z)$ is regular and hence single-valued in the body. Integration of this equation gives

$$\varphi(z) = \sum_{k=1}^{m} A_k\big[(z - z_k)\ln(z - z_k) - (z - z_k)\big] + \int_A^P \varphi_1^*(z)\, dz + C$$

The second term on the right-hand side of the above equation, that is, the integration of a regular function, is not always single-valued. If the integral $\int_A^P \varphi_1^*(z)\, dz$ undergoes an increase $2\pi i D_k$ for one circuit around some contour L'_k containing one of the boundary contours L_k, the above equation becomes

$$\varphi(z) = \sum_{k=1}^{m} A_k\big[(z - z_k)\ln(z - z_k) - (z - z_k)\big] + \sum_{k=1}^{m} D_k \ln(z - z_k) + \varphi_2^*(z) + C$$

where D_k are real constants, and the function $\varphi_2^*(z)$ is regular in the body. This equation can be rewritten more simply as

$$\varphi(z) = z\sum_{k=1}^{m} A_k \ln(z - z_k) + \sum_{k=1}^{m} B_k \ln(z - z_k) + \varphi^*(z) \tag{5.101}$$

where B_k are complex constants, and the function $\varphi^*(z)$ is regular in the body.

We consider the second equation of Eqs. (5.93)

$$\sigma_{yy} - \sigma_{xx} + 2i\sigma_{xy} = 2[\bar{z}\varphi''(z) + \psi'(z)] + \left(\frac{\partial}{\partial x} - i\frac{\partial}{\partial y}\right)^2 \chi_p$$

When the components of stress and the function $((\partial/\partial x) - i(\partial/\partial y))^2\chi_p$ are single-valued, the function $\varphi''(z)$ is regular from Eq. (5.101) so that a complex function $\psi'(z)$ must be regular. Then the complex stress function $\psi(z)$ can be written as

$$\psi(z) = \sum_{k=1}^{m} C_k \ln(z - z_k) + \psi^*(z) \tag{5.102}$$

where C_k are complex constants, and the function $\psi^*(z)$ is regular in the body.

Next, we consider the condition of single-valuedness of displacements. From Eq. (5.95) the displacements are

$$u_x + iu_y = \frac{1}{2G}\left[\left(\frac{3 - \nu^*}{1 + \nu^*}\right)\varphi(z) - z\overline{\varphi'(z)} - \overline{\psi(z)} - \left(\frac{\partial \chi_p}{\partial x} + i\frac{\partial \chi_p}{\partial y}\right)\right] - c^*(x + iy)$$

When $[(\partial\chi_p/\partial x)+i(\partial\chi_p/\partial y)]$ is single-valued, it is found by substitution of Eqs. (5.101) and (5.102) into this equation that the dislocation $\Delta u_x+i\Delta u_y$ which denotes an increasing displacement for one counter-clockwise circuit around some contour L'_k containing one of the boundary contours L_k, becomes

$$\Delta u_x+i\Delta u_y=\frac{1}{2G}\left[2\pi i\left(\frac{4}{1+\nu^*}A_k z+\frac{3-\nu^*}{1+\nu^*}B_k+\overline{C}_k\right)-\Delta\left(\frac{\partial\chi_p}{\partial x}+i\frac{\partial\chi_p}{\partial y}\right)_{L_k}\right] \tag{5.103}$$

where $\Delta[(\partial\chi_p/\partial x)+i(\partial\chi_p/\partial y)]_{L_k}$ denotes the increase undergone by the equation $[(\partial\chi_p/\partial x)+i(\partial\chi_p/\partial y)]$ for one counter-clockwise circuit around some contour L'_k. Hence the conditions of single-valuedness of displacements are given by

$$A_k=0,\quad \left(\frac{3-\nu^*}{1+\nu^*}\right)B_k+\overline{C}_k=\frac{1}{2\pi i}\Delta\left(\frac{\partial\chi_p}{\partial x}+i\frac{\partial\chi_p}{\partial y}\right)_{L_k}\quad (k=1,\ldots,m) \tag{5.104}$$

Next, let us consider the resultant force $(P_x+iP_y)_{L_k}$ due to external forces acting on any boundary contour L_k. The resultant force $(P_x+iP_y)_{L_k}$ defined as

$$(P_x+iP_y)_{L_k}=\oint_{L_k}(p_{nx}+ip_{ny})\,ds \tag{5.105}$$

is given by Eq. (5.96)

$$(P_x+iP_y)_{L_k}=2\pi\left[B_k-\overline{C_k}\right]-i\Delta\left(\frac{\partial\chi_p}{\partial x}+i\frac{\partial\chi_p}{\partial y}\right)_{L_k} \tag{5.106}$$

It is easily found from Eqs. (5.104) and (5.106) that B_k and $\overline{C}_k$ are given by

$$\begin{aligned}B_k&=\frac{1+\nu^*}{8\pi}(P_x+iP_y)_{L_k}\\ \overline{C}_k&=\frac{1}{2\pi i}\Delta\left(\frac{\partial\chi_p}{\partial x}+i\frac{\partial\chi_p}{\partial y}\right)_{L_k}-\frac{3-\nu^*}{8\pi}(P_x+iP_y)_{L_k}\end{aligned} \tag{5.107}$$

The two complex stress functions may finally be expressed by

$$\begin{aligned}\varphi(z)&=\frac{1+\nu^*}{8\pi}\sum_{k=1}^{m}(P_x+iP_y)_{L_k}\ln(z-z_k)+\varphi^*(z)\\ \psi(z)&=-\sum_{k=1}^{m}\left[\frac{1}{2\pi i}\Delta\left(\frac{\partial\chi_p}{\partial x}-i\frac{\partial\chi_p}{\partial y}\right)_{L_k}+\frac{3-\nu^*}{8\pi}(P_x-iP_y)_{L_k}\right]\\ &\quad\times\ln(z-z_k)+\psi^*(z)\end{aligned} \tag{5.108}$$

where $\varphi^*(z)$ and $\psi^*(z)$ are regular functions.

When mechanical forces do not apply a body, Eqs. (5.108) reduce to

$$\varphi(z)=\varphi^*(z)$$
$$\psi(z)=-\frac{1}{2\pi i}\sum_{k=1}^{m}\Delta\left(\frac{\partial\chi_p}{\partial x}-i\frac{\partial\chi_p}{\partial y}\right)_{L_k}\ln(z-z_k)+\psi^*(z) \tag{5.109}$$

5.3.4 Conformal Mapping of the Formulae of Plane Theory

In many cases it is convenient to express displacements and stresses by use of conformal mapping. Let us consider the functional relationship

$$z=\omega(\zeta) \tag{5.110}$$

between the points $z=x+iy$ of the complex z-plane and $\zeta=\xi+i\eta$ of the complex ζ-plane. If the function $\omega(\zeta)$ is a regular function in the ζ-plane, Eq. (5.110) is called conformal mapping of the complex z-plane into the ζ-plane. The mapping performed by the function $\omega(\zeta)$ is conformal at all points.

Let us translate a given region S in the complex z-plane into a region Σ in the complex ζ-plane by use of the conformal mapping function $\omega(\zeta)$. The lines $\xi=$constant and $\eta=$constant in the ζ-plane correspond to certain curves in the z-plane which are related by the conformal mapping function $\omega(\zeta)$

$$z=x+iy=\omega(\zeta)=\omega(\xi+i\eta) \tag{5.111}$$

The curvilinear coordinate system corresponding to lines $\xi=$constant and $\eta=$constant are dotted in the z-plane shown in Fig. 5.6. The curvilinear coordinate system is defined by the comformal mapping function in such a way that the curvilinear coordinate system is orthogonal. The direction cosines between the Cartesian coordinate axes (x, y) and the curvilinear coordinate axes which are normal to lines $\xi=$constant and $\eta=$constant are

$$l_1=m_2=\cos\alpha,\quad m_1=-l_2=\sin\alpha,\quad n_3=1,\quad n_1=n_2=l_3=m_3=0 \tag{a}$$

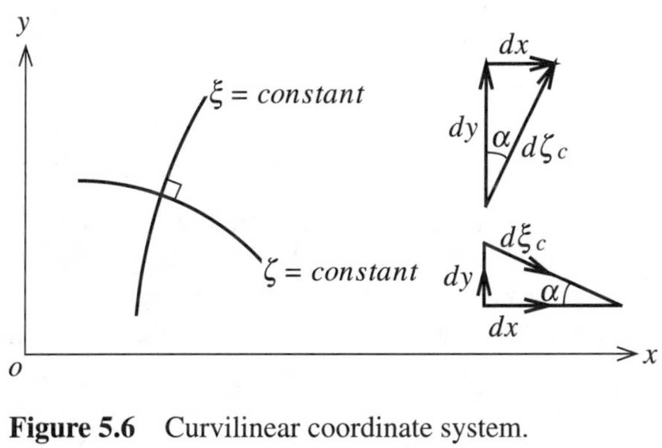

Figure 5.6 Curvilinear coordinate system.

Therefore, the relation between the Cartesian coordinate axes (x, y) and the curvilinear coordinate axes (ξ_c, η_c) are

$$x+iy=(\xi_c\cos\alpha-\eta_c\sin\alpha)+i(\xi_c\sin\alpha+\eta_c\cos\alpha)=e^{i\alpha}(\xi_c+i\eta_c) \tag{5.112}$$

$$\xi_c+i\eta_c=e^{-i\alpha}(x+iy) \tag{5.112$'$}$$

The components (u_{ξ_c}, u_{η_c}) of the displacement vector $\boldsymbol{u}$ in the z-plane referred to the curvilinear coordinate system (ξ_c, η_c) can be expressed by the components (u_x, u_y) of the displacement vector $\boldsymbol{u}$ in the z-plane referred to the Cartesian coordinate system (x, y)

$$u_\xi+iu_\eta\equiv u_{\xi_c}+iu_{\eta_c}=e^{-i\alpha}(u_x+iu_y) \tag{5.113}$$

The components of stress in plane problems referred to the curvilinear coordinate system (ξ_c, η_c) can be expressed by the components referred to the Cartesian coordinate system (x, y) by use of Eqs. (4.6) and (a)

$$\begin{aligned}
\sigma_{\xi\xi}&=\sigma_{xx}\cos^2\alpha+\sigma_{yy}\sin^2\alpha+\sigma_{xy}\sin 2\alpha\\
\sigma_{\eta\eta}&=\sigma_{xx}\sin^2\alpha+\sigma_{yy}\cos^2\alpha-\sigma_{xy}\sin 2\alpha\\
\sigma_{\xi\eta}&=-\tfrac{1}{2}(\sigma_{xx}-\sigma_{yy})\sin 2\alpha+\sigma_{xy}\cos 2\alpha
\end{aligned} \tag{b}$$

Then

$$\begin{aligned}
\sigma_{\xi\xi}+\sigma_{\eta\eta}&=\sigma_{xx}+\sigma_{yy}\\
\sigma_{\eta\eta}-\sigma_{\xi\xi}+2i\sigma_{\xi\eta}&=e^{2i\alpha}(\sigma_{yy}-\sigma_{xx}+2i\sigma_{xy})
\end{aligned} \tag{5.114}$$

When a given region S in the complex z-plane is translated into a circle, a circular ring or an infinite plane with a circular hole in the complex ζ-plane by use of the conformal mapping function $\omega(\zeta)$, it is natural to select the polar coordinate system (ρ, θ) in the ζ-plane. The circles $\rho=$constant and radii $\theta=$constant in the ζ-plane correspond to certain curves in the z-plane which are related by the conformal mapping function $\omega(\zeta)$

$$z=x+iy=\omega(\zeta)=\omega(\rho e^{i\theta}), \qquad \zeta=\rho e^{i\theta} \tag{5.115}$$

The curvilinear coordinate system (ρ_c, θ_c) consists of curves $\rho=$constant and radii $\theta=$ constant. The components $(u_{\rho_c}, u_{\theta_c})$ of displacement vector $\boldsymbol{u}$ in the z-plane referred to the curvilinear coordinate system (ρ_c, θ_c) can be expressed by the components (u_x, u_y) of displacement vector $\boldsymbol{u}$ in the z-plane referred to the Cartesian coordinate system (x, y)

$$u_\rho+iu_\theta\equiv u_{\rho_c}+iu_{\theta_c}=e^{-i\alpha}(u_x+iu_y) \tag{5.116}$$

where α denotes the angle between x axis and ρ_c axis. The components of stress in plane problems referred to the curvilinear coordinate system (ρ_c, θ_c) can be expressed by the components referred to the Cartesian coordinate system (x, y)

$$\begin{aligned}\sigma_{\rho\rho}+\sigma_{\theta\theta}&=\sigma_{xx}+\sigma_{yy}\\ \sigma_{\theta\theta}-\sigma_{\rho\rho}+2i\sigma_{\rho\theta}&=e^{2i\alpha}(\sigma_{yy}-\sigma_{xx}+2i\sigma_{xy})\end{aligned} \tag{5.117}$$

When a small displacement dz is produced in the tangential direction ρ at the point z, a corresponding point ζ will undergo a small displacement $d\zeta$ in the radial direction in the ζ plane

$$dz=|dz|e^{i\alpha}, \qquad d\zeta=|d\zeta|e^{i\theta} \tag{c}$$

Then,

$$\begin{aligned}e^{i\alpha}&=\frac{dz}{|dz|}=\frac{\omega'(\zeta)\,d\zeta}{|\omega'(\zeta)|\cdot|d\zeta|}=e^{i\theta}\frac{\omega'(\zeta)}{|\omega'(\zeta)|}=\frac{\zeta}{\rho}\frac{\omega'(\zeta)}{|\omega'(\zeta)|}\\ e^{-i\alpha}&=e^{-i\theta}\frac{\overline{\omega'(\zeta)}}{|\omega'(\zeta)|}=\frac{\bar{\zeta}}{\rho}\frac{\overline{\omega'(\zeta)}}{|\omega'(\zeta)|}\end{aligned} \tag{5.118}$$

$$e^{2i\alpha}=\left[\frac{\zeta}{\rho}\frac{\omega'(\zeta)}{|\omega'(\zeta)|}\right]^2=\frac{\zeta^2}{\rho^2}\frac{\omega'(\zeta)\omega'(\zeta)}{\omega'(\zeta)\overline{\omega'(\zeta)}}=\frac{\zeta^2}{\rho^2}\frac{\omega'(\zeta)}{\overline{\omega'(\zeta)}} \tag{5.119}$$

Next, we introduce the new notation

$$\varphi(z)=\varphi(\omega(\zeta))\equiv\phi(\zeta), \qquad \psi(z)=\psi(\omega(\zeta))\equiv\Psi(\zeta)$$

$$\varphi'(z)=\frac{d\varphi(z)}{dz}=\frac{d\phi(\zeta)}{d\zeta}\frac{d\zeta}{dz}=\frac{1}{\omega'(\zeta)}\frac{d\phi(\zeta)}{d\zeta}=\frac{\phi'(\zeta)}{\omega'(\zeta)}$$

With this notation, and using Eq. (5.95) with $c^*=0$, Eq. (5.116) becomes

$$\begin{aligned}u_\rho+iu_\theta=\frac{1}{2G}\frac{\bar{\zeta}}{\rho}\frac{\overline{\omega'(\zeta)}}{|\omega'(\zeta)|}\Bigg[&\frac{3-\nu^*}{1+\nu^*}\phi(\zeta)-\frac{\omega(\zeta)}{\overline{\omega'(\zeta)}}\overline{\phi'(\zeta)}-\overline{\Psi(\zeta)}\\ &-\frac{\zeta}{\rho}\frac{1}{\omega'(\zeta)}\left(\frac{\partial\chi_p}{\partial\rho}+i\frac{1}{\rho}\frac{\partial\chi_p}{\partial\theta}\right)\Bigg]\end{aligned} \tag{5.120}$$

where χ_p must satisfy the governing equation

$$\left(\frac{\partial^2}{\partial\xi^2}+\frac{\partial^2}{\partial\eta^2}\right)\chi_p=-\alpha^*E^*\omega'(\zeta)\overline{\omega'(\zeta)}\tau \tag{5.121}$$

The stress fields (5.117) reduce to

$$\sigma_{\rho\rho}+\sigma_{\theta\theta}=4\,\mathrm{Re}\left[\frac{\phi'(\zeta)}{\omega'(\zeta)}\right]-\alpha^*E^*\tau$$

$$\sigma_{\theta\theta}-\sigma_{\rho\rho}+2i\sigma_{\rho\theta}=\frac{2\zeta^2}{\rho^2\overline{\omega'(\zeta)}}\left\{\overline{\omega(\zeta)}\left[\frac{\phi'(\zeta)}{\omega'(\zeta)}\right]'+\Psi'(\zeta)\right\}$$
$$+\frac{4\zeta^2}{\rho^2\overline{\omega'(\zeta)}}\left\{\frac{\partial^2\chi_p}{\partial\zeta^2}\frac{1}{\omega'(\zeta)}-\frac{\partial\chi_p}{\partial\zeta}\frac{\omega''(\zeta)}{[\omega'(\zeta)]^2}\right\} \tag{5.122}$$

Example 5.4 When the conformal mapping function is given by $z=\omega(\zeta)=\zeta+(a/\zeta)$, derive the relation between the z plane and the ζ plane.

Solution The conformal mapping function is

$$z=x+iy=\omega(\zeta)=\zeta+\frac{a}{\zeta}$$
$$=\rho e^{i\alpha}+\frac{a}{\rho e^{i\alpha}}=\left(\rho+\frac{a}{\rho}\right)\cos\alpha+i\left(\rho-\frac{a}{\rho}\right)\sin\alpha \tag{a}$$

Equation (a) gives

$$x=\left(\rho+\frac{a}{\rho}\right)\cos\alpha,\qquad y=\left(\rho-\frac{a}{\rho}\right)\sin\alpha \tag{b}$$

Elimination of α or ρ from Eqs. (b) gives, respectively

$$\left(\frac{x}{\rho+a\rho^{-1}}\right)^2+\left(\frac{y}{\rho-a\rho^{-1}}\right)^2=1 \tag{c}$$

$$\left(\frac{x}{\cos\alpha}\right)^2-\left(\frac{y}{\sin\alpha}\right)^2=4a \tag{d}$$

The line of $\rho=$constant in the ζ plane yields an ellipse in the z plane, and the line of $\alpha=$constant in the ζ plane yields a hyperbola in the z plane.

5.4 Potential Method

The Navier's equations (5.30) which are the governing equations expressed by displacement components for the plane problem can be rewritten as

$$\begin{aligned}\mu\nabla^2u_x+(\lambda^*+\mu)\left(\frac{\partial^2u_x}{\partial x^2}+\frac{\partial^2u_y}{\partial x\partial y}\right)-\beta^*\frac{\partial\tau}{\partial x}=0\\ \mu\nabla^2u_y+(\lambda^*+\mu)\left(\frac{\partial^2u_x}{\partial x\partial y}+\frac{\partial^2u_y}{\partial y^2}\right)-\beta^*\frac{\partial\tau}{\partial y}=0\end{aligned} \tag{5.123}$$

The general solutions of Navier's equations (5.123) for the plane problem can be expressed as the sum of the complementary solutions u_x^c and u_y^c, and the particular solutions u_x^p and u_y^p

$$u_x = u_x^c + u_x^p, \qquad u_y = u_y^c + u_y^p \tag{5.124}$$

where the complementary solutions u_x^c and u_y^c, and the particular solutions u_x^p and u_y^p are governed by linear homogeneous differential equations and linear non-homogeneous differential equations, respectively:

$$\begin{aligned}
&\mu\nabla^2 u_x^c + (\lambda^* + \mu)\left(\frac{\partial^2 u_x^c}{\partial x^2} + \frac{\partial^2 u_y^c}{\partial x \partial y}\right) = 0\\
&\mu\nabla^2 u_y^c + (\lambda^* + \mu)\left(\frac{\partial^2 u_x^c}{\partial x \partial y} + \frac{\partial^2 u_y^c}{\partial y^2}\right) = 0
\end{aligned} \tag{5.125}$$

$$\begin{aligned}
&\mu\nabla^2 u_x^p + (\lambda^* + \mu)\left(\frac{\partial^2 u_x^p}{\partial x^2} + \frac{\partial^2 u_y^p}{\partial x \partial y}\right) = \beta^* \frac{\partial \tau}{\partial x}\\
&\mu\nabla^2 u_y^p + (\lambda^* + \mu)\left(\frac{\partial^2 u_x^p}{\partial x \partial y} + \frac{\partial^2 u_y^p}{\partial y^2}\right) = \beta^* \frac{\partial \tau}{\partial y}
\end{aligned} \tag{5.126}$$

The particular solutions u_x^p and u_y^p can be expressed in terms of the Goodier's thermoelastic potential function Φ as follows:

$$u_x^p = \Phi_{,x}, \qquad u_y^p = \Phi_{,y} \tag{5.127}$$

Substituting Eqs. (5.127) into Eqs. (5.126), we get

$$\nabla^2 \Phi = K\tau \tag{5.128}$$

where

$$K = \frac{\beta^*}{\lambda^* + 2\mu} = (1 + \nu^*)\alpha^* \tag{5.129}$$

One solution of Eq. (5.128) can be obtained from the potential theory in the form

$$\Phi(x, y, t) = \frac{K}{2\pi} \iint \tau(x', y', t) \ln\sqrt{(x - x')^2 + (y - y')^2}\, dx'\, dy' \tag{5.130}$$

When transient thermoelasticity in a body without heat generation is treated, that is, the heat conduction equation for the temperature change is governed by

$$\frac{\partial \tau}{\partial t} = \kappa \nabla^2 \tau \tag{5.131}$$

the thermoelastic displacement potential may be determined in the same way as in Subsection 4.4.1

$$\Phi(x,y,t)=\kappa K\int_{t_r}^{t}\tau(x,y,t)dt+\Phi_r(x,y,t_r)+(t-t_r)\Phi_0(x,y) \tag{5.132}$$

where t_r denotes the reference time, and Φ_r and Φ_0 denote solutions of a following Poisson and harmonic equation, respectively:

$$\nabla^2\Phi_r(x,y,t_r)=K\tau_r(x,y,t_r) \tag{5.133}$$

$$\nabla^2\Phi_0(x,y)=0 \tag{5.134}$$

The function Φ_0 is usually selected in such a way that a thermoelastic displacement potential Φ has the finite value at an arbitrary time if the thermal stresses are to be finite at the arbitrary time. The function Φ_r is a thermoelastic displacement potential with temperature change τ_r at reference time (usually, at the initial time or final time).

Next, we consider the complementary solutions for Navier's equations (5.125). The displacement components may generally be expressed as the sum of the displacement components, one without dilatation and one without rotation. Then the displacement components u_x^c and u_y^c can be expressed as

$$u_x^c=\frac{\partial A}{\partial x}+\frac{\partial B}{\partial y},\qquad u_y^c=\frac{\partial A}{\partial y}-\frac{\partial B}{\partial x} \tag{a}$$

Substitution of Eqs. (a) into Eqs. (5.125) yields

$$\mu\nabla^2\left(\frac{\lambda^*+2\mu}{\mu}\frac{\partial A}{\partial x}+\frac{\partial B}{\partial y}\right)=0,\qquad \mu\nabla^2\left(\frac{\lambda^*+2\mu}{\mu}\frac{\partial A}{\partial y}-\frac{\partial B}{\partial x}\right)=0 \tag{b}$$

The solution of Eqs. (b) can be expressed as

$$\left(\frac{\lambda^*+2\mu}{\mu}\right)\frac{\partial A}{\partial x}+\frac{\partial B}{\partial y}=\varphi_1,\qquad \left(\frac{\lambda^*+2\mu}{\mu}\right)\frac{\partial A}{\partial y}-\frac{\partial B}{\partial x}=\varphi_2 \tag{c}$$

where functions φ_1 and φ_2 are harmonic functions:

$$\nabla^2\varphi_1=0,\qquad \nabla^2\varphi_2=0 \tag{d}$$

After differentiating the first equation and the second equation of Eqs. (c) with respect to x and y, respectively, the sum of these equations gives

$$\left(\frac{\lambda^*+2\mu}{\mu}\right)\nabla^2 A=\frac{\partial\varphi_1}{\partial x}+\frac{\partial\varphi_2}{\partial y} \tag{e}$$

The solution of Eq. (e) is

$$A=\frac{\mu}{2(\lambda^*+2\mu)}(x\varphi_1+y\varphi_2+\psi) \tag{f}$$

where ψ is a plane harmonic function. From Eqs. (c) and (f) we get

$$\begin{aligned}\frac{\partial B}{\partial y}&=\frac{1}{2}\left(\varphi_1-x\frac{\partial\varphi_1}{\partial x}-y\frac{\partial\varphi_2}{\partial x}-\frac{\partial\psi}{\partial x}\right)\\ \frac{\partial B}{\partial x}&=-\frac{1}{2}\left(\varphi_2-x\frac{\partial\varphi_1}{\partial y}-y\frac{\partial\varphi_2}{\partial y}-\frac{\partial\psi}{\partial y}\right)\end{aligned}\tag{g}$$

Therefore, the displacement components u_x^c and u_y^c can be expressed as

$$\begin{aligned}u_x^c&=\frac{\lambda^*+\mu}{2(\lambda^*+2\mu)}\left[\left(\frac{\lambda^*+3\mu}{\lambda^*+\mu}\right)\varphi_1-x\frac{\partial\varphi_1}{\partial x}-y\frac{\partial\varphi_2}{\partial x}-\frac{\partial\psi}{\partial x}\right]\\ u_y^c&=\frac{\lambda^*+\mu}{2(\lambda^*+2\mu)}\left[\left(\frac{\lambda^*+3\mu}{\lambda^*+\mu}\right)\varphi_2-x\frac{\partial\varphi_1}{\partial y}-y\frac{\partial\varphi_2}{\partial y}-\frac{\partial\psi}{\partial y}\right]\end{aligned}\tag{h}$$

Renaming the functions φ_1, φ_2, and ψ as:

$$\frac{\lambda^*+\mu}{2(\lambda^*+2\mu)}\varphi_1\Rightarrow\varphi_1,\qquad\frac{\lambda^*+\mu}{2(\lambda^*+2\mu)}\varphi_2\Rightarrow\varphi_2,\qquad\frac{\lambda^*+\mu}{2(\lambda^*+2\mu)}\psi\Rightarrow\psi\tag{i}$$

Eqs. (h) reduce to

$$\begin{aligned}u_x^c&=\frac{\lambda^*+3\mu}{\lambda^*+\mu}\varphi_1-x\frac{\partial\varphi_1}{\partial x}-y\frac{\partial\varphi_2}{\partial x}-\frac{\partial\psi}{\partial x}\\ u_y^c&=\frac{\lambda^*+3\mu}{\lambda^*+\mu}\varphi_2-x\frac{\partial\varphi_1}{\partial y}-y\frac{\partial\varphi_2}{\partial y}-\frac{\partial\psi}{\partial y}\end{aligned}\tag{5.135}$$

and their alternative forms are

$$\begin{aligned}u_x^c&=\frac{3-\nu^*}{1+\nu^*}\varphi_1-x\frac{\partial\varphi_1}{\partial x}-y\frac{\partial\varphi_2}{\partial x}-\frac{\partial\psi}{\partial x}\\ u_y^c&=\frac{3-\nu^*}{1+\nu^*}\varphi_2-x\frac{\partial\varphi_1}{\partial y}-y\frac{\partial\varphi_2}{\partial y}-\frac{\partial\psi}{\partial y}\end{aligned}\tag{5.136}$$

where the functions φ_1, φ_2 and ψ must satisfy the Laplace equation

$$\nabla^2\varphi_1=0,\qquad\nabla^2\varphi_2=0,\qquad\nabla^2\psi=0\tag{5.137}$$

The displacement components u_x^c and u_y^c are expressed by three plane harmonic functions. However, it is possible that the displacement components u_x^c and u_y^c may be expressed by two plane harmonic functions. We take three plane harmonic functions φ_1, φ_2, and ψ by employing three other functions ϕ, ϕ_1, and ϕ_2:

$$\varphi_1=\phi_1+\phi_{,x},\qquad\varphi_2=\phi_2+\phi_{,y},\qquad\psi=\frac{4}{1+\nu^*}\phi-x\phi_{,x}-y\phi_{,y}\tag{5.138}$$

Substituting Eqs. (5.138) into Eqs. (5.136) and (5.137), the displacement components u_x^c and u_y^c are expressed as

$$u_x^c = \frac{3-\nu^*}{1+\nu^*}\phi_1 - x\frac{\partial\phi_1}{\partial x} - y\frac{\partial\phi_2}{\partial x}, \qquad u_y^c = \frac{3-\nu^*}{1+\nu^*}\phi_2 - x\frac{\partial\phi_1}{\partial y} - y\frac{\partial\phi_2}{\partial y} \tag{5.139}$$

where two functions ϕ_1 and ϕ_2 are harmonic:

$$\nabla^2\phi_1 = 0, \qquad \nabla^2\phi_2 = 0 \tag{5.140}$$

Problems

5.1 Prove that the arbitrary integration constants C_1, C_2, and C_3 in Eq. (5.48) may be taken as zero for a simply connected body.

5.2 Prove that for the integration constants C_{1i}, C_{2i}, and C_{3i} on the boundary contour L_i $(i = 1, 2, \ldots, m)$ in Eq. (5.49), the integration constants on only one contour can be taken as zero.

5.3 Prove that no thermal stress results in a strip due to a temperature given by Eq. (3.35).

5.4 Evaluate the thermal stress in a rectangular body with $l \times b$ when temperature is given by Eq. (3.34).

Answer No thermal stress.

5.5 Find displacements in a strip when a temperature is given by Eq. (3.35).

Answer

$$u_x = \frac{C}{E^*} - c^* x + \alpha^*\left[T_a x + (T_b - T_a)\frac{x^2}{2l}\right]$$

$$u_y = \frac{B}{E^*} - c^* y + \alpha^*\left[T_a y + (T_b - T_a)\frac{xy}{l}\right]$$

5.6 Derive Eq. (5.120).

5.7 Derive Eq. (5.122).

5.8 The Airy's stress function F related to the components of stress

$$\sigma_{xx} = \frac{\partial^2 F}{\partial y^2}, \qquad \sigma_{yy} = \frac{\partial^2 F}{\partial x^2}, \qquad \sigma_{xy} = -\frac{\partial^2 F}{\partial x \partial y}$$

is usually used in isothermal plane problems, where a governing equation of F is $\nabla^4 F = 0$. Prove that the thermal stress in plane problems can be expressed by

$$\sigma_{xx} = \frac{\partial^2}{\partial y^2}(F - 2\mu\Phi), \qquad \sigma_{yy} = \frac{\partial^2}{\partial x^2}(F - 2\mu\Phi), \qquad \sigma_{xy} = -\frac{\partial^2}{\partial x \partial y}(F - 2\mu\Phi)$$

where Φ is the Goodier's thermoelastic potential function and F is the Airy's stress function.

5.9 Prove that the components of thermal stress in plane problems can be expressed by

$$\sigma_{xx} = 2\mu\left[-\Phi_{,yy} + x\phi_{1,yy} + y\phi_{2,yy} + \frac{2}{1+\nu^*}(\phi_{1,x} + \nu^*\phi_{2,y})\right]$$

$$\sigma_{yy} = 2\mu\left[-\Phi_{,xx} + x\phi_{1,xx} + y\phi_{2,xx} + \frac{2}{1+\nu^*}(\phi_{2,y} + \nu^*\phi_{1,x})\right]$$

$$\sigma_{xy} = 2\mu\left[\Phi_{,xy} - (x\phi_1 + y\phi_2)_{,xy} + \frac{1-\nu^*}{1+\nu^*}(\phi_{1,y} + \phi_{2,x})\right]$$

where the Goodier's thermoelastic potential function Φ in Eq. (5.127) and two harmonic functions ϕ_1, ϕ_2 in Eq. (5.139) are used.

Chapter 6

THERMAL STRESSES IN CIRCULAR CYLINDERS

Various techniques are introduced to determine steady and transient thermal stresses in 1D, 2D, and 3D solid and hollow cylinders. The displacement technique, which is one of the most common techniques for thermal stress field determination, is applied to the solution of 1D problems of cylindrical bodies. In this context, the thermal stress function, the thermoelastic potential technique, the complex variable method, the reciprocal theory, and the dislocation techniques are presented. The steady thermal stress field of infinitely long solid cylinders and the transient thermal stress for circular plates are obtained with the same thermal stress function technique. Next, 2D (axisymmetric) problems, both steady and transient, are treated with the Goodier's thermoelastic displacement potential and the Michell's function. Lastly, 3D problems in cylindrical coordinates are expressed in terms of the Goodier's thermoelastic displacement potential and the Boussinesq harmonic functions.

6.1 One-Dimensional Problems

In Chapter 5, many solution techniques for plane thermoelasticity were considered. This section treats one-dimensional problems of cylindrical bodies due to axi-symmetric temperature fields in order to demonstrate how to use the solution techniques and to compare various techniques. Some techniques are familiar, but it may be difficult to apply them to special problems. On the other hand, other less familiar techniques are useful in dealing with special problems.

6.1.1 Displacement Technique

The displacement technique is perhaps the most familiar technique applied to the solution of one-dimensional problems of cylindrical bodies. The one-dimensional equilibrium equation in the radial direction is

$$\frac{d\sigma_{rr}}{dr} + \frac{\sigma_{rr} - \sigma_{\theta\theta}}{r} = 0 \tag{6.1}$$

The generalized Hooke's law for plane problems in a cylindrical coordinate system is

$$\begin{aligned}
\epsilon_{rr} &= \frac{1}{E^*}\left(\sigma_{rr} - \nu^*\sigma_{\theta\theta}\right) + \alpha^*\tau - c^* \\
\epsilon_{\theta\theta} &= \frac{1}{E^*}\left(\sigma_{\theta\theta} - \nu^*\sigma_{rr}\right) + \alpha^*\tau - c^* \\
\epsilon_{r\theta} &= \frac{1}{2G}\sigma_{r\theta}
\end{aligned} \tag{6.2}$$

where

$$\begin{aligned}
E^* &= \begin{cases} \dfrac{E}{1-\nu^2} & \text{for plane strain} \\ E & \text{for plane stress} \end{cases} \\
\nu^* &= \begin{cases} \dfrac{\nu}{1-\nu} & \text{for plane strain} \\ \nu & \text{for plane stress} \end{cases} \\
\alpha^* &= \begin{cases} (1+\nu)\alpha & \text{for plane strain} \\ \alpha & \text{for plane stress} \end{cases} \\
c^* &= \begin{cases} \nu\epsilon_0 & \text{for plane strain} \\ 0 & \text{for plane stress} \end{cases}
\end{aligned} \tag{6.3}$$

in which the axial stress is zero ($\sigma_{zz} = 0$) for plane stress, and the constant axial strain ($\epsilon_{zz} = \epsilon_0$) may occur for a plane strain. When all strains and stresses are functions of r, the strain–displacement relations are

$$\epsilon_{rr} = \frac{du}{dr}, \qquad \epsilon_{\theta\theta} = \frac{u}{r}, \qquad \epsilon_{r\theta} = 0 \tag{6.4}$$

where u is the radial displacement. By use of Eqs. (6.2) and (6.4), the components of stress are

$$\begin{aligned}
\sigma_{rr} &= \frac{E^*}{1-\nu^{*2}}\left[\frac{du}{dr} + \nu^*\frac{u}{r} - (1+\nu^*)\alpha^*\tau + (1+\nu^*)c^*\right] \\
\sigma_{\theta\theta} &= \frac{E^*}{1-\nu^{*2}}\left[\nu^*\frac{du}{dr} + \frac{u}{r} - (1+\nu^*)\alpha^*\tau + (1+\nu^*)c^*\right] \\
\sigma_{r\theta} &= 0
\end{aligned} \tag{6.5}$$

Substitution of Eqs. (6.5) into Eq. (6.1) yields

$$\frac{d}{dr}\left[\frac{1}{r}\frac{d(ru)}{dr}\right] = (1+\nu^*)\alpha^*\frac{d\tau}{dr} \tag{6.6}$$

The general solution of Eq. (6.6) is

$$u = (1+\nu^*)\alpha^*\frac{1}{r}\int_r \tau r\,dr + C_1 r + \frac{C_2}{r} \tag{6.7}$$

where C_1 and C_2 are integration constants which may be determined from the boundary conditions. By use of Eqs. (6.5) and (6.7), the stresses are

$$\begin{aligned}
\sigma_{rr} &= -\frac{\alpha^* E^*}{r^2}\int_r \tau r\,dr + \frac{E^*}{1-\nu^*}C_1 - \frac{E^*}{1+\nu^*}\frac{C_2}{r^2} + \frac{E^*}{1-\nu^*}c^* \\
\sigma_{\theta\theta} &= \frac{\alpha^* E^*}{r^2}\int_r \tau r\,dr - \alpha^* E^*\tau + \frac{E^*}{1-\nu^*}C_1 + \frac{E^*}{1+\nu^*}\frac{C_2}{r^2} + \frac{E^*}{1-\nu^*}c^* \\
\sigma_{r\theta} &= 0
\end{aligned} \tag{6.8}$$

Let us consider a hollow circular cylinder with an inner radius a and an outer radius b. The boundary conditions for traction free surfaces are

$$\sigma_{rr} = 0 \qquad \text{at } r = a \text{ and } r = b \tag{6.9}$$

The integration constants C_1 and C_2 can be determined from Eq. (6.9)

$$\begin{aligned}
C_1 &= \alpha^*(1-\nu^*)\frac{1}{b^2-a^2}\int_a^b \tau r\,dr - c^* \\
C_2 &= \alpha^*(1+\nu^*)\frac{a^2}{b^2-a^2}\int_a^b \tau r\,dr
\end{aligned}$$

Thus, the stresses are

$$\begin{aligned}
\sigma_{rr} &= \alpha^* E^*\left[-\frac{1}{r^2}\int_a^r \tau r\,dr + \frac{r^2-a^2}{r^2(b^2-a^2)}\int_a^b \tau r\,dr\right] \\
\sigma_{\theta\theta} &= \alpha^* E^*\left[\frac{1}{r^2}\int_a^r \tau r\,dr + \frac{r^2+a^2}{r^2(b^2-a^2)}\int_a^b \tau r\,dr - \tau\right] \\
\sigma_{zz} &= \begin{cases} 0 & \text{for plane stress} \\ \dfrac{\alpha E}{1-\nu}\left(\dfrac{2\nu}{b^2-a^2}\displaystyle\int_a^b \tau r\,dr - \tau\right) + E\epsilon_0 & \text{for plane strain} \end{cases}
\end{aligned} \tag{6.10}$$

and the radial displacement is

$$u = (1+\nu^*)\alpha^*\left[\frac{1}{r}\int_a^r \tau r\,dr + \left(\frac{1-\nu^*}{1+\nu^*}r + \frac{a^2}{r}\right)\frac{1}{b^2-a^2}\int_a^b \tau r\,dr\right] - c^* r \tag{6.11}$$

The constant axial strain ϵ_0 for plane strain can be determined from the condition that the axial force is zero

$$2\pi \int_a^b \sigma_{zz} r\, dr = 0 \tag{6.12}$$

Substituting the third of Eqs. (6.10) into above equation, the constant axial strain ϵ_0 is given by

$$\epsilon_0 = \frac{2\alpha}{b^2 - a^2} \int_a^b \tau r\, dr \tag{6.13}$$

The axial stress and the radial displacement for plane strain with constant axial strain are

$$\begin{aligned}
\sigma_{zz} &= \frac{\alpha E}{1-\nu}\left(\frac{2}{b^2-a^2}\int_a^b \tau r\, dr - \tau\right) \\
u &= \frac{1+\nu}{1-\nu}\alpha\left[\frac{1}{r}\int_a^r \tau r\, dr + \left(\frac{1-3\nu}{1+\nu}r + \frac{a^2}{r}\right)\frac{1}{b^2-a^2}\int_a^b \tau r\, dr\right]
\end{aligned} \tag{6.14}$$

Therefore, the stresses and radial displacement for a circular cylinder are for plane stress (thin hollow circular disk):

$$\begin{aligned}
u &= (1+\nu)\alpha\left[\frac{1}{r}\int_a^r \tau r\, dr + \left(\frac{1-\nu}{1+\nu}r + \frac{a^2}{r}\right)\frac{1}{b^2-a^2}\int_a^b \tau r\, dr\right] \\
\sigma_{rr} &= \alpha E\left[-\frac{1}{r^2}\int_a^r \tau r\, dr + \frac{r^2-a^2}{r^2(b^2-a^2)}\int_a^b \tau r\, dr\right] \\
\sigma_{\theta\theta} &= \alpha E\left[\frac{1}{r^2}\int_a^r \tau r\, dr + \frac{r^2+a^2}{r^2(b^2-a^2)}\int_a^b \tau r\, dr - \tau\right] \\
\sigma_{zz} &= \sigma_{r\theta} = 0
\end{aligned} \tag{6.15}$$

for plane strain (long hollow circular cylinder):

$$u = \begin{cases}
\dfrac{1+\nu}{1-\nu}\alpha\left\{\dfrac{1}{r}\displaystyle\int_a^r \tau r\, dr + \left[(1-2\nu)r + \dfrac{a^2}{r}\right]\dfrac{1}{b^2-a^2}\displaystyle\int_a^b \tau r\, dr\right\} \\
\text{for } \epsilon_{zz} = 0 \\
\dfrac{1+\nu}{1-\nu}\alpha\left[\dfrac{1}{r}\displaystyle\int_a^r \tau r\, dr + \left(\dfrac{1-3\nu}{1+\nu}r + \dfrac{a^2}{r}\right)\dfrac{1}{b^2-a^2}\displaystyle\int_a^b \tau r\, dr\right] \\
\text{for } \epsilon_{zz} = \epsilon_0
\end{cases}$$

$$\sigma_{rr} = \frac{\alpha E}{1-\nu}\left[-\frac{1}{r^2}\int_a^r \tau r\, dr + \frac{r^2-a^2}{r^2(b^2-a^2)}\int_a^b \tau r\, dr\right] \tag{6.16}$$

$$\sigma_{\theta\theta} = \frac{\alpha E}{1-\nu}\left[\frac{1}{r^2}\int_a^r \tau r\,dr + \frac{r^2+a^2}{r^2(b^2-a^2)}\int_a^b \tau r\,dr - \tau\right]$$

$$\sigma_{zz} = \begin{cases} \dfrac{\alpha E}{1-\nu}\left(\dfrac{2\nu}{b^2-a^2}\displaystyle\int_a^b \tau r\,dr - \tau\right) & \text{for } \epsilon_{zz} = 0 \\ \dfrac{\alpha E}{1-\nu}\left(\dfrac{2}{b^2-a^2}\displaystyle\int_a^b \tau r\,dr - \tau\right) & \text{for } \epsilon_{zz} = \epsilon_0 \end{cases}$$

$$\sigma_{r\theta} = 0$$

The above solutions place no restriction on the temperature field. Hence, the formulae for radial displacement and stresses are valid for both steady and transient temperature fields in the hollow circular cylinder.

Next, we consider a solid circular cylinder with a radius a. The radial displacement in this case is given by

$$u = (1+\nu^*)\alpha^*\frac{1}{r}\int_0^r \tau r\,dr + C_1 r + \frac{C_2}{r} \tag{6.17}$$

Since the displacement is finite when r tends to zero, C_2 must be zero, so that the displacement reduces to

$$u = (1+\nu^*)\alpha^*\frac{1}{r}\int_0^r \tau r\,dr + C_1 r \tag{6.18}$$

Requiring that the surface be traction free gives

$$C_1 = \frac{1-\nu^*}{a^2}\alpha^*\int_0^a \tau r\,dr - c^*$$

The displacement and stresses are

$$\begin{aligned} u &= (1+\nu^*)\alpha^*\left(\frac{1}{r}\int_0^r \tau r\,dr + \frac{1-\nu^*}{1+\nu^*}\frac{r}{a^2}\int_0^a \tau r\,dr\right) - c^* r \\ \sigma_{rr} &= \alpha^* E^*\left(-\frac{1}{r^2}\int_0^r \tau r\,dr + \frac{1}{a^2}\int_0^a \tau r\,dr\right) \\ \sigma_{\theta\theta} &= \alpha^* E^*\left(\frac{1}{r^2}\int_0^r \tau r\,dr + \frac{1}{a^2}\int_0^a \tau r\,dr - \tau\right) \\ \sigma_{zz} &= \begin{cases} 0 & \text{for plane stress} \\ \dfrac{\alpha E}{1-\nu}\left(\dfrac{2\nu}{a^2}\displaystyle\int_0^a \tau r\,dr - \tau\right) + E\epsilon_0 & \text{for plane strain} \end{cases} \end{aligned} \tag{6.19}$$

The constant axial strain ϵ_0 for plane strain can be determined from the condition that the axial force is zero

$$2\pi\int_0^a \sigma_{zz} r\,dr = 0 \tag{6.20}$$

Substituting the fourth of Eqs. (6.19) into the above equation gives constant axial strain ϵ_0 for plane strain. Then we have for plane stress (thin solid circular disk):

$$
\begin{aligned}
u &= (1+\nu)\alpha\left(\frac{1}{r}\int_0^r \tau r\,dr + \frac{1-\nu}{1+\nu}\frac{r}{a^2}\int_0^a \tau r\,dr\right)\\
\sigma_{rr} &= \alpha E\left(-\frac{1}{r^2}\int_0^r \tau r\,dr + \frac{1}{a^2}\int_0^a \tau r\,dr\right)\\
\sigma_{\theta\theta} &= \alpha E\left(\frac{1}{r^2}\int_0^r \tau r\,dr + \frac{1}{a^2}\int_0^a \tau r\,dr - \tau\right)\\
\sigma_{zz} &= \sigma_{r\theta} = 0
\end{aligned}
\tag{6.21}
$$

for plane strain (long solid circular cylinder):

$$
\begin{aligned}
u &= \begin{cases} \dfrac{1+\nu}{1-\nu}\alpha\left[\dfrac{1}{r}\displaystyle\int_0^r \tau r\,dr + (1-2\nu)\dfrac{r}{a^2}\displaystyle\int_0^a \tau r\,dr\right] & \text{for } \epsilon_{zz} = 0\\ \dfrac{1+\nu}{1-\nu}\alpha\left(\dfrac{1}{r}\displaystyle\int_0^r \tau r\,dr + \dfrac{1-3\nu}{1+\nu}\dfrac{r}{a^2}\displaystyle\int_0^a \tau r\,dr\right) & \text{for } \epsilon_{zz} = \epsilon_0 \end{cases}\\
\sigma_{rr} &= \frac{\alpha E}{1-\nu}\left(-\frac{1}{r^2}\int_0^r \tau r\,dr + \frac{1}{a^2}\int_0^a \tau r\,dr\right)\\
\sigma_{\theta\theta} &= \frac{\alpha E}{1-\nu}\left(\frac{1}{r^2}\int_0^r \tau r\,dr + \frac{1}{a^2}\int_0^a \tau r\,dr - \tau\right)\\
\sigma_{zz} &= \begin{cases} \dfrac{\alpha E}{1-\nu}\left(\dfrac{2\nu}{a^2}\displaystyle\int_0^a \tau r\,dr - \tau\right) & \text{for } \epsilon_{zz} = 0\\ \dfrac{\alpha E}{1-\nu}\left(\dfrac{2}{a^2}\displaystyle\int_0^a \tau r\,dr - \tau\right) & \text{for } \epsilon_{zz} = \epsilon_0 \end{cases}\\
\sigma_{r\theta} &= 0
\end{aligned}
\tag{6.22}
$$

Example 6.1 When both the inner surface ($r = a$) and the outer surface ($r = b$) of a hollow circular cylinder at initial temperature T_i are heated to T_a and T_b, respectively, find the thermal stresses in the cylinder under a steady temperature field.

Solution The temperature is given by Eq. (3.82), so that temperature difference τ is

$$
\tau = T_a - T_i + (T_b - T_a)\frac{\ln(r/a)}{\ln(b/a)} \tag{6.23}
$$

Substitution of Eq. (6.23) into Eqs. (6.15) and (6.16) yields

for plane stress:

$$
\begin{aligned}
u &= \alpha(T_a - T_i)r + \alpha(T_b - T_a)r\left\{\frac{1+\nu}{2}\frac{\ln(r/a)}{\ln(b/a)}\right.\\
&\quad \left. + \left[1-\nu+(1+\nu)\frac{a^2}{r^2}\right]\frac{b^2}{2(b^2-a^2)} - \frac{1}{2\ln(b/a)}\right\}\\
\sigma_{rr} &= \frac{\alpha E}{2}(T_b - T_a)\left[-\frac{\ln(r/a)}{\ln(b/a)} + \left(1-\frac{a^2}{r^2}\right)\frac{b^2}{b^2-a^2}\right]\\
\sigma_{\theta\theta} &= \frac{\alpha E}{2}(T_b - T_a)\left[-\frac{1+\ln(r/a)}{\ln(b/a)} + \left(1+\frac{a^2}{r^2}\right)\frac{b^2}{b^2-a^2}\right]\\
\sigma_{zz} &= \sigma_{r\theta} = 0
\end{aligned}
\tag{6.24}
$$

for plane strain:

$$
\begin{aligned}
u &= \begin{cases}
(1+\nu)\alpha(T_a - T_i)r + \dfrac{1+\nu}{1-\nu}\alpha(T_b - T_a)r\left[\dfrac{\ln(r/a)}{2\ln(b/a)}\right.\\
\quad \left. + \left(1-2\nu+\dfrac{a^2}{r^2}\right)\dfrac{b^2}{2(b^2-a^2)} - \dfrac{1-\nu}{2\ln(b/a)}\right] & \text{for } \epsilon_{zz} = 0\\
\alpha(T_a - T_i)r + \dfrac{1+\nu}{1-\nu}\alpha(T_b - T_a)r\left[\dfrac{\ln(r/a)}{2\ln(b/a)}\right.\\
\quad \left. + \left(\dfrac{1-3\nu}{1+\nu}+\dfrac{a^2}{r^2}\right)\dfrac{b^2}{2(b^2-a^2)} - \dfrac{1-\nu}{1+\nu}\dfrac{1}{2\ln(b/a)}\right] & \text{for } \epsilon_{zz} = \epsilon_0
\end{cases}\\
\sigma_{rr} &= \frac{\alpha E}{2(1-\nu)}(T_b - T_a)\left[-\frac{\ln(r/a)}{\ln(b/a)} + \left(1-\frac{a^2}{r^2}\right)\frac{b^2}{b^2-a^2}\right]\\
\sigma_{\theta\theta} &= \frac{\alpha E}{2(1-\nu)}(T_b - T_a)\left[-\frac{1+\ln(r/a)}{\ln(b/a)} + \left(1+\frac{a^2}{r^2}\right)\frac{b^2}{b^2-a^2}\right]\\
\sigma_{zz} &= \begin{cases}
-\alpha E(T_a - T_i) + \dfrac{\alpha E}{2(1-\nu)}(T_b - T_a)\left[-\dfrac{\nu+2\ln(r/a)}{\ln(b/a)} + \dfrac{2\nu b^2}{b^2-a^2}\right]\\
\hfill \text{for } \epsilon_{zz} = 0\\
\dfrac{\alpha E}{2(1-\nu)}(T_b - T_a)\left[-\dfrac{1+2\ln(r/a)}{\ln(b/a)} + \dfrac{2b^2}{b^2-a^2}\right] \quad \text{for } \epsilon_{zz} = \epsilon_0
\end{cases}\\
\sigma_{r\theta} &= 0
\end{aligned}
\tag{6.25}
$$

Figure 6.1 shows the non-dimensional thermal stresses $\widehat{rr}$, $\widehat{\theta\theta}$ and $\widehat{zz}$, in which

$$
\widehat{rr} = \sigma_{rr}/\alpha^* E^*(T_b - T_a), \quad \widehat{\theta\theta} = \sigma_{\theta\theta}/\alpha^* E^*(T_b - T_a), \quad \widehat{zz} = \sigma_{zz}/\alpha^* E^*(T_b - T_a)
$$

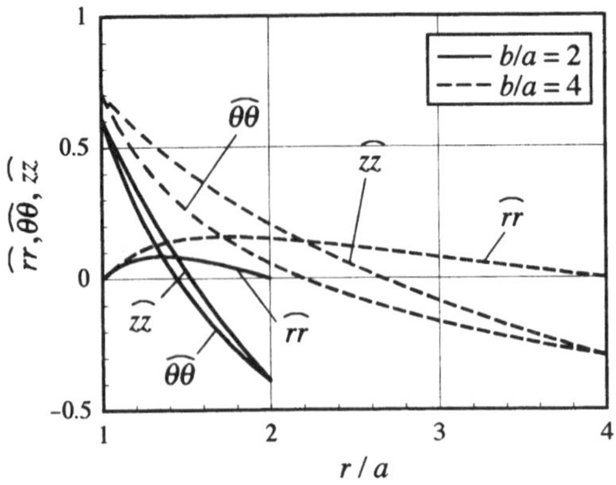

Figure 6.1 Stresses in a hollow circular cylinder.

Of course, the non-dimensional axial stress $\widehat{zz}$ in Fig. 6.1 is shown for plane strain with $\epsilon_{zz} = \epsilon_0$, but the axial stress is zero for the plane stress.

The radial and axial thermal stresses at $r = a$ and $r = b$ for the plane strain with $\epsilon_{zz} = \epsilon_0$ reduce to

$$
\begin{aligned}
(\sigma_{\theta\theta})_{r=a} = (\sigma_{zz})_{r=a} &= \frac{\alpha E}{2(1-\nu)}(T_b - T_a)\left(-\frac{1}{\ln(b/a)} + \frac{2b^2}{b^2 - a^2}\right) \\
(\sigma_{\theta\theta})_{r=b} = (\sigma_{zz})_{r=b} &= \frac{\alpha E}{2(1-\nu)}(T_b - T_a)\left(-\frac{1}{\ln(b/a)} + \frac{2a^2}{b^2 - a^2}\right)
\end{aligned}
\tag{6.26}
$$

When the thickness of the hollow circular cylinder is very small, it is found

$$
2\ln\frac{b}{a} = \ln\left(\frac{b}{a}\right)^2 = \ln\left[\frac{1 + (b^2 - a^2)/(b^2 + a^2)}{1 - (b^2 - a^2)/(b^2 + a^2)}\right] = \ln\frac{1+\delta}{1-\delta}
$$

where $\delta = (b^2 - a^2)/(b^2 + a^2)$. When δ is very small, we find

$$
\ln\frac{1+\delta}{1-\delta} = 2\left(\delta + \frac{\delta^3}{3} + \cdots\right) \cong 2\delta
$$

Then,

$$
2\ln\frac{b}{a} \cong 2\frac{b^2 - a^2}{b^2 + a^2}
$$

Substitution of this equation into Eq. (6.26) gives the thermal stresses at the boundary surfaces

$$
\begin{aligned}
(\sigma_{\theta\theta})_{r=a} = (\sigma_{zz})_{r=a} &= \frac{\alpha E}{2(1-\nu)}(T_b - T_a) \\
(\sigma_{\theta\theta})_{r=b} = (\sigma_{zz})_{r=b} &= -\frac{\alpha E}{2(1-\nu)}(T_b - T_a)
\end{aligned}
\tag{6.27}
$$

On the other hand, the thermal stresses at the boundary surfaces for very thick hollow circular cylinder are

$$
\begin{aligned}
(\sigma_{\theta\theta})_{r=a} &= (\sigma_{zz})_{r=a} = \frac{\alpha E}{1-\nu}(T_b - T_a) \\
(\sigma_{\theta\theta})_{r=b} &= (\sigma_{zz})_{r=b} = 0
\end{aligned}
\tag{6.28}
$$

It is found that the stresses at the inner surface of a very thick hollow circular cylinder are two times as large as those for very thin hollow circular cylinder.

Example 6.2 When the temperature at the boundary surface of a long solid circular cylinder of radius a with the initial temperature T_i is raised to T_a, find the thermal stresses in the cylinder under a transient temperature field.

Solution The temperature is given by Eq. (3.92), so that temperature difference τ is

$$
\tau = (T_a - T_i)\left[1 - 2\sum_{n=1}^{\infty} \frac{J_0(s_n r)}{s_n a J_1(s_n a)} e^{-\kappa s_n^2 t}\right] \tag{6.29}
$$

where $J_i(x)$ denotes the Bessel function of the first kind of order i, and s_n are roots of the equation $J_0(s_n a) = 0$. Using the integral formula:

$$
\int_0^r r J_0(\alpha r)\, dr = \frac{r}{\alpha} J_1(\alpha r)
$$

we find

$$
\begin{aligned}
\int_0^r \tau r\, dr &= (T_a - T_i)\left[\frac{r^2}{2} - 2\sum_{n=1}^{\infty} \frac{r J_1(s_n r)}{s_n^2 a J_1(s_n a)} e^{-\kappa s_n^2 t}\right] \\
\int_0^a \tau r\, dr &= (T_a - T_i)\left[\frac{a^2}{2} - 2\sum_{n=1}^{\infty} \frac{1}{s_n^2} e^{-\kappa s_n^2 t}\right]
\end{aligned}
$$

Substitution of these equations into Eqs. (6.22) gives the thermal stresses for plane strain with $\epsilon_{zz} = \epsilon_0$:

$$
\begin{aligned}
\sigma_{rr} &= -\frac{2\alpha E}{1-\nu}(T_a - T_i)\sum_{n=1}^{\infty}\left[\frac{1}{s_n^2 a^2} - \frac{J_1(s_n r)}{s_n^2 a r J_1(s_n a)}\right] e^{-\kappa s_n^2 t} \\
\sigma_{\theta\theta} &= -\frac{2\alpha E}{1-\nu}(T_a - T_i)\sum_{n=1}^{\infty}\left[\frac{1}{s_n^2 a^2} + \frac{J_1(s_n r) - s_n r J_0(s_n r)}{s_n^2 a r J_1(s_n a)}\right] e^{-\kappa s_n^2 t} \\
\sigma_{zz} &= -\frac{2\alpha E}{1-\nu}(T_a - T_i)\sum_{n=1}^{\infty}\left[\frac{2}{s_n^2 a^2} - \frac{J_0(s_n r)}{s_n a J_1(s_n a)}\right] e^{-\kappa s_n^2 t}
\end{aligned}
\tag{6.30}
$$

Figure 6.2 shows transient non-dimensional thermal stresses $\widehat{rr}$, $\widehat{\theta\theta}$, and $\widehat{zz}$ for various non-dimensional times $F(=\kappa t/a^2)$, where

$$\widehat{rr} = \frac{\sigma_{rr}}{\dfrac{\alpha E}{1-\nu}(T_a - T_i)}, \qquad \widehat{\theta\theta} = \frac{\sigma_{\theta\theta}}{\dfrac{\alpha E}{1-\nu}(T_a - T_i)}, \qquad \text{and} \qquad \widehat{zz} = \frac{\sigma_{zz}}{\dfrac{\alpha E}{1-\nu}(T_a - T_i)}$$

Example 6.3 When both the inner surface ($r = a$) and the outer surface ($r = b$) of a hollow circular cylinder at constant initial temperature T_i are heated to T_a and T_b, respectively, find the thermal stress in the hollow cylinder under transient temperature field (See Eq. (c) in Example 3.9).

Solution The temperature is given by Eq. (c) in Example 3.9, so that the temperature change τ is

$$\begin{aligned} \tau &= (T_a - T_i) + (T_b - T_a)\left[\frac{\ln(r/a)}{\ln(b/a)}\right] \\ &\quad - \pi \sum_{n=1}^{\infty} \frac{(T_a - T_i)J_0(s_n b) - (T_b - T_i)J_0(s_n a)}{J_0^2(s_n b) - J_0^2(s_n a)} J_0(s_n b) f(s_n, r) e^{-\kappa s_n^2 t} \\ &= (T_a - T_i) + (T_b - T_a)\left[\frac{\ln(r/a)}{\ln(b/a)}\right] + \sum_{n=1}^{\infty} F_n(t) f(s_n, r) \end{aligned} \tag{a}$$

where

$$\begin{aligned} f(s_n, r) &= Y_0(s_n a)J_0(s_n r) - J_0(s_n a)Y_0(s_n r) \\ F_n(t) &= -\pi \frac{(T_a - T_i)J_0(s_n b) - (T_b - T_i)J_0(s_n a)}{J_0^2(s_n b) - J_0^2(s_n a)} J_0(s_n b) e^{-\kappa s_n^2 t} \end{aligned} \tag{b}$$

and s_n are the positive roots of the equation $f(s_n, b) = 0$. First, we calculate an integral:

$$\begin{aligned} \int_a^r f(s_n, r) r\, dr &= \frac{1}{s_n}\left[r Y_0(s_n a)J_1(s_n r) - r J_0(s_n a)Y_1(s_n r)\right]_a^r \\ &= \frac{r}{s_n}\left[Y_0(s_n a)J_1(s_n r) - J_0(s_n a)Y_1(s_n r)\right] - \frac{a}{s_n}\frac{2}{\pi s_n a} \\ &= \frac{r}{s_n} f_1(s_n, r) - \frac{2}{\pi s_n^2} \end{aligned} \tag{c}$$

where

$$f_1(s_n, r) = Y_0(s_n a)J_1(s_n r) - J_0(s_n a)Y_1(s_n r) \tag{d}$$

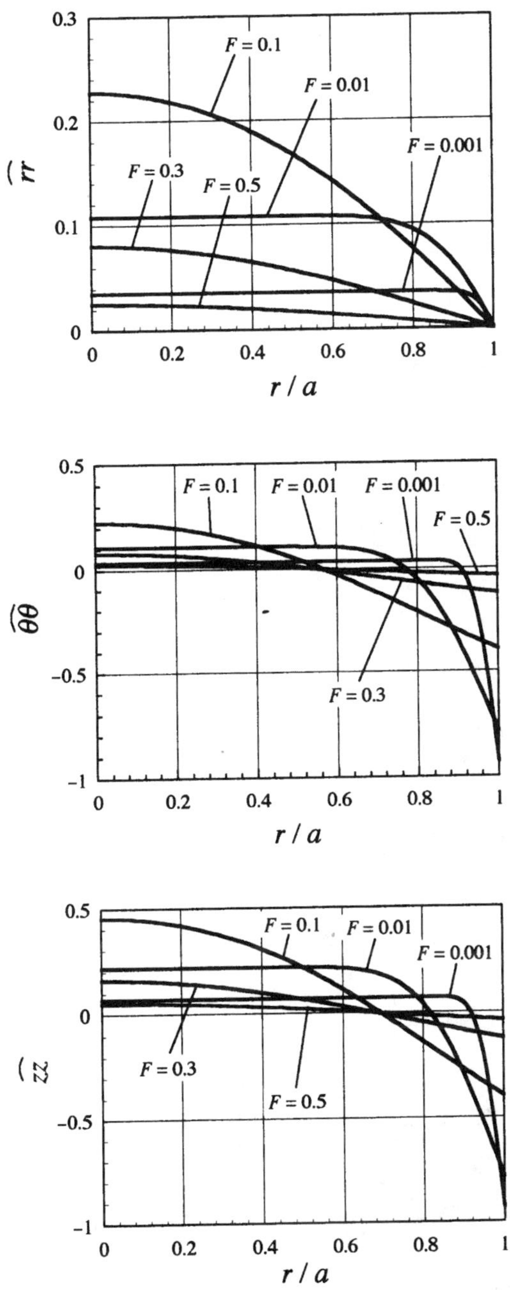

Figure 6.2 Transient thermal stresses in a long solid circular cylinder.

On the other hand, from Eqs. (b), (c), and (d) we get

$$\int_a^b f(s_n, r) r\, dr = \frac{2}{\pi s_n^2} \frac{J_0(s_n a) - J_0(s_n b)}{J_0(s_n b)} \tag{e}$$

Taking Eqs. (c) and (e) into consideration, the substitution of Eq. (a) into Eqs. (6.10) gives

$$\begin{aligned}
\sigma_{rr} &= \alpha^* E^* \left[-\frac{1}{r^2} \int_a^r \tau r\, dr + \frac{r^2 - a^2}{r^2 (b^2 - a^2)} \int_a^b \tau r\, dr \right] \\
&= \frac{1}{2} \alpha^* E^* (T_b - T_a) \left[-\frac{\ln(r/a)}{\ln(b/a)} + \left(1 - \frac{a^2}{r^2}\right) \frac{b^2}{b^2 - a^2} \right] \\
&\quad + \alpha^* E^* \left\{ -\frac{1}{r^2} \sum_{n=1}^{\infty} F_n(t) \left[\frac{r}{s_n} f_1(s_n, r) - \frac{2}{\pi s_n^2} \right] \right. \\
&\quad \left. + \frac{r^2 - a^2}{r^2 (b^2 - a^2)} \sum_{n=1}^{\infty} F_n(t) \frac{2}{\pi s_n^2} \frac{J_0(s_n a) - J_0(s_n b)}{J_0(s_n b)} \right\} \\
&= \frac{1}{2} \alpha^* E^* (T_b - T_a) \left[-\frac{\ln(r/a)}{\ln(b/a)} + \left(1 - \frac{a^2}{r^2}\right) \frac{b^2}{b^2 - a^2} \right] \\
&\quad + \alpha^* E^* \sum_{n=1}^{\infty} F_n(t) \left[-\frac{1}{r s_n} f_1(s_n, r) + \frac{2}{\pi s_n^2} \frac{(r^2 - a^2) J_0(s_n a) + (b^2 - r^2) J_0(s_n b)}{r^2 (b^2 - a^2) J_0(s_n b)} \right]
\end{aligned} \tag{f}$$

$$\begin{aligned}
\sigma_{\theta\theta} &= \alpha^* E^* \left[\frac{1}{r^2} \int_a^r \tau r\, dr + \frac{r^2 + a^2}{r^2 (b^2 - a^2)} \int_a^b \tau r\, dr - \tau \right] \\
&= \frac{1}{2} \alpha^* E^* (T_b - T_a) \left[-\frac{1 + \ln(r/a)}{\ln(b/a)} + \left(1 + \frac{a^2}{r^2}\right) \frac{b^2}{b^2 - a^2} \right] \\
&\quad + \alpha^* E^* \left\{ \frac{1}{r^2} \sum_{n=1}^{\infty} F_n(t) \left[\frac{r}{s_n} f_1(s_n, r) - \frac{2}{\pi s_n^2} \right] + \frac{r^2 + a^2}{r^2 (b^2 - a^2)} \right. \\
&\quad \left. \times \sum_{n=1}^{\infty} F_n(t) \frac{2}{\pi s_n^2} \frac{J_0(s_n a) - J_0(s_n b)}{J_0(s_n b)} - \sum_{n=1}^{\infty} F_n(t) f(s_n, r) \right\} \\
&= \frac{1}{2} \alpha^* E^* (T_b - T_a) \left[-\frac{1 + \ln(r/a)}{\ln(b/a)} + \left(1 + \frac{a^2}{r^2}\right) \frac{b^2}{b^2 - a^2} \right] \\
&\quad + \alpha^* E^* \sum_{n=1}^{\infty} F_n(t) \left[\frac{1}{r s_n} f_1(s_n, r) - f(s_n, r) \right. \\
&\quad \left. + \frac{2}{\pi s_n^2} \frac{(r^2 + a^2) J_0(s_n a) - (b^2 + r^2) J_0(s_n b)}{r^2 (b^2 - a^2) J_0(s_n b)} \right]
\end{aligned} \tag{g}$$

6.1.2 Thermal Stress Function Technique

The equilibrium equation (6.1) is satisfied if the stress is expressed in terms of the thermal stress function χ as

$$\sigma_{rr} = \frac{1}{r}\frac{d\chi}{dr}, \qquad \sigma_{\theta\theta} = \frac{d^2\chi}{dr^2} \tag{6.31}$$

The compatibility equation is given by

$$\frac{d\epsilon_{\theta\theta}}{dr} = \frac{\epsilon_{rr} - \epsilon_{\theta\theta}}{r} \tag{6.32}$$

Substitution of Eqs. (6.2) into Eq. (6.32) gives

$$\frac{d}{dr}\left[\frac{1}{r}\frac{d}{dr}\left(r\frac{d\chi}{dr}\right)\right] = -\alpha^* E^* \frac{d\tau}{dr} \tag{6.33}$$

where Eqs. (6.31) have been used. The general solution of Eq. (6.33) is

$$\chi = -\alpha^* E^* \int \frac{1}{r} \int \tau r \, dr \, dr + C_1 + C_2 \ln r + C_3 r^2 \tag{6.34}$$

Substitution of Eq. (6.34) into Eqs. (6.31) gives the following radial and hoop stresses:

$$\begin{aligned} \sigma_{rr} &= -\alpha^* E^* \frac{1}{r^2} \int \tau r \, dr + \frac{C_2}{r^2} + 2C_3 \\ \sigma_{\theta\theta} &= \alpha^* E^* \left(\frac{1}{r^2} \int \tau r \, dr - \tau\right) - \frac{C_2}{r^2} + 2C_3 \end{aligned} \tag{6.35}$$

When we consider a hollow circular cylinder, the unknown coefficients C_2 and C_3 can be determined from the boundary conditions (6.9)

$$C_2 = -\alpha^* E^* \frac{a^2}{b^2 - a^2} \int_a^b \tau r \, dr, \qquad C_3 = \frac{\alpha^* E^*}{2(b^2 - a^2)} \int_a^b \tau r \, dr \tag{6.36}$$

Thus, the thermal stresses are

$$\begin{aligned} \sigma_{rr} &= \alpha^* E^* \left[-\frac{1}{r^2} \int_a^r \tau r \, dr + \frac{r^2 - a^2}{r^2(b^2 - a^2)} \int_a^b \tau r \, dr\right] \\ \sigma_{\theta\theta} &= \alpha^* E^* \left[\frac{1}{r^2} \int_a^r \tau r \, dr + \frac{r^2 + a^2}{r^2(b^2 - a^2)} \int_a^b \tau r \, dr - \tau\right] \end{aligned} \tag{6.37}$$

Notice that Eqs. (6.37) agree with Eqs. (6.10) derived by the displacement technique.

Next, we treat the same problem by another method, which also makes use of the thermal stress function technique. The governing equation (5.33) for axisymmetric cylinder reduces to

$$\frac{1}{r}\frac{d}{dr}\left\{r\frac{d}{dr}\left[\frac{1}{r}\frac{d}{dr}\left(r\frac{d\chi}{dr}\right)\right]\right\} = -\alpha^* E^* \frac{1}{r}\frac{d}{dr}\left(r\frac{d\tau}{dr}\right) \tag{6.38}$$

The general solution of Eq. (6.38) is

$$\chi = -\alpha^* E^* \int \frac{1}{r} \int \tau r \, dr \, dr + C_1 + C_2 \ln r + C_3 r^2 + C_4 r^2 \ln r \tag{6.39}$$

Substitution of Eq. (6.39) into Eqs. (6.31) gives the following radial and hoop stresses:

$$\begin{aligned} \sigma_{rr} &= -\alpha^* E^* \frac{1}{r^2} \int \tau r \, dr \, dr + \frac{C_2}{r^2} + 2C_3 + C_4 (1 + 2 \ln r) \\ \sigma_{\theta\theta} &= \alpha^* E^* \left(\frac{1}{r^2} \int \tau r \, dr - \tau \right) - \frac{C_2}{r^2} + 2C_3 + C_4 (3 + 2 \ln r) \end{aligned} \tag{6.40}$$

Thermal stresses (6.40) have three unknown coefficients, but there are only two boundary conditions to determine three unknown coefficients. Therefore, it is necessary to add another condition of single-valuedness of rotation and displacement as a complementary condition. The condition of single-valuedness of the rotation (5.62) for the cylindrical coordinate system may be expressed as

$$\int_0^{2\pi} \frac{d}{dr} \left[\frac{1}{r} \frac{d}{dr} \left(r \frac{d\chi}{dr} \right) + \alpha^* E^* \tau \right] r \, d\theta = 0 \tag{6.41}$$

Substitution of Eq. (6.39) into Eq. (6.41) gives $C_4 = 0$. Therefore, the thermal stresses are

$$\begin{aligned} \sigma_{rr} &= -\alpha^* E^* \frac{1}{r^2} \int \tau r \, dr + \frac{C_2}{r^2} + 2C_3 \\ \sigma_{\theta\theta} &= \alpha^* E^* \left(\frac{1}{r^2} \int \tau r \, dr - \tau \right) - \frac{C_2}{r^2} + 2C_3 \end{aligned} \tag{6.42}$$

Equations (6.42) agree with Eqs. (6.35) so that we can easily obtain the thermal stresses (6.37).

6.1.3 Thermoelastic Potential Technique

We define the Goodier's thermoelastic potential Φ related to the radial displacement u from Eq. (4.39) as follows:

$$u = \frac{d\Phi}{dr} \tag{6.43}$$

Substitution of Eq. (6.43) into the equilibrium equation (6.6) expressed by the radial displacement u yields the governing equation for the Goodier's thermoelastic potential Φ

$$\frac{1}{r} \frac{d}{dr} \left(r \frac{d\Phi}{dr} \right) = (1 + \nu^*) \alpha^* \tau \tag{6.44}$$

A particular solution of Eq. (6.44) is

$$\Phi = (1+\nu^*)\alpha^* \int \frac{1}{r} \int \tau r \, dr dr \tag{6.45}$$

A radial displacement corresponding to the Goodier's thermoelastic potential Φ is

$$\bar{u} = (1+\nu^*)\alpha^* \frac{1}{r} \int \tau r \, dr \tag{6.46}$$

to within an additive constant due to integration. Substitution of Eq. (6.46) into Eqs. (6.5) gives the stresses corresponding to the Goodier's thermoelastic potential Φ:

$$\begin{aligned} \bar{\sigma}_{rr} &= -\alpha^* E^* \frac{1}{r^2} \int \tau r \, dr + \frac{E^* c^*}{1-\nu^*} \\ \bar{\sigma}_{\theta\theta} &= \alpha^* E^* \left(\frac{1}{r^2} \int \tau r \, dr - \tau \right) + \frac{E^* c^*}{1-\nu^*} \end{aligned} \tag{6.47}$$

In general, the stress field corresponding to the Goodier's thermoelastic potential Φ does not satisfy the boundary conditions.

As a concrete example, a hollow circular cylinder which is traction free is considered. Rewriting the integral $\int \tau r \, dr$ in Eqs. (6.47) as $\int_a^r \tau r \, dr$, the radial stresses $\bar{\sigma}_{rr}$ at the inner surface ($r = a$) and outer surface ($r = b$) are

$$(\bar{\sigma}_{rr})_{r=a} = \frac{E^* c^*}{1-\nu^*}, \qquad (\bar{\sigma}_{rr})_{r=b} = \frac{E^* c^*}{1-\nu^*} - \alpha^* E^* \frac{1}{b^2} \int_a^b \tau r \, dr \tag{6.48}$$

In order to satisfy the boundary conditions we should consider an isothermal problem with boundary conditions as follows:

$$\begin{aligned} \left(\bar{\bar{\sigma}}_{rr}\right)_{r=a} &= -\frac{E^* c^*}{1-\nu^*} \equiv -p_a \\ \left(\bar{\bar{\sigma}}_{rr}\right)_{r=b} &= -\frac{E^* c^*}{1-\nu^*} + \alpha^* E^* \frac{1}{b^2} \int_a^b \tau r \, dr \equiv -p_b \end{aligned} \tag{6.49}$$

The isothermal problem can be solved with the Airy stress function $\bar{\bar{\chi}}$ which is the thermal stress function for the isothermal problem. Taking the term with respect to the temperature change in Eq. (6.34) as zero, the Airy stress function $\bar{\bar{\chi}}$ is

$$\bar{\bar{\chi}} = \bar{\bar{C}}_1 + \bar{\bar{C}}_2 \ln r + \bar{\bar{C}}_3 r^2 \tag{6.50}$$

The stresses due to $\bar{\bar{\chi}}$ are

$$\bar{\bar{\sigma}}_{rr} = \frac{\bar{\bar{C}}_2}{r^2} + 2\bar{\bar{C}}_3, \qquad \bar{\bar{\sigma}}_{\theta\theta} = -\frac{\bar{\bar{C}}_2}{r^2} + 2\bar{\bar{C}}_3 \tag{6.51}$$

The boundary conditions (6.49) give the coefficients $\overline{\overline{C}}_2$ and $\overline{\overline{C}}_3$

$$\begin{aligned}\overline{\overline{C}}_2 &= \frac{a^2b^2(p_b - p_a)}{b^2 - a^2} = -\alpha^* E^* \frac{a^2}{b^2 - a^2}\int_a^b \tau r\, dr \\ 2\overline{\overline{C}}_3 &= \frac{p_a a^2 - p_b b^2}{b^2 - a^2} = -\frac{E^* c^*}{1 - \nu^*} + \alpha^* E^* \frac{1}{b^2 - a^2}\int_a^b \tau r\, dr\end{aligned} \tag{6.52}$$

The stresses for the isothermal problem are

$$\begin{aligned}\bar{\bar{\sigma}}_{rr} &= \alpha^* E^* \frac{r^2 - a^2}{r^2(b^2 - a^2)}\int_a^b \tau r\, dr - \frac{E^* c^*}{1 - \nu^*} \\ \bar{\bar{\sigma}}_{\theta\theta} &= \alpha^* E^* \frac{r^2 + a^2}{r^2(b^2 - a^2)}\int_a^b \tau r\, dr - \frac{E^* c^*}{1 - \nu^*}\end{aligned} \tag{6.53}$$

Adding Eqs. (6.47) and (6.53), the thermal stresses satisfying the boundary conditions can be obtained as

$$\begin{aligned}\sigma_{rr} &= \bar{\sigma}_{rr} + \bar{\bar{\sigma}}_{rr} = \alpha^* E^* \left[-\frac{1}{r^2}\int_a^r \tau r\, dr + \frac{r^2 - a^2}{r^2(b^2 - a^2)}\int_a^b \tau r\, dr \right] \\ \sigma_{\theta\theta} &= \bar{\sigma}_{\theta\theta} + \bar{\bar{\sigma}}_{\theta\theta} = \alpha^* E^* \left[\frac{1}{r^2}\int_a^r \tau r\, dr + \frac{r^2 + a^2}{r^2(b^2 - a^2)}\int_a^b \tau r\, dr - \tau \right]\end{aligned} \tag{6.54}$$

6.1.4 Complex Variable Technique

The complex variable technique is a powerful method for treating thermal stress problems in bodies with complex shapes. First, we resolve the thermal stress function χ into two parts:

$$\chi = \chi_c + \chi_p \tag{6.55}$$

where a complementary solution χ_c and a particular solution χ_p must satisfy the following equations:

$$\nabla^4 \chi_c = 0, \qquad \nabla^2 \chi_p = -\alpha^* E^* \tau \tag{6.56}$$

By use of the complex variable method for the plane problem in Section 5.3, the general expression of stresses and displacements are given by Eqs. (5.93) and (5.95) in a Cartesian coordinate system:

$$\begin{aligned}&\sigma_{xx} + \sigma_{yy} = 4\mathrm{Re}\left[\varphi'(z)\right] - \alpha^* E^* \tau \\ &\sigma_{yy} - \sigma_{xx} + 2i\sigma_{xy} = 2\left[\bar{z}\varphi''(z) + \psi'(z)\right] + \left(\frac{\partial^2 \chi_p}{\partial x^2} - \frac{\partial^2 \chi_p}{\partial y^2} - 2i\frac{\partial^2 \chi_p}{\partial x \partial y} \right) \\ &u_x + iu_y = \frac{1}{2G}\left[\frac{3 - \nu^*}{1 + \nu^*}\varphi(z) - z\overline{\varphi'(z)} - \overline{\psi(z)} - \left(\frac{\partial \chi_p}{\partial x} + i\frac{\partial \chi_p}{\partial y} \right) \right] - c^*(x + iy)\end{aligned} \tag{6.57}$$

Using the relationships (4.70) and (4.74) between the stress and the displacement in a Cartesian and cylindrical coordinate systems, we get for the plane problem

$$\begin{aligned}\sigma_{rr}+\sigma_{\theta\theta}&=\sigma_{xx}+\sigma_{yy}\\ \sigma_{\theta\theta}-\sigma_{rr}+2i\sigma_{r\theta}&=e^{2i\theta}(\sigma_{yy}-\sigma_{xx}+2i\sigma_{xy})\\ u_r+iu_\theta&=e^{-i\theta}(u_x+iu_y)\end{aligned}\tag{6.58}$$

Substitution of Eqs. (6.57) into Eqs. (6.58) gives

$$\begin{aligned}\sigma_{rr}+\sigma_{\theta\theta}&=4\mathrm{Re}\left[\varphi'(z)\right]-\alpha^*E^*\tau\\ \sigma_{\theta\theta}-\sigma_{rr}+2i\sigma_{r\theta}&=2e^{2i\theta}\left[\bar{z}\varphi''(z)+\psi'(z)\right]+e^{2i\theta}\left(\frac{\partial^2\chi_p}{\partial x^2}-\frac{\partial^2\chi_p}{\partial y^2}-2i\frac{\partial^2\chi_p}{\partial x\partial y}\right)\\ u_r+iu_\theta&=\frac{1}{2G}e^{-i\theta}\left[\frac{3-\nu^*}{1+\nu^*}\varphi(z)-z\overline{\varphi'(z)}-\overline{\psi(z)}-\left(\frac{\partial\chi_p}{\partial x}+i\frac{\partial\chi_p}{\partial y}\right)\right]-e^{-i\theta}c^*z\end{aligned}\tag{6.59}$$

The above equations still contain the terms expressed by the Cartesian coordinate system, so that these terms must be expressed by the cylindrical coordinate system. The relations between the Cartesian and the cylindrical coordinate systems are

$$x=r\cos\theta,\qquad y=r\sin\theta,\qquad r^2=x^2+y^2,\qquad \theta=\tan^{-1}\left(\frac{y}{x}\right)$$

so that it is clear that

$$\frac{\partial r}{\partial x}=\cos\theta,\qquad \frac{\partial r}{\partial y}=\sin\theta,\qquad \frac{\partial\theta}{\partial x}=-\frac{\sin\theta}{r},\qquad \frac{\partial\theta}{\partial y}=\frac{\cos\theta}{r}$$

Using these relations, we have

$$\begin{aligned}&\frac{\partial}{\partial x}=\cos\theta\frac{\partial}{\partial r}-\frac{\sin\theta}{r}\frac{\partial}{\partial\theta},\quad \frac{\partial}{\partial y}=\sin\theta\frac{\partial}{\partial r}+\frac{\cos\theta}{r}\frac{\partial}{\partial\theta}\\ &\frac{\partial^2}{\partial x^2}=\cos^2\theta\frac{\partial^2}{\partial r^2}+\sin^2\theta\left(\frac{1}{r}\frac{\partial}{\partial r}+\frac{1}{r^2}\frac{\partial^2}{\partial\theta^2}\right)-2\sin\theta\cos\theta\frac{\partial}{\partial r}\left(\frac{1}{r}\frac{\partial}{\partial\theta}\right)\\ &\frac{\partial^2}{\partial y^2}=\sin^2\theta\frac{\partial^2}{\partial r^2}+\cos^2\theta\left(\frac{1}{r}\frac{\partial}{\partial r}+\frac{1}{r^2}\frac{\partial^2}{\partial\theta^2}\right)+2\sin\theta\cos\theta\frac{\partial}{\partial r}\left(\frac{1}{r}\frac{\partial}{\partial\theta}\right)\\ &\frac{\partial^2}{\partial x\partial y}=-\sin\theta\cos\theta\left(\frac{1}{r}\frac{\partial}{\partial r}+\frac{1}{r^2}\frac{\partial^2}{\partial\theta^2}-\frac{\partial^2}{\partial r^2}\right)+(\cos^2\theta-\sin^2\theta)\frac{\partial}{\partial r}\left(\frac{1}{r}\frac{\partial}{\partial\theta}\right)\end{aligned}\tag{6.60}$$

The term in the particular solution on the right-hand side of the second equation of Eqs. (6.59) may be expressed in cylindrical coordinates through

$$\frac{\partial^2\chi_p}{\partial x^2}-\frac{\partial^2\chi_p}{\partial y^2}-2i\frac{\partial^2\chi_p}{\partial x\partial y}=e^{-2i\theta}\left[\left(\frac{\partial^2\chi_p}{\partial r^2}-\frac{1}{r}\frac{\partial\chi_p}{\partial r}-\frac{1}{r^2}\frac{\partial^2\chi_p}{\partial\theta^2}\right)-2i\frac{\partial}{\partial r}\left(\frac{1}{r}\frac{\partial\chi_p}{\partial\theta}\right)\right]$$

We get the stresses and displacements in the cylindrical coordinate system:

$$\begin{aligned}
\sigma_{rr} + \sigma_{\theta\theta} &= 4\mathrm{Re}\left[\varphi'(z)\right] - \alpha^* E^* \tau \\
\sigma_{\theta\theta} - \sigma_{rr} + 2i\sigma_{r\theta} &= 2e^{2i\theta}[\bar{z}\varphi''(z) + \psi'(z)] \\
&\quad + \left(\frac{\partial^2 \chi_p}{\partial r^2} - \frac{1}{r}\frac{\partial \chi_p}{\partial r} - \frac{1}{r^2}\frac{\partial^2 \chi_p}{\partial \theta^2}\right) - 2i\frac{\partial}{\partial r}\left(\frac{1}{r}\frac{\partial \chi_p}{\partial \theta}\right) \\
u_r + iu_\theta &= \frac{1+\nu^*}{E^*}\left\{e^{-i\theta}\left[\frac{3-\nu^*}{1+\nu^*}\varphi(z) - z\overline{\varphi'(z)} - \overline{\psi(z)}\right]\right. \\
&\quad \left. - \left(\frac{\partial \chi_p}{\partial r} + i\frac{1}{r}\frac{\partial \chi_p}{\partial \theta}\right)\right\} - e^{-i\theta}c^* z
\end{aligned} \tag{6.61}$$

Let us consider a one-dimensional problem in the long circular cylinder. The second of Eqs. (6.56) reduces to

$$\frac{1}{r}\frac{d}{dr}\left(r\frac{d\chi_p}{dr}\right) = -\alpha^* E^* \tau$$

The particular solution of the above equation is

$$\chi_p = -\alpha^* E^* \int \frac{1}{r} \int \tau r\, dr\, dr \tag{6.62}$$

When we take the functions $\varphi(z)$ and $\psi(z)$ as follows:

$$\varphi(z) = \sum_{n=-\infty}^{\infty} A_n z^n, \qquad \psi(z) = \sum_{n=-\infty}^{\infty} B_n z^n \tag{6.63}$$

Equations (6.61) reduce to

$$\begin{aligned}
\sigma_{rr} + \sigma_{\theta\theta} &= 2\sum_{n=-\infty}^{\infty} n\left(A_n z^{n-1} + \overline{A}_n \bar{z}^{n-1}\right) - \alpha^* E^* \tau \\
\sigma_{\theta\theta} - \sigma_{rr} + 2i\sigma_{r\theta} &= 2e^{2i\theta}\sum_{n=-\infty}^{\infty} n\left[(n-1)A_n \bar{z} z^{n-2} + B_n z^{n-1}\right] + \alpha^* E^*\left(\frac{2}{r^2}\int \tau r\, dr - \tau\right) \\
u_r + iu_\theta &= \frac{1+\nu^*}{E^*}\left[e^{-i\theta}\sum_{n=-\infty}^{\infty}\left(\frac{3-\nu^*}{1+\nu^*}A_n z^n - n\overline{A}_n z\bar{z}^{n-1} - \overline{B}_n \bar{z}^n\right)\right] \\
&\quad + (1+\nu^*)\alpha^* \frac{1}{r}\int \tau r\, dr - e^{-i\theta}c^* z
\end{aligned} \tag{6.64}$$

In the one-dimensional problem of a circular cylinder, the stress and the displacement are independent of θ. Thus, through substitution of $z = re^{i\theta}$ into Eqs. (6.64) gives

$$A_n = 0 \quad \text{for } n \neq 1, \qquad B_n = 0 \quad \text{for } n \neq -1, \qquad B_{-1} = \text{real number} \tag{6.65}$$

The radial and hoop stresses and displacements reduce to

$$\begin{aligned}
\sigma_{rr} &= (A_1 + \overline{A}_1) + B_{-1}\frac{1}{r^2} - \alpha^* E^* \frac{1}{r^2}\int \tau r\, dr \\
\sigma_{\theta\theta} &= (A_1 + \overline{A}_1) - B_{-1}\frac{1}{r^2} + \alpha^* E^* \left(\frac{1}{r^2}\int \tau r\, dr - \tau\right) \\
u_r &= \frac{1+\nu^*}{E^*}\left(2\frac{1-\nu^*}{1+\nu^*}\mathrm{Re}[A_1]r - B_{-1}r^{-1}\right) + (1+\nu^*)\alpha^*\frac{1}{r}\int \tau r\, dr - c^* r \\
u_\theta &= \frac{4}{E^*}\mathrm{Im}[A_1]r
\end{aligned} \tag{6.66}$$

When a hollow circular cylinder with traction free surfaces at inner radius a and outer radius b is considered, the boundary conditions become

$$A_1 + \overline{A}_1 = \alpha^* E^* \frac{1}{b^2 - a^2}\int_a^b \tau r\, dr, \qquad B_{-1} = -\alpha^* E^* \frac{a^2}{b^2 - a^2}\int_a^b \tau r\, dr \tag{6.67}$$

The thermal stresses and displacements are

$$\begin{aligned}
\sigma_{rr} &= \alpha^* E^* \left[-\frac{1}{r^2}\int_a^r \tau r\, dr + \frac{r^2 - a^2}{r^2(b^2 - a^2)}\int_a^b \tau r\, dr\right] \\
\sigma_{\theta\theta} &= \alpha^* E^* \left[\frac{1}{r^2}\int_a^r \tau r\, dr + \frac{r^2 + a^2}{r^2(b^2 - a^2)}\int_a^b \tau r\, dr - \tau\right] \\
u_r &= (1+\nu^*)\alpha^* \left[\frac{1}{r}\int_a^r \tau r\, dr + \left(\frac{1-\nu^*}{1+\nu^*}r + \frac{a^2}{r}\right)\frac{1}{b^2 - a^2}\int_a^b \tau r\, dr\right] - c^* r \\
u_\theta &= \frac{4}{E^*}\mathrm{Im}[A_1]r
\end{aligned} \tag{6.68}$$

The results (6.68) agree with Eqs. (6.10) and (6.11). The rotation with respect to the axial direction is given by

$$\omega_z = \frac{1}{2r}\left[\frac{\partial(ru_\theta)}{\partial r} - \frac{\partial u_r}{\partial \theta}\right] = \frac{4}{E^*}\mathrm{Im}[A_1] \tag{6.69}$$

When the rigid rotation is omitted, the circumferential displacement u_θ reduces to zero.

6.1.5 A Note on Reciprocal Theory Technique

For completeness, we mention here the method called Reciprocal Theory Technique. The method is based on the theory which will be presented in Chapter 10. Therefore, the method is described later in the book, see Subsection 10.7.2.

6.1.6 Dislocation Technique

When a temperature field is in the steady state without internal heat generation, then it satisfies the Laplace equation. In this case, thermal stresses do not occur in simply connected bodies, but they occur in multiply connected bodies. The thermal stresses in a multiply connected body due to a steady temperature field without internal heat generation can be determined by the use of difference in these phenomena. Consider two rings, one that has a slit and another one that does not have a slit. The doubly connected ring is called ring 1 and the ring with a slit which is a simply connected body is called ring 2. When two rings are subjected to a steady temperature field with no internal heat generation, ring 1 keeps the same shape, but thermal stresses are produced in the ring. On the other hand, ring 2 changes shape from the ring with a slit to an open ring in Fig. 6.3. Thermal dislocation occurs in ring 2, but thermal stresses are not produced because ring 2 is a simply connected body. If we properly apply forces to the open ring in order to close the gap, that is, eliminate the dislocation, the open ring becomes a ring with no gap. If an isothermal elastic problem to eliminate the thermal dislocation can be solved, the thermal stress in the ring can be determined.

We consider steady thermal stresses in a hollow circular cylinder with inner and outer radii a and b by the dislocation technique, obviously a one-dimensional problem. Using the definition of the rotation

$$\omega = \frac{1}{2}\left(\frac{\partial v}{\partial x} - \frac{\partial u}{\partial y}\right) \tag{6.70}$$

the differential of rotation is

$$\begin{aligned} d\omega &= \frac{1}{2}\left[\left(\frac{\partial^2 v}{\partial x^2} - \frac{\partial^2 u}{\partial x \partial y}\right) dx + \left(\frac{\partial^2 v}{\partial x \partial y} - \frac{\partial^2 u}{\partial y^2}\right) dy\right] \\ &= \frac{\partial \epsilon_{xy}}{\partial x} dx - \frac{\partial \epsilon_{xy}}{\partial y} dy + \left(-\frac{\partial \epsilon_{xx}}{\partial y} dx + \frac{\partial \epsilon_{yy}}{\partial x} dy\right) \end{aligned}$$

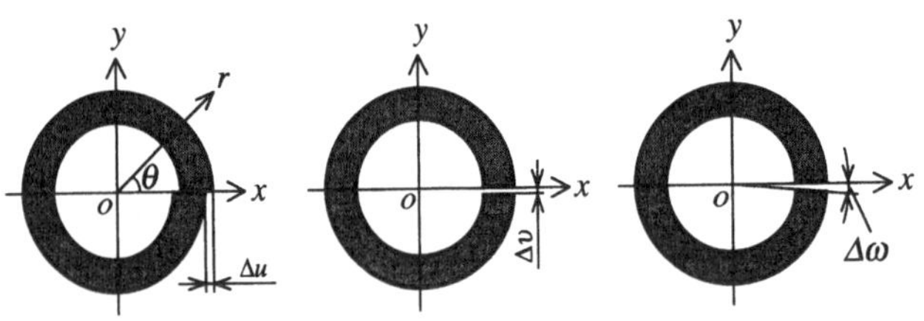

Figure 6.3 Dislocation.

If a free thermal expansion is considered, we have

$$\epsilon_{xx} = \epsilon_{yy} = \alpha^*\tau, \qquad \epsilon_{xy} = 0$$

The substitution of these relations into $d\omega$ yields

$$d\omega = \left[-\frac{\partial(\alpha^*\tau)}{\partial y}dx + \frac{\partial(\alpha^*\tau)}{\partial x}dy\right] = \frac{\partial}{\partial n}(\alpha^*\tau)\,ds \tag{6.71}$$

where n denotes the normal direction, s denotes the circumferential direction. Integrating Eq. (6.71) from some fixed point P_0 to a variable point P along the circular contour, the increment of rotation is

$$\omega - \omega_0 = \int_{\omega_0}^{\omega} d\omega = \int_{P_0}^{P} \frac{\partial}{\partial n}(\alpha^*\tau)\,ds \tag{6.72}$$

In a similar way, we have

$$\begin{aligned} du &= d(x\alpha^*\tau) - d(y\omega) \\ &\quad - \left\{x\left[\frac{\partial(\alpha^*\tau)}{\partial x}dx + \frac{\partial(\alpha^*\tau)}{\partial y}dy\right] - y\left[-\frac{\partial(\alpha^*\tau)}{\partial y}dx + \frac{\partial(\alpha^*\tau)}{\partial x}dy\right]\right\} \\ dv &= d(x\omega) + d(y\alpha^*\tau) \\ &\quad - \left\{x\left[-\frac{\partial(\alpha^*\tau)}{\partial y}dx + \frac{\partial(\alpha^*\tau)}{\partial x}dy\right] + y\left[\frac{\partial(\alpha^*\tau)}{\partial x}dx + \frac{\partial(\alpha^*\tau)}{\partial y}dy\right]\right\} \end{aligned} \tag{6.73}$$

Using the relations of direction cosines between Cartesian coordinate system (x, y) and the local coordinate system (n, s)

$$\cos(n, x) = \cos(s, y) = \frac{dy}{ds} = \frac{dx}{dn}, \qquad \cos(n, y) = -\cos(s, x) = -\frac{dx}{ds} = \frac{dy}{dn}$$

increments of rotation and displacements from $\theta = 0$ to $\theta = 2\pi$ along the circle of radius r are

$$\begin{aligned} \Delta\omega &= \omega_{\theta=2\pi} - \omega_{\theta=0} = \int_0^{2\pi} \frac{\partial(\alpha^*\tau)}{\partial r} r\,d\theta \\ \Delta u &= u_{\theta=2\pi} - u_{\theta=0} = -\int_0^{2\pi} \left[\cos\theta\frac{\partial(\alpha^*\tau)}{\partial\theta} - r\sin\theta\frac{\partial(\alpha^*\tau)}{\partial r}\right] r\,d\theta \\ \Delta v &= v_{\theta=2\pi} - v_{\theta=0} \\ &= r\int_0^{2\pi} \frac{\partial(\alpha^*\tau)}{\partial r} r\,d\theta - \int_0^{2\pi} \left[r\cos\theta\frac{\partial(\alpha^*\tau)}{\partial r} + \sin\theta\frac{\partial(\alpha^*\tau)}{\partial\theta}\right] r\,d\theta \end{aligned} \tag{6.74}$$

The general solution of the steady temperature field is given by

$$\tau(r,\theta) = A_0 + B_0 \ln r + \left(A_1 r + B_1 r^{-1}\right)\cos\theta + \left(C_1 r + D_1 r^{-1}\right)\sin\theta$$
$$+ \sum_{n=2}^{\infty}\left[\left(A_n r^n + B_n r^{-n}\right)\cos n\theta + \left(C_n r^n + D_n r^{-n}\right)\sin n\theta\right] \qquad (6.75)$$

Substituting τ in Eq. (6.75) into Eqs. (6.74), the increments of rotation and displacements due to free thermal expansion are obtained as

$$\begin{aligned} \Delta\omega_\tau &= 2\pi\alpha^* B_0 \\ \Delta u_\tau &= -2\pi\alpha^* D_1 \\ \Delta v_\tau &= 2\pi\alpha^* B_0 r + 2\pi\alpha^* B_1 \end{aligned} \qquad (6.76)$$

or

$$\Delta u_\tau + i\Delta v_\tau = 2\pi i\alpha^* B_0 r + 2\pi\alpha^*(-D_1 + iB_1) \qquad (6.76')$$

where the subscript τ means the increments of rotation and displacements due to free thermal expansion.

Next, we consider an isothermal problem for a traction free ring to eliminate the increments of rotation and displacements (6.76):

$$\begin{aligned} &\Delta u_r + i\Delta u_\theta = -(\Delta u_\tau + i\Delta v_\tau) = -[2\pi i\alpha^* B_0 r + 2\pi\alpha^*(-D_1 + iB_1)] \\ &\sigma_{rr} = 0 \quad \text{and} \quad \sigma_{r\theta} = 0 \qquad \text{on } r = a \text{ and } r = b \end{aligned} \qquad (6.77)$$

The governing equations of the complex variable method for the isothermal problem are reduced from Eqs. (6.61) to the form

$$\begin{aligned} \sigma_{rr} + \sigma_{\theta\theta} &= 4\,\mathrm{Re}\,[\varphi'(z)] \\ \sigma_{\theta\theta} - \sigma_{rr} + 2i\sigma_{r\theta} &= 2e^{2i\theta}[\bar{z}\varphi''(z) + \psi'(z)] \\ u_r + iu_\theta &= \frac{1+\nu^*}{E^*}e^{-i\theta}\left[\frac{3-\nu^*}{1+\nu^*}\varphi(z) - z\overline{\varphi'(z)} - \overline{\psi(z)}\right] - e^{-i\theta}c^* z \end{aligned} \qquad (6.78)$$

We adopt the complex functions $\varphi(z)$ and $\psi(z)$ as follows:

$$\begin{aligned} \varphi(z) &= G_0 \ln z + G_1 z \ln z + E_1 z + E_2 z^2 \\ \psi(z) &= \overline{G_0} \ln z + F_1 z^{-1} + F_2 z^{-2} \end{aligned} \qquad (6.79)$$

in which G_0, G_1, E_1, E_2, F_1 and F_2 are complex constants. Substituting complex functions $\varphi(z)$ and $\psi(z)$ into Eq. (6.78), we obtain the following stresses and increments of displacements

$$\begin{aligned} \sigma_{rr} = {} & \left[2E_1^R + F_1^R r^{-2} + G_1^R(2\ln r + 1) - 2\theta G_1^I\right] \\ & + 2\left(E_2^R r + F_2^R r^{-3} + G_0^R r^{-1}\right)\cos\theta \\ & + 2\left(-E_2^I r + F_2^I r^{-3} + G_0^I r^{-1}\right)\sin\theta \end{aligned}$$

$$
\begin{aligned}
\sigma_{\theta\theta} &= \left[2E_1^R - F_1^R r^{-2} + G_1^R(2\ln r + 3) - 2\theta G_1^I\right] \\
&\quad + 2\left(3E_2^R r - F_2^R r^{-3} + G_0^R r^{-1}\right)\cos\theta \\
&\quad + 2\left(-3E_2^I r - F_2^I r^{-3} + G_0^I r^{-1}\right)\sin\theta \qquad (6.80)\\
\sigma_{r\theta} &= \left(G_1^I - F_1^I r^{-2}\right) + 2\left(E_2^I r - F_2^I r^{-3} - G_0^I r^{-1}\right)\cos\theta \\
&\quad + 2\left(E_2^R r + F_2^R r^{-3} + G_0^R r^{-1}\right)\sin\theta \\
\Delta u_r + i\Delta u_\theta &= [u_r + iu_\theta]_{\theta=0}^{\theta=2\pi} \\
&= 2\pi i\frac{1+\nu^*}{E^*}\left[\frac{4}{1+\nu^*}G_0 + \left(\frac{3-\nu^*}{1+\nu^*}G_1 + \overline{G_1}\right)r\right]
\end{aligned}
$$

where the superscript R and I denote the real and imaginary parts, respectively, of the complex constants. We can get these real constants from the boundary conditions for traction free surfaces:

$$
G_0^R = -\frac{\alpha^* E^*}{4}B_1, \qquad G_0^I = -\frac{\alpha^* E^*}{4}D_1, \qquad G_1^R = -\frac{\alpha^* E^*}{4}B_0
$$

$$
G_1^I = 0, \qquad F_1^I = 0
$$

$$
E_1^R = \frac{\alpha^* E^* B_0}{8(b^2-a^2)}\left(2b^2\ln b - 2a^2\ln a + b^2 - a^2\right)
$$

$$
E_2^R = \frac{\alpha^* E^* B_1}{4(a^2+b^2)}, \qquad E_2^I = -\frac{\alpha^* E^* D_1}{4(a^2+b^2)}
$$

$$
F_1^R = -\frac{\alpha^* E^* B_0 a^2 b^2}{2(b^2-a^2)}\ln\frac{b}{a}, \qquad F_2^R = \frac{\alpha^* E^* B_1 a^2 b^2}{4(a^2+b^2)}, \qquad F_2^I = \frac{\alpha^* E^* D_1 a^2 b^2}{4(a^2+b^2)}
$$

Substituting these constants into Eqs. (6.80), the thermal stresses are obtained as the stresses required to close the dislocation in the ring:

$$
\begin{aligned}
\sigma_{rr} &= -\frac{\alpha^* E^* B_0}{2}\left[\ln\frac{r}{a} - \frac{b^2}{b^2-a^2}\left(1-\frac{a^2}{r^2}\right)\ln\frac{b}{a}\right] \\
&\quad - \frac{\alpha^* E^*}{2(a^2+b^2)}\left\{B_1[(a^2+b^2)r^{-1} - r - a^2b^2r^{-3}]\cos\theta\right. \\
&\qquad \left. + D_1[(a^2+b^2)r^{-1} - r - a^2b^2r^{-3}]\sin\theta\right\} \\
\sigma_{\theta\theta} &= -\frac{\alpha^* E^* B_0}{2}\left[1 + \ln\frac{r}{a} - \frac{b^2}{b^2-a^2}\left(1+\frac{a^2}{r^2}\right)\ln\frac{b}{a}\right] \\
&\quad - \frac{\alpha^* E^*}{2(a^2+b^2)}\left\{B_1[(a^2+b^2)r^{-1} - 3r + a^2b^2r^{-3}]\cos\theta\right. \qquad (6.81)\\
&\qquad \left. + D_1[(a^2+b^2)r^{-1} - 3r + a^2b^2r^{-3}]\sin\theta\right\}
\end{aligned}
$$

$$\sigma_{r\theta} = -\frac{\alpha^* E^*}{2(a^2+b^2)}\{-D_1[(a^2+b^2)r^{-1} - r - a^2b^2r^{-3}]\cos\theta$$
$$+B_1[(a^2+b^2)r^{-1} - r - a^2b^2r^{-3}]\sin\theta\}$$

When the temperature in the case of one-dimensional temperature field may be expressed by Eq. (3.82):

$$T = T_a + (T_b - T_a)\frac{\ln(r/a)}{\ln(b/a)}$$

then B_0 is given by

$$B_0 = \frac{T_b - T_a}{\ln(b/a)}$$

so that thermal stresses are

$$\begin{aligned}\sigma_{rr} &= \frac{\alpha^* E^*(T_b - T_a)}{2\ln(b/a)}\left[-\ln\frac{r}{a} + \frac{b^2}{b^2-a^2}\left(1-\frac{a^2}{r^2}\right)\ln(b/a)\right] \\ \sigma_{\theta\theta} &= \frac{\alpha^* E^*(T_b - T_a)}{2\ln(b/a)}\left[-1-\ln\frac{r}{a} + \frac{b^2}{b^2-a^2}\left(1+\frac{a^2}{r^2}\right)\ln(b/a)\right]\end{aligned} \tag{6.82}$$

6.2 Plane Problems

6.2.1 Steady State Thermal Stresses in a Hollow Cylinder

Consider a long hollow cylinder with an inner radius a and an outer radius b. The initial temperature $T_i(r)$ in the cylinder is given by a constant temperature T_i and heat is exchanged at the inner and outer surfaces. The temperatures of the surrounding media $T_a(\theta)$ and $T_b(\theta)$ at the inner and outer surfaces, respectively, depend on the circumferential direction θ. The heat conduction problem for the steady state is

Governing equation:

$$\frac{\partial^2 T}{\partial r^2} + \frac{1}{r}\frac{\partial T}{\partial r} + \frac{1}{r^2}\frac{\partial^2 T}{\partial \theta^2} = 0 \tag{a}$$

Boundary conditions:

$$\lambda\frac{\partial T}{\partial r} = h_a[T - T_a(\theta)] \qquad \text{on } r = a \tag{b}$$

$$-\lambda\frac{\partial T}{\partial r} = h_b[T - T_b(\theta)] \qquad \text{on } r = b \tag{c}$$

where

$$T_a(\theta) = T_{a0} + \sum_{n=1}^{\infty} \left(T_{an} \cos n\theta + T'_{an} \sin n\theta\right) \tag{d}$$

$$T_b(\theta) = T_{b0} + \sum_{n=1}^{\infty} \left(T_{bn} \cos n\theta + T'_{bn} \sin n\theta\right) \tag{e}$$

Initial condition:

$$T = T_i \qquad \text{at } t = 0 \tag{f}$$

We assume that the temperature can be expressed by

$$T(r,\theta) = T_0(r) + \sum_{n=1}^{\infty} \left[T_n(r) \cos n\theta + T'_n(r) \sin n\theta\right] \tag{g}$$

Substituting Eq. (g) into Eq. (a), functions $T_0(r)$, $T_n(r)$ and $T'_n(r)$ should satisfy

$$\left(\frac{d^2}{dr^2} + \frac{1}{r}\frac{d}{dr}\right) T_0(r) = 0, \qquad \left(\frac{d^2}{dr^2} + \frac{1}{r}\frac{d}{dr} - \frac{n^2}{r^2}\right)\left[T_n(r), T'_n(r)\right] = 0 \tag{h}$$

The general solutions of Eqs. (h) are

$$T_0(r) = A_0 + B_0 \ln r, \quad T_n(r) = A_n r^n + B_n r^{-n}, \quad T'_n(r) = A'_n r^n + B'_n r^{-n} \tag{i}$$

where A_0, B_0, A_n, B_n, A'_n, and B'_n are unknown coefficients. Substituting Eqs. (i) into Eqs. (b) and (c), we have A_0, B_0, A_n, B_n, A'_n, and B'_n

$$A_0 = T_{a0} + (T_{b0} - T_{a0}) \frac{\dfrac{\lambda}{h_a a} - \ln a}{\dfrac{\lambda}{h_a a} + \dfrac{\lambda}{h_b b} + \ln(b/a)}, \qquad B_0 = \frac{T_{b0} - T_{a0}}{\dfrac{\lambda}{h_a a} + \dfrac{\lambda}{h_b b} + \ln(b/a)}$$

$$\begin{aligned}
\left(A_n, A'_n\right) &= \frac{1}{D_n}\left[-\left(T_{an}, T'_{an}\right) h_a a (h_b b - \lambda n) b^{-n}\right. \\
&\qquad \left. + \left(T_{bn}, T'_{bn}\right) h_b b (h_a a + \lambda n) a^{-n}\right] \\
\left(B_n, B'_n\right) &= \frac{1}{D_n}\left[\left(T_{an}, T'_{an}\right) h_a a (h_b b + \lambda n) b^{n} - \left(T_{bn}, T'_{bn}\right) h_b b (h_a a - \lambda n) a^{n}\right]
\end{aligned} \tag{j}$$

where

$$D_n = (h_a a + \lambda n)(h_b b + \lambda n)\left(\frac{b}{a}\right)^n - (h_a a - \lambda n)(h_b b - \lambda n)\left(\frac{b}{a}\right)^{-n}$$

The temperature change from the initial temperature is given by

$$
\begin{aligned}
\tau &= T - T_i \\
&= (T_{a0} - T_i) + (T_{b0} - T_{a0}) \frac{\dfrac{\lambda}{h_a a} + \ln(r/a)}{\dfrac{\lambda}{h_a a} + \dfrac{\lambda}{h_b b} + \ln(b/a)} \\
&+ \sum_{n=1}^{\infty} \frac{1}{D_n} \left(\left\{ T_{bn} h_b b \left[(h_a a + \lambda n) \left(\frac{r}{a}\right)^{n} - (h_a a - \lambda n) \left(\frac{r}{a}\right)^{-n} \right] \right. \right. \\
&\left. - T_{an} h_a a \left[(h_b b - \lambda n) \left(\frac{r}{b}\right)^{n} - (h_b b + \lambda n) \left(\frac{r}{b}\right)^{-n} \right] \right\} \cos n\theta \\
&+ \left\{ T'_{bn} h_b b \left[(h_a a + \lambda n) \left(\frac{r}{a}\right)^{n} - (h_a a - \lambda n) \left(\frac{r}{a}\right)^{-n} \right] \right. \\
&\left. \left. - T'_{an} h_a a \left[(h_b b - \lambda n) \left(\frac{r}{b}\right)^{n} - (h_b b + \lambda n) \left(\frac{r}{b}\right)^{-n} \right] \right\} \sin n\theta \right)
\end{aligned} \tag{6.83}
$$

We solve the thermal stress problem by the thermal stress function technique which was discussed in Section 5.2. Let us introduce a thermal stress function χ related to the components of stress in the cylindrical coordinate system

$$
\sigma_{rr} = \frac{1}{r}\frac{\partial \chi}{\partial r} + \frac{1}{r^2}\frac{\partial^2 \chi}{\partial \theta^2}, \qquad \sigma_{\theta\theta} = \frac{\partial^2 \chi}{\partial r^2}, \qquad \sigma_{r\theta} = -\frac{\partial}{\partial r}\left(\frac{1}{r}\frac{\partial \chi}{\partial \theta}\right) \tag{6.84}
$$

The governing equation for χ is (from Eq. (5.33))

$$
\nabla^4 \chi = -\alpha^* E^* \nabla^2 \tau \tag{6.85}
$$

where

$$
\nabla^4 = \nabla^2 \nabla^2 = \left(\frac{\partial^2}{\partial r^2} + \frac{1}{r}\frac{\partial}{\partial r} + \frac{1}{r^2}\frac{\partial^2}{\partial \theta^2} \right)^2
$$

The conditions of single-valuedness of rotation and displacements (5.62), (5.63) and (5.64) for the cylindrical coordinate system reduce to

$$
\begin{aligned}
&\int_0^{2\pi} r \frac{\partial}{\partial r} (\nabla^2 \chi + \alpha^* E^* \tau) \, d\theta = 0 \\
&\int_0^{2\pi} r \left(\cos\theta \frac{\partial}{\partial \theta} - r \sin\theta \frac{\partial}{\partial r} \right) (\nabla^2 \chi + \alpha^* E^* \tau) \, d\theta = 0 \\
&\int_0^{2\pi} r \left(r\cos\theta \frac{\partial}{\partial r} + \sin\theta \frac{\partial}{\partial \theta} \right) (\nabla^2 \chi + \alpha^* E^* \tau) \, d\theta = 0
\end{aligned} \tag{6.86}
$$

The boundary conditions for traction free surfaces are

$$
\sigma_{rr} = 0, \quad \sigma_{r\theta} = 0 \qquad \text{on } r = a \text{ and } r = b \tag{6.87}
$$

Substitution of the temperature change (6.83) into the governing equation (6.85) reduces to

$$\nabla^4 \chi = 0 \tag{k}$$

General solutions of Eq. (k) are from Eq. (5.82)

$$\begin{aligned}
\chi = {} & E_0 + F_0 \ln r + G_0 r^2 + H_0 r^2 \ln r \\
& + \left(E_1 r + F_1 r^{-1} + G_1 r^3 + H_1 r \ln r\right)\cos\theta \\
& + \left(E_1' r + F_1' r^{-1} + G_1' r^3 + H_1' r \ln r\right)\sin\theta \\
& + \sum_{n=2}^{\infty}\Big[\left(E_n r^n + F_n r^{-n} + G_n r^{n+2} + H_n r^{-n+2}\right)\cos n\theta \\
& \qquad + \left(E_n' r^n + F_n' r^{-n} + G_n' r^{n+2} + H_n' r^{-n+2}\right)\sin n\theta\Big]
\end{aligned} \tag{6.88}$$

Substituting Eq. (6.88) into Eqs. (6.84), we obtain the following thermal stress components:

$$\begin{aligned}
\sigma_{rr} = {} & F_0 r^{-2} + 2G_0 + H_0(2\ln r + 1) + \left(-2F_1 r^{-3} + 2G_1 r + H_1 r^{-1}\right)\cos\theta \\
& + \left(-2F_1' r^{-3} + 2G_1' r + H_1' r^{-1}\right)\sin\theta \\
& - \sum_{n=2}^{\infty}\Big\{\big[n(n-1)E_n r^{n-2} + n(n+1)F_n r^{-n-2} \\
& \qquad + (n-2)(n+1)G_n r^n + (n-1)(n+2)H_n r^{-n}\big]\cos n\theta \\
& \qquad + \big[n(n-1)E_n' r^{n-2} + n(n+1)F_n' r^{-n-2} \\
& \qquad + (n-2)(n+1)G_n' r^n + (n-1)(n+2)H_n' r^{-n}\big]\sin n\theta\Big\}
\end{aligned}$$

$$\begin{aligned}
\sigma_{\theta\theta} = {} & -F_0 r^{-2} + 2G_0 + H_0(2\ln r + 3) + \left(2F_1 r^{-3} + 6G_1 r + H_1 r^{-1}\right)\cos\theta \\
& + \left(2F_1' r^{-3} + 6G_1' r + H_1' r^{-1}\right)\sin\theta \\
& + \sum_{n=2}^{\infty}\Big\{\big[n(n-1)E_n r^{n-2} + n(n+1)F_n r^{-n-2} \\
& \qquad + (n+1)(n+2)G_n r^n + (n-1)(n-2)H_n r^{-n}\big]\cos n\theta \\
& \qquad + \big[n(n-1)E_n' r^{n-2} + n(n+1)F_n' r^{-n-2} \\
& \qquad + (n+1)(n+2)G_n' r^n + (n-1)(n-2)H_n' r^{-n}\big]\sin n\theta\Big\}
\end{aligned}$$

$$\sigma_{r\theta} = \left(-2F_1 r^{-3} + 2G_1 r + H_1 r^{-1}\right)\sin\theta - \left(-2F_1' r^{-3} + 2G_1' r + H_1' r^{-1}\right)\cos\theta$$
$$+ \sum_{n=2}^{\infty} n\Big\{\big[(n-1)E_n r^{n-2} - (n+1)F_n r^{-n-2}$$
$$+ (n+1)G_n r^n - (n-1)H_n r^{-n}\big]\sin n\theta$$
$$- \big[(n-1)E_n' r^{n-2} - (n+1)F_n' r^{-n-2}$$
$$+ (n+1)G_n' r^n - (n-1)H_n' r^{-n}\big]\cos n\theta\Big\} \tag{6.89}$$

Using the boundary conditions (6.87), we obtain

$$\begin{aligned}
F_0 a^{-2} + 2G_0 + H_0(2\ln a + 1) &= 0\\
F_0 b^{-2} + 2G_0 + H_0(2\ln b + 1) &= 0\\
-2F_1 a^{-3} + 2G_1 a + H_1 a^{-1} &= 0\\
-2F_1 b^{-3} + 2G_1 b + H_1 b^{-1} &= 0\\
-2F_1' a^{-3} + 2G_1' a + H_1' a^{-1} &= 0\\
-2F_1' b^{-3} + 2G_1' b + H_1' b^{-1} &= 0\\
E_n = F_n = G_n = H_n = E_n' = F_n' = G_n' = H_n' &= 0 \qquad (n = 2, 3, \ldots)
\end{aligned} \tag{l}$$

From Eqs. (l) the coefficients F_0, G_0, F_1, G_1, F_1' and G_1' can be obtained

$$\begin{aligned}
F_0 &= \left(\frac{2H_0}{b^2 - a^2}\right) a^2 b^2 \ln\frac{b}{a}\\
G_0 &= -\frac{H_0}{2(b^2 - a^2)}\left[(b^2 - a^2) + 2b^2\ln b - 2a^2\ln a\right]\\
F_1 &= \frac{a^2 b^2 H_1}{2(a^2 + b^2)}, \qquad G_1 = -\frac{H_1}{2(a^2 + b^2)}\\
F_1' &= \frac{a^2 b^2 H_1'}{2(a^2 + b^2)}, \qquad G_1' = -\frac{H_1'}{2(a^2 + b^2)}
\end{aligned} \tag{m}$$

Introduction of Eqs. (6.83) and (6.88) into the conditions of single-valuedness of rotation and displacements (6.86) gives

$$\begin{aligned}
H_0 &= -\frac{1}{4}\alpha^* E^* B_0 = -\frac{\alpha^* E^*}{4}\frac{T_{b0} - T_{a0}}{\dfrac{\lambda}{h_a a} + \dfrac{\lambda}{h_b b} + \ln(b/a)}\\
H_1 &= -\frac{1}{2}\alpha^* E^* B_1 = -\frac{\alpha^* E^*}{2D_1}\left[T_{a1} h_a a(h_b b + \lambda)b - T_{b1} h_b b(h_a a - \lambda)a\right]\\
H_1' &= -\frac{1}{2}\alpha^* E^* B_1' = -\frac{\alpha^* E^*}{2D_1}\left[T_{a1}' h_a a(h_b b + \lambda)b - T_{b1}' h_b b(h_a a - \lambda)a\right]
\end{aligned} \tag{6.90}$$

We have thermal stress components from Eqs. (6.89), (m) and (6.90)

$$\sigma_{rr} = \frac{\alpha^* E^*}{2} \frac{T_{b0} - T_{a0}}{\dfrac{\lambda}{h_a a} + \dfrac{\lambda}{h_b b} + \ln(b/a)} \left[-\ln\frac{r}{a} + \frac{b^2}{b^2 - a^2}\left(1 - \frac{a^2}{r^2}\right)\ln\frac{b}{a} \right]$$
$$+ \frac{\alpha^* E^*}{2(a^2 + b^2)D_1} \Big\{ \left[T_{a1} h_a a(h_b b + \lambda)b - T_{b1} h_b b(h_a a - \lambda)a \right]$$
$$\times \left(r + \frac{a^2 b^2}{r^3} - \frac{a^2 + b^2}{r} \right) \cos\theta + \left[T'_{a1} h_a a(h_b b + \lambda)b \right.$$
$$\left. - T'_{b1} h_b b(h_a a - \lambda)a \right] \left(r + \frac{a^2 b^2}{r^3} - \frac{a^2 + b^2}{r} \right) \sin\theta \Big\}$$

$$\sigma_{\theta\theta} = \frac{\alpha^* E^*}{2} \frac{T_{b0} - T_{a0}}{\dfrac{\lambda}{h_a a} + \dfrac{\lambda}{h_b b} + \ln(b/a)} \left[-\left(1 + \ln\frac{r}{a}\right) + \frac{b^2}{b^2 - a^2}\left(1 + \frac{a^2}{r^2}\right)\ln\frac{b}{a} \right]$$
$$- \frac{\alpha^* E^*}{2(a^2 + b^2)D_1} \Big\{ \left[T_{a1} h_a a(h_b b + \lambda)b - T_{b1} h_b b(h_a a - \lambda)a \right]$$
$$\times \left(-3r + \frac{a^2 b^2}{r^3} + \frac{a^2 + b^2}{r} \right) \cos\theta + \left[T'_{a1} h_a a(h_b b + \lambda)b \right.$$
$$\left. - T'_{b1} h_b b(h_a a - \lambda)a \right] \left(-3r + \frac{a^2 b^2}{r^3} + \frac{a^2 + b^2}{r} \right) \sin\theta \Big\}$$

$$\sigma_{r\theta} = \frac{\alpha^* E^*}{2(a^2 + b^2)D_1} \Big\{ \left[T_{a1} h_a a(h_b b + \lambda)b - T_{b1} h_b b(h_a a - \lambda)a \right]$$
$$\times \left(r + \frac{a^2 b^2}{r^3} - \frac{a^2 + b^2}{r} \right) \sin\theta - \left[T'_{a1} h_a a(h_b b + \lambda)b \right.$$
$$\left. - T'_{b1} h_b b(h_a a - \lambda)a \right] \left(r + \frac{a^2 b^2}{r^3} - \frac{a^2 + b^2}{r} \right) \cos\theta \Big\} \tag{6.91}$$

When steady state temperature distribution without heat generation is considered in the plane problem, the components of $\cos n\theta$ and $\sin n\theta$ for $n \geq 2$ in the temperature do not affect the thermal stress. Only three components, that is, the component independent of θ and the components with $\cos\theta$ and $\sin\theta$ in the temperature produce the thermal stress.

Example 6.4 When both the inner surface ($r = a$) and the outer surface ($r = b$) of a hollow circular cylinder at constant initial temperature T_i are heated to $T_a(\theta)$ and $T_b(\theta)$, respectively, find the steady thermal stress in the hollow cylinder.

Solution The temperature conditions are

$$T = T_a(\theta) \quad \text{on } r = a$$
$$T = T_b(\theta) \quad \text{on } r = b \tag{a}$$

The temperature conditions (b) and (c) in Subsection 6.2.1 are

$$T - T_a(\theta) = \frac{\lambda}{h_a}\frac{\partial T}{\partial r} \qquad \text{on } r = a \tag{b}$$

$$T - T_b(\theta) = -\frac{\lambda}{h_a}\frac{\partial T}{\partial r} \qquad \text{on } r = b \tag{c}$$

Application of the limits $h_a \to \infty$ and $h_b \to \infty$ to the boundary conditions (b) and (c) gives Eqs. (a). Therefore, the thermal stresses may be obtained by the application of $h_a \to \infty$ and $h_b \to \infty$ to the thermal stresses (6.91) in Subsection 6.2.1. We receive

$$\begin{aligned}
\sigma_{rr} &= \frac{\alpha^* E^*}{2}(T_{b0} - T_{a0})\left[-\frac{\ln(r/a)}{\ln(b/a)} + \left(1 - \frac{a^2}{r^2}\right)\frac{b^2}{b^2 - a^2}\right] \\
&\quad + \frac{\alpha^* E^* ab}{2(b^2 + a^2)(b^2 - a^2)}\left(r - \frac{b^2 + a^2}{r} + \frac{a^2 b^2}{r^3}\right) \\
&\quad \times \left[(T_{a1}b - T_{b1}a)\cos\theta + \left(T'_{a1}b - T'_{b1}a\right)\sin\theta\right] \\
\sigma_{\theta\theta} &= \frac{\alpha^* E^*}{2}(T_{b0} - T_{a0})\left[-\frac{1 + \ln(r/a)}{\ln(b/a)} + \left(1 + \frac{a^2}{r^2}\right)\frac{b^2}{b^2 - a^2}\right] \\
&\quad - \frac{\alpha^* E^* ab}{2(b^2 + a^2)(b^2 - a^2)}\left(-3r + \frac{b^2 + a^2}{r} + \frac{a^2 b^2}{r^3}\right) \\
&\quad \times \left[(T_{a1}b - T_{b1}a)\cos\theta + \left(T'_{a1}b - T'_{b1}a\right)\sin\theta\right] \\
\sigma_{r\theta} &= \frac{\alpha^* E^* ab}{2(b^2 + a^2)(b^2 - a^2)}\left(r - \frac{b^2 + a^2}{r} + \frac{a^2 b^2}{r^3}\right) \\
&\quad \times \left[(T_{a1}b - T_{b1}a)\sin\theta - \left(T'_{a1}b - T'_{b1}a\right)\cos\theta\right]
\end{aligned} \tag{d}$$

6.2.2 Transient Thermal Stresses in a Solid Cylinder

Consider transient thermal stresses in a long solid cylinder with constant initial temperature T_i and the heat is exchanged by the heat transfer between the boundary surface of radius a and the surrounding medium with temperature $T_a(\theta)$ which depends on the tangential direction θ. The equations to be solved are

Governing equation:

$$\frac{1}{\kappa}\frac{\partial T}{\partial t} = \frac{\partial^2 T}{\partial r^2} + \frac{1}{r}\frac{\partial T}{\partial r} + \frac{1}{r^2}\frac{\partial^2 T}{\partial \theta^2} \tag{a}$$

Boundary condition:

$$-\lambda\frac{\partial T}{\partial r} = h_a[T - T_a(\theta)] \qquad \text{on } r = a \tag{b}$$

where

$$T_a(\theta) = T_{a0} + \sum_{n=1}^{\infty} \left(T_{an} \cos n\theta + T'_{an} \sin n\theta\right) \tag{c}$$

Initial condition:

$$T = T_i \qquad \text{at } t = 0 \tag{d}$$

We assume that the temperature can be expressed by

$$T(r, \theta, t) = T_0(r) e^{-\kappa s_0^2 t} + \sum_{n=1}^{\infty} \left[T_n(r) \cos n\theta + T'_n(r) \sin n\theta\right] e^{-\kappa s_n^2 t} \tag{e}$$

Substituting Eq. (e) into Eq. (a), functions $T_0(r)$, $T_n(r)$, and $T'_n(r)$ should satisfy

$$\begin{aligned} &\left(\frac{d^2}{dr^2} + \frac{1}{r}\frac{d}{dr} + s_0^2\right) T_0(r) = 0 && \text{for } s_0 \neq 0 \\ &\left[\frac{d^2}{dr^2} + \frac{1}{r}\frac{d}{dr} + \left(s_n^2 - \frac{n^2}{r^2}\right)\right]\left[T_n(r), T'_n(r)\right] = 0 && \text{for } s_n \neq 0 \end{aligned} \tag{f}$$

$$\begin{aligned} &\left(\frac{d^2}{dr^2} + \frac{1}{r}\frac{d}{dr}\right) T_0(r) = 0 && \text{for } s_0 = 0 \\ &\left[\frac{d^2}{dr^2} + \frac{1}{r}\frac{d}{dr} - \frac{n^2}{r^2}\right]\left[T_n(r), T'_n(r)\right] = 0 && \text{for } s_n = 0 \end{aligned} \tag{f$'$}$$

Equations (f) are Bessel's differential equations, the general solutions of which are

$$\begin{aligned} T_0(r) &= A_0 J_0(s_0 r) + B_0 Y_0(s_0 r) && \text{for } s_0 \neq 0 \\ T_n(r) &= A_n J_n(s_n r) + B_n Y_n(s_n r) && \text{for } s_n \neq 0 \\ T'_n(r) &= A'_n J_n(s_n r) + B'_n Y_n(s_n r) && \text{for } s_n \neq 0 \end{aligned} \tag{g}$$

The general solutions of Eqs. (f′) are

$$\begin{aligned} &T_0(r) = \overline{A}_0 + \overline{B}_0 \ln r && \text{for } s_0 = 0 \\ &T_n(r) = \overline{A}_n r^n + \overline{B}_n r^{-n}, \quad T'_n(r) = \overline{A}'_n r^n + \overline{B}'_n r^{-n} && \text{for } s_n = 0 \end{aligned} \tag{g$'$}$$

where A_0, B_0, A_n, B_n, A'_n, B'_n, $\overline{A}_0$, $\overline{B}_0$, $\overline{A}_n$, $\overline{B}_n$, $\overline{A}'_n$, and $\overline{B}'_n$ are unknown coefficients, and $J_n(s_n r)$ and $Y_n(s_n r)$ are Bessel functions of the first and second kind of order n, respectively. As Bessel functions of the second kind of order n tend to negative infinity when r tends to zero, the general solution of Eq. (a) which is suitable for this problem is given by

$$T = \overline{A}_0 + \sum_{n=1}^{\infty} \left(\overline{A}_n r^n \cos n\theta + \overline{A}'_n r^n \sin n\theta \right) + A_0 J_0(s_0 r) e^{-\kappa s_0^2 t}$$
$$+ \sum_{n=1}^{\infty} \left[A_n J_n(s_n r) \cos n\theta + A'_n J_n(s_n r) \sin n\theta \right] e^{-\kappa s_n^2 t} \tag{h}$$

Substituting Eq. (h) into Eq. (b) yields

$$\overline{A}_0 = T_{a0}, \qquad \overline{A}_n = T_{an} \frac{\dfrac{h_a a}{\lambda}}{\left(n + \dfrac{h_a a}{\lambda} \right) a^n}, \qquad \overline{A}'_n = T'_{an} \frac{\dfrac{h_a a}{\lambda}}{\left(n + \dfrac{h_a a}{\lambda} \right) a^n} \tag{i}$$

and s_0 and s_n are the roots of the equations

$$\frac{h_a a}{\lambda} J_0(s_0 a) - s_0 a J_1(s_0 a) = 0 \tag{j}$$

and

$$\left(\frac{h_a a}{\lambda} - n \right) J_n(s_n a) + s_n a J_{n-1}(s_n a) = 0 \tag{j'}$$

respectively. As the number of the roots of Eqs. (j) and (j′) is infinite, the temperature (h) can be expressed by

$$T = \overline{A}_0 + \sum_{n=1}^{\infty} \left(\overline{A}_n r^n \cos n\theta + \overline{A}'_n r^n \sin n\theta \right) + \sum_{k=1}^{\infty} A_{0k} J_0(s_{0k} r) e^{-\kappa s_{0k}^2 t}$$
$$+ \sum_{n=1}^{\infty} \sum_{k=1}^{\infty} \left[A_{nk} J_n(s_{nk} r) \cos n\theta + A'_{nk} J_n(s_{nk} r) \sin n\theta \right] e^{-\kappa s_{nk}^2 t} \tag{k}$$

Substitution of Eq. (k) into the initial condition (d) yields

$$\overline{A}_0 + \sum_{k=1}^{\infty} A_{0k} J_0(s_{0k} r) + \sum_{n=1}^{\infty} \left\{ \left[\overline{A}_n r^n + \sum_{k=1}^{\infty} A_{nk} J_n(s_{nk} r) \right] \cos n\theta \right.$$
$$\left. + \left[\overline{A}'_n r^n + \sum_{k=1}^{\infty} A'_{nk} J_n(s_{nk} r) \right] \sin n\theta \right\} = T_i$$

Then,

$$\sum_{k=1}^{\infty} A_{0k} J_0(s_{0k} r) = T_i - \overline{A}_0$$
$$\sum_{k=1}^{\infty} A_{nk} J_n(s_{nk} r) = -\overline{A}_n r^n, \qquad \sum_{k=1}^{\infty} A'_{nk} J_n(s_{nk} r) = -\overline{A}'_n r^n \tag{l}$$

Multiplying $r J_0(s_{0i} r)$, $r J_n(s_{ni} r)$, and $r J_n(s_{ni} r)$ by the first, second, and third equations in Eqs. (l), respectively, and integrating from zero to a we find

$$\sum_{k=1}^{\infty} A_{0k} \int_0^a r J_0(s_{0k} r) J_0(s_{0i} r)\, dr = (T_i - \overline{A}_0) \int_0^a r J_0(s_{0i} r)\, dr$$
$$\sum_{k=1}^{\infty} A_{nk} \int_0^a r J_n(s_{nk} r) J_n(s_{ni} r)\, dr = -\overline{A}_n \int_0^a r^{n+1} J_n(s_{ni} r)\, dr \tag{m}$$
$$\sum_{k=1}^{\infty} A'_{nk} \int_0^a r J_n(s_{nk} r) J_n(s_{ni} r)\, dr = -\overline{A}'_n \int_0^a r^{n+1} J_n(s_{ni} r)\, dr$$

By use of Eqs. (B.31) and (B.32) in Appendix B:

$$\int_0^z J_\nu(kz) J_\nu(lz) z\, dz = \frac{z}{k^2 - l^2}[k J_\nu(lz) J_{\nu+1}(kz) - l J_\nu(kz) J_{\nu+1}(lz)]$$
$$= \frac{z}{k^2 - l^2}[l J_{\nu-1}(lz) J_\nu(kz) - k J_{\nu-1}(kz) J_\nu(lz)] \qquad k \neq l \tag{n}$$
$$\int_0^z J_\nu^2(kz) z\, dz = \frac{z^2}{2}\left[J_\nu^2(kz) - J_{\nu-1}(kz) J_{\nu+1}(kz)\right]$$

and using Eqs. (j) and (j′), we have from Eqs. (m)

$$A_{0k} = \left(T_i - \overline{A}_0\right) \frac{2}{a^2\left[J_0^2(s_{0k} a) + J_1^2(s_{0k} a)\right]} \int_0^a r J_0(s_{0k} r)\, dr$$
$$= (T_i - T_{a0}) \frac{2 J_1(s_{0k} a)}{s_{0k} a\left[J_0^2(s_{0k} a) + J_1^2(s_{0k} a)\right]}$$
$$\left(A_{nk}, A'_{nk}\right) = -\left(\overline{A}_n, \overline{A}'_n\right) \frac{2}{a^2\left[J_n^2(s_{nk} a) - J_{n-1}(s_{nk} a) J_{n+1}(s_{nk} a)\right]} \int_0^a r^{n+1} J_n(s_{nk} r)\, dr$$
$$= -\left(T_{an}, T'_{an}\right) \frac{2\dfrac{h_a a}{\lambda} J_{n+1}(s_{nk} a)}{\left(\dfrac{h_a a}{\lambda} + n\right) s_{nk} a}$$
$$\times \frac{1}{\left[J_n^2(s_{nk} a) - \dfrac{2n}{s_{nk} a} J_n(s_{nk} a) J_{n-1}(s_{nk} a) + J_{n-1}^2(s_{nk} a)\right]} \tag{o}$$

The temperature change from the initial temperature is given by

$$\begin{aligned}\tau &= T - T_i \\ &= T_{a0} - T_i + \frac{h_a a}{\lambda}\sum_{n=1}^{\infty}\frac{1}{\left(\dfrac{h_a a}{\lambda}+n\right)}\left(\frac{r}{a}\right)^n\left(T_{an}\cos n\theta + T'_{an}\sin n\theta\right) \\ &\quad + (T_{a0} - T_i)\sum_{k=1}^{\infty}F_{0k}J_0(s_{0k}r)e^{-\kappa s_{0k}^2 t} \\ &\quad + \sum_{n=1}^{\infty}\left(T_{an}\cos n\theta + T'_{an}\sin n\theta\right)\sum_{k=1}^{\infty}F_{nk}J_n(s_{nk}r)e^{-\kappa s_{nk}^2 t}\end{aligned} \tag{6.92}$$

where

$$\begin{aligned}F_{0k} &= -\frac{2J_1(s_{0k}a)}{s_{0k}a\left[J_0^2(s_{0k}a) + J_1^2(s_{0k}a)\right]} \\ F_{nk} &= -\frac{2\dfrac{h_a a}{\lambda}}{\left(\dfrac{h_a a}{\lambda}+n\right)s_{nk}a}\frac{\left[\dfrac{2n}{s_n a}J_n(s_n a) - J_{n-1}(s_n a)\right]}{\left[J_n^2(s_{nk}a) - \dfrac{2n}{s_{nk}a}J_n(s_{nk}a)J_{n-1}(s_{nk}a) + J_{n-1}^2(s_{nk}a)\right]}\end{aligned} \tag{p}$$

Taking Eqs. (j) and (j′) into consideration, Eqs. (p) reduce to

$$\begin{aligned}F_{0k} &= -\frac{2\dfrac{h_a a}{\lambda}}{\left(s_{0k}^2a^2 + \dfrac{h_a^2a^2}{\lambda^2}\right)J_0(s_{0k}a)} \\ F_{nk} &= -\frac{2\dfrac{h_a a}{\lambda}}{\left(s_{nk}^2a^2 + \dfrac{h_a^2a^2}{\lambda^2} - n^2\right)J_n(s_{nk}a)}\end{aligned} \tag{p′}$$

Let us introduce a thermal stress function χ related to the components of stress in the cylindrical coordinate system by Eqs. (6.84). The governing equation for χ is given by Eq. (6.85). It is not necessary to consider the conditions of single-valuedness of rotation and displacements (6.86) because we consider a simply connected body. The boundary conditions for a traction free body are

$$\sigma_{rr} = 0, \quad \sigma_{r\theta} = 0 \qquad \text{on } r = a \tag{6.93}$$

Resolving the stress function χ into a complementary solution χ_c and a particular solution χ_p, the governing equations for χ_c and χ_p are

$$\nabla^4\chi_c = 0, \qquad \nabla^4\chi_p = -\alpha^* E^* \nabla^2\tau \tag{6.94}$$

The particular solution χ_p is

$$\chi_p = \alpha^* E^* \left[(T_{a0} - T_i) \sum_{k=1}^{\infty} \frac{F_{0k}}{s_{0k}^2} J_0(s_{0k} r) e^{-\kappa s_{0k}^2 t} + \sum_{n=1}^{\infty} \sum_{k=1}^{\infty} \frac{F_{nk}}{s_{nk}^2} J_n(s_{nk} r) \left(T_{an} \cos n\theta + T'_{an} \sin n\theta \right) e^{-\kappa s_{nk}^2 t} \right] \tag{q}$$

A general form of the complementary solution χ_c suitable for this problem is, from Eq. (6.88),

$$\chi_c = C_0 + D_0 r^2 + \sum_{n=1}^{\infty} \left[\left(C_n r^n + D_n r^{n+2} \right) \cos n\theta + \left(C'_n r^n + D'_n r^{n+2} \right) \sin n\theta \right] \tag{r}$$

Thus, the stress function χ is expressed by

$$\begin{aligned} \chi = {} & C_0 + D_0 r^2 + \sum_{n=1}^{\infty} \left[\left(C_n r^n + D_n r^{n+2} \right) \cos n\theta + \left(C'_n r^n + D'_n r^{n+2} \right) \sin n\theta \right] \\ & + \alpha^* E^* \left[(T_{a0} - T_i) \sum_{k=1}^{\infty} \frac{F_{0k}}{s_{0k}^2} J_0(s_{0k} r) e^{-\kappa s_{0k}^2 t} \right. \\ & \left. + \sum_{n=1}^{\infty} \sum_{k=1}^{\infty} \frac{F_{nk}}{s_{nk}^2} J_n(s_{nk} r) \left(T_{an} \cos n\theta + T'_{an} \sin n\theta \right) e^{-\kappa s_{nk}^2 t} \right] \end{aligned} \tag{6.95}$$

Substituting Eq. (6.95) into Eqs. (6.84) gives the following thermal stress components:

$$\begin{aligned} \sigma_{rr} = {} & 2D_0 - \sum_{n=1}^{\infty} \left\{ \left[n(n-1) C_n r^{n-2} + (n-2)(n+1) D_n r^n \right] \cos n\theta \right. \\ & \left. + \left[n(n-1) C'_n r^{n-2} + (n-2)(n+1) D'_n r^n \right] \sin n\theta \right\} \\ & + \alpha^* E^* \left\{ -(T_{a0} - T_i) \sum_{k=1}^{\infty} \frac{F_{0k}}{s_{0k} r} J_1(s_{0k} r) e^{-\kappa s_{0k}^2 t} \right. \\ & + \sum_{n=1}^{\infty} \sum_{k=1}^{\infty} \frac{F_{nk}}{s_{nk}^2 r^2} \left[s_{nk} r J_{n-1}(s_{nk} r) - n(n+1) J_n(s_{nk} r) \right] \\ & \left. \times \left(T_{an} \cos n\theta + T'_{an} \sin n\theta \right) e^{-\kappa s_{nk}^2 t} \right\} \end{aligned}$$

$$\sigma_{\theta\theta} = 2D_0 + \sum_{n=1}^{\infty}\{[n(n-1)C_n r^{n-2} + (n+2)(n+1)D_n r^n]\cos n\theta$$

$$+ [n(n-1)C'_n r^{n-2} + (n+2)(n+1)D'_n r^n]\sin n\theta\}$$

$$+ \alpha^* E^* \left\{ (T_{a0} - T_i)\sum_{k=1}^{\infty}\frac{F_{0k}}{s_{0k} r}[J_1(s_{0k} r) - s_{0k} r J_0(s_{0k} r)]e^{-\kappa s_{0k}^2 t} \right.$$

$$+ \sum_{n=1}^{\infty}\sum_{k=1}^{\infty}\frac{F_{nk}}{s_{nk}^2 r^2}\left[\left(n^2 + n - s_{nk}^2 r^2\right)J_n(s_{nk} r) - s_{nk} r J_{n-1}(s_{nk} r)\right]$$

$$\left. \times \left(T_{an}\cos n\theta + T'_{an}\sin n\theta\right)e^{-\kappa s_{nk}^2 t} \right\}$$

$$\sigma_{r\theta} = \sum_{n=1}^{\infty} n \left\{ [(n-1)C_n r^{n-2} + (n+1)D_n r^n]\sin n\theta \right.$$

$$- [(n-1)C'_n r^{n-2} + (n+1)D'_n r^n]\cos n\theta + \alpha^* E^* \sum_{k=1}^{\infty}\frac{F_{nk}}{s_{nk}^2 r^2}[s_{nk} r J_{n-1}(s_{nk} r)$$

$$\left. - (n+1)J_n(s_{nk} r)] \times \left(T_{an}\sin n\theta - T'_{an}\cos n\theta\right)e^{-\kappa s_{nk}^2 t} \right\} \quad (6.96)$$

Substitution of thermal stress components (6.96) into the boundary conditions (6.93) allows us to determine the unknown coefficients:

$$D_0 = \frac{\alpha^* E^*}{2}(T_{a0} - T_i)\sum_{k=1}^{\infty}\frac{F_{0k} J_1(s_{0k} a)}{s_{0k} a}e^{-\kappa s_{0k}^2 t}$$

$$\left(C_n, C'_n\right) = \left(T_{an}, T'_{an}\right)\frac{\alpha^* E^*}{2}\sum_{k=1}^{\infty}\frac{F_{nk}}{s_{nk}^2 a^n}[s_{nk} a J_{n-1}(s_{nk} a) - 2(n+1)J_n(s_{nk} a)]e^{-\kappa s_{nk}^2 t}$$

$$\left(D_n, D'_n\right) = -\left(T_{an}, T'_{an}\right)\frac{\alpha^* E^*}{2}\sum_{k=1}^{\infty}\frac{F_{nk}}{s_{nk}^2 a^{n+2}}[s_{nk} a J_{n-1}(s_{nk} a) - 2n J_n(s_{nk} a)]e^{-\kappa s_{nk}^2 t}$$

$$(6.97)$$

From Eqs. (6.96) and (6.97) we have the components of thermal stress

$$\sigma_{rr} = \alpha^* E^* (T_{a0} - T_i)\sum_{k=1}^{\infty}\frac{F_{0k}}{s_{0k} a r}[r J_1(s_{0k} a) - a J_1(s_{0k} r)]e^{-\kappa s_{0k}^2 t}$$

$$- \frac{\alpha^* E^*}{2}\sum_{n=1}^{\infty}\left(T_{an}\cos n\theta + T'_{an}\sin n\theta\right)$$

$$\times \sum_{k=1}^{\infty} \frac{F_{nk}}{s_{nk}^2 a^2} \left\{ n(n-1)\left(\frac{r}{a}\right)^{n-2} [s_{nk}aJ_{n-1}(s_{nk}a) - 2(n+1)J_n(s_{nk}a)] \right.$$

$$- (n-2)(n+1)\left(\frac{r}{a}\right)^{n} [s_{nk}aJ_{n-1}(s_{nk}a) - 2nJ_n(s_{nk}a)]$$

$$\left. +2\left(\frac{a}{r}\right)^2 [n(n+1)J_n(s_{nk}r) - s_{nk}rJ_{n-1}(s_{nk}r)] \right\} e^{-\kappa s_{nk}^2 t}$$

$$\sigma_{\theta\theta} = \alpha^* E^* (T_{a0} - T_i) \sum_{k=1}^{\infty} \frac{F_{0k}}{s_{0k}a} \left[J_1(s_{0k}a) - s_{0k}aJ_0(s_{0k}r) + \frac{a}{r}J_1(s_{0k}r) \right] e^{-\kappa s_{0k}^2 t}$$

$$- \frac{\alpha^* E^*}{2} \sum_{n=1}^{\infty} \left(T_{an}\cos n\theta + T'_{an}\sin n\theta\right)$$

$$\times \sum_{k=1}^{\infty} \frac{F_{nk}}{s_{nk}^2 a^2} \left\{ n(n-1)\left(\frac{r}{a}\right)^{n-2} [-s_{nk}aJ_{n-1}(s_{nk}a) + 2(n+1)J_n(s_{nk}a)] \right.$$

$$+ (n+1)(n+2)\left(\frac{r}{a}\right)^{n} [s_{nk}aJ_{n-1}(s_{nk}a) - 2nJ_n(s_{nk}a)]$$

$$\left. -2\left(\frac{a}{r}\right)^2 \left[(n^2 + n - s_{nk}^2 r^2)J_n(s_{nk}r) - s_{nk}rJ_{n-1}(s_{nk}r)\right] \right\} e^{-\kappa s_{nk}^2 t}$$

$$\sigma_{r\theta} = -\frac{\alpha^* E^*}{2} \sum_{n=1}^{\infty} n\left(T_{an}\sin n\theta - T'_{an}\cos n\theta\right)$$

$$\times \sum_{k=1}^{\infty} \frac{F_{nk}}{s_{nk}^2 a^2} \left\{ (n-1)\left(\frac{r}{a}\right)^{n-2} [-s_{nk}aJ_{n-1}(s_{nk}a) + 2(n+1)J_n(s_{nk}a)] \right.$$

$$+ (n+1)\left(\frac{r}{a}\right)^{n} [s_{nk}aJ_{n-1}(s_{nk}a) - 2nJ_n(s_{nk}a)]$$

$$\left. -2\left(\frac{a}{r}\right)^2 [s_{nk}rJ_{n-1}(s_{nk}r) - (n+1)J_n(s_{nk}r)] \right\} e^{-\kappa s_{nk}^2 t} \tag{6.98}$$

Example 6.5 When the surface $r = a$ of a solid cylinder with the constant initial temperature T_i is heated to $T_a(\theta)$, find the transient thermal stress in the cylinder.

Solution The temperature condition is

$$T = T_a(\theta) \qquad \text{on } r = a \tag{a}$$

The temperature condition from Subsection 6.2.2 is

$$T - T_a(\theta) = -\frac{\lambda}{h_a}\frac{\partial T}{\partial r} \qquad \text{on } r = a \tag{b}$$

Application of the limit $h_a \to \infty$ to the boundary condition (b) gives Eq. (a). Therefore, the transient thermal stress can be obtained by the application of $h_a \to \infty$ to the thermal stresses (6.98). From Eqs. (j) an (j′) in Subsection 6.2.2, s_{0k} and s_{nk} are the roots of the equations $J_0(s_{0k}a) = 0$ and $J_n(s_{nk}a) = 0$, respectively. Equation (p) in Subsection 6.2.2 reduces to

$$F_{0k} = -\frac{2}{s_{0k}a J_1(s_{0k}a)}, \qquad F_{nk} = \frac{2}{s_{nk}a J_{n-1}(s_{nk}a)} \tag{c}$$

The temperature and the thermal stresses are

$$\begin{aligned}
\tau &= (T_{a0} - T_i)\left[1 - 2\sum_{k=1}^{\infty}\frac{J_0(s_{0k}r)}{s_{0k}a J_1(s_{0k}a)}e^{-\kappa s_{0k}^2 t}\right] \\
&\quad + \sum_{n=1}^{\infty}\left(T_{an}\cos n\theta + T'_{an}\sin n\theta\right)\left[\left(\frac{r}{a}\right)^n + 2\sum_{k=1}^{\infty}\frac{J_n(s_{nk}r)}{s_{nk}a J_{n-1}(s_{nk}a)}e^{-\kappa s_{nk}^2 t}\right] \\
\sigma_{rr} &= -2\alpha^* E^*(T_{a0} - T_i)\sum_{k=1}^{\infty}\frac{r J_1(s_{0k}a) - a J_1(s_{0k}r)}{s_{0k}^2 a^2 r J_1(s_{0k}a)}e^{-\kappa s_{0k}^2 t} \\
&\quad - \alpha^* E^*\sum_{n=1}^{\infty}\left(T_{an}\cos n\theta + T'_{an}\sin n\theta\right) \\
&\quad \times \sum_{k=1}^{\infty}\frac{1}{s_{nk}^2 a^2}\left\{n(n-1)\left(\frac{r}{a}\right)^{n-2} - (n-2)(n+1)\left(\frac{r}{a}\right)^n\right. \\
&\quad \left. + \frac{2a}{s_{nk}r^2 J_{n-1}(s_{nk}a)}\left[n(n+1)J_n(s_{nk}r) - s_{nk}r J_{n-1}(s_{nk}r)\right]\right\} e^{-\kappa s_{nk}^2 t} \\
\sigma_{\theta\theta} &= -2\alpha^* E^*(T_{a0} - T_i)\sum_{k=1}^{\infty}\frac{r J_1(s_{0k}a) - s_{0k}ar J_0(s_{0k}r) + a J_1(s_{0k}r)}{s_{0k}^2 a^2 r J_1(s_{0k}a)}e^{-\kappa s_{0k}^2 t} \\
&\quad - \alpha^* E^*\sum_{n=1}^{\infty}\left(T_{an}\cos n\theta + T'_{an}\sin n\theta\right) \\
&\quad \times \sum_{k=1}^{\infty}\frac{1}{s_{nk}^2 a^2}\left\{-n(n-1)\left(\frac{r}{a}\right)^{n-2} + (n+1)(n+2)\left(\frac{r}{a}\right)^n - \frac{2a}{s_{nk}r^2 J_{n-1}(s_{nk}a)}\right. \\
&\quad \left. \times \left[\left(n^2 + n - s_{nk}^2 r^2\right)J_n(s_{nk}r) - s_{nk}r J_{n-1}(s_{nk}r)\right]\right\} e^{-\kappa s_{nk}^2 t} \\
\sigma_{r\theta} &= -\alpha^* E^*\sum_{n=1}^{\infty} n\left(T_{an}\cos n\theta - T'_{an}\sin n\theta\right) \\
&\quad \times \sum_{k=1}^{\infty}\frac{1}{s_{nk}^2 a^2}\left\{-(n-1)\left(\frac{r}{a}\right)^{n-2} + (n+1)\left(\frac{r}{a}\right)^n\right. \\
&\quad \left. - \frac{2a}{s_{nk}r^2 J_{n-1}(s_{nk}a)}\left[s_{nk}r J_{n-1}(s_{nk}r) - (n+1)J_n(s_{nk}r)\right]\right\} e^{-\kappa s_{nk}^2 t}
\end{aligned} \tag{d}$$

6.3 Two-Dimensional Problems (Axisymmetric Problems)

6.3.1 Fundamental Equations for Two-Dimensional Problems

We now consider axisymmetric problems of a cylindrical body. It is assumed that the body is deformed symmetrically with respect to the coordinate axis z. Making use of the cylindrical coordinate system (r, θ, z), for axially symmetric problems, the equilibrium of the forces in the directions r and z are obtained from Eqs. (4.69)

$$\begin{aligned} \frac{\partial \sigma_{rr}}{\partial r} + \frac{\partial \sigma_{zr}}{\partial z} + \frac{\sigma_{rr} - \sigma_{\theta\theta}}{r} + F_r = 0 \\ \frac{\partial \sigma_{rz}}{\partial r} + \frac{\partial \sigma_{zz}}{\partial z} + \frac{\sigma_{rz}}{r} + F_z = 0 \end{aligned} \tag{6.99}$$

The constitutive equations for a homogeneous, isotropic body in the cylindrical coordinate system, or the generalized Hooke's law, are obtained from Eqs. (4.76)

$$\begin{aligned} \epsilon_{rr} &= \frac{1}{E}[\sigma_{rr} - \nu(\sigma_{\theta\theta} + \sigma_{zz})] + \alpha\tau = \frac{1}{2G}\left(\sigma_{rr} - \frac{\nu}{1+\nu}\Theta\right) + \alpha\tau \\ \epsilon_{\theta\theta} &= \frac{1}{E}[\sigma_{\theta\theta} - \nu(\sigma_{zz} + \sigma_{rr})] + \alpha\tau = \frac{1}{2G}\left(\sigma_{\theta\theta} - \frac{\nu}{1+\nu}\Theta\right) + \alpha\tau \\ \epsilon_{zz} &= \frac{1}{E}[\sigma_{zz} - \nu(\sigma_{rr} + \sigma_{\theta\theta})] + \alpha\tau = \frac{1}{2G}\left(\sigma_{zz} - \frac{\nu}{1+\nu}\Theta\right) + \alpha\tau \\ \epsilon_{zr} &= \frac{\sigma_{zr}}{2G} \end{aligned} \tag{6.100}$$

where $\Theta = \sigma_{rr} + \sigma_{\theta\theta} + \sigma_{zz}$. An alternative form is

$$\begin{aligned} \sigma_{rr} &= 2\mu\epsilon_{rr} + \lambda e - \beta\tau \\ \sigma_{\theta\theta} &= 2\mu\epsilon_{\theta\theta} + \lambda e - \beta\tau \\ \sigma_{zz} &= 2\mu\epsilon_{zz} + \lambda e - \beta\tau \\ \sigma_{zr} &= 2\mu\epsilon_{zr} \end{aligned} \tag{6.101}$$

where $e = \epsilon_{rr} + \epsilon_{\theta\theta} + \epsilon_{zz}$, $\mu = G$, $\lambda = [2\nu G/(1-2\nu)]$, and $\beta = \alpha(3\lambda + 2\mu) = [\alpha E/(1-2\nu)]$.

The components of strain in the cylindrical coordinate system are given by Eqs. (4.71) for the three-dimensional deformation. Therefore, the components of strain for an axisymmetric deformation are

$$\begin{aligned} \epsilon_{rr} &= \frac{\partial u_r}{\partial r}, \qquad \epsilon_{\theta\theta} = \frac{u_r}{r}, \qquad \epsilon_{zz} = \frac{\partial u_z}{\partial z} \\ \epsilon_{zr} &= \frac{1}{2}\left(\frac{\partial u_r}{\partial z} + \frac{\partial u_z}{\partial r}\right) \\ e &= \frac{\partial u_r}{\partial r} + \frac{u_r}{r} + \frac{\partial u_z}{\partial z} \end{aligned} \tag{6.102}$$

where u_r and u_z are the components of displacement in the direction of r and z, respectively.

Substituting Eqs. (6.101) and (6.102) into Eqs. (6.99), Navier's equations for axisymmetric thermoelastic problems can be expressed as

$$\begin{aligned} \nabla^2 u_r - \frac{u_r}{r^2} + \frac{1}{1-2\nu}\frac{\partial e}{\partial r} - 2\frac{1+\nu}{1-2\nu}\alpha\frac{\partial \tau}{\partial r} + \frac{2(1+\nu)}{E}F_r = 0 \\ \nabla^2 u_z + \frac{1}{1-2\nu}\frac{\partial e}{\partial z} - 2\frac{1+\nu}{1-2\nu}\alpha\frac{\partial \tau}{\partial z} + \frac{2(1+\nu)}{E}F_z = 0 \end{aligned} \tag{6.103}$$

in which ∇^2 in axisymmetrical case denotes

$$\nabla^2 = \frac{\partial^2}{\partial r^2} + \frac{1}{r}\frac{\partial}{\partial r} + \frac{\partial^2}{\partial z^2} \tag{6.104}$$

The solution to the Navier's equations (6.103) without the body forces can be expressed, for example, by Goodier's thermoelastic displacement potential Φ and Boussinesq harmonic functions φ and ψ discussed in Section 4.4. Referring to Eqs. (4.80) under the axisymmetric condition, the displacements can be expressed by

$$\begin{aligned} u_r &= \frac{\partial \Phi}{\partial r} + \frac{\partial \varphi}{\partial r} + z\frac{\partial \psi}{\partial r} \\ u_z &= \frac{\partial \Phi}{\partial z} + \frac{\partial \varphi}{\partial z} + z\frac{\partial \psi}{\partial z} - (3-4\nu)\psi \end{aligned} \tag{6.105}$$

in which Goodier's thermoelastic displacement potential Φ and Boussinesq harmonic functions φ, ψ must satisfy the following governing equations

$$\nabla^2\Phi = K\tau, \qquad \nabla^2\varphi = 0, \qquad \nabla^2\psi = 0 \tag{6.106}$$

The components of stress are

$$\begin{aligned} \sigma_{rr} &= 2G\left(\frac{\partial^2\Phi}{\partial r^2} - K\tau + \frac{\partial^2\varphi}{\partial r^2} + z\frac{\partial^2\psi}{\partial r^2} - 2\nu\frac{\partial\psi}{\partial z}\right) \\ \sigma_{\theta\theta} &= 2G\left(\frac{1}{r}\frac{\partial\Phi}{\partial r} - K\tau + \frac{1}{r}\frac{\partial\varphi}{\partial r} + \frac{z}{r}\frac{\partial\psi}{\partial r} - 2\nu\frac{\partial\psi}{\partial z}\right) \\ \sigma_{zz} &= 2G\left[\frac{\partial^2\Phi}{\partial z^2} - K\tau + \frac{\partial^2\varphi}{\partial z^2} + z\frac{\partial^2\psi}{\partial z^2} - 2(1-\nu)\frac{\partial\psi}{\partial z}\right] \\ \sigma_{rz} &= 2G\left[\frac{\partial^2\Phi}{\partial r\partial z} + \frac{\partial^2\varphi}{\partial r\partial z} + z\frac{\partial^2\psi}{\partial r\partial z} - (1-2\nu)\frac{\partial\psi}{\partial r}\right] \end{aligned} \tag{6.107}$$

When deformations in the cylindrical coordinate system are discussed, Michell's function M instead of Boussinesq harmonic functions φ and ψ is often used. We take

$$M = -\int(\varphi + z\psi)\,dz \tag{6.108}$$

The displacements in the cylindrical coordinate system are represented by the Michell's function M

$$u_r = \frac{\partial \Phi}{\partial r} - \frac{\partial^2 M}{\partial r \partial z}$$
$$u_z = \frac{\partial \Phi}{\partial z} + 2(1-\nu)\nabla^2 M - \frac{\partial^2 M}{\partial z^2} \tag{6.109}$$

where the Michell's function M must satisfy the equation

$$\nabla^2 \nabla^2 M = 0 \tag{6.110}$$

The components of the stress are represented by the thermoelastic displacement potential Φ and the Michell's function M:

$$\sigma_{rr} = 2G\left[\frac{\partial^2 \Phi}{\partial r^2} - K\tau + \frac{\partial}{\partial z}\left(\nu\nabla^2 M - \frac{\partial^2 M}{\partial r^2}\right)\right]$$
$$\sigma_{\theta\theta} = 2G\left[\frac{1}{r}\frac{\partial \Phi}{\partial r} - K\tau + \frac{\partial}{\partial z}\left(\nu\nabla^2 M - \frac{1}{r}\frac{\partial M}{\partial r}\right)\right]$$
$$\sigma_{zz} = 2G\left\{\frac{\partial^2 \Phi}{\partial z^2} - K\tau + \frac{\partial}{\partial z}\left[(2-\nu)\nabla^2 M - \frac{\partial^2 M}{\partial z^2}\right]\right\}$$
$$\sigma_{rz} = 2G\left\{\frac{\partial^2 \Phi}{\partial r \partial z} + \frac{\partial}{\partial r}\left[(1-\nu)\nabla^2 M - \frac{\partial^2 M}{\partial z^2}\right]\right\} \tag{6.111}$$

The general solutions of the Eq. (6.110) for the axisymmetric case of the cylindrical coordinate system are

$$\begin{matrix}
\begin{pmatrix}1\\ \ln r\end{pmatrix}\begin{pmatrix}1\\ z\end{pmatrix}, & \begin{pmatrix}J_0(ar)\\ Y_0(ar)\end{pmatrix}\begin{pmatrix}e^{az}\\ e^{-az}\end{pmatrix}, & \begin{pmatrix}I_0(ar)\\ K_0(ar)\end{pmatrix}\begin{pmatrix}\sin az\\ \cos az\end{pmatrix}\\
\begin{pmatrix}r^2\\ r^2\ln r\end{pmatrix}\begin{pmatrix}1\\ z\end{pmatrix}, & \begin{pmatrix}rJ_1(ar)\\ rY_1(ar)\end{pmatrix}\begin{pmatrix}e^{az}\\ e^{-az}\end{pmatrix}, & \begin{pmatrix}rI_1(ar)\\ rK_1(ar)\end{pmatrix}\begin{pmatrix}\sin az\\ \cos az\end{pmatrix}\\
\begin{pmatrix}1\\ \ln r\end{pmatrix}\begin{pmatrix}z^2\\ z^3\end{pmatrix}, & \begin{pmatrix}J_0(ar)\\ Y_0(ar)\end{pmatrix}\begin{pmatrix}ze^{az}\\ ze^{-az}\end{pmatrix}, & \begin{pmatrix}I_0(ar)\\ K_0(ar)\end{pmatrix}\begin{pmatrix}z\sin az\\ z\cos az\end{pmatrix}
\end{matrix} \tag{6.112}$$

where a is an arbitrary constant, $J_n(ar)$ and $Y_n(ar)$ are the Bessel functions, $I_n(ar)$ and $K_n(ar)$ are the modified Bessel functions, and $\binom{\sinh az}{\cosh az}$ can be used instead of $\binom{e^{az}}{e^{-az}}$.

6.3.2 Steady State Thermal Stresses in a Semi-Infinite Body

Consider a semi-infinite body subjected to a steady temperature field depending on radial and axial directions of the cylindrical coordinate system shown in Fig. 6.4. The initial

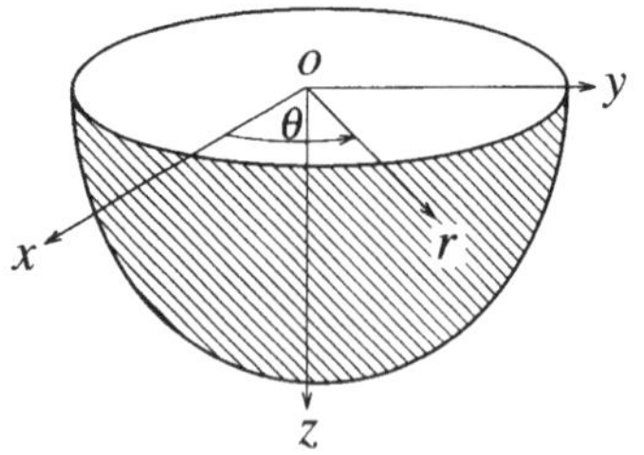

Figure 6.4 A semi-infinite body.

temperature $T_i(r, z)$ in the semi-infinite body is given by a constant temperature T_i and heat is exchanged through heat transfer at the boundary surface. The temperature of the surrounding medium $T_a(r)$ at the boundary surface depends on the radial direction r. The fundamental equation system to be solved for the steady state is

Governing equation:

$$\frac{\partial^2 T}{\partial r^2} + \frac{1}{r}\frac{\partial T}{\partial r} + \frac{\partial^2 T}{\partial z^2} = 0 \tag{a}$$

Boundary condition:

$$\lambda \frac{\partial T}{\partial z} = h[T - T_a(r)] \qquad \text{on } z = 0 \tag{b}$$

$$T = T_i \qquad \text{at } r, z \to \infty \tag{c}$$

Initial condition:

$$T = T_i \qquad \text{at } t = 0 \tag{d}$$

Referring to the general solution of the harmonic function (4.82), the temperature suitable for this problem is

$$T(r, z) = A_0 + \int_0^\infty A(s) J_0(sr) e^{-sz}\, ds \tag{e}$$

in which A_0 and $A(s)$ are unknown functions and $J_0(sr)$ denotes the Bessel function of the first kind of order zero. The boundary condition (c) gives

$$A_0 = T_i \tag{f}$$

From the boundary condition (b), we have

$$\int_0^\infty (h + \lambda s) A(s) J_0(sr)\, ds = h[T_a(r) - T_i] \tag{g}$$

Now we introduce the Hankel transform* which is defined by

$$\bar{f}(s) = \int_0^\infty f(r) r J_\nu(sr)\, dr \tag{6.113}$$

where $\bar{f}(s)$ is called the Hankel transform of order ν of an arbitrary function $f(r)$. The Hankel inversion theorem is defined as

$$f(r) = \int_0^\infty \bar{f}(s) s J_\nu(sr)\, ds \tag{6.114}$$

The right-hand side of Eq. (g) can be rewritten by use of the Hankel transform

$$T_a(r) - T_i = \int_0^\infty \bar{f}(s) s J_0(sr)\, ds \tag{h}$$

where

$$\bar{f}(s) = \int_0^\infty [T_a(r) - T_i] r J_0(sr)\, dr \tag{i}$$

From Eqs. (g) and (h), the unknown function $A(s)$ is

$$A(s) = \frac{hs\bar{f}(s)}{h + \lambda s} \tag{j}$$

Therefore, the temperature field is obtained as

$$T = T_i + \int_0^\infty \frac{hs\bar{f}(s)}{h + \lambda s} J_0(sr) e^{-sz}\, ds \tag{6.115}$$

The temperature change τ is

$$\tau = T - T_i = \int_0^\infty \frac{hs\bar{f}(s)}{h + \lambda s} J_0(sr) e^{-sz}\, ds \tag{6.116}$$

Let us introduce Goodier's thermoelastic displacement potential Φ and the Boussinesq harmonic functions φ and ψ discussed in Subsection 6.3.1. Referring to Eqs. (4.82), the functions φ and ψ can be expressed by use of integrals suitable for the semi-infinite body under the axial symmetric condition

$$\begin{aligned} \varphi &= \int_0^\infty B(s) J_0(sr) e^{-sz}\, ds \\ \psi &= \int_0^\infty C(s) J_0(sr) e^{-sz}\, ds \end{aligned} \tag{k}$$

*See: I. N. Sneddon, *Fourier Transforms*, McGraw-Hill, New York (1951), p. 48, and I. N. Sneddon, *The Use of Integral Transforms*, McGraw-Hill, New York (1972), p. 298.

The potential Φ is obtained from Eqs. (6.106) and (6.116) as

$$\Phi = -\frac{K}{2}\int_0^\infty A(s)J_0(sr)\frac{z}{s}e^{-sz}\,ds \tag{l}$$

Substituting Eqs. (k) and (l) into Eqs. (6.105) and (6.107), the displacements and the thermal stresses are

$$\begin{aligned}
u_r &= -\int_0^\infty sJ_1(sr)\left[-K\frac{A(s)}{2s}z + B(s) + C(s)z\right]e^{-sz}\,ds\\
u_z &= -\int_0^\infty J_0(sr)\left[K\frac{A(s)}{2s}(1-sz) + sB(s) + (3-4\nu+sz)C(s)\right]e^{-sz}\,ds\\
\sigma_{rr} &= 2G\int_0^\infty \left\{\left[-s^2J_0(sr) + \frac{s}{r}J_1(sr)\right]\left[-K\frac{A(s)}{2s}z + B(s) + C(s)z\right]\right.\\
&\qquad \left. -\,[KA(s) - 2\nu sC(s)]J_0(sr)\right\}e^{-sz}\,ds\\
\sigma_{\theta\theta} &= -2G\int_0^\infty \left\{\frac{s}{r}J_1(sr)\left[-K\frac{A(s)}{2s}z + B(s) + C(s)z\right]\right.\\
&\qquad \left. +\,[KA(s) - 2\nu sC(s)]J_0(sr)\right\}e^{-sz}\,ds\\
\sigma_{zz} &= 2G\int_0^\infty sJ_0(sr)\left[-K\frac{A(s)}{2s}sz + sB(s) + (2-2\nu+sz)C(s)\right]e^{-sz}\,ds\\
\sigma_{rz} &= 2G\int_0^\infty sJ_1(sr)\left[K\frac{A(s)}{2s}(1-sz) + sB(s) + (1-2\nu+sz)C(s)\right]e^{-sz}\,ds
\end{aligned} \tag{6.117}$$

The boundary conditions on the traction free surface are

$$\sigma_{zz} = 0, \quad \sigma_{zr} = 0 \qquad \text{on } z = 0 \tag{m}$$

The unknown functions $B(s)$ and $C(s)$ can be determined by the boundary conditions (m)

$$B(s) = -(1-\nu)K\frac{A(s)}{s^2}, \qquad C(s) = K\frac{A(s)}{2s} \tag{n}$$

Substituting Eqs. (n) into Eqs. (6.117), the displacements and the thermal stresses are

$$\begin{aligned}
u_r &= (1+\nu)\alpha\int_0^\infty \frac{A(s)}{s}J_1(sr)e^{-sz}\,ds\\
u_z &= -(1+\nu)\alpha\int_0^\infty \frac{A(s)}{s}J_0(sr)e^{-sz}\,ds\\
\sigma_{rr} &= -\alpha E\int_0^\infty \frac{A(s)}{sr}J_1(sr)e^{-sz}\,ds
\end{aligned} \tag{6.118}$$

$$\sigma_{\theta\theta} = -\alpha E \int_0^\infty A(s)\left[J_0(sr) - \frac{J_1(sr)}{sr}\right] e^{-sz}\, ds$$

$$\sigma_{zz} = \sigma_{zr} = 0$$

where $A(s)$ is given by Eq. (j).

Example 6.6 Derive the thermal stress in a semi-infinite body under steady temperature field, when the surface of the semi-infinite body is partially heated by the temperature $T_a(r) = T_i + T_a H(a-r)$, where $H(a-r)$ denotes the Heaviside's step function.

Solution The boundary conditions are

$$\begin{aligned} T &= T_a(r) = T_i + T_a H(a-r) && \text{on } z = 0 \\ T &= T_i && \text{at } r, z \to \infty \end{aligned} \tag{a}$$

The temperature is given by Eq. (e) in Subsection 6.3.2

$$T(r, z) = A + \int_0^\infty A(s) J_0(sr) e^{-sz}\, ds \tag{b}$$

The boundary conditions give

$$A = T_i, \qquad \int_0^\infty A(s) J_0(sr)\, ds = T_a H(a-r) \tag{c}$$

Equation (c) determines the unknown coefficient $A(s)$:

$$\begin{gathered} \frac{A(s)}{s} = T_a \int_0^\infty H(a-r) r J_0(sr)\, ds = T_a \int_0^a r J_0(sr)\, ds = T_a \frac{a}{s} J_1(sa) \\ A(s) = T_a a J_1(sa) \end{gathered} \tag{d}$$

We now obtain the temperature change τ:

$$\begin{aligned} \tau(r, z) &= T(r, z) - T_i \\ &= \int_0^\infty A(s) J_0(sr) e^{-sz}\, ds = T_a \int_0^\infty a J_1(sa) J_0(sr) e^{-sz}\, ds \end{aligned} \tag{e}$$

Substitution of Eq. (d) into Eqs. (6.118) gives

$$\begin{aligned} u_r &= (1+\nu)\alpha T_a \int_0^\infty \frac{a}{s} J_1(sa) J_1(sr) e^{-sz}\, ds \\ u_z &= -(1+\nu)\alpha T_a \int_0^\infty \frac{a}{s} J_1(sa) J_0(sr) e^{-sz}\, ds \\ \sigma_{rr} &= -\alpha E T_a \int_0^\infty \frac{a}{sr} J_1(sa) J_1(sr) e^{-sz}\, ds \end{aligned} \tag{f}$$

$$\sigma_{\theta\theta} = -\alpha E T_a \int_0^\infty a J_1(sa) \left[J_0(sr) - \frac{J_1(sr)}{sr} \right] e^{-sz}\, ds$$

$$\sigma_{zz} = \sigma_{zr} = 0$$

6.3.3 Steady State Thermal Stresses in a Long Circular Cylinder

Consider an infinitely long solid circular cylinder of radius a subjected to a steady axisymmetric temperature field. The initial temperature $T_i(r, z)$ in the cylinder is given by a constant temperature T_i and heat is exchanged through heat transfer at the boundary surface. The temperature of the surrounding medium $T_a(z)$ at the boundary surface depends on the axial direction z. The fundamental equation system to be solved for the steady state is

Governing equation:

$$\frac{\partial^2 T}{\partial r^2} + \frac{1}{r}\frac{\partial T}{\partial r} + \frac{\partial^2 T}{\partial z^2} = 0 \tag{a}$$

Boundary condition:

$$-\lambda \frac{\partial T}{\partial r} = h[T - T_a(z)] \qquad \text{on } r = a \tag{b}$$

$$T = T_i \qquad \text{at } |z| \to \infty \tag{c}$$

Initial condition:

$$T = T_i \qquad \text{at } t = 0 \tag{d}$$

in which we assume $T_a(z) = T_a(-z)$ for simplicity. Referring to the general solution of the harmonic function (4.82), the temperature suitable for this problem is

$$T(r, z) = A_0 + \int_0^\infty A(s) I_0(sr) \cos sz\, ds \tag{e}$$

in which A_0 and $A(s)$ are unknown functions and $I_0(sr)$ denotes the modified Bessel function of the first kind of order zero. The boundary condition (c) gives

$$A_0 = T_i \tag{f}$$

From the boundary condition (b), we have

$$\int_0^\infty A(s)\, [h I_0(sa) + \lambda s I_1(sa)] \cos sz\, ds = h\, [T_a(z) - T_i] \tag{g}$$

Now we introduce the Fourier cosine transform† which is defined by

$$\bar{f}(s) = \frac{2}{\pi}\int_0^\infty f(z)\cos sz\,dz \tag{6.119}$$

where $\bar{f}(s)$ is called the Fourier cosine transform of an arbitrary function $f(z)$. The Fourier cosine inversion is defined as

$$f(z) = \int_0^\infty \bar{f}(s)\cos sz\,ds \tag{6.120}$$

The right-hand side of Eq. (g) can be rewritten by use of the Fourier cosine transform

$$T_a(z) - T_i = \int_0^\infty \bar{f}(s)\cos sz\,ds \tag{h}$$

where

$$\bar{f}(s) = \frac{2}{\pi}\int_0^\infty [T_a(z) - T_i]\cos sz\,dz \tag{i}$$

From Eqs. (g) and (h), the unknown function $A(s)$ is

$$A(s) = \frac{h\bar{f}(s)}{hI_0(sa) + \lambda sI_1(sa)} \tag{j}$$

Thus, the temperature field is obtained as

$$T = T_i + \int_0^\infty \frac{h\bar{f}(s)}{hI_0(sa) + \lambda sI_1(sa)} I_0(sr)\cos sz\,ds \tag{6.121}$$

The temperature change τ is

$$\tau = T - T_i = \int_0^\infty \frac{h\bar{f}(s)}{hI_0(sa) + \lambda sI_1(sa)} I_0(sr)\cos sz\,ds \tag{6.122}$$

Let us introduce the Goodier's thermoelastic displacement potential Φ and the Michell's function M discussed in Subsection 6.3.1. Referring to Eqs. (6.112), the Michell's function M can be expressed by use of a form suitable for the infinitely long cylinder under the axisymmetric condition

$$M = \int_0^\infty [B(s)I_0(sr) + C(s)rI_1(sr)]\sin sz\,ds \tag{k}$$

The potential Φ is obtained from Eqs. (6.106) and (6.122) as

$$\Phi = \frac{K}{2}\int_0^\infty A(s)\frac{r}{s}I_1(sr)\cos sz\,ds \tag{l}$$

---

†See: I. N. Sneddon, *Fourier Transforms*, McGraw-Hill, New York (1951), pp. 5, 17, and I. N. Sneddon, *The Use of Integral Transforms*, McGraw-Hill, New York (1972), p. 42.

Substituting Eqs. (k) and (l) into Eqs. (6.109) and (6.111), the displacements and the thermal stresses are

$$u_r = \int_0^\infty \left[K\frac{A(s)}{2} r I_0(sr) - s^2 B(s) I_1(sr) - s^2 C(s) r I_0(sr) \right] \cos sz\, ds$$

$$u_z = \int_0^\infty \left\{ -K\frac{A(s)}{2} r I_1(sr) + s^2 B(s) I_0(sr) \right.$$

$$\left. + sC(s)\,[4(1-\nu) I_0(sr) + sr I_1(sr)] \right\} \sin sz\, ds$$

$$\sigma_{rr} = 2G \int_0^\infty \left\{ -K\frac{A(s)}{2}\,[I_0(sr) - sr I_1(sr)] + s^3 B(s) \left[\frac{I_1(sr)}{sr} - I_0(sr) \right] \right.$$

$$\left. - s^2 C(s)\,[(1-2\nu) I_0(sr) + sr I_1(sr)] \right\} \cos sz\, ds$$

$$\sigma_{\theta\theta} = -2G \int_0^\infty \left[K\frac{A(s)}{2} I_0(sr) + s^3 B(s) \frac{I_1(sr)}{sr} + (1-2\nu) s^2 C(s) I_0(sr) \right] \cos sz\, ds$$

$$\sigma_{zz} = 2G \int_0^\infty \left\{ -K A(s) \left[I_0(sr) + \frac{sr}{2} I_1(sr) \right] + s^3 B(s) I_0(sr) \right.$$

$$\left. + s^2 C(s)\,[2(2-\nu) I_0(sr) + sr I_1(sr)] \right\} \cos sz\, ds$$

$$\sigma_{rz} = 2G \int_0^\infty \left\{ -K\frac{A(s)}{2} sr I_0(sr) + s^3 B(s) I_1(sr) \right.$$

$$\left. + s^2 C(s)\,[2(1-\nu) I_1(sr) + sr I_0(sr)] \right\} \sin sz\, ds \qquad (6.123)$$

The boundary conditions on the traction free surface are

$$\sigma_{rr} = 0, \quad \sigma_{rz} = 0 \qquad \text{on } r = a \qquad \text{(m)}$$

The unknown functions $B(s)$ and $C(s)$ can be determined from the boundary conditions (m)

$$B(s) = (1-\nu) K A(s) \frac{a}{s^2} \frac{sa I_0^2(sa) - sa I_1^2(sa) + I_0(sa) I_1(sa)}{(2 - 2\nu + s^2a^2) I_1^2(sa) - s^2a^2 I_0^2(sa)}$$

$$C(s) = K A(s) \frac{a^2}{2} \frac{I_1^2(sa) - I_0^2(sa)}{(2 - 2\nu + s^2a^2) I_1^2(sa) - s^2a^2 I_0^2(sa)} \qquad \text{(n)}$$

Substituting Eqs. (n) into Eqs. (6.123), the displacements and the thermal stresses can be obtained.

Example 6.7 Derive the steady thermal stress in a long circular cylinder when its surface is partially heated by the temperature $T_a(z) = T_i + T_a H(a - |z|)$, where $H(a - |z|)$ denotes the Heaviside's step function.

Solution The boundary conditions are

$$\begin{aligned} T &= T_a(z) = T_i + T_a H(a - |z|) && \text{on } r = a \\ T &= T_i && \text{at } r, z \to \infty \end{aligned} \tag{a}$$

The temperature is given by Eq. (e) in Subsection 6.3.3

$$T(r, z) = A + \int_0^\infty A(s) I_0(sr) \cos sz \, ds \tag{b}$$

The boundary conditions give

$$A = T_i, \qquad \int_0^\infty A(s) I_0(sa) \cos sz \, ds = T_a H(a - |z|) \tag{c}$$

Equation (c) determines the unknown coefficient $A(s)$:

$$A(s) = \frac{2T_a}{\pi I_0(sa)} \int_0^\infty H(a - |z|) \cos sz \, ds = \frac{2T_a}{\pi I_0(sa)} \int_0^a \cos sz \, ds = \frac{2T_a \sin sa}{\pi s I_0(sa)} \tag{d}$$

We now obtain the temperature change τ:

$$\begin{aligned} \tau(r, z) &= T(r, z) - T_i \\ &= \int_0^\infty A(s) I_0(sr) \cos sz \, ds = \frac{2}{\pi} T_a \int_0^\infty \frac{\sin sa}{s I_0(sa)} I_0(sr) \cos sz \, ds \end{aligned} \tag{e}$$

Substitution of Eq. (d) into Eqs. (6.123) gives the displacement and the thermal stress.

6.3.4 Transient Thermal Stresses in a Thick Plate

Consider a thick plate of thickness $2b$ subjected to a transient axisymmetric temperature field dependent on the radial and axial directions of the cylindrical coordinate system. The initial temperature $T_i(r, z)$ in the thick plate is given by a constant temperature T_i and the same heat flux $QF(r)$ is prescribed on the upper and lower boundary surfaces. The fundamental equation system to be solved for the unsteady state is

Governing equation:

$$\frac{\partial^2 T}{\partial r^2} + \frac{1}{r}\frac{\partial T}{\partial r} + \frac{\partial^2 T}{\partial z^2} = \frac{1}{\kappa}\frac{\partial T}{\partial t} \tag{a}$$

Boundary conditions:

$$\lambda \frac{\partial T}{\partial z} = \pm QF(r) \qquad \text{on } z = \pm b \tag{b}$$

$$T = T_i \qquad \text{at } r \to \infty \tag{c}$$

Initial condition:

$$T = T_i \qquad \text{at } t = 0 \tag{d}$$

Introducing the temperature change $\tau = T - T_i$ into Eqs. (a)–(d) and making use of the Laplace transform with respect to the time t, the governing equation (a) and the boundary condition (b) reduce to

$$\frac{\partial^2 \bar{\tau}}{\partial r^2} + \frac{1}{r}\frac{\partial \bar{\tau}}{\partial r} + \frac{\partial^2 \bar{\tau}}{\partial z^2} = \frac{p}{\kappa}\bar{\tau} \tag{e}$$

$$\lambda \frac{\partial \bar{\tau}}{\partial z} = \pm \frac{Q}{p} F(r) \qquad \text{on } z = \pm b \tag{f}$$

where p is the Laplace transform parameter. Considering the symmetry of the boundary condition, the general solution of Eq. (e) suitable for this problem is

$$\bar{\tau} = \int_0^\infty A(s) J_0(sr) \cosh \sqrt{s^2 + q^2}\, z \, ds \tag{g}$$

in which $q^2 = (p/\kappa)$, $A(s)$ is an unknown function, and $J_0(sr)$ denotes the Bessel function of the first kind of order zero. The boundary condition (f) gives

$$A(s) = \frac{Qsf(s)}{\lambda p \sqrt{s^2 + q^2} \sinh \sqrt{s^2 + q^2}\, b} \tag{h}$$

where

$$f(s) = \int_0^\infty F(r) r J_0(sr)\, dr, \qquad F(r) = \int_0^\infty f(s) s J_0(sr)\, ds \tag{i}$$

Then, the temperature field in the transform domain is obtained as

$$\bar{\tau} = \frac{Q}{\lambda} \int_0^\infty f(s) \frac{\cosh \sqrt{s^2 + q^2}\, z}{p\sqrt{s^2 + q^2} \sinh \sqrt{s^2 + q^2}\, b} s J_0(sr)\, ds \tag{j}$$

After inverting the Laplace transform, we have the temperature change τ

$$\tau = \frac{Q}{\lambda} \int_0^\infty f(s) J_0(sr) \left[\frac{\cosh sz}{\sinh sb} - \frac{1}{sb} e^{-\kappa s^2 t} + 2sb \sum_{n=1}^{\infty} \frac{(-1)^{n+1} \cos n\pi (z/b)}{n^2\pi^2 + s^2 b^2} e^{-\kappa (n^2\pi^2 + s^2 b^2) t / b^2} \right] ds \tag{6.124}$$

Let us introduce the Goodier's thermoelastic displacement potential Φ and the Michell's function M discussed in Subsection 6.3.1. Referring to Eqs. (6.112) the

Michell's function M can be expressed by use of a form suitable for the thick plate under the axial symmetric condition

$$M = \frac{Q}{\lambda} K \int_0^\infty f(s) J_0(sr) \left[B(s) \sinh sz + C(s) sz \cosh sz \right] ds \tag{k}$$

The potential Φ is obtained from Eqs. (6.106) and (6.124) as

$$\Phi = \frac{Q}{\lambda} K \int_0^\infty f(s) J_0(sr) \left[\frac{z \sinh sz}{2s \sinh sb} + \frac{1}{s^3 b} e^{-\kappa s^2 t} \right.$$
$$\left. + 2sb^3 \sum_{n=1}^{\infty} \frac{(-1)^n \cos n\pi(z/b)}{(n^2\pi^2 + s^2 b^2)^2} e^{-\kappa(n^2\pi^2 + s^2 b^2)t/b^2} \right] ds \tag{l}$$

Substituting the potential Φ and the Michell's function M into Eqs. (6.109) and (6.111), the radial and axial displacements, and the thermal stresses are

$$u_r = \frac{Q}{\lambda} K \int_0^\infty f(s) s J_1(sr) \left\{ B(s) s \cosh sz + C(s) s (\cosh sz + sz \sinh sz) \right.$$
$$\left. - \left[\frac{z \sinh sz}{2s \sinh sb} + \frac{1}{s^3 b} e^{-\kappa s^2 t} + 2sb^3 \sum_{n=1}^{\infty} \frac{(-1)^n \cos n\pi(z/b)}{(n^2\pi^2 + s^2 b^2)^2} e^{-\kappa(n^2\pi^2 + s^2 b^2)t/b^2} \right] \right\} ds$$

$$u_z = -\frac{Q}{\lambda} K \int_0^\infty f(s) J_0(sr) \left\{ B(s) s^2 \sinh sz - C(s) s^2 [2(1-2\nu) \sinh sz - sz \cosh sz] \right.$$
$$\left. - \frac{\sinh sz + sz \cosh sz}{2s \sinh sb} + 2sb^2 \pi \sum_{n=1}^{\infty} \frac{n(-1)^n \sin n\pi(z/b)}{(n^2\pi^2 + s^2 b^2)^2} e^{-\kappa(n^2\pi^2 + s^2 b^2)t/b^2} \right\} ds$$

$$\sigma_{rr} = 2\frac{Q}{\lambda} KG \int_0^\infty f(s) \left(B(s) s^3 \left[J_0(sr) - \frac{J_1(sr)}{sr} \right] \cosh sz \right.$$
$$+ C(s) s^3 \left\{ 2\nu J_0(sr) \cosh sz + \left[J_0(sr) - \frac{J_1(sr)}{sr} \right] (\cosh sz + sz \sinh sz) \right\}$$
$$- \left[J_0(sr) - \frac{J_1(sr)}{sr} \right] \frac{sz \sinh sz}{2 \sinh sb} - J_0(sr) \frac{\cosh sz}{\sinh sb} + \frac{J_1(sr)}{srsb} e^{-\kappa s^2 t}$$
$$\left. + 2sb \sum_{n=1}^{\infty} \frac{(\ 1)^n \cos n\pi(z/b)}{(n^2\pi^2 + s^2 b^2)^2} \left[\frac{s^2 b^2}{sr} J_1(sr) + n^2 \pi^2 J_0(sr) \right] e^{-\kappa(n^2\pi^2 + s^2 b^2)t/b^2} \right) ds$$

$$\sigma_{\theta\theta} = 2\frac{Q}{\lambda} KG \int_0^\infty f(s) \left\{ B(s) s^3 \frac{J_1(sr)}{sr} \cosh sz + C(s) s^3 \left[2\nu J_0(sr) \cosh sz \right. \right.$$
$$\left. + \frac{J_1(sr)}{sr} (\cosh sz + sz \sinh sz) \right] - J_0(sr) \frac{\cosh sz}{\sinh sb} - \frac{J_1(sr)}{sr} \frac{sz \sinh sz}{2 \sinh sb}$$

$$+\frac{1}{sb}\left[J_0(sr)-\frac{J_1(sr)}{sr}\right]e^{-\kappa s^2 t}+2sb\sum_{n=1}^{\infty}\frac{(-1)^n\cos n\pi(z/b)}{(n^2\pi^2+s^2b^2)^2}$$

$$\times\left[(n^2\pi^2+s^2b^2)J_0(sr)-\frac{s^2b^2}{sr}J_1(sr)\right]e^{-\kappa(n^2\pi^2+s^2b^2)t/b^2}\Bigg\}ds$$

$$\sigma_{zz}=2\frac{Q}{\lambda}KG\int_0^{\infty}f(s)J_0(sr)\Bigg\{-B(s)s^3\cosh sz$$

$$+C(s)s^3[(1-2\nu)\cosh sz-sz\sinh sz]+\frac{sz\sinh sz}{2\sinh sb}$$

$$+\frac{1}{sb}e^{-\kappa s^2 t}+2s^3b^3\sum_{n=1}^{\infty}\frac{(-1)^n\cos n\pi(z/b)}{(n^2\pi^2+s^2b^2)^2}e^{-\kappa(n^2\pi^2+s^2b^2)t/b^2}\Bigg\}ds$$

$$\sigma_{rz}=2\frac{Q}{\lambda}KG\int_0^{\infty}f(s)J_1(sr)\Bigg[B(s)s^3\sinh sz+C(s)s^3(2\nu\sinh sz+sz\cosh sz)$$

$$-\frac{\sinh sz+sz\cosh sz}{2\sinh sb}+2s^2b^2\pi\sum_{n=1}^{\infty}\frac{n(-1)^n\sin n\pi(z/b)}{(n^2\pi^2+s^2b^2)^2}e^{-\kappa(n^2\pi^2+s^2b^2)t/b^2}\Bigg]ds$$

(6.125)

The boundary conditions on the traction free surfaces are

$$\sigma_{zz}=0,\quad \sigma_{rz}=0\qquad \text{on } z=\pm b \tag{m}$$

The unknown function $B(s)$ and $C(s)$ can be determined from the boundary conditions (m)

$$B(s)=\frac{1-2\nu}{2s^3\sinh sb}+\frac{2\nu\sinh sb+sb\cosh sb}{s^3(\sinh sb\cosh sb+sb)}$$

$$\times\left[\frac{1}{sb}e^{-\kappa s^2 t}+2s^3b^3\sum_{n=1}^{\infty}\frac{e^{-\kappa(n^2\pi^2+s^2b^2)t/b^2}}{(n^2\pi^2+s^2b^2)^2}\right]$$

$$C(s)=\frac{1}{2s^3\sinh sb}-\frac{\sinh sb}{s^3(\sinh sb\cosh sb+sb)}$$

$$\times\left[\frac{1}{sb}e^{-\kappa s^2 t}+2s^3b^3\sum_{n=1}^{\infty}\frac{e^{-\kappa(n^2\pi^2+s^2b^2)t/b^2}}{(n^2\pi^2+s^2b^2)^2}\right]$$

(n)

Substituting Eqs. (n) into Eqs. (6.125), the displacements and the thermal stresses can be obtained.

6.4 Three-Dimensional Problems

6.4.1 Solution Methods in Three-Dimensional Problems

We now consider a cylindrical body subjected to a three-dimensional temperature field in the cylindrical coordinate system (r, θ, z). The equilibrium equations in the cylindrical coordinate system are given by Eqs. (4.69)

$$\begin{aligned}
\frac{\partial \sigma_{rr}}{\partial r} + \frac{1}{r}\frac{\partial \sigma_{\theta r}}{\partial \theta} + \frac{\partial \sigma_{zr}}{\partial z} + \frac{\sigma_{rr} - \sigma_{\theta\theta}}{r} + F_r = 0 \\
\frac{\partial \sigma_{r\theta}}{\partial r} + \frac{1}{r}\frac{\partial \sigma_{\theta\theta}}{\partial \theta} + \frac{\partial \sigma_{z\theta}}{\partial z} + 2\frac{\sigma_{r\theta}}{r} + F_\theta = 0 \\
\frac{\partial \sigma_{rz}}{\partial r} + \frac{1}{r}\frac{\partial \sigma_{\theta z}}{\partial \theta} + \frac{\partial \sigma_{zz}}{\partial z} + \frac{\sigma_{rz}}{r} + F_z = 0
\end{aligned} \tag{6.126}$$

The constitutive equations for a homogeneous, isotropic body in the cylindrical coordinate system, or the generalized Hooke's law, are Eqs. (4.69), which are given by

$$\begin{aligned}
\epsilon_{rr} &= \frac{1}{E}[\sigma_{rr} - \nu(\sigma_{\theta\theta} + \sigma_{zz})] + \alpha\tau = \frac{1}{2G}\left(\sigma_{rr} - \frac{\nu}{1+\nu}\Theta\right) + \alpha\tau \\
\epsilon_{\theta\theta} &= \frac{1}{E}[\sigma_{\theta\theta} - \nu(\sigma_{zz} + \sigma_{rr})] + \alpha\tau = \frac{1}{2G}\left(\sigma_{\theta\theta} - \frac{\nu}{1+\nu}\Theta\right) + \alpha\tau \\
\epsilon_{zz} &= \frac{1}{E}[\sigma_{zz} - \nu(\sigma_{rr} + \sigma_{\theta\theta})] + \alpha\tau = \frac{1}{2G}\left(\sigma_{zz} - \frac{\nu}{1+\nu}\Theta\right) + \alpha\tau \\
\epsilon_{r\theta} &= \frac{\sigma_{r\theta}}{2G}, \qquad \epsilon_{\theta z} = \frac{\sigma_{\theta z}}{2G}, \qquad \epsilon_{zr} = \frac{\sigma_{zr}}{2G}
\end{aligned} \tag{6.127}$$

where $\Theta = \sigma_{rr} + \sigma_{\theta\theta} + \sigma_{zz}$. The alternative form is

$$\begin{aligned}
\sigma_{rr} &= 2\mu\epsilon_{rr} + \lambda e - \beta\tau, \qquad \sigma_{r\theta} = 2\mu\epsilon_{r\theta} \\
\sigma_{\theta\theta} &= 2\mu\epsilon_{\theta\theta} + \lambda e - \beta\tau, \qquad \sigma_{\theta z} = 2\mu\epsilon_{\theta z} \\
\sigma_{zz} &= 2\mu\epsilon_{zz} + \lambda e - \beta\tau, \qquad \sigma_{zr} = 2\mu\epsilon_{zr}
\end{aligned} \tag{6.128}$$

where $e = \epsilon_{rr} + \epsilon_{\theta\theta} + \epsilon_{zz}$.

The components of strain in the cylindrical coordinate system are given by Eqs. (4.71) for three-dimensional deformation

$$\begin{aligned}
\epsilon_{rr} &= \frac{\partial u_r}{\partial r}, \qquad \epsilon_{\theta\theta} = \frac{u_r}{r} + \frac{1}{r}\frac{\partial u_\theta}{\partial \theta}, \qquad \epsilon_{zz} = \frac{\partial u_z}{\partial z} \\
\epsilon_{r\theta} &= \frac{1}{2}\left(\frac{1}{r}\frac{\partial u_r}{\partial \theta} + \frac{\partial u_\theta}{\partial r} - \frac{u_\theta}{r}\right), \qquad \epsilon_{\theta z} = \frac{1}{2}\left(\frac{\partial u_\theta}{\partial z} + \frac{1}{r}\frac{\partial u_z}{\partial \theta}\right) \\
\epsilon_{zr} &= \frac{1}{2}\left(\frac{\partial u_r}{\partial z} + \frac{\partial u_z}{\partial r}\right) \\
e &= \frac{\partial u_r}{\partial r} + \frac{u_r}{r} + \frac{1}{r}\frac{\partial u_\theta}{\partial \theta} + \frac{\partial u_z}{\partial z}
\end{aligned} \tag{6.129}$$

where u_r, u_θ, u_z are the components of displacement in the r, θ, z directions, respectively. Substituting Eqs. (6.129) and (6.128) into Eqs. (6.126), the Navier's equations for three-dimensional thermoelastic problems can be expressed as

$$
\begin{aligned}
&(\lambda+2\mu)\frac{\partial e}{\partial r} - \mu\left[\frac{1}{r^2}\frac{\partial^2(ru_\theta)}{\partial r\partial\theta} - \frac{1}{r^2}\frac{\partial^2 u_r}{\partial\theta^2} - \frac{\partial^2 u_r}{\partial z^2} + \frac{\partial^2 u_z}{\partial r\partial z}\right] - \beta\frac{\partial\tau}{\partial r} + F_r = 0\\
&(\lambda+2\mu)\frac{1}{r}\frac{\partial e}{\partial\theta} - \mu\left[\frac{1}{r}\frac{\partial^2 u_z}{\partial z\partial\theta} - \frac{\partial^2 u_\theta}{\partial z^2} + \frac{1}{r^2}\frac{\partial(ru_\theta)}{\partial r}\right.\\
&\quad\left. - \frac{1}{r}\frac{\partial^2(ru_\theta)}{\partial r^2} - \frac{1}{r^2}\frac{\partial u_r}{\partial\theta} + \frac{1}{r}\frac{\partial^2 u_r}{\partial r\partial\theta}\right] - \beta\frac{1}{r}\frac{\partial\tau}{\partial\theta} + F_\theta = 0\\
&(\lambda+2\mu)\frac{\partial e}{\partial z} - \mu\left[\frac{1}{r}\frac{\partial u_r}{\partial z} + \frac{\partial^2 u_r}{\partial r\partial z} - \frac{\partial^2 u_z}{\partial r^2} - \frac{1}{r}\frac{\partial u_z}{\partial r} - \frac{1}{r^2}\frac{\partial^2 u_z}{\partial\theta^2} + \frac{1}{r}\frac{\partial^2 u_\theta}{\partial z\partial\theta}\right]\\
&\quad - \beta\frac{\partial\tau}{\partial z} + F_z = 0
\end{aligned}
\tag{6.130}
$$

The solution of the Navier's equations (6.130) without body forces in the cylindrical coordinate system can be expressed by the Goodier's thermoelastic displacement potential Φ and the Boussinesq harmonic functions φ, ϑ, and ψ:

$$
\begin{aligned}
u_r &= \frac{\partial\Phi}{\partial r} + \frac{\partial\varphi}{\partial r} + \frac{2}{r}\frac{\partial\vartheta}{\partial\theta} + z\frac{\partial\psi}{\partial r}\\
u_\theta &= \frac{1}{r}\frac{\partial\Phi}{\partial\theta} + \frac{1}{r}\frac{\partial\varphi}{\partial\theta} - 2\frac{\partial\vartheta}{\partial r} + \frac{z}{r}\frac{\partial\psi}{\partial\theta}\\
u_z &= \frac{\partial\Phi}{\partial z} + \frac{\partial\varphi}{\partial z} + z\frac{\partial\psi}{\partial z} - (3-4\nu)\psi
\end{aligned}
\tag{6.131}
$$

in which the four functions must satisfy

$$\nabla^2\Phi = K\tau, \qquad \nabla^2\varphi = \nabla^2\vartheta = \nabla^2\psi = 0 \tag{6.132}$$

where $\nabla^2 = (\partial^2/\partial r^2) + (1/r)(\partial/\partial r) + (1/r^2)(\partial^2/\partial\theta^2) + (\partial^2/\partial z^2)$. The components of stress can be expressed by use of the potential Φ and the harmonic functions φ, ϑ, and ψ as

$$
\begin{aligned}
\sigma_{rr} &= 2G\left[\frac{\partial^2\Phi}{\partial r^2} - K\tau + \frac{\partial^2\varphi}{\partial r^2} + 2\frac{\partial}{\partial r}\left(\frac{1}{r}\frac{\partial\vartheta}{\partial\theta}\right) + z\frac{\partial^2\psi}{\partial r^2} - 2\nu\frac{\partial\psi}{\partial z}\right]\\
\sigma_{\theta\theta} &= 2G\left[\frac{1}{r}\frac{\partial\Phi}{\partial r} + \frac{1}{r^2}\frac{\partial^2\Phi}{\partial\theta^2} - K\tau + \frac{1}{r}\frac{\partial\varphi}{\partial r} + \frac{1}{r^2}\frac{\partial^2\varphi}{\partial\theta^2} - 2\frac{\partial}{\partial r}\left(\frac{1}{r}\frac{\partial\vartheta}{\partial\theta}\right)\right.\\
&\quad\left. + \frac{z}{r}\frac{\partial\psi}{\partial r} + \frac{z}{r^2}\frac{\partial^2\psi}{\partial\theta^2} - 2\nu\frac{\partial\psi}{\partial z}\right]\\
\sigma_{zz} &= 2G\left[\frac{\partial^2\Phi}{\partial z^2} - K\tau + \frac{\partial^2\varphi}{\partial z^2} + z\frac{\partial^2\psi}{\partial z^2} - 2(1-\nu)\frac{\partial\psi}{\partial z}\right]
\end{aligned}
$$

$$\sigma_{rz} = 2G\left[\frac{\partial^2\Phi}{\partial r\partial z} + \frac{\partial^2\varphi}{\partial r\partial z} + \frac{1}{r}\frac{\partial^2\vartheta}{\partial\theta\partial z} + z\frac{\partial^2\psi}{\partial r\partial z} - (1-2\nu)\frac{\partial\psi}{\partial r}\right]$$

$$\sigma_{r\theta} = 2G\left[\frac{\partial}{\partial r}\left(\frac{1}{r}\frac{\partial\Phi}{\partial\theta}\right) + \frac{\partial}{\partial r}\left(\frac{1}{r}\frac{\partial\varphi}{\partial\theta}\right) - \frac{\partial^2\vartheta}{\partial r^2} + \frac{1}{r}\frac{\partial\vartheta}{\partial r} + \frac{1}{r^2}\frac{\partial^2\vartheta}{\partial\theta^2} + z\frac{\partial}{\partial r}\left(\frac{1}{r}\frac{\partial\psi}{\partial\theta}\right)\right]$$

$$\sigma_{\theta z} = 2G\left(\frac{1}{r}\frac{\partial^2\Phi}{\partial\theta\partial z} + \frac{1}{r}\frac{\partial^2\varphi}{\partial\theta\partial z} - \frac{\partial^2\vartheta}{\partial r\partial z} + \frac{z}{r}\frac{\partial^2\psi}{\partial\theta\partial z} - \frac{1-2\nu}{r}\frac{\partial\psi}{\partial\theta}\right) \tag{6.133}$$

Now, we recall another displacement function which is the Michell's function M, and which is related to the Boussinesq harmonic functions φ and ψ, see Eq. (6.108)

$$M = -\int(\varphi + z\psi)\,dz \tag{6.134}$$

The displacements in the cylindrical coordinate system are represented by the potential Φ, the Michell's function M, and the Boussinesq harmonic function ϑ

$$\begin{aligned}
u_r &= \frac{\partial\Phi}{\partial r} - \frac{\partial^2 M}{\partial r\partial z} + \frac{2}{r}\frac{\partial\vartheta}{\partial\theta}\\
u_\theta &= \frac{1}{r}\frac{\partial\Phi}{\partial\theta} - \frac{1}{r}\frac{\partial^2 M}{\partial z\partial\theta} - 2\frac{\partial\vartheta}{\partial r}\\
u_z &= \frac{\partial\Phi}{\partial z} + 2(1-\nu)\nabla^2 M - \frac{\partial^2 M}{\partial z^2}
\end{aligned} \tag{6.135}$$

The Michell's function M must satisfy the equation

$$\nabla^2\nabla^2 M = 0 \tag{6.136}$$

The components of stress are represented by the potential Φ, the Michell's function M, and the Boussinesq harmonic function ϑ

$$\sigma_{rr} = 2G\left[\frac{\partial^2\Phi}{\partial r^2} - K\tau + \frac{\partial}{\partial z}\left(\nu\nabla^2 M - \frac{\partial^2 M}{\partial r^2}\right) + \frac{2}{r}\frac{\partial^2\vartheta}{\partial r\partial\theta} - \frac{2}{r^2}\frac{\partial\vartheta}{\partial\theta}\right]$$

$$\begin{aligned}
\sigma_{\theta\theta} = 2G\Bigg[&\frac{1}{r}\frac{\partial\Phi}{\partial r} + \frac{1}{r^2}\frac{\partial^2\Phi}{\partial\theta^2} - K\tau + \frac{\partial}{\partial z}\left(\nu\nabla^2 M - \frac{1}{r}\frac{\partial M}{\partial r} - \frac{1}{r^2}\frac{\partial^2 M}{\partial\theta^2}\right)\\
&+\frac{2}{r^2}\frac{\partial\vartheta}{\partial\theta} - \frac{2}{r}\frac{\partial^2\vartheta}{\partial r\partial\theta}\Bigg]
\end{aligned}$$

$$\sigma_{zz} = 2G\left\{\frac{\partial^2\Phi}{\partial z^2} - K\tau + \frac{\partial}{\partial z}\left[(2-\nu)\nabla^2 M - \frac{\partial^2 M}{\partial z^2}\right]\right\}$$

$$\sigma_{rz} = 2G\left\{\frac{\partial^2\Phi}{\partial r\partial z} + \frac{\partial}{\partial r}\left[(1-\nu)\nabla^2 M - \frac{\partial^2 M}{\partial z^2}\right] + \frac{1}{r}\frac{\partial^2\vartheta}{\partial\theta\partial z}\right\}$$

$$\sigma_{r\theta} = 2G\left[\frac{\partial}{\partial r}\left(\frac{1}{r}\frac{\partial\Phi}{\partial\theta}\right) - \frac{\partial}{\partial r}\left(\frac{1}{r}\frac{\partial^2 M}{\partial\theta\partial z}\right) - \frac{\partial^2\vartheta}{\partial r^2} + \frac{1}{r}\frac{\partial\vartheta}{\partial r} + \frac{1}{r^2}\frac{\partial^2\vartheta}{\partial\theta^2}\right]$$

$$\sigma_{\theta z} = 2G\left\{\frac{1}{r}\frac{\partial^2 \Phi}{\partial\theta\partial z} + \frac{1}{r}\frac{\partial}{\partial\theta}\left[(1-\nu)\nabla^2 M - \frac{\partial^2 M}{\partial z^2}\right] - \frac{\partial^2 \vartheta}{\partial r\partial z}\right\} \tag{6.137}$$

The general solutions of the Laplace equation (6.132) in the cylindrical coordinate system are

$$\begin{aligned}
&\begin{pmatrix}1\\ \ln r\end{pmatrix}\begin{pmatrix}1\\ \theta\end{pmatrix}\begin{pmatrix}1\\ z\end{pmatrix}, &&\begin{pmatrix}J_0(ar)\\ Y_0(ar)\end{pmatrix}\begin{pmatrix}1\\ \theta\end{pmatrix}\begin{pmatrix}e^{az}\\ e^{-az}\end{pmatrix}\\
&\begin{pmatrix}I_0(ar)\\ K_0(ar)\end{pmatrix}\begin{pmatrix}1\\ \theta\end{pmatrix}\begin{pmatrix}\sin az\\ \cos az\end{pmatrix}, &&\begin{pmatrix}r^n\\ r^{-n}\end{pmatrix}\begin{pmatrix}\cos n\theta\\ \sin n\theta\end{pmatrix}\begin{pmatrix}1\\ z\end{pmatrix}\\
&\begin{pmatrix}J_n(ar)\\ Y_n(ar)\end{pmatrix}\begin{pmatrix}\cos n\theta\\ \sin n\theta\end{pmatrix}\begin{pmatrix}e^{az}\\ e^{-az}\end{pmatrix}, &&\begin{pmatrix}I_n(ar)\\ K_n(ar)\end{pmatrix}\begin{pmatrix}\cos n\theta\\ \sin n\theta\end{pmatrix}\begin{pmatrix}\cos az\\ \sin az\end{pmatrix}
\end{aligned} \tag{6.138}$$

The general solutions of Eq. (6.136) in the cylindrical coordinate system are given by Eqs. (6.138) and the following equations:

$$\begin{aligned}
&\begin{pmatrix}r^2\\ r^2\ln r\end{pmatrix}\begin{pmatrix}1\\ \theta\end{pmatrix}\begin{pmatrix}1\\ z\end{pmatrix}, &&\begin{pmatrix}1\\ \ln r\end{pmatrix}\begin{pmatrix}1\\ \theta\end{pmatrix}\begin{pmatrix}z^2\\ z^3\end{pmatrix}\\
&\begin{pmatrix}rJ_1(ar)\\ rY_1(ar)\end{pmatrix}\begin{pmatrix}1\\ \theta\end{pmatrix}\begin{pmatrix}e^{az}\\ e^{-az}\end{pmatrix}, &&\begin{pmatrix}J_0(ar)\\ Y_0(ar)\end{pmatrix}\begin{pmatrix}1\\ \theta\end{pmatrix}\begin{pmatrix}ze^{az}\\ ze^{-az}\end{pmatrix}\\
&\begin{pmatrix}rI_1(ar)\\ rK_1(ar)\end{pmatrix}\begin{pmatrix}1\\ \theta\end{pmatrix}\begin{pmatrix}\sin az\\ \cos az\end{pmatrix}, &&\begin{pmatrix}I_0(ar)\\ K_0(ar)\end{pmatrix}\begin{pmatrix}1\\ \theta\end{pmatrix}\begin{pmatrix}z\sin az\\ z\cos az\end{pmatrix}\\
&\begin{pmatrix}0\\ r\end{pmatrix}\begin{pmatrix}\theta\cos\theta\\ \theta\sin\theta\end{pmatrix}\begin{pmatrix}1\\ z\end{pmatrix}, &&\begin{pmatrix}r^3\\ r\ln r\end{pmatrix}\begin{pmatrix}\cos\theta\\ \sin\theta\end{pmatrix}\begin{pmatrix}1\\ z\end{pmatrix}\\
&\begin{pmatrix}r^{n+2}\\ r^{-n+2}\end{pmatrix}\begin{pmatrix}\cos n\theta\\ \sin n\theta\end{pmatrix}\begin{pmatrix}1\\ z\end{pmatrix}, &&\begin{pmatrix}r^n\\ r^{-n}\end{pmatrix}\begin{pmatrix}\cos n\theta\\ \sin n\theta\end{pmatrix}\begin{pmatrix}z^2\\ z^3\end{pmatrix}\\
&\begin{pmatrix}rJ_{n+1}(ar)\\ rY_{n+1}(ar)\end{pmatrix}\begin{pmatrix}\cos n\theta\\ \sin n\theta\end{pmatrix}\begin{pmatrix}e^{az}\\ e^{-az}\end{pmatrix}, &&\begin{pmatrix}J_n(ar)\\ Y_n(ar)\end{pmatrix}\begin{pmatrix}\cos n\theta\\ \sin n\theta\end{pmatrix}\begin{pmatrix}ze^{az}\\ ze^{-az}\end{pmatrix}\\
&\begin{pmatrix}rI_{n+1}(ar)\\ rK_{n+1}(ar)\end{pmatrix}\begin{pmatrix}\cos n\theta\\ \sin n\theta\end{pmatrix}\begin{pmatrix}\cos az\\ \sin az\end{pmatrix}, &&\begin{pmatrix}I_n(ar)\\ K_n(ar)\end{pmatrix}\begin{pmatrix}\cos n\theta\\ \sin n\theta\end{pmatrix}\begin{pmatrix}z\cos az\\ z\sin az\end{pmatrix}
\end{aligned} \tag{6.139}$$

where a is an arbitrary constant, $J_n(ar)$ and $Y_n(ar)$ are Bessel functions, $I_n(ar)$ and $K_n(ar)$ are the modified Bessel functions. Moreover, functions $\binom{\sinh az}{\cosh az}$ can be used instead of $\binom{e^{az}}{e^{-az}}$.

6.4.2 Steady State Thermal Stresses in a Semi-Infinite Body

Consider a semi-infinite body subjected to a steady, non-axisymmetric temperature field dependent on the radial, circumferential, and axial directions of the cylindrical coordinate system shown in Fig. 6.4. The initial temperature $T_i(r, \theta, z)$ in the semi-infinite body is a constant T_i and heat is exchanged by heat transfer at the boundary surface. The temperature of the surrounding medium $T_a(r, \theta)$ at the boundary surface depends on the radial direction r and the circumferential direction θ. The fundamental equation system to be solved for the steady state is

Governing equation:

$$\frac{\partial^2 T}{\partial r^2} + \frac{1}{r}\frac{\partial T}{\partial r} + \frac{1}{r^2}\frac{\partial^2 T}{\partial \theta^2} + \frac{\partial^2 T}{\partial z^2} = 0 \tag{a}$$

Boundary conditions:

$$\lambda \frac{\partial T}{\partial z} = h\,[T - T_a(r,\theta)] \qquad \text{on } z = 0 \tag{b}$$

$$T = T_i \qquad \text{at } r, z \to \infty \tag{c}$$

Initial condition:

$$T = T_i \qquad \text{at } t = 0 \tag{d}$$

Referring to the general solution of the harmonic function (6.138), the temperature suitable for this problem is

$$\begin{aligned} T(r,\theta,z) &= B_0 + \int_0^\infty A_0(s) J_0(sr) e^{-sz}\,ds \\ &\quad + \sum_{n=1}^{\infty} \int_0^\infty J_n(sr)\,[A_n(s)\cos n\theta + B_n(s)\sin n\theta]\,e^{-sz}\,ds \end{aligned} \tag{e}$$

in which B_0, $A_0(s)$, $A_n(s)$, and $B_n(s)$ are unknown functions, and $J_n(sr)$ denotes the Bessel function of the first kind of order n. The boundary condition (c) gives

$$B_0 = T_i \tag{f}$$

The temperature of the surrounding medium in boundary condition (b) is assumed to be expressed by

$$\begin{aligned} T_a(r,\theta) &= T_i + \int_0^\infty T_{a0}(s) s J_0(sr)\,ds \\ &\quad + \sum_{n=1}^{\infty} \int_0^\infty s J_n(sr) \left[T_{an}(s)\cos n\theta + T_{an}^*(s)\sin n\theta\right] ds \end{aligned} \tag{g}$$

where

$$T_{a0}(s) = \frac{1}{2\pi}\int_0^{2\pi}\int_0^\infty [T_a(r,\theta) - T_i] r J_0(sr)\,dr\,d\theta$$

$$T_{an}(s) = \frac{1}{\pi}\int_0^{2\pi} \cos n\theta \int_0^\infty [T_a(r,\theta) - T_i] r J_n(sr)\,dr\,d\theta$$

$$T_{an}^*(s) = \frac{1}{\pi}\int_0^{2\pi} \sin n\theta \int_0^\infty [T_a(r,\theta) - T_i] r J_n(sr)\,dr\,d\theta$$

Boundary condition (b) gives unknown functions $A_0(s)$, $A_n(s)$, and $B_n(s)$.

$$A_0(s) = \frac{hsT_{a0}(s)}{h+\lambda s}, \qquad A_n(s) = \frac{hsT_{an}(s)}{h+\lambda s}, \qquad B_n(s) = \frac{hsT_{an}^*(s)}{h+\lambda s} \tag{h}$$

Therefore, the temperature field is

$$T = T_i + \int_0^\infty T_{a0}(s)\frac{hs}{h+\lambda s}J_0(sr)e^{-sz}\,ds + \sum_{n=1}^\infty \int_0^\infty J_n(sr)\frac{hs}{h+\lambda s}\left[T_{an}(s)\cos n\theta + T_{an}^*(s)\sin n\theta\right]e^{-sz}\,ds \tag{6.140}$$

The temperature change τ is

$$\tau = T - T_i = \int_0^\infty T_{a0}(s)\frac{hs}{h+\lambda s}J_0(sr)e^{-sz}\,ds + \sum_{n=1}^\infty \int_0^\infty J_n(sr)\frac{hs}{h+\lambda s}\left[T_{an}(s)\cos n\theta + T_{an}^*(s)\sin n\theta\right]e^{-sz}\,ds \tag{6.141}$$

Let us employ the Goodier's thermoelastic displacement potential Φ and the Boussinesq harmonic functions φ, ϑ, and ψ. Referring to Eqs. (6.138), the functions φ, ϑ, and ψ can be expressed by forms suitable for the semi-infinite body

$$\begin{aligned}
\varphi(r,\theta,z) &= \int_0^\infty C_0(s)J_0(sr)e^{-sz}\,ds + \sum_{n=1}^\infty \int_0^\infty J_n(sr)\left[C_n(s)\cos n\theta + C_n^*(s)\sin n\theta\right]e^{-sz}\,ds \\
\vartheta(r,\theta,z) &= \sum_{n=1}^\infty \int_0^\infty J_n(sr)\left[D_n(s)\cos n\theta + D_n^*(s)\sin n\theta\right]e^{-sz}\,ds \\
\psi(r,\theta,z) &= \int_0^\infty E_0(s)J_0(sr)e^{-sz}\,ds + \sum_{n=1}^\infty \int_0^\infty J_n(sr)\left[E_n(s)\cos n\theta + E_n^*(s)\sin n\theta\right]e^{-sz}\,ds
\end{aligned} \tag{i}$$

and the potential Φ is obtained from Eqs. (6.132) and (6.141) as

$$\Phi(r,\theta,z) = -\frac{K}{2}\int_0^\infty A_0(s)J_0(sr)\frac{z}{s}e^{-sz}\,ds - \frac{K}{2}\sum_{n=1}^\infty \int_0^\infty J_n(sr)\left[A_n(s)\cos n\theta + B_n(s)\sin n\theta\right]\frac{z}{s}e^{-sz}\,ds \tag{j}$$

Substituting the potential Φ and the functions φ, ϑ, and ψ given by Eqs. (j) and (i) into Eqs. (6.131) and (6.133), the displacements and the thermal stresses are

$$u_r = \int_0^\infty sJ_1(sr)\left[K\frac{A_0}{2s}z - C_0 - E_0 z\right]e^{-sz}\,ds$$

$$+\sum_{n=1}^{\infty}\cos n\theta \int_0^\infty \left\{\left[sJ_{n-1}(sr) - \frac{n}{r}J_n(sr)\right]\right.$$

$$\left.\times\left[-K\frac{A_n}{2s}z + C_n + E_n z\right] + \frac{2n}{r}J_n(sr)D_n^*\right\}e^{-sz}\,ds$$

$$+\sum_{n=1}^{\infty}\sin n\theta \int_0^\infty \left\{\left[sJ_{n-1}(sr) - \frac{n}{r}J_n(sr)\right]\right.$$

$$\left.\times\left[-K\frac{B_n}{2s}z + C_n^* + E_n^* z\right] - \frac{2n}{r}J_n(sr)D_n\right\}e^{-sz}\,ds$$

$$u_\theta = -\sum_{n=1}^{\infty}\sin n\theta \int_0^\infty \left\{\frac{n}{r}J_n(sr)\left[-K\frac{A_n}{2s}z + C_n + E_n z\right]\right.$$

$$\left. + 2\left[sJ_{n-1}(sr) - \frac{n}{r}J_n(sr)\right]D_n^*\right\}e^{-sz}\,ds + \sum_{n=1}^{\infty}\cos n\theta \int_0^\infty \left\{\frac{n}{r}J_n(sr)\right.$$

$$\left.\times\left[-K\frac{B_n}{2s}z + C_n^* + E_n^* z\right] - 2\left[sJ_{n-1}(sr) - \frac{n}{r}J_n(sr)\right]D_n\right\}e^{-sz}\,ds$$

$$u_z = -\int_0^\infty J_0(sr)\left[K\frac{A_0}{2s}(1 - sz) + sC_0 + (3 - 4\nu + sz)E_0\right]e^{-sz}\,ds$$

$$-\sum_{n=1}^{\infty}\cos n\theta \int_0^\infty J_n(sr)\left[K\frac{A_n}{2s}(1 - sz) + sC_n + (3 - 4\nu + sz)E_n\right]e^{-sz}\,ds$$

$$-\sum_{n=1}^{\infty}\sin n\theta \int_0^\infty J_n(sr)\left[K\frac{B_n}{2s}(1 - sz) + sC_n^* + (3 - 4\nu + sz)E_n^*\right]e^{-sz}\,ds$$

$$\frac{\sigma_{rr}}{2G} = \int_0^\infty \left\{\left[-s^2J_0(sr) + \frac{s}{r}J_1(sr)\right]\left(-K\frac{A_0}{2s}z + C_0 + E_0 z\right)\right.$$

$$\left. - (KA_0 - 2\nu sE_0)\,J_0(sr)\right\}e^{-sz}\,ds$$

$$+\sum_{n=1}^{\infty}\cos n\theta \int_0^\infty \left\{\left[\frac{n^2 + n - s^2r^2}{r^2}J_n(sr) - \frac{s}{r}J_{n-1}(sr)\right]\right.$$

$$\times\left(-K\frac{A_n}{2s}z + C_n + E_n z\right) - (KA_n - 2\nu sE_n)J_n(sr)$$

$$+\frac{2n}{r}\left[sJ_{n-1}(sr)-\frac{n+1}{r}J_n(sr)\right]D_n^*\Bigg\}e^{-sz}ds$$

$$+\sum_{n=1}^{\infty}\sin n\theta\int_0^{\infty}\left\{\left[\frac{n^2+n-s^2r^2}{r^2}J_n(sr)-\frac{s}{r}J_{n-1}(sr)\right]\left(-K\frac{B_n}{2s}z+C_n^*+E_n^*z\right)\right.$$

$$\left.-(KB_n-2\nu sE_n^*)J_n(sr)-\frac{2n}{r}\left[sJ_{n-1}(sr)-\frac{n+1}{r}J_n(sr)\right]D_n\right\}e^{-sz}ds$$

$$\frac{\sigma_{\theta\theta}}{2G}=-\int_0^{\infty}\left\{\frac{s}{r}J_1(sr)\left(-K\frac{A_0}{2s}z+C_0+E_0z\right)+(KA_0-2\nu sE_0)J_0(sr)\right\}e^{-sz}ds$$

$$+\sum_{n=1}^{\infty}\cos n\theta\int_0^{\infty}\left\{\left[\frac{s}{r}J_{n-1}(sr)-\frac{n}{r^2}J_n(sr)\right]\left(-K\frac{A_n}{2s}z+C_n-2nD_n^*+E_nz\right)\right.$$

$$\left.+\frac{1}{r^2}J_n(sr)\left[K\frac{n^2A_n}{2s}z-Kr^2A_n-n^2C_n+2nD_n^*+(2\nu sr^2-n^2z)E_n\right]\right\}e^{-sz}ds$$

$$+\sum_{n=1}^{\infty}\sin n\theta\int_0^{\infty}\left\{\left[\frac{s}{r}J_{n-1}(sr)-\frac{n}{r^2}J_n(sr)\right]\left(-K\frac{B_n}{2s}z+C_n^*+2nD_n+E_n^*z\right)\right.$$

$$\left.+\frac{1}{r^2}J_n(sr)\left[K\frac{n^2B_n}{2s}z-Kr^2B_n-n^2C_n^*-2nD_n+(2\nu sr^2-n^2z)E_n^*\right]\right\}e^{-sz}ds$$

$$\frac{\sigma_{zz}}{2G}=\int_0^{\infty}J_0(sr)\left[-K\frac{A_0}{2}sz+s^2C_0+s(2-2\nu+sz)E_0\right]e^{-sz}ds$$

$$+\sum_{n=1}^{\infty}\cos n\theta\int_0^{\infty}J_n(sr)\left[-K\frac{A_n}{2}sz+s^2C_n+s(2-2\nu+sz)E_n\right]e^{-sz}ds$$

$$+\sum_{n=1}^{\infty}\sin n\theta\int_0^{\infty}J_n(sr)\left[-K\frac{B_n}{2}sz+s^2C_n^*+s(2-2\nu+sz)E_n^*\right]e^{-sz}ds$$

$$\frac{\sigma_{rz}}{2G}=\int_0^{\infty}sJ_1(sr)\left[K\frac{A_0}{2s}(1-sz)+sC_0+(1-2\nu+sz)E_0\right]e^{-sz}ds$$

$$-\sum_{n=1}^{\infty}\cos n\theta\int_0^{\infty}\left\{\left[sJ_{n-1}(sr)-\frac{n}{r}J_n(sr)\right]\right.$$

$$\left.\times\left[K\frac{A_n}{2s}(1-sz)+sC_n+(1-2\nu+sz)E_n\right]+\frac{ns}{r}J_n(sr)D_n^*\right\}e^{-sz}ds$$

$$-\sum_{n=1}^{\infty}\sin n\theta\int_0^{\infty}\left\{\left[sJ_{n-1}(sr)-\frac{n}{r}J_n(sr)\right]\right.$$

$$\left.\times\left[K\frac{B_n}{2s}(1-sz)+sC_n^*+(1-2\nu+sz)E_n^*\right]-\frac{ns}{r}J_n(sr)D_n\right\}e^{-sz}ds$$

$$\frac{\sigma_{\theta z}}{2G}=\sum_{n=1}^{\infty}\sin n\theta\int_0^{\infty}\left\{\frac{n}{r}J_n(sr)\left[K\frac{A_n}{2s}(1-sz)+sC_n+(1-2\nu+sz)E_n\right]\right.$$
$$\left.+\left[sJ_{n-1}(sr)-\frac{n}{r}J_n(sr)\right]sD_n^*\right\}e^{-sz}ds$$
$$-\sum_{n=1}^{\infty}\cos n\theta\int_0^{\infty}\left\{\frac{n}{r}J_n(sr)\left[K\frac{B_n}{2s}(1-sz)+sC_n^*+(1-2\nu+sz)E_n^*\right]\right.$$
$$\left.-\left[sJ_{n-1}(sr)-\frac{n}{r}J_n(sr)\right]sD_n\right\}e^{-sz}ds$$

$$\frac{\sigma_{r\theta}}{2G}=-\sum_{n=1}^{\infty}\sin n\theta\int_0^{\infty}\left\{\frac{n}{r}\left[sJ_{n-1}(sr)-\frac{n+1}{r}J_n(sr)\right]\left[-K\frac{A_n}{2s}z+C_n+E_nz\right]\right.$$
$$\left.+\left[\frac{2n^2+2n-s^2r^2}{r^2}J_n(sr)-\frac{2s}{r}J_{n-1}(sr)\right]D_n^*\right\}e^{-sz}ds$$
$$+\sum_{n=1}^{\infty}\cos n\theta\int_0^{\infty}\left\{\frac{n}{r}\left[sJ_{n-1}(sr)-\frac{n+1}{r}J_n(sr)\right]\left[-K\frac{B_n}{2s}z+C_n^*+E_n^*z\right]\right.$$
$$\left.-\left[\frac{2n^2+2n-s^2r^2}{r^2}J_n(sr)-\frac{2s}{r}J_{n-1}(sr)\right]D_n\right\}e^{-sz}ds \tag{6.142}$$

The boundary conditions on the traction free surface are

$$\sigma_{zz}=0,\quad \sigma_{zr}=0,\quad \sigma_{z\theta}=0\qquad \text{on } z=0 \tag{k}$$

The first of Eqs. (k) gives

$$\begin{aligned} sC_0+2(1-\nu)E_0&=0\\ sC_n+2(1-\nu)E_n&=0\\ sC_n^*+2(1-\nu)E_n^*&=0 \end{aligned} \tag{l}$$

The second of Eqs. (k) gives

$$sC_0+(1-2\nu)E_0=-K\frac{A_0}{2s} \tag{m}$$

$$\int_0^{\infty}\left\{\left[sJ_{n-1}(sr)-\frac{n}{r}J_n(sr)\right]\left[K\frac{A_n}{2s}+sC_n+(1-2\nu)E_n\right]+\frac{ns}{r}J_n(sr)D_n^*\right\}ds=0 \tag{n}$$

$$\int_0^{\infty}\left\{\left[sJ_{n-1}(sr)-\frac{n}{r}J_n(sr)\right]\left[K\frac{B_n}{2s}+sC_n^*+(1-2\nu)E_n^*\right]-\frac{ns}{r}J_n(sr)D_n\right\}ds=0 \tag{o}$$

The third of Eqs. (k) gives

$$\int_0^\infty \left\{ \frac{n}{r} J_n(sr) \left[K\frac{A_n}{2s} + sC_n + (1-2\nu)E_n \right] + \left[sJ_{n-1}(sr) - \frac{n}{r} J_n(sr) \right] sD_n^* \right\} ds = 0 \tag{p}$$

$$\int_0^\infty \left\{ \frac{n}{r} J_n(sr) \left[K\frac{B_n}{2s} + sC_n^* + (1-2\nu)E_n^* \right] - \left[sJ_{n-1}(sr) - \frac{n}{r} J_n(sr) \right] sD_n \right\} ds = 0 \tag{q}$$

Equation (n) added to Eq. (p) gives

$$\int_0^\infty sJ_{n-1}(sr) \left[K\frac{A_n}{2s} + sC_n + (1-2\nu)E_n + sD_n^* \right] ds = 0 \tag{r}$$

Equation (n) subtracted from Eq. (p) gives

$$\int_0^\infty \left[sJ_{n-1}(sr) - \frac{2n}{r} J_n(sr) \right] \left[K\frac{A_n}{2s} + sC_n + (1-2\nu)E_n - sD_n^* \right] ds = 0 \tag{s}$$

Equation (o) added to Eq. (q) gives

$$\int_0^\infty sJ_{n-1}(sr) \left[K\frac{B_n}{2s} + sC_n^* + (1-2\nu)E_n^* - sD_n \right] ds = 0 \tag{t}$$

Equation (o) subtracted from Eq. (q) gives

$$\int_0^\infty \left[sJ_{n-1}(sr) - \frac{2n}{r} J_n(sr) \right] \left[K\frac{B_n}{2s} + sC_n^* + (1-2\nu)E_n^* + sD_n \right] ds = 0 \tag{u}$$

As Eq. (r) to Eq. (u) are available for an arbitrary value of r, we have

$$\begin{aligned} sC_n + (1-2\nu)E_n + sD_n^* &= -K\frac{A_n}{2s}, \qquad & sC_n + (1-2\nu)E_n - sD_n^* &= -K\frac{A_n}{2s} \\ sC_n^* + (1-2\nu)E_n^* - sD_n &= -K\frac{B_n}{2s}, \qquad & sC_n^* + (1-2\nu)E_n^* + sD_n &= -K\frac{B_n}{2s} \end{aligned} \tag{v}$$

From Eqs. (l), (m), and (v), the unknown functions are

$$\begin{aligned} C_0 &= -(1-\nu)K\frac{A_0}{s^2}, \qquad & E_0 &= K\frac{A_0}{2s} \\ C_n &= -(1-\nu)K\frac{A_n}{s^2}, \qquad & E_n &= K\frac{A_n}{2s}, \qquad & D_n &= 0 \\ C_n^* &= -(1-\nu)K\frac{B_n}{s^2}, \qquad & E_n^* &= K\frac{B_n}{2s}, \qquad & D_n^* &= 0 \end{aligned} \tag{w}$$

Substituting Eqs. (w) into Eqs. (6.142), the displacements and the thermal stresses are

$$
\begin{aligned}
u_r &= (1+\nu)\alpha \int_0^\infty \frac{1}{s} A_0(s) J_1(sr) e^{-sz}\, ds + (1+\nu)\alpha \\
&\quad \times \sum_{n=1}^{\infty} \int_0^\infty \frac{1}{s}\left[\frac{n}{rs} J_n(sr) - J_{n-1}(sr)\right][A_n(s)\cos n\theta + B_n(s)\sin n\theta]e^{-sz}\, ds \\
u_\theta &= (1+\nu)\alpha \sum_{n=1}^{\infty} \int_0^\infty \frac{n}{s^2 r} J_n(sr)[A_n(s)\sin n\theta - B_n(s)\cos n\theta]e^{-sz}\, ds \\
u_z &= -(1+\nu)\alpha \int_0^\infty \frac{1}{s} A_0(s) J_0(sr) e^{-sz}\, ds \\
&\quad -(1+\nu)\alpha \sum_{n=1}^{\infty} \int_0^\infty \frac{1}{s} J_n(sr)[A_n(s)\cos n\theta + B_n(s)\sin n\theta]e^{-sz}\, ds \\
\sigma_{rr} &= -\alpha E \int_0^\infty \frac{A_0(s)}{sr} J_1(sr) e^{-sz}\, ds - \alpha E \sum_{n=1}^{\infty} \int_0^\infty \left[\frac{n(n+1)}{s^2 r^2} J_n(sr) - \frac{1}{sr} J_{n-1}(sr)\right] \\
&\quad \times [A_n(s)\cos n\theta + B_n(s)\sin n\theta]e^{-sz}\, ds \\
\sigma_{\theta\theta} &= -\alpha E \int_0^\infty A_0(s)\left[J_0(sr) - \frac{J_1(sr)}{sr}\right] e^{-sz}\, ds \\
&\quad -\alpha E \sum_{n=1}^{\infty} \int_0^\infty \left[\frac{1}{sr} J_{n-1}(sr) + \frac{s^2 r^2 - n - n^2}{s^2 r^2} J_n(sr)\right] \\
&\quad \times [A_n(s)\cos n\theta + B_n(s)\sin n\theta]e^{-sz}\, ds \\
\sigma_{r\theta} &= \alpha E \sum_{n=1}^{\infty} \int_0^\infty \frac{n}{sr}\left[J_{n-1}(sr) - \frac{n+1}{sr} J_n(sr)\right][A_n(s)\sin n\theta - B_n(s)\cos n\theta]e^{-sz}\, ds \\
\sigma_{zz} &= \sigma_{zr} = \sigma_{z\theta} = 0
\end{aligned} \tag{6.143}
$$

where $A_n(s)$ and $B_n(s)$ are given by Eqs. (h).

Example 6.8 Derive the steady thermal stress in a semi-infinite body when the temperature on the surface of the semi-infinite body is prescribed by $T_{z=0}(r,\theta) = T_i + T_a(r,\theta)$.

Solution The boundary conditions are

$$
\begin{aligned}
T &= T_i + T_a(r,\theta) && \text{on } z = 0 \\
T &= T_i && \text{at } r, z \to \infty
\end{aligned} \tag{a}
$$

where

$$T_a(r, \theta) = \int_0^\infty T_{a0}(s) s J_0(sr)\, ds + \sum_{n=1}^{\infty} \int_0^\infty s J_n(sr) \left[T_{an}(s) \cos n\theta + T_{an}^*(s) \sin n\theta\right] ds \tag{b}$$

The temperature is given by Eq. (e) in Subsection 6.4.2.

$$T(r, \theta, z) = B_0 + \int_0^\infty A_0(s) J_0(sr) e^{-sz}\, ds + \sum_{n=1}^{\infty} \int_0^\infty J_n(sr) \left[A_n(s) \cos n\theta + B_n(s) \sin n\theta\right] e^{-sz}\, ds \tag{c}$$

The boundary conditions give

$$B_0 = T_i, \qquad A_0(s) = sT_{a0}(s), \qquad A_n(s) = sT_{an}(s), \qquad B_n(s) = sT_{an}^*(s) \tag{d}$$

We now obtain the temperature change τ:

$$\begin{aligned} \tau(r, \theta, z) &= T(r, \theta, z) - T_i \\ &= \int_0^\infty sT_{a0}(s) J_0(sr) e^{-sz}\, dr \\ &\quad + \sum_{n=1}^{\infty} \int_0^\infty s J_n(sr) \left[T_{an}(s) \cos n\theta + T_{an}^*(s) \sin n\theta\right] e^{-sz}\, ds \end{aligned} \tag{e}$$

Substitution of Eq. (d) into Eqs. (6.143) gives the displacement and the thermal stress.

6.4.3 Steady State Thermal Stresses in a Long Circular Cylinder

Consider an infinitely long solid circular cylinder of radius a subjected to a steady, non-axisymmetric temperature field. The initial temperature $T_i(r, \theta, z)$ in the cylinder is assumed to be a constant T_i and heat is exchanged by heat transfer at the boundary surface. The temperature of the surrounding medium $T_a(\theta, z)$ at the boundary surface depends on the directions θ and z, and satisfies the condition $T_a(\theta, z) = T_a(\theta, -z)$ for simplicity. The fundamental equation system to be solved for the steady state is

Governing equation:

$$\frac{\partial^2 T}{\partial r^2} + \frac{1}{r}\frac{\partial T}{\partial r} + \frac{1}{r^2}\frac{\partial^2 T}{\partial \theta^2} + \frac{\partial^2 T}{\partial z^2} = 0 \tag{a}$$

Boundary conditions:

$$-\lambda \frac{\partial T}{\partial r} = h\left[T - T_a(\theta, z)\right] \qquad \text{on } r = a \tag{b}$$

$$T = T_i \qquad \text{at } |z| \to \infty \tag{c}$$

Initial condition:

$$T = T_i \qquad \text{at } t = 0 \tag{d}$$

Referring to the general solution of the harmonic function (6.138), the temperature suitable for this problem is

$$\begin{aligned} T(r, \theta, z) &= B_0 + \int_0^\infty A_0(s) I_0(sr) \cos sz \, ds \\ &\quad + \sum_{n=1}^{\infty} \int_0^\infty I_n(sr)\left[A_n(s) \cos n\theta + B_n(s) \sin n\theta\right] \cos sz \, ds \end{aligned} \tag{e}$$

in which B_0, $A_0(s)$, $A_n(s)$, and $B_n(s)$ are unknown functions and $I_n(sr)$ denotes the modified Bessel function of the first kind of order n. Boundary condition (c) gives

$$B_0 = T_i \tag{f}$$

The temperature of the surrounding medium in boundary condition (b) is assumed to be expressed by

$$\begin{aligned} T_a(\theta, z) &= T_i + \int_0^\infty T_{a0}(s) \cos sz \, ds \\ &\quad + \sum_{n=1}^{\infty} \int_0^\infty \left[T_{an}(s) \cos n\theta + T_{an}^*(s) \sin n\theta\right] \cos sz \, ds \end{aligned} \tag{g}$$

where

$$T_{a0}(s) = \frac{1}{\pi^2} \int_0^{2\pi} \int_0^\infty \left[T_a(r, \theta) - T_i\right] \cos sz \, dz \, d\theta$$

$$T_{an}(s) = \frac{2}{\pi^2} \int_0^{2\pi} \cos n\theta \int_0^\infty \left[T_a(r, \theta) - T_i\right] \cos sz \, dz \, d\theta$$

$$T_{an}^*(s) = \frac{2}{\pi^2} \int_0^{2\pi} \sin n\theta \int_0^\infty \left[T_a(r, \theta) - T_i\right] \cos sz \, dz \, d\theta$$

From the boundary condition (b), the unknown functions $A_n(s)$ and $B_n(s)$ can be determined as

$$A_0(s) = \frac{h}{h I_0(sa) + \lambda s I_1(sa)} T_{a0}(s)$$

$$A_n(s) = \frac{ha}{(ha - \lambda n)I_n(sa) + \lambda sa I_{n-1}(sa)} T_{an}(s) \tag{h}$$
$$B_n(s) = \frac{ha}{(ha - \lambda n)I_n(sa) + \lambda sa I_{n-1}(sa)} T^*_{an}(s)$$

Therefore, the temperature field is obtained as

$$\begin{aligned} T = T_i &+ \int_0^\infty \frac{h}{hI_0(sa) + \lambda s I_1(sa)} T_{a0}(s) I_0(sr) \cos sz\, ds \\ &+ \sum_{n=1}^\infty \int_0^\infty \frac{ha}{(ha - \lambda n)I_n(sa) + \lambda sa I_{n-1}(sa)} I_n(sr) \\ &\times \left[T_{an}(s)\cos n\theta + T^*_{an}(s)\sin n\theta\right] \cos sz\, ds \end{aligned} \tag{6.144}$$

The temperature change τ is

$$\begin{aligned} \tau &= T - T_i \\ &= \int_0^\infty A_0(s) I_0(sr) \cos sz\, ds \\ &\quad + \sum_{n=1}^\infty \int_0^\infty I_n(sr)\left[A_n(s)\cos n\theta + B_n(s)\sin n\theta\right] \cos sz\, ds \\ &= \int_0^\infty \frac{h}{hI_0(sa) + \lambda s I_1(sa)} T_{a0}(s) I_0(sr) \cos sz\, ds \\ &\quad + \sum_{n=1}^\infty \int_0^\infty \frac{ha}{(ha - \lambda n)I_n(sa) + \lambda sa I_{n-1}(sa)} I_n(sr) \\ &\quad \times \left[T_{an}(s)\cos n\theta + T^*_{an}(s)\sin n\theta\right] \cos sz\, ds \end{aligned} \tag{6.145}$$

Let us employ the Goodier's thermoelastic displacement potential Φ, the Boussinesq harmonic function ϑ, and the Michell's function M discussed in Subsection 6.4.1. Referring to Eqs. (6.138) and (6.139), the Michell's function M and the Boussinesq function ϑ can be expressed for the infinitely long cylinder as

$$\begin{aligned} M(r,\theta,z) &= \int_0^\infty \left[C_0(s) I_0(sr) + D_0(s) r I_1(sr)\right] \sin sz\, ds \\ &\quad + \sum_{n=1}^\infty \cos n\theta \int_0^\infty \left[C_n(s) I_n(sr) + D_n(s) r I_{n+1}(sr)\right] \sin sz\, ds \\ &\quad + \sum_{n=1}^\infty \sin n\theta \int_0^\infty \left[C^*_n(s) I_n(sr) + D^*_n(s) r I_{n+1}(sr)\right] \sin sz\, ds \end{aligned} \tag{i}$$

$$\vartheta(r,\theta,z) = \sum_{n=1}^\infty \int_0^\infty I_n(sr)\left[E_n(s)\cos n\theta + E^*_n(s)\sin n\theta\right] \cos sz\, ds \tag{j}$$

and the potential Φ is obtained from Eqs. (6.106) and (6.145) as

$$\Phi(r,\theta,z) = \frac{K}{2}\int_0^\infty A_0(s)\frac{r}{s}I_1(sr)\cos sz\,ds + \frac{K}{2}\sum_{n=1}^{\infty}\int_0^\infty \frac{r}{s}I_{n+1}(sr)$$
$$\times [A_n(s)\cos n\theta + B_n(s)\sin n\theta]\cos sz\,ds \tag{k}$$

Substituting Eqs. (k), (i), and (j) into Eqs. (6.135) and (6.137), the displacements and the thermal stresses are

$$u_r = -\int_0^\infty \left[s^2 C_0(s)I_1(sr) + s^2 D_0(s) r I_0(sr) - K\frac{A_0(s)}{2} r I_0(sr)\right]\cos sz\,ds$$
$$-\sum_{n=1}^{\infty}\cos n\theta \int_0^\infty \left\{s^2 C_n(s)\left[I_{n+1}(sr) + \frac{n}{sr}I_n(sr)\right]\right.$$
$$+ sD_n(s)[srI_n(sr) - nI_{n+1}(sr)] - E_n^*(s)\frac{2n}{r}I_n(sr)$$
$$\left. - K\frac{A_n(s)}{2s}[srI_n(sr) - nI_{n+1}(sr)]\right\}\cos sz\,ds$$
$$-\sum_{n=1}^{\infty}\sin n\theta \int_0^\infty \left\{s^2 C_n^*(s)\left[I_{n+1}(sr) + \frac{n}{sr}I_n(sr)\right]\right.$$
$$+ sD_n^*(s)[srI_n(sr) - nI_{n+1}(sr)] + E_n(s)\frac{2n}{r}I_n(sr)$$
$$\left. - K\frac{B_n(s)}{2s}[srI_n(sr) - nI_{n+1}(sr)]\right\}\cos sz\,ds$$

$$u_\theta = \sum_{n=1}^{\infty}\sin n\theta \int_0^\infty \left\{sC_n(s)\frac{n}{r}I_n(sr) + nsD_n(s)I_{n+1}(sr)\right.$$
$$\left. - 2E_n^*(s)\left[sI_{n+1}(sr) + \frac{n}{r}I_n(sr)\right] - nK\frac{A_n(s)}{2s}I_{n+1}(sr)\right\}\cos sz\,ds$$
$$-\sum_{n=1}^{\infty}\cos n\theta \int_0^\infty \left\{sC_n^*(s)\frac{n}{r}I_n(sr) + nsD_n^*(s)I_{n+1}(sr)\right.$$
$$\left. + 2E_n(s)\left[sI_{n+1}(sr) + \frac{n}{r}I_n(sr)\right] - nK\frac{B_n(s)}{2s}I_{n+1}(sr)\right\}\cos sz\,ds$$

$$u_z = \int_0^\infty \left\{s^2 C_0(s)I_0(sr) + sD_0(s)[4(1-\nu)I_0(sr) + srI_1(sr)]\right.$$
$$\left. - K\frac{A_0(s)}{2} r I_1(sr)\right\}\sin sz\,ds$$

$$+\sum_{n=1}^{\infty}\cos n\theta\int_0^\infty\left\{s^2C_n(s)I_n(sr)+sD_n(s)[4(1-\nu)I_n(sr)+srI_{n+1}(sr)]\right.$$

$$\left.-K\frac{A_n(s)}{2}rI_{n+1}(sr)\right\}\sin sz\,ds+\sum_{n=1}^{\infty}\sin n\theta\int_0^\infty\left\{s^2C_n^*(s)I_n(sr)\right.$$

$$\left.+sD_n^*(s)[4(1-\nu)I_n(sr)+srI_{n+1}(sr)]-K\frac{B_n(s)}{2}rI_{n+1}(sr)\right\}\sin sz\,ds$$

$$\frac{\sigma_{rr}}{2G}=-\int_0^\infty\left\{s^3C_0(s)\left[I_0(sr)-\frac{1}{sr}I_1(sr)\right]+s^2D_0(s)[(1-2\nu)I_0(sr)+srI_1(sr)]\right.$$

$$\left.+K\frac{A_0(s)}{2}[I_0(sr)-srI_1(sr)]\right\}\cos sz\,ds$$

$$-\sum_{n=1}^{\infty}\cos n\theta\int_0^\infty\left\{s^3C_n(s)\left[\frac{n^2-n+s^2r^2}{s^2r^2}I_n(sr)-\frac{1}{sr}I_{n+1}(sr)\right]\right.$$

$$+s^2D_n(s)\left[(1-2\nu)I_n(sr)+\frac{n^2+n+s^2r^2}{sr}I_{n+1}(sr)\right]$$

$$-\frac{2n}{r}sE_n^*(s)\left[I_{n+1}(sr)+\frac{n-1}{sr}I_n(sr)\right]$$

$$\left.+K\frac{A_n(s)}{2}\left[I_n(sr)-\frac{n^2+n+s^2r^2}{sr}I_{n+1}(sr)\right]\right\}\cos sz\,ds$$

$$-\sum_{n=1}^{\infty}\sin n\theta\int_0^\infty\left\{s^3C_n^*(s)\left[\frac{n^2-n+s^2r^2}{s^2r^2}I_n(sr)-\frac{1}{sr}I_{n+1}(sr)\right]\right.$$

$$+s^2D_n^*(s)\left[(1-2\nu)I_n(sr)+\frac{n^2+n+s^2r^2}{sr}I_{n+1}(sr)\right]$$

$$+\frac{2n}{r}sE_n(s)\left[I_{n+1}(sr)+\frac{n-1}{sr}I_n(sr)\right]$$

$$\left.+K\frac{B_n(s)}{2}\left[I_n(sr)-\frac{n^2+n+s^2r^2}{sr}I_{n+1}(sr)\right]\right\}\cos sz\,ds$$

$$\frac{\sigma_{\theta\theta}}{2G}=-\int_0^\infty\left[s^3C_0(s)\frac{I_1(sr)}{sr}+s^2D_0(s)(1-2\nu)I_0(sr)+K\frac{A_0(s)}{2}I_0(sr)\right]\cos sz\,ds$$

$$-\sum_{n=1}^{\infty}\cos n\theta\int_0^\infty\left\{s^3C_n(s)\left[\frac{1}{sr}I_{n+1}(sr)-\frac{n^2-n}{s^2r^2}I_n(sr)\right]\right.$$

$$+s^2D_n(s)\left[(1-2\nu)I_n(sr)-\frac{n^2+n}{sr}I_{n+1}(sr)\right]$$

$$+\frac{2n}{r}sE_n^*(s)\left[I_{n+1}(sr)+\frac{n-1}{sr}I_n(sr)\right]$$

$$\left.+K\frac{A_n(s)}{2}\left[\frac{n^2+n}{sr}I_{n+1}(sr)+I_n(sr)\right]\right\}\cos sz\,ds$$

$$-\sum_{n=1}^{\infty}\sin n\theta\int_0^{\infty}\left\{s^3C_n^*(s)\left[\frac{1}{sr}I_{n+1}(sr)-\frac{n^2-n}{s^2r^2}I_n(sr)\right]\right.$$

$$+s^2D_n^*(s)\left[(1-2\nu)I_n(sr)-\frac{n^2+n}{sr}I_{n+1}(sr)\right]$$

$$-\frac{2n}{r}sE_n(s)\left[I_{n+1}(sr)+\frac{n-1}{sr}I_n(sr)\right]$$

$$\left.+K\frac{B_n(s)}{2}\left[\frac{n^2+n}{sr}I_{n+1}(sr)+I_n(sr)\right]\right\}\cos sz\,ds$$

$$\frac{\sigma_{zz}}{2G}=\int_0^{\infty}\left\{s^3C_0(s)I_0(sr)+s^2D_0(s)[2(2-\nu)I_0(sr)+srI_1(sr)]\right.$$

$$\left.-KA_0(s)\left[I_0(sr)+\frac{sr}{2}I_1(sr)\right]\right\}\cos sz\,ds$$

$$+\sum_{n=1}^{\infty}\cos n\theta\int_0^{\infty}\left\{s^3C_n(s)I_n(sr)+s^2D_n(s)[2(2-\nu)I_n(sr)+srI_{n+1}(sr)]\right.$$

$$\left.-KA_n(s)\left[I_n(sr)+\frac{sr}{2}I_{n+1}(sr)\right]\right\}\cos sz\,ds$$

$$+\sum_{n=1}^{\infty}\sin n\theta\int_0^{\infty}\left\{s^3C_n^*(s)I_n(sr)+s^2D_n^*(s)[2(2-\nu)I_n(sr)+srI_{n+1}(sr)]\right.$$

$$\left.-KB_n(s)\left[I_n(sr)+\frac{sr}{2}I_{n+1}(sr)\right]\right\}\cos sz\,ds$$

$$\frac{\sigma_{rz}}{2G}=\int_0^{\infty}\left\{s^3C_0(s)I_1(sr)+s^2D_0(s)[2(1-\nu)I_1(sr)+srI_0(sr)]\right.$$

$$\left.-K\frac{A_0(s)}{2}srI_0(sr)\right\}\sin sz\,ds$$

$$+\sum_{n=1}^{\infty}\cos n\theta\int_0^{\infty}\left\{s^3C_n(s)\left[I_{n+1}(sr)+\frac{n}{sr}I_n(sr)\right]\right.$$

$$+s^2D_n(s)\left[\frac{2n(1-\nu)+s^2r^2}{sr}I_n(sr)+(2-2\nu-n)I_{n+1}(sr)\right]$$

$$\left.-\frac{n}{r}sE_n^*(s)I_n(sr)-K\frac{A_n(s)}{2}[srI_n(sr)-nI_{n+1}(sr)]\right\}\sin sz\,ds$$

$$
\begin{aligned}
&+\sum_{n=1}^{\infty}\sin n\theta\int_0^{\infty}\left\{s^3C_n^*(s)\left[I_{n+1}(sr)+\frac{n}{sr}I_n(sr)\right]\right.\\
&+s^2D_n^*(s)\left[\frac{2n(1-\nu)+s^2r^2}{sr}I_n(sr)+(2-2\nu-n)I_{n+1}(sr)\right]\\
&\left.+\frac{n}{r}sE_n(s)I_n(sr)-K\frac{B_n(s)}{2}[srI_n(sr)-nI_{n+1}(sr)]\right\}\sin sz\,ds
\end{aligned}
$$

$$
\begin{aligned}
\frac{\sigma_{r\theta}}{2G}=&\sum_{n=1}^{\infty}\sin n\theta\int_0^{\infty}\left\{ns^3C_n(s)\left[\frac{1}{sr}I_{n+1}(sr)+\frac{n-1}{s^2r^2}I_n(sr)\right]\right.\\
&+ns^2D_n(s)\left[I_n(sr)-\frac{n+1}{sr}I_{n+1}(sr)\right]\\
&-s^2E_n^*(s)\left[\frac{2n^2-2n+s^2r^2}{s^2r^2}I_n(sr)-\frac{2}{sr}I_{n+1}(sr)\right]\\
&\left.-nK\frac{A_n(s)}{2}\left[I_n(sr)-\frac{n+1}{sr}I_{n+1}(sr)\right]\right\}\cos sz\,ds\\
&-\sum_{n=1}^{\infty}\cos n\theta\int_0^{\infty}\left\{ns^3C_n^*(s)\left[\frac{1}{sr}I_{n+1}(sr)+\frac{n-1}{s^2r^2}I_n(sr)\right]\right.\\
&+ns^2D_n^*(s)\left[I_n(sr)-\frac{n+1}{sr}I_{n+1}(sr)\right]\\
&+s^2E_n(s)\left[\frac{2n^2-2n+s^2r^2}{s^2r^2}I_n(sr)-\frac{2}{sr}I_{n+1}(sr)\right]\\
&\left.-nK\frac{B_n(s)}{2}\left[I_n(sr)-\frac{n+1}{sr}I_{n+1}(sr)\right]\right\}\cos sz\,ds
\end{aligned}
$$

$$
\begin{aligned}
\frac{\sigma_{z\theta}}{2G}=&-\sum_{n=1}^{\infty}\sin n\theta\int_0^{\infty}\left\{ns^3C_n(s)\frac{1}{sr}I_n(sr)+ns^2D_n(s)\left[\frac{2(1-\nu)}{sr}I_n(sr)+I_{n+1}(sr)\right]\right.\\
&\left.-s^2E_n^*(s)\left[I_{n+1}(sr)+\frac{n}{sr}I_n(sr)\right]-nK\frac{A_n(s)}{2}I_{n+1}(sr)\right\}\sin sz\,ds\\
&+\sum_{n=1}^{\infty}\cos n\theta\int_0^{\infty}\left\{ns^3C_n^*(s)\frac{1}{sr}I_n(sr)\right.\\
&+ns^2D_n^*(s)\left[\frac{2(1-\nu)}{sr}I_n(sr)+I_{n+1}(sr)\right]+s^2E_n(s)\left[I_{n+1}(sr)+\frac{n}{sr}I_n(sr)\right]\\
&\left.-nK\frac{B_n(s)}{2}I_{n+1}(sr)\right\}\sin sz\,ds
\end{aligned}
\tag{6.146}
$$

The boundary conditions on the traction free surface are

$$\sigma_{rr} = 0, \quad \sigma_{rz} = 0, \quad \sigma_{r\theta} = 0 \qquad \text{on } r = a \tag{l}$$

The unknown functions $C_0(s)$ and $D_0(s)$ can be determined from the boundary conditions (l)

$$C_0(s) = (1-\nu)KA_0(s)\frac{a}{s^2}\frac{saI_0^2(sa) - saI_1^2(sa) + I_0(sa)I_1(sa)}{(2-2\nu+s^2a^2)I_1^2(sa) - s^2a^2I_0^2(sa)}$$

$$D_0(s) = KA_0(s)\frac{a^2}{2}\frac{I_1^2(sa) - I_0^2(sa)}{(2-2\nu+s^2a^2)I_1^2(sa) - s^2a^2I_0^2(sa)} \tag{m}$$

The unknown functions $C_n(s)$, $D_n(s)$, and $E_n^*(s)$ can be determined by solving the system of three simultaneous algebraic equations

$$s^3C_n(s)\left[\frac{n^2 - n + s^2a^2}{s^2a^2}I_n(sa) - \frac{1}{sa}I_{n+1}(sa)\right]$$
$$+ s^2D_n(s)\left[(1-2\nu)I_n(sa) + \frac{n^2+n+s^2a^2}{sa}I_{n+1}(sa)\right]$$
$$- \frac{2n}{a}sE_n^*(s)\left[I_{n+1}(sa) + \frac{n-1}{sa}I_n(sa)\right]$$
$$= -K\frac{A_n(s)}{2}\left[I_n(sa) - \frac{n^2+n+s^2a^2}{sa}I_{n+1}(sa)\right]$$

$$s^3C_n(s)\left[I_{n+1}(sa) + \frac{n}{sa}I_n(sa)\right]$$
$$+ s^2D_n(s)\left[\frac{2n(1-\nu)+s^2a^2}{sa}I_n(sa) + (2-2\nu-n)I_{n+1}(sa)\right]$$
$$- \frac{n}{a}sE_n^*(s)I_n(sa) = K\frac{A_n(s)}{2}[saI_n(sa) - nI_{n+1}(sa)]$$

$$ns^3C_n(s)\left[\frac{1}{sa}I_{n+1}(sa) + \frac{n-1}{s^2a^2}I_n(sa)\right] + ns^2D_n(s)\left[I_n(sa) - \frac{n+1}{sa}I_{n+1}(sa)\right]$$
$$- s^2E_n^*(s)\left[\frac{2n^2 - 2n + s^2a^2}{s^2a^2}I_n(sa) - \frac{2}{sa}I_{n+1}(sa)\right]$$
$$= nK\frac{A_n(s)}{2}\left[I_n(sa) - \frac{n+1}{sa}I_{n+1}(sa)\right] \tag{n}$$

The unknown functions $C_n^*(s)$, $D_n^*(s)$, and $E_n(s)$ can be determined by solving the system of three simultaneous algebraic equations

$$
\begin{aligned}
& s^3 C_n^*(s)\left[\frac{n^2-n+s^2a^2}{s^2a^2} I_n(sa) - \frac{1}{sa} I_{n+1}(sa)\right] \\
& \quad + s^2 D_n^*(s)\left[(1-2\nu) I_n(sa) + \frac{n^2+n+s^2a^2}{sa} I_{n+1}(sa)\right] \\
& \quad + \frac{2n}{a} s E_n(s)\left[I_{n+1}(sa) + \frac{n-1}{sa} I_n(sa)\right] \\
& \quad = -K\frac{B_n(s)}{2}\left[I_n(sa) - \frac{n^2+n+s^2a^2}{sa} I_{n+1}(sa)\right] \\
& s^3 C_n^*(s)\left[I_{n+1}(sa) + \frac{n}{sa} I_n(sa)\right] \\
& \quad + s^2 D_n^*(s)\left[\frac{2n(1-\nu)+s^2a^2}{sa} I_n(sa) + (2-2\nu-n) I_{n+1}(sa)\right] \\
& \quad + \frac{n}{a} s E_n(s) I_n(sa) = K\frac{B_n(s)}{2}[sa I_n(sa) - n I_{n+1}(sa)] \\
& ns^3 C_n^*(s)\left[\frac{1}{sa} I_{n+1}(sa) + \frac{n-1}{s^2a^2} I_n(sa)\right] + ns^2 D_n^*(s)\left[I_n(sa) - \frac{n+1}{sa} I_{n+1}(sa)\right] \\
& \quad + s^2 E_n(s)\left[\frac{2n^2-2n+s^2a^2}{s^2a^2} I_n(sa) - \frac{2}{sa} I_{n+1}(sa)\right] \\
& \quad = nK\frac{B_n(s)}{2}\left[I_n(sa) - \frac{n+1}{sa} I_{n+1}(sa)\right] \qquad \text{(o)}
\end{aligned}
$$

Substituting those determined functions into Eqs. (6.146), the displacements and the thermal stresses can be obtained.

Problems

6.1 Derive Eq. (6.85).

6.2 Derive Eq. (6.86).

6.3 Derive Eq. (6.89).

6.4 Prove that the Navier's equation in the cylindrical coordinate system in plane problems can be expressed by

$$(\lambda^* + 2\mu)\frac{\partial e}{\partial r} - \frac{2\mu}{r}\frac{\partial \omega}{\partial \theta} = \beta^* \frac{\partial \tau}{\partial r}$$

$$(\lambda^* + 2\mu)\frac{1}{r}\frac{\partial e}{\partial \theta} + 2\mu\frac{\partial \omega}{\partial r} = \beta^* \frac{1}{r}\frac{\partial \tau}{\partial \theta}$$

where

$$e = \frac{\partial u_r}{\partial r} + \frac{u_r}{r} + \frac{1}{r}\frac{\partial u_\theta}{\partial \theta}, \qquad \omega = \frac{1}{2r}\left[\frac{\partial (r u_\theta)}{\partial r} - \frac{\partial u_r}{\partial \theta}\right]$$

6.5 Prove that the governing equation in Problem 6.4 can be solved by use of the Goodier's thermoelastic potential function Φ and two harmonic functions φ, ϑ related to the components of displacement as follows

$$u_r = \frac{\partial \Phi}{\partial r} + \frac{\partial \varphi}{\partial r} + \frac{2}{r}\frac{\partial \vartheta}{\partial \theta}$$

$$u_\theta = \frac{1}{r}\frac{\partial \Phi}{\partial \theta} + \frac{1}{r}\frac{\partial \varphi}{\partial \theta} - 2\frac{\partial \vartheta}{\partial r}$$

where the Goodier's thermoelastic potential function Φ and two harmonic functions φ, ϑ must satisfy the equations

$$\nabla^2 \Phi = \frac{\beta^*}{\lambda^* + 2\mu}\tau, \qquad \nabla^2 \varphi = 0, \qquad \nabla^2 \vartheta = 0$$

in which

$$\nabla^2 = \frac{\partial^2}{\partial r^2} + \frac{1}{r}\frac{\partial}{\partial r} + \frac{1}{r^2}\frac{\partial^2}{\partial \theta^2}$$

6.6 Derive Eq. (6.109).

6.7 Derive Eq. (6.111).

6.8 Derive Eq. (6.123).

6.9 Derive Eq. (6.139).

Chapter 7

THERMAL STRESSES IN SPHERICAL BODIES

Thermoelastic problems in spherical coordinates are presented. One-dimensional steady state and transient thermal stresses are discussed for the cases of a hollow and a solid sphere, and for infinite bodies with a spherical cavity. Moreover, two-dimensional (axisymmetric) problems are expressed in terms of the Goodier's thermoelastic displacement potential and the Boussinesq harmonic functions and, in this context, the solid sphere, the hollow sphere, and the infinite body with a spherical cavity are discussed for a steady temperature field.

7.1 Thermal Stresses in Spherical Bodies

In this chapter, one-, two-, and three-dimensional thermoelastic problems are treated in the spherical coordinate system (r, θ, ϕ) shown in Fig. 3.3. Some illustrative examples cover a solid sphere, a hollow sphere, and an infinite body with a spherical cavity or a spherical inclusion.

7.2 One-Dimensional Problems

For one-dimensional problems in the spherical coordinate system, which means spherically symmetric problems, the familiar displacement technique is extensively used. Because of the spherical symmetry, the stress and strain components in the θ and ϕ directions are identical, and the shearing stress and strain components are zero:

$$\begin{aligned} &\sigma_{\theta\theta} = \sigma_{\phi\phi}, \qquad \epsilon_{\theta\theta} = \epsilon_{\phi\phi} \\ &\sigma_{r\theta} = \sigma_{r\phi} = \sigma_{\theta\phi} = 0, \qquad \epsilon_{r\theta} = \epsilon_{r\phi} = \epsilon_{\theta\phi} = 0 \end{aligned} \tag{7.1}$$

The equilibrium equation without the body force for a one-dimensional problem in the spherical coordinate system is obtained from Eqs. (4.83):

$$\frac{d\sigma_{rr}}{dr} + \frac{2}{r}(\sigma_{rr} - \sigma_{\theta\theta}) = 0 \tag{7.2}$$

The stress–strain relations are derived from Eqs. (4.91):

$$\begin{aligned} \sigma_{rr} &= 2\mu\epsilon_{rr} + \lambda e - \beta\tau \\ \sigma_{\theta\theta} &= \sigma_{\phi\phi} = 2\mu\epsilon_{\theta\theta} + \lambda e - \beta\tau \end{aligned} \tag{7.3}$$

where $e = \epsilon_{rr} + 2\epsilon_{\theta\theta}$. The strain components in terms of the radial displacement u_r are determined from Eqs. (4.85):

$$\epsilon_{rr} = \frac{du_r}{dr}, \qquad \epsilon_{\theta\theta} = \epsilon_{\phi\phi} = \frac{u_r}{r} \tag{7.4}$$

The stress components in terms of the displacement component u_r are represented by use of Eqs. (7.3), (7.4) and (4.20)

$$\begin{aligned} \sigma_{rr} &= \frac{E}{(1+\nu)(1-2\nu)}\left[(1-\nu)\frac{du_r}{dr} + 2\nu\frac{u_r}{r} - (1+\nu)\alpha\tau\right] \\ \sigma_{\theta\theta} = \sigma_{\phi\phi} &= \frac{E}{(1+\nu)(1-2\nu)}\left[\nu\frac{du_r}{dr} + \frac{u_r}{r} - (1+\nu)\alpha\tau\right] \end{aligned} \tag{7.5}$$

Making use of Eqs. (7.2) and (7.5), the equilibrium equation reduces to

$$\frac{d}{dr}\left[\frac{1}{r^2}\frac{d}{dr}(r^2 u_r)\right] = \frac{1+\nu}{1-\nu}\alpha\frac{d\tau}{dr} \tag{7.6}$$

The general solution of Eq. (7.6) is

$$u_r = \frac{1+\nu}{1-\nu}\alpha\frac{1}{r^2}\int \tau r^2\,dr + C_1 r + C_2\frac{1}{r^2} \tag{7.7}$$

where C_1 and C_2 are integration constants. From Eqs. (7.5) and (7.7), the stresses are written as

$$\begin{aligned}\sigma_{rr} &= \frac{E}{(1+\nu)(1-2\nu)}\left[-2\frac{(1+\nu)(1-2\nu)}{1-\nu}\alpha\frac{1}{r^3}\int \tau r^2\,dr\right.\\ &\qquad\left. +(1+\nu)C_1 - 2(1-2\nu)C_2\frac{1}{r^3}\right]\\ \sigma_{\theta\theta} = \sigma_{\phi\phi} &= \frac{E}{(1+\nu)(1-2\nu)}\left[\frac{(1+\nu)(1-2\nu)}{1-\nu}\alpha\left(\frac{1}{r^3}\int \tau r^2\,dr - \tau\right)\right.\\ &\qquad\left. +(1+\nu)C_1 + (1-2\nu)C_2\frac{1}{r^3}\right]\end{aligned} \tag{7.8}$$

The integration constants C_1 and C_2 must be determined from the boundary conditions.

7.2.1 A Hollow Sphere

Let us consider a hollow sphere with an inner radius a and an outer radius b. The integral involving the temperature change, which appears in Eqs. (7.7) and (7.8), has the following limits:

$$\int \tau r^2\,dr = \int_a^r \tau r^2\,dr \tag{7.9}$$

The boundary conditions on the traction free surface are

$$\sigma_{rr} = 0 \qquad \text{at } r = a \text{ and } r = b \tag{7.10}$$

Thus, the integration constants C_1 and C_2 can be determined by Eqs. (7.10)

$$\begin{aligned}C_1 &= \frac{2(1-2\nu)}{1-\nu}\alpha\frac{1}{b^3-a^3}\int_a^b \tau r^2\,dr\\ C_2 &= \frac{1+\nu}{1-\nu}\alpha\frac{a^3}{b^3-a^3}\int_a^b \tau r^2\,dr\end{aligned} \tag{7.11}$$

By use of Eqs. (7.7), (7.8), and (7.11), the displacement and stress components are given by

$$\begin{aligned}u_r &= \frac{1+\nu}{1-\nu}\alpha\left[\frac{1}{r^2}\int_a^r \tau r^2\,dr + \frac{2(1-2\nu)}{1+\nu}\frac{r}{b^3-a^3}\int_a^b \tau r^2\,dr + \frac{a^3}{b^3-a^3}\frac{1}{r^2}\int_a^b \tau r^2\,dr\right]\\ \sigma_{rr} &= \frac{\alpha E}{1-\nu}\left[\frac{2(r^3-a^3)}{r^3(b^3-a^3)}\int_a^b \tau r^2\,dr - \frac{2}{r^3}\int_a^r \tau r^2\,dr\right]\\ \sigma_{\theta\theta} = \sigma_{\phi\phi} &= \frac{\alpha E}{1-\nu}\left[\frac{2r^3+a^3}{r^3(b^3-a^3)}\int_a^b \tau r^2\,dr + \frac{1}{r^3}\int_a^r \tau r^2\,dr - \tau\right]\end{aligned} \tag{7.12}$$

The radial displacement and the stress components in Eqs. (7.12) are valid for steady and transient states of spherically symmetric temperature fields in the hollow sphere.

Example 7.1 Both the inner surface ($r = a$) and the outer surface ($r = b$) of a hollow sphere with constant initial temperature T_i are maintained at constant temperatures T_a and T_b, respectively. Find the thermal displacement and thermal stresses in the hollow sphere under a steady temperature field.

Solution The temperature T is given by Eq. (3.111), so that a temperature change τ is

$$\tau = T_a - T_i + (T_b - T_a)\left[\frac{1 - \dfrac{a}{r}}{1 - \dfrac{a}{b}}\right] \tag{7.13}$$

We find

$$\int \tau r^2\,dr = \frac{1}{3}(T_a - T_i)r^3 + \frac{1}{6}(T_b - T_a)\frac{r^2(2r - 3a)}{1 - \dfrac{a}{b}} \tag{7.14}$$

Substituting Eqs. (7.13) and (7.14) into Eqs. (7.12), the radial displacement and stress components are

$$\begin{aligned}
u_r &= (T_a - T_i)\alpha r + \frac{1+\nu}{2(1-\nu)}(T_b - T_a)\alpha\left(\frac{b}{b^3 - a^3}\right) \\
&\quad\times\left\{\frac{2r}{1+\nu}\left[(1-\nu)b^2 + \nu(ab + a^2)\right] - a(b^2 + ab + a^2) + \frac{a^3b^2}{r^2}\right\} \\
\sigma_{rr} &= -\frac{\alpha E}{1-\nu}(T_b - T_a)\frac{ab}{b^3 - a^3}\left(1 - \frac{a}{r}\right)\left(1 - \frac{b}{r}\right)\left(b + a + \frac{ab}{r}\right) \\
\sigma_{\theta\theta} = \sigma_{\phi\phi} &= -\frac{\alpha E}{1-\nu}(T_b - T_a)\frac{ab}{b^3 - a^3} \\
&\quad\times\left[b + a - (b^2 + ab + a^2)\frac{1}{2r} - \frac{a^2b^2}{2r^3}\right]
\end{aligned} \tag{7.15}$$

Making use of Eqs. (7.15), it can be seen that the extremal value of σ_{rr} appears at the position $r = ((3a^2b^2)/(b^2 + ab + a^2))^{1/2}$, which is

$$(\sigma_{rr})_{\text{extremum}} = -\frac{\alpha E}{1-\nu}(T_b - T_a)\left(\frac{ab}{b^3 - a^3}\right)\left[a + b - \frac{2\sqrt{3}}{9ab}(a^2 + ab + b^2)^{3/2}\right] \tag{7.16}$$

Furthermore, the extremal values of $\sigma_{\theta\theta}(=\sigma_{\phi\phi})$ appear at the inner and outer surfaces, and they are

$$\begin{aligned}\sigma_{\theta\theta}|_{r=a} &= \frac{\alpha E}{2(1-\nu)}(T_b - T_a)\frac{b(2b+a)}{b^2+ab+a^2} \\ \sigma_{\theta\theta}|_{r=b} &= -\frac{\alpha E}{2(1-\nu)}(T_b - T_a)\frac{a(b+2a)}{b^2+ab+a^2}\end{aligned} \tag{7.17}$$

Now, let $t = b - a$ be the thickness of the hollow sphere. When the thickness t is very small, the thermal stresses at the boundary surfaces are given from Eqs. (7.17) as

$$\begin{aligned}\sigma_{\theta\theta}|_{r=a} &= \sigma_{\phi\phi}|_{r=a} \\ &= \frac{\alpha E}{2(1-\nu)}(T_b - T_a)\frac{(a+t)(3a+2t)}{3a^2+3at+t^2} \\ &= \frac{\alpha E}{2(1-\nu)}(T_b - T_a)\frac{\left[1+\left(\frac{t}{a}\right)\right]\left[3+2\left(\frac{t}{a}\right)\right]}{3+3\left(\frac{t}{a}\right)+\left(\frac{t}{a}\right)^2} \cong \frac{\alpha E}{2(1-\nu)}(T_b - T_a) \\ \sigma_{\theta\theta}|_{r=b} &= \sigma_{\phi\phi}|_{r=b} \\ &= -\frac{\alpha E}{2(1-\nu)}(T_b - T_a)\frac{a(3a+t)}{3a^2+3at+t^2} \\ &= -\frac{\alpha E}{2(1-\nu)}(T_b - T_a)\frac{3+\left(\frac{t}{a}\right)}{3+3\left(\frac{t}{a}\right)+\left(\frac{t}{a}\right)^2} \cong -\frac{\alpha E}{2(1-\nu)}(T_b - T_a)\end{aligned} \tag{7.18}$$

On the other hand, when the hollow sphere is very thick, the thermal stresses at the boundary surfaces are given from Eqs. (7.17) as

$$\begin{aligned}\sigma_{\theta\theta}|_{r=a} = \sigma_{\phi\phi}|_{r=a} &= \frac{\alpha E}{2(1-\nu)}(T_b - T_a)\left[\frac{2+\left(\frac{a}{b}\right)}{1+\left(\frac{a}{b}\right)+\left(\frac{a}{b}\right)^2}\right] \\ &\cong \frac{\alpha E}{1-\nu}(T_b - T_a) \\ \sigma_{\theta\theta}|_{r=b} = \sigma_{\phi\phi}|_{r=b} &= -\frac{\alpha E}{2(1-\nu)}(T_b - T_a)\left\{\frac{\left(\frac{a}{b}\right)\left[1+2\left(\frac{a}{b}\right)\right]}{1+\left(\frac{a}{b}\right)+\left(\frac{a}{b}\right)^2}\right\} \cong 0\end{aligned} \tag{7.19}$$

It is found that the stress at the inner surface for a very thick hollow sphere is twice as large as that for a very thin hollow sphere. Furthermore, comparing Eqs. (7.18) and (7.19) with Eqs. (6.27) and (6.28), the thermal stresses at the boundary surfaces for very thin or very thick hollow spheres show identical values to those (see Eqs. (6.27) and (6.28)) at the boundary surfaces for very thin or very thick hollow circular cylinders, respectively.

Example 7.2 When the temperature in Example 7.1 represents the transient state, find the thermal displacement and thermal stresses in the hollow sphere.

Solution The temperature T is calculated via Eq. (3.123), which yields

$$T = T_a + (T_b - T_a)\left[\frac{1 - \dfrac{a}{r}}{1 - \dfrac{a}{b}}\right] - \frac{2}{\pi r}\sum_{n=1}^{\infty}\frac{1}{n}\sin\left[n\pi\left(\frac{r-a}{b-a}\right)\right]e^{-\kappa n^2\pi^2 t/(b-a)^2}$$
$$\times\left[(T_i - T_b)(-1)^n b + (T_b - T_a)a\right] \tag{7.20}$$

Then the temperature change τ is

$$\tau = (T_a - T_i) + (T_b - T_a)\left[\frac{1 - \dfrac{a}{r}}{1 - \dfrac{a}{b}}\right]$$
$$- \frac{2}{\pi r}\sum_{n=1}^{\infty}\frac{1}{n}\sin\left[n\pi\left(\frac{r-a}{b-a}\right)\right]e^{-\kappa n^2\pi^2 t/(b-a)^2}$$
$$\times\left[(T_i - T_b)(-1)^n b + (T_b - T_a)a\right] \tag{a}$$

We find the following relation

$$\int \tau r^2\,dr = \frac{1}{3}(T_a - T_i)r^3 + (T_b - T_a)\frac{b}{b-a}\frac{1}{6}r^2(2r - 3a)$$
$$+ \frac{2(b-a)}{\pi^3}\sum_{n=1}^{\infty}\frac{1}{n^3}e^{-\kappa n^2\pi^2 t/(b-a)^2}\left[(T_i - T_b)b(-1)^n + (T_a - T_i)a\right]$$
$$\times\left\{n\pi r\cos\left[n\pi\left(\frac{r-a}{b-a}\right)\right] - (b-a)\sin\left[n\pi\left(\frac{r-a}{b-a}\right)\right]\right\} \tag{b}$$

Substituting Eqs. (a) and (b) into Eqs. (7.12), the radial displacement and stresses are given as

$$u_r = \frac{1+\nu}{1-\nu}\alpha\left\{(T_a - T_i)\frac{1-\nu}{1+\nu}r + (T_b - T_a)\frac{b}{2(b^3 - a^3)}\right.$$
$$\times\left\{\frac{2r}{1+\nu}\left[(1-\nu)b^2 + \nu(ab + a^2)\right] - a(b^2 + ab + a^2) + \frac{a^3b^2}{r^2}\right\}$$

$$
+ \frac{2(b-a)}{\pi^3} \sum_{n=1}^{\infty} \frac{1}{n^3} e^{-\kappa n^2 \pi^2 t/(b-a)^2} \left[(T_i - T_b)(-1)^n b + (T_a - T_i)a\right]
$$

$$
\times \left(\frac{2(1-2\nu)}{1+\nu} r \frac{n\pi}{b^3 - a^3}((-1)^n b - a) + \frac{n\pi}{r} \cos\left[n\pi\left(\frac{r-a}{b-a}\right)\right] \right.
$$

$$
\left. \left. + \frac{1}{r^2} \left\{ -(b-a)\sin\left[n\pi\left(\frac{r-a}{b-a}\right)\right] + \frac{n\pi ab}{b^3 - a^3}(a^2(-1)^n - b^2) \right\} \right) \right\}
$$

$$
\sigma_{rr} = \frac{\alpha E}{1-\nu} \left\{ -(T_b - T_a) \frac{ab}{b^3 - a^3} \left(1 - \frac{a}{r}\right)\left(1 - \frac{b}{r}\right)\left(b + a + \frac{ab}{r}\right) \right.
$$

$$
+ \frac{4(b-a)}{\pi^3} \sum_{n=1}^{\infty} \frac{1}{n^3} e^{-\kappa n^2 \pi^2 t/(b-a)^2} \left[(T_i - T_b)(-1)^n b + (T_a - T_i)a\right]
$$

$$
\times \left(\frac{(r^3 - a^3) n\pi}{r^3 (b^3 - a^3)}((-1)^n b - a) \right. \tag{7.21}
$$

$$
\left. \left. - \frac{1}{r^3} \left\{ n\pi r \cos\left[n\pi\left(\frac{r-a}{b-a}\right)\right] - (b-a)\sin\left[n\pi\left(\frac{r-a}{b-a}\right)\right] - n\pi a \right\} \right) \right\}
$$

$$
\sigma_{\theta\theta} = \sigma_{\phi\phi}
$$

$$
= \frac{\alpha E}{1-\nu} \left\{ -(T_b - T_a) \frac{ab}{b^3 - a^3} \left[b + a - (b^2 + ab + a^2)\frac{1}{2r} - \frac{a^2 b^2}{2r^3} \right] \right.
$$

$$
+ \frac{2(b-a)}{\pi^3} \sum_{n=1}^{\infty} \frac{1}{n^3} e^{-\kappa n^2 \pi^2 t/(b-a)^2} \left[(T_i - T_b)(-1)^n b + (T_a - T_i)a\right]
$$

$$
\times \left(\frac{(2r^3 + a^3) n\pi}{r^3 (b^3 - a^3)}((-1)^n b - a) + \frac{1}{r^3} \left\{ n\pi r \cos\left[n\pi\left(\frac{r-a}{b-a}\right)\right] \right. \right.
$$

$$
\left. \left. \left. -(b-a)\sin\left[n\pi\left(\frac{r-a}{b-a}\right)\right] - n\pi a \right\} + \frac{n^2\pi^2}{b-a}\frac{1}{r}\sin\left[n\pi\left(\frac{r-a}{b-a}\right)\right] \right) \right\}
$$

7.2.2 A Solid Sphere

Let us consider a solid sphere with a radius a. The integral $\int \tau r^2\, dr$ involving the temperature change in Eqs. (7.7) and (7.8) means $\int_0^r \tau r^2\, dr$ for the solid sphere. The radial displacement and stress components of the solid sphere are given from Eqs. (7.7) and (7.8) as

$$
u_r = \frac{1+\nu}{1-\nu} \alpha \frac{1}{r^2} \int_0^r \tau r^2\, dr + C_1 r + C_2 \frac{1}{r^2}
$$

$$\sigma_{rr} = \frac{E}{(1+\nu)(1-2\nu)}\left[-2\frac{(1+\nu)(1-2\nu)}{1-\nu}\alpha\frac{1}{r^3}\int_0^r \tau r^2\,dr + (1+\nu)C_1 - 2(1-2\nu)C_2\frac{1}{r^3}\right] \tag{7.22}$$

$$\sigma_{\theta\theta} = \sigma_{\phi\phi} = \frac{E}{(1+\nu)(1-2\nu)}\left[\frac{(1+\nu)(1-2\nu)}{1-\nu}\alpha\left(\frac{1}{r^3}\int_0^r \tau r^2\,dr - \tau\right) + (1+\nu)C_1 + (1-2\nu)C_2\frac{1}{r^3}\right]$$

When the temperature change has a finite value in the whole region of the solid sphere, the corresponding displacement and stresses have to take finite values. The displacement and stresses expressed by Eqs. (7.22) have infinite values at the origin $r = 0$ if the coefficient C_2 is not zero. Therefore, the coefficient C_2 must be zero for the displacement and stresses to be finite. The boundary condition on the traction free surface is

$$\sigma_{rr} = 0 \qquad \text{at } r = a \tag{7.23}$$

which gives

$$C_1 = \frac{2(1-2\nu)}{1-\nu}\alpha\frac{1}{a^3}\int_0^a \tau r^2\,dr \tag{7.24}$$

Therefore, the displacement and stresses for the solid sphere are

$$\begin{aligned}
u_r &= \frac{\alpha}{1-\nu}\left[(1+\nu)\frac{1}{r^2}\int_0^r \tau r^2\,dr + 2(1-2\nu)\frac{r}{a^3}\int_0^a \tau r^2\,dr\right]\\
\sigma_{rr} &= \frac{\alpha E}{1-\nu}\left[\frac{2}{a^3}\int_0^a \tau r^2\,dr - \frac{2}{r^3}\int_0^r \tau r^2\,dr\right]\\
\sigma_{\theta\theta} = \sigma_{\phi\phi} &= \frac{\alpha E}{1-\nu}\left[\frac{2}{a^3}\int_0^a \tau r^2\,dr + \frac{1}{r^3}\int_0^r \tau r^2\,dr - \tau\right]
\end{aligned} \tag{7.25}$$

Example 7.3 When the boundary surface ($r = a$) of a solid sphere with constant initial temperature T_i is heated by the surrounding medium at a constant temperature T_a with heat transfer coefficient h_a, find the radial displacement and thermal stresses in the solid sphere under a steady temperature field.

Solution The heat conduction equation and the thermal boundary condition are

$$\frac{d^2T}{dr^2} + \frac{2}{r}\frac{dT}{dr} = 0 \tag{a}$$

$$-\lambda\frac{dT}{dr} = h_a(T - T_a) \qquad \text{on } r = a \tag{b}$$

The general solution of Eq. (a) is

$$T = A + Br^{-1} \tag{c}$$

Since T is finite at $r = 0$, the temperature solution satisfying Eqs. (b) and (c) is given as

$$T = T_a \tag{d}$$

Thus, the temperature change τ becomes

$$\tau = T_a - T_i \tag{e}$$

Therefore, the following integral relation is obtained:

$$\int \tau r^2\, dr = \frac{1}{3}(T_a - T_i)r^3 \tag{f}$$

Substitution of Eq. (f) into Eqs. (7.25) gives the radial displacement and stresses as

$$\begin{gathered} u_r = \alpha(T_a - T_i)r \\ \sigma_{rr} = 0, \qquad \sigma_{\theta\theta} = \sigma_{\phi\phi} = 0 \end{gathered} \tag{7.26}$$

Example 7.4 When the temperature in Example 7.3 represents the transient state, find the radial displacement and thermal stresses in the solid sphere.

Solution The temperature T is calculated by Eq. (3.120), so that the temperature change τ is

$$\tau = (T_a - T_i)\left[1 - 2\sum_{n=1}^{\infty} \frac{\lambda h_a a^2 \sin(s_n r)}{r\left(\lambda^2 s_n^2 a^2 + h_a^2 a^2 - \lambda h_a a\right)\sin(s_n a)} e^{-\kappa s_n^2 t}\right] \tag{a}$$

in which λ is the thermal conductivity, and s_n are positive roots of the transcendental equation

$$\tan(sa) = \frac{\lambda s a}{\lambda - h_a a} \tag{b}$$

Now, the following relation is derived from Eq. (a):

$$\begin{aligned} \int \tau r^2\, dr = (T_a - T_i)\Bigg[\frac{1}{3}r^3 - 2\sum_{n=1}^{\infty} &\frac{\lambda h_a a^2}{\left(\lambda^2 s_n^2 a^2 + h_a^2 a^2 - \lambda h_a a\right)\sin(s_n a)} e^{-\kappa s_n^2 t} \\ &\times \frac{1}{s_n^2}(\sin(s_n r) - s_n r\cos(s_n r))\Bigg] \end{aligned} \tag{c}$$

Substitution of Eqs. (a) and (c) into Eqs. (7.25) gives the radial displacement and stresses:

$$u_r = \frac{\alpha}{1-\nu}(T_a - T_i)$$
$$\times \left\{ (1-\nu)r - 2\sum_{n=1}^{\infty} \frac{\lambda h_a a^2}{(\lambda^2 s_n^2 a^2 + h_a^2 a^2 - \lambda h_a a) s_n^2 \sin(s_n a)} e^{-\kappa s_n^2 t} \right.$$
$$\left. \times \left[(1+\nu)\frac{1}{r^2}(\sin(s_n r) - s_n r\cos(s_n r)) + 2(1-2\nu) r \frac{h_a a}{\lambda a^3}\sin(s_n a) \right] \right\}$$

$$\sigma_{rr} = -\frac{4\alpha E}{1-\nu}(T_a - T_i)\sum_{n=1}^{\infty} \frac{\lambda h_a a^2}{(\lambda^2 s_n^2 a^2 + h_a^2 a^2 - \lambda h_a a) s_n^2 \sin(s_n a)} e^{-\kappa s_n^2 t} \tag{7.27}$$
$$\times \left[\frac{h_a a}{\lambda a^3}\sin(s_n a) - \frac{1}{r^3}(\sin(s_n r) - s_n r\cos(s_n r)) \right]$$

$$\sigma_{\theta\theta} = \sigma_{\phi\phi} = -\frac{2\alpha E}{1-\nu}(T_a - T_i)\sum_{n=1}^{\infty} \frac{\lambda h_a a^2}{(\lambda^2 s_n^2 a^2 + h_a^2 a^2 - \lambda h_a a) s_n^2 \sin(s_n a)} e^{-\kappa s_n^2 t}$$
$$\times \left[\frac{2h_a a}{\lambda a^3}\sin(s_n a) + \frac{1}{r^3}(\sin(s_n r) - s_n r\cos(s_n r)) - \frac{s_n^2}{r}\sin(s_n r) \right]$$

7.2.3 Infinite Bodies

Let us consider an infinite body with a spherical cavity of radius a. Referring to Eqs. (7.7) and (7.8), the radial displacement and stress components in the infinite body with the spherical cavity are given by

$$u_r = \frac{1+\nu}{1-\nu}\alpha\frac{1}{r^2}\int_a^r \tau r^2\,dr + C_1 r + C_2\frac{1}{r^2}$$
$$\sigma_{rr} = \frac{E}{(1+\nu)(1-2\nu)}\left[-2\frac{(1+\nu)(1-2\nu)}{1-\nu}\alpha\frac{1}{r^3}\int_a^r \tau r^2\,dr \right.$$
$$\left. + (1+\nu)C_1 - 2(1-2\nu)C_2\frac{1}{r^3} \right] \tag{7.28}$$
$$\sigma_{\theta\theta} = \sigma_{\phi\phi} = \frac{E}{(1+\nu)(1-2\nu)}\left[\frac{(1+\nu)(1-2\nu)}{1-\nu}\alpha\left(\frac{1}{r^3}\int_a^r \tau r^2\,dr - \tau \right) \right.$$
$$\left. + (1+\nu)C_1 + (1-2\nu)C_2\frac{1}{r^3} \right]$$

As is the case for the solid sphere, if the temperature change has a finite value in the whole region of the infinite body, the corresponding displacement and stresses have to be finite all over the region including infinity, i.e., $r = \infty$, which means that the coefficient

C_1 must be zero. Moreover, the boundary condition on the traction free surface is given by Eqs. (7.23), which yields

$$C_2 = 0 \tag{7.29}$$

Therefore, the radial displacement and stresses for the infinite body with a spherical cavity are

$$\begin{aligned} u_r &= \frac{1+\nu}{1-\nu}\alpha\frac{1}{r^2}\int_a^r \tau r^2\,dr \\ \sigma_{rr} &= -\frac{2\alpha E}{1-\nu}\frac{1}{r^3}\int_a^r \tau r^2\,dr \\ \sigma_{\theta\theta} = \sigma_{\phi\phi} &= \frac{\alpha E}{1-\nu}\left(\frac{1}{r^3}\int_a^r \tau r^2\,dr - \tau\right) \end{aligned} \tag{7.30}$$

Let us consider the case of an infinite body without a cavity. In such case, we may consider that the radius a of the spherical cavity tends to zero, and the displacement and stresses for the infinite body without a cavity are, referring to Eqs. (7.30),

$$\begin{aligned} u_r &= \frac{1+\nu}{1-\nu}\alpha\frac{1}{r^2}\int_0^r \tau r^2\,dr \\ \sigma_{rr} &= -\frac{2\alpha E}{1-\nu}\frac{1}{r^3}\int_0^r \tau r^2\,dr \\ \sigma_{\theta\theta} = \sigma_{\phi\phi} &= \frac{\alpha E}{1-\nu}\left(\frac{1}{r^3}\int_0^r \tau r^2\,dr - \tau\right) \end{aligned} \tag{7.31}$$

Example 7.5 When the boundary surface ($r = a$) of an infinite body with a spherical cavity at initial temperature T_i is heated by the surrounding medium at a constant temperature T_a with heat transfer coefficient h_a, find the radial displacement and thermal stresses under a steady temperature field.

Solution The heat conduction equation and thermal boundary conditions are

$$\frac{d^2T}{dr^2} + \frac{2}{r}\frac{dT}{dr} = 0 \tag{a}$$

$$\begin{aligned} \lambda\frac{dT}{dr} &= h_a(T - T_a) \qquad \text{on } r = a \\ T &= T_i \qquad \text{at } r \to \infty \end{aligned} \tag{b}$$

The temperature solution satisfying Eqs. (a) and (b) is given as

$$T = \frac{h_a a}{\lambda + h_a a}\frac{a}{r}(T_a - T_i) + T_i \tag{c}$$

The temperature change τ becomes

$$\tau = \frac{h_a a}{\lambda + h_a a} \frac{a}{r}(T_a - T_i) \tag{d}$$

Making use of Eq. (d), the following integral relation is obtained:

$$\int \tau r^2 \, dr = \frac{h_a a}{\lambda + h_a a}(T_a - T_i)\frac{ar^2}{2} \tag{e}$$

Substitution of Eq. (e) into Eqs. (7.30) gives the radial displacement and stresses as

$$\begin{aligned}
u_r &= \frac{1+\nu}{1-\nu}\alpha \frac{h_a a}{\lambda + h_a a}(T_a - T_i)\frac{a}{2}\left(1 - \frac{a^2}{r^2}\right) \\
\sigma_{rr} &= -\frac{\alpha E}{1-\nu}\frac{h_a a}{\lambda + h_a a}(T_a - T_i)\frac{a}{r}\left(1 - \frac{a^2}{r^2}\right) \\
\sigma_{\theta\theta} = \sigma_{\phi\phi} &= -\frac{\alpha E}{1-\nu}\frac{h_a a}{\lambda + h_a a}(T_a - T_i)\frac{1}{2}\frac{a}{r}\left(1 + \frac{a^2}{r^2}\right)
\end{aligned} \tag{7.32}$$

For the special case that the boundary surface of the spherical cavity is maintained at a constant temperature T_a, the temperature change, the radial displacement and stresses can be obtained from Eqs. (d) and (7.32) by equating the term $(h_a a)/(\lambda + h_a a)$ to 1.

Example 7.6 When an infinite body at initial temperature T_i is heated by a partially distributed internal heat generation Q_0 within the region $r \le a$, find the radial displacement and thermal stresses in the infinite body under a steady temperature field.

Solution Referring to Eqs. (3.104), the heat conduction equation is given as

$$\frac{d^2T}{dr^2} + \frac{2}{r}\frac{dT}{dr} + \frac{Q_0}{\lambda}H(a - r) = 0 \tag{a}$$

where H is the Heaviside step function and the boundary conditions are

$$\begin{aligned}
T &= T_i \qquad \text{at } r \to \infty \\
T &\neq \infty \qquad \text{at } r \to 0
\end{aligned} \tag{b}$$

The temperature solution satisfying Eqs. (a) and (b) is given as

$$T = \begin{cases} -\dfrac{Q_0}{6\lambda}r^2 + C_1 & \text{for } 0 \le r \le a \\[2ex] T_i + \dfrac{C_2}{r} & \text{for } r \ge a \end{cases} \tag{c}$$

where the integration constants C_1 and C_2 are determined so as to satisfy the condition of continuity of temperature and heat flux at $r = a$. Hence, the temperature solution is represented as

$$T = \begin{cases} \dfrac{Q_0}{6\lambda}(3a^2 - r^2) + T_i & \text{for } 0 \le r \le a \\ \dfrac{Q_0}{6\lambda}\dfrac{2a^3}{r} + T_i & \text{for } r \ge a \end{cases} \tag{7.33}$$

Thus, the temperature change τ takes the form

$$\tau = \begin{cases} \dfrac{Q_0}{6\lambda}(3a^2 - r^2) & \text{for } 0 \le r \le a \\ \dfrac{Q_0}{6\lambda}\dfrac{2a^3}{r} & \text{for } r \ge a \end{cases} \tag{d}$$

Equation (d) gives an integral relation:

$$\int \tau r^2\, dr = \begin{cases} \dfrac{Q_0}{30\lambda} r^3 (5a^2 - r^2) & \text{for } 0 \le r \le a \\ \dfrac{Q_0}{6\lambda} a^3 r^2 & \text{for } r \ge a \end{cases} \tag{e}$$

Substitution of Eq. (e) into Eqs. (7.31) gives the radial displacement and stresses as

$$u_r = \begin{cases} \dfrac{1+\nu}{1-\nu}\alpha \dfrac{Q_0}{30\lambda} r(5a^2 - r^2) & \text{for } 0 \le r \le a \\ \dfrac{1+\nu}{1-\nu}\alpha \dfrac{Q_0}{30\lambda}\dfrac{a^3}{r^2}(5r^2 - a^2) & \text{for } r \ge a \end{cases}$$

$$\sigma_{rr} = \begin{cases} -\dfrac{\alpha E}{1-\nu}\dfrac{Q_0}{15\lambda}(5a^2 - r^2) & \text{for } 0 \le r \le a \\ -\dfrac{\alpha E}{1-\nu}\dfrac{Q_0}{15\lambda}\dfrac{a^3}{r^3}(5r^2 - a^2) & \text{for } r \ge a \end{cases} \tag{7.34}$$

$$\sigma_{\theta\theta} = \sigma_{\phi\phi} = \begin{cases} -\dfrac{\alpha E}{1-\nu}\dfrac{Q_0}{15\lambda}(5a^2 - 2r^2) & \text{for } 0 \le r \le a \\ -\dfrac{\alpha E}{1-\nu}\dfrac{Q_0}{30\lambda}\dfrac{a^3}{r^3}(5r^2 + a^2) & \text{for } r \ge a \end{cases}$$

It is found from Eqs. (7.34) that the expressions for the radial displacement and stresses are different for the inner region of internal heat generation and for the outer region in spite of the fact that the continuity conditions of the radial displacement u_r and stress σ_{rr} at the boundary surface $(r = a)$ of the internal heat generation are satisfied.

Example 7.7 When the temperature in Example 7.5 is in a transient state, find the radial displacement and thermal stresses in the infinite body with a spherical cavity.

Solution Referring to Eqs. (3.105), the heat conduction equation is given as

$$\frac{\partial T}{\partial t} = \kappa \left(\frac{\partial^2 T}{\partial r^2} + \frac{2}{r}\frac{\partial T}{\partial r} \right) \tag{a}$$

and the initial and boundary conditions are

$$T = T_i \qquad \text{at } t = 0 \tag{b}$$

$$\lambda \frac{\partial T}{\partial r} = h_a (T - T_a) \qquad \text{on } r = a \tag{c}$$

Introducing new variables

$$\Theta = rT, \qquad R = r - a \tag{d}$$

the fundamental equation system with the new variables may be obtained as

$$\frac{\partial \Theta}{\partial t} = \kappa \frac{\partial^2 \Theta}{\partial R^2} \tag{a$'$}$$

$$\Theta = T_i (R + a) \qquad \text{at } t = 0 \tag{b$'$}$$

$$\frac{\partial \Theta}{\partial R} - \left(\frac{h_a}{\lambda} + \frac{1}{R + a} \right) \Theta = -\frac{h_a}{\lambda} T_a (R + a) \qquad \text{on } R = 0 \tag{c$'$}$$

Making use of the Laplace transform, the solution $\overline{\Theta}$ in the transform domain is given as

$$\overline{\Theta} = \frac{h_a a}{\lambda}(T_a - T_i) \frac{e^{-R\sqrt{\frac{p}{\kappa}}}}{p\left(\sqrt{\frac{p}{\kappa}} + \frac{h_a}{\lambda} + \frac{1}{a} \right)} + \frac{1}{p} T_i (R + a) \tag{e}$$

in which p is the Laplace transform parameter. Inversion of the Laplace transform gives

$$\begin{aligned}\Theta = {} & \frac{h_a a}{\lambda}(T_a - T_i) \left(\frac{h_a}{\lambda} + \frac{1}{a} \right)^{-1} \\ & \times \left\{ \operatorname{erfc}\left(\frac{R}{2\sqrt{\kappa t}} \right) - e^{\left(\frac{h_a}{\lambda} + \frac{1}{a} \right) R + \kappa t \left(\frac{h_a}{\lambda} + \frac{1}{a} \right)^2} \right. \\ & \left. \times \operatorname{erfc}\left[\frac{R}{2\sqrt{\kappa t}} + \left(\frac{h_a}{\lambda} + \frac{1}{a} \right) \sqrt{\kappa t} \right] \right\} + T_i (R + a)\end{aligned} \tag{f}$$

in which erfc (x) means the complementary error function. Thus, the temperature change τ reduces to

$$\begin{aligned}\tau &= T - T_i \\ &= \frac{h_a a}{\lambda + h_a a} \frac{a}{r} (T_a - T_i)\end{aligned}$$

$$\times\left\{-e^{\left(\frac{h_a}{\lambda}+\frac{1}{a}\right)(r-a)+\kappa t\left(\frac{h_a}{\lambda}+\frac{1}{a}\right)^2}\right.$$

$$\left.\times\,\mathrm{erfc}\left[\frac{r-a}{2\sqrt{\kappa t}}+\left(\frac{h_a}{\lambda}+\frac{1}{a}\right)\sqrt{\kappa t}\right]+\mathrm{erfc}\left(\frac{r-a}{2\sqrt{\kappa t}}\right)\right\} \tag{7.35}$$

Substitution of Eq. (7.35) into Eqs. (7.30) gives the radial displacement and stresses as

$$\begin{aligned} u_r &= \frac{1+\nu}{1-\nu}\alpha\frac{1}{r^2}F(r,t) \\ \sigma_{rr} &= -\frac{2\alpha E}{1-\nu}\frac{1}{r^3}F(r,t) \\ \sigma_{\theta\theta} &= \sigma_{\phi\phi} = \frac{\alpha E}{1-\nu}\left[\frac{1}{r^3}F(r,t)-\tau(r,t)\right] \end{aligned} \tag{g}$$

in which

$$\begin{aligned} F(r,t) &= \int_a^r \tau r^2 dr \\ &= \frac{h_a a}{\lambda}(T_a-T_i)\left(\frac{h_a}{\lambda}+\frac{1}{a}\right)^{-1}\left(\frac{1}{2}(r^2-a^2)\,\mathrm{erfc}\left(\frac{r-a}{2\sqrt{\kappa t}}\right)\right. \\ &\quad+\kappa t\left[\mathrm{erf}\left(\frac{r-a}{2\sqrt{\kappa t}}\right)-\frac{2}{\sqrt{\pi}}\frac{r+a}{2\sqrt{\kappa t}}e^{-\frac{(r-a)^2}{4\kappa t}}\right] \\ &\quad+\frac{2}{\sqrt{\pi}}\left(\frac{h_a}{\lambda}+\frac{1}{a}\right)^{-1}\sqrt{\kappa t}\left[\frac{h_a a}{\lambda}-e^{-\frac{(r-a)^2}{4\kappa t}}\right] \\ &\quad-\frac{h_a a}{\lambda}\left(\frac{h_a}{\lambda}+\frac{1}{a}\right)^{-2}\mathrm{erf}\left(\frac{r-a}{2\sqrt{\kappa t}}\right)+\left(\frac{h_a}{\lambda}+\frac{1}{a}\right)^{-2}e^{\kappa t\left(\frac{h_a}{\lambda}+\frac{1}{a}\right)^2} \\ &\quad\times\left\{\frac{h_a a}{\lambda}\mathrm{erfc}\left[\left(\frac{h_a}{\lambda}+\frac{1}{a}\right)\sqrt{\kappa t}\right]-\left[\left(\frac{h_a}{\lambda}+\frac{1}{a}\right)r-1\right]e^{\left(\frac{h_a}{\lambda}+\frac{1}{a}\right)(r-a)}\right. \\ &\quad\left.\left.\times\,\mathrm{erfc}\left[\left(\frac{r-a}{2\sqrt{\kappa t}}\right)+\left(\frac{h_a}{\lambda}+\frac{1}{a}\right)\sqrt{\kappa t}\right]\right\}\right) \end{aligned} \tag{7.36}$$

7.3 Two-Dimensional Problems (Axisymmetric Problems)

We now consider two-dimensional axisymmetric problems of a spherical body. It is assumed that the body is deformed symmetrically with respect to the coordinate axis z. Making use of the spherical coordinate system (r, θ, ϕ), the force equilibrium equations in the directions of r and θ are obtained form Eq. (4.83)

$$\begin{aligned}
&\frac{\partial \sigma_{rr}}{\partial r} + \frac{1}{r}\frac{\partial \sigma_{\theta r}}{\partial \theta} + \frac{1}{r}(2\sigma_{rr} - \sigma_{\theta\theta} - \sigma_{\phi\phi} + \sigma_{\theta r}\cot\theta) + F_r = 0 \\
&\frac{\partial \sigma_{r\theta}}{\partial r} + \frac{1}{r}\frac{\partial \sigma_{\theta\theta}}{\partial \theta} + \frac{1}{r}[(\sigma_{\theta\theta} - \sigma_{\phi\phi})\cot\theta + 3\sigma_{r\theta}] + F_\theta = 0
\end{aligned} \tag{7.37}$$

The constitutive equations in the spherical coordinate system, or the generalized Hooke's law, are Eq. (4.90), which are given by:

$$\begin{aligned}
\epsilon_{rr} &= \frac{1}{E}[\sigma_{rr} - \nu(\sigma_{\theta\theta} + \sigma_{\phi\phi})] + \alpha\tau = \frac{1}{2G}\left(\sigma_{rr} - \frac{\nu}{1+\nu}\Theta\right) + \alpha\tau \\
\epsilon_{\theta\theta} &= \frac{1}{E}[\sigma_{\theta\theta} - \nu(\sigma_{\phi\phi} + \sigma_{rr})] + \alpha\tau = \frac{1}{2G}\left(\sigma_{\theta\theta} - \frac{\nu}{1+\nu}\Theta\right) + \alpha\tau \\
\epsilon_{\phi\phi} &= \frac{1}{E}[\sigma_{\phi\phi} - \nu(\sigma_{rr} + \sigma_{\theta\theta})] + \alpha\tau = \frac{1}{2G}\left(\sigma_{\phi\phi} - \frac{\nu}{1+\nu}\Theta\right) + \alpha\tau \\
\epsilon_{r\theta} &= \frac{1}{2G}\sigma_{r\theta}
\end{aligned} \tag{7.38}$$

where $\Theta = \sigma_{rr} + \sigma_{\theta\theta} + \sigma_{\phi\phi}$. Alternative forms are

$$\begin{aligned}
\sigma_{rr} &= 2\mu\epsilon_{rr} + \lambda e - \beta\tau \\
\sigma_{\theta\theta} &= 2\mu\epsilon_{\theta\theta} + \lambda e - \beta\tau \\
\sigma_{\phi\phi} &= 2\mu\epsilon_{\phi\phi} + \lambda e - \beta\tau \\
\sigma_{r\theta} &= 2\mu\epsilon_{r\theta}
\end{aligned} \tag{7.39}$$

where $e = \epsilon_{rr} + \epsilon_{\theta\theta} + \epsilon_{\phi\phi}$.

The components of strain in the spherical coordinate system are given by Eq. (4.85) for three-dimensional deformation. Therefore, the components of strain for an axisymmetric deformation are

$$\begin{aligned}
\epsilon_{rr} &= \frac{\partial u_r}{\partial r} \\
\epsilon_{\theta\theta} &= \frac{u_r}{r} + \frac{1}{r}\frac{\partial u_\theta}{\partial \theta} \\
\epsilon_{\phi\phi} &= \frac{u_r}{r} + \cot\theta\frac{u_\theta}{r} \\
\epsilon_{r\theta} &= \frac{1}{2}\left(\frac{1}{r}\frac{\partial u_r}{\partial \theta} + \frac{\partial u_\theta}{\partial r} - \frac{u_\theta}{r}\right) \\
e &= \frac{\partial u_r}{\partial r} + 2\frac{u_r}{r} + \cot\theta\frac{u_\theta}{r} + \frac{1}{r}\frac{\partial u_\theta}{\partial \theta}
\end{aligned} \tag{7.40}$$

Substituting Eqs. (7.39) and (7.40) into Eq. (7.37), Navier's equations of thermoelasticity for axisymmetric problems may be expressed as

$$
\begin{aligned}
(\lambda+2\mu)\frac{\partial e}{\partial r} - \frac{2\mu}{r\sin\theta}\frac{\partial(\omega_\phi \sin\theta)}{\partial\theta} - \beta\frac{\partial\tau}{\partial r} + F_r = 0 \\
(\lambda+2\mu)\frac{1}{r}\frac{\partial e}{\partial\theta} + \frac{2\mu}{r}\frac{\partial(r\omega_\phi)}{\partial r} - \beta\frac{1}{r}\frac{\partial\tau}{\partial\theta} + F_\theta = 0
\end{aligned}
\tag{7.41}
$$

where

$$
\omega_\phi = \frac{1}{2r}\left[\frac{\partial(ru_\theta)}{\partial r} - \frac{\partial u_r}{\partial\theta}\right] \tag{7.42}
$$

The solution of the Navier's equations (7.41) without the body force for axisymmetric problems in the spherical coordinate system can be expressed, for example, by the Goodier's thermoelastic displacement potential Φ and the Boussinesq harmonic functions φ and ψ

$$
\begin{aligned}
u_r &= \frac{\partial\Phi}{\partial r} + \frac{\partial\varphi}{\partial r} + r\cos\theta\frac{\partial\psi}{\partial r} - (3-4\nu)\,\psi\cos\theta \\
u_\theta &= \frac{1}{r}\frac{\partial\Phi}{\partial\theta} + \frac{1}{r}\frac{\partial\varphi}{\partial\theta} + \cos\theta\frac{\partial\psi}{\partial\theta} + (3-4\nu)\,\psi\sin\theta
\end{aligned}
\tag{7.43}
$$

where the three functions must satisfy the equations

$$
\nabla^2\Phi = K\tau, \qquad \nabla^2\varphi = 0, \qquad \nabla^2\psi = 0 \tag{7.44}
$$

in which

$$
\begin{aligned}
\nabla^2 &= \frac{\partial^2}{\partial r^2} + \frac{2}{r}\frac{\partial}{\partial r} + \frac{1}{r^2}\frac{\partial^2}{\partial\theta^2} + \frac{1}{r^2}\cot\theta\frac{\partial}{\partial\theta} \\
K &= \frac{1+\nu}{1-\nu}\alpha
\end{aligned}
\tag{7.45}
$$

Making use of the components of displacement defined by Eq. (7.43), the strain components are represented as

$$
\begin{aligned}
\epsilon_{rr} &= \frac{\partial^2\Phi}{\partial r^2} + \frac{\partial^2\varphi}{\partial r^2} + r\cos\theta\frac{\partial^2\psi}{\partial r^2} - 2(1-2\nu)\cos\theta\frac{\partial\psi}{\partial r} \\
\epsilon_{\theta\theta} &= \frac{1}{r}\frac{\partial\Phi}{\partial r} + \frac{1}{r^2}\frac{\partial^2\Phi}{\partial\theta^2} + \frac{1}{r}\frac{\partial\varphi}{\partial r} + \frac{1}{r^2}\frac{\partial^2\varphi}{\partial\theta^2} \\
&\quad + \cos\theta\frac{\partial\psi}{\partial r} + \frac{1}{r}\cos\theta\frac{\partial^2\psi}{\partial\theta^2} + 2(1-2\nu)\frac{1}{r}\sin\theta\frac{\partial\psi}{\partial\theta} \\
\epsilon_{\phi\phi} &= \frac{1}{r}\frac{\partial\Phi}{\partial r} + \cot\theta\frac{1}{r^2}\frac{\partial\Phi}{\partial\theta} + \frac{1}{r}\frac{\partial\varphi}{\partial r} + \cot\theta\frac{1}{r^2}\frac{\partial\varphi}{\partial\theta} + \cos\theta\frac{\partial\psi}{\partial r} + \frac{\cos^2\theta}{\sin\theta}\frac{1}{r}\frac{\partial\psi}{\partial\theta}
\end{aligned}
$$

$$
\begin{aligned}
\epsilon_{r\theta} &= \frac{\partial^2}{\partial r \partial \theta}\left(\frac{\Phi}{r}\right) + \frac{\partial^2}{\partial r \partial \theta}\left(\frac{\varphi}{r}\right) \\
&\quad + (1-2\nu)\sin\theta \frac{\partial \psi}{\partial r} + \cos\theta \frac{\partial^2 \psi}{\partial r \partial \theta} - 2(1-\nu)\frac{1}{r}\cos\theta \frac{\partial \psi}{\partial \theta} \\
e &= \nabla^2 \Phi + \nabla^2 \varphi + r\cos\theta \nabla^2 \psi - 2(1-2\nu)\left(\cos\theta \frac{\partial \psi}{\partial r} - \frac{1}{r}\sin\theta \frac{\partial \psi}{\partial \theta}\right) \\
&= K\tau - 2(1-2\nu)\left(\cos\theta \frac{\partial \psi}{\partial r} - \frac{1}{r}\sin\theta \frac{\partial \psi}{\partial \theta}\right)
\end{aligned} \tag{7.46}
$$

Substituting Eqs. (7.46) into Eqs. (7.39), the stress components in the spherical coordinate system are represented in terms of three functions Φ, φ, and ψ:

$$
\begin{aligned}
\sigma_{rr} &= 2G\left[\frac{\partial^2 \Phi}{\partial r^2} + \frac{\partial^2 \varphi}{\partial r^2} + r\cos\theta \frac{\partial^2 \psi}{\partial r^2} - 2(1-\nu)\cos\theta \frac{\partial \psi}{\partial r}\right. \\
&\quad \left. + 2\nu \frac{1}{r}\sin\theta \frac{\partial \psi}{\partial \theta} - K\tau\right] \\
\sigma_{\theta\theta} &= 2G\left[\frac{1}{r}\frac{\partial \Phi}{\partial r} + \frac{1}{r^2}\frac{\partial^2 \Phi}{\partial \theta^2} + \frac{1}{r}\frac{\partial \varphi}{\partial r} + \frac{1}{r^2}\frac{\partial^2 \varphi}{\partial \theta^2}\right. \\
&\quad \left. + (1-2\nu)\cos\theta \frac{\partial \psi}{\partial r} + \frac{1}{r}\cos\theta \frac{\partial^2 \psi}{\partial \theta^2} + 2(1-\nu)\frac{1}{r}\sin\theta \frac{\partial \psi}{\partial \theta} - K\tau\right] \\
\sigma_{\phi\phi} &= 2G\left[\frac{1}{r}\frac{\partial \Phi}{\partial r} + \cot\theta \frac{1}{r^2}\frac{\partial \Phi}{\partial \theta} + \frac{1}{r}\frac{\partial \varphi}{\partial r} + \cot\theta \frac{1}{r^2}\frac{\partial \varphi}{\partial \theta}\right. \\
&\quad \left. + (1-2\nu)\cos\theta \frac{\partial \psi}{\partial r} + (\cos\theta\cot\theta + 2\nu\sin\theta)\frac{1}{r}\frac{\partial \psi}{\partial \theta} - K\tau\right] \\
\sigma_{r\theta} &= 2G\left[\frac{\partial^2}{\partial r \partial \theta}\left(\frac{\Phi}{r}\right) + \frac{\partial^2}{\partial r \partial \theta}\left(\frac{\varphi}{r}\right)\right. \\
&\quad \left. + (1-2\nu)\sin\theta \frac{\partial \psi}{\partial r} + \cos\theta \frac{\partial^2 \psi}{\partial r \partial \theta} - 2(1-\nu)\frac{1}{r}\cos\theta \frac{\partial \psi}{\partial \theta}\right]
\end{aligned} \tag{7.47}
$$

We now consider the general solutions for displacement functions φ and ψ in Eqs. (7.44). It can be seen from Eqs. (7.44) that the fundamental equations for φ and ψ are given by the Laplace equation, namely

$$
\left(\frac{\partial^2}{\partial r^2} + \frac{2}{r}\frac{\partial}{\partial r} + \frac{1}{r^2}\frac{\partial^2}{\partial \theta^2} + \frac{1}{r^2}\cot\theta \frac{\partial}{\partial \theta}\right)\begin{pmatrix}\varphi \\ \psi\end{pmatrix} = 0 \tag{7.48}
$$

Introducing the variable transformation defined by

$$
\mu = \cos\theta \tag{7.49}
$$

the fundamental equation (7.48) reduces to

$$\left\{\frac{\partial^2}{\partial r^2}+\frac{2}{r}\frac{\partial}{\partial r}+\frac{1}{r^2}\frac{\partial}{\partial \mu}\left[(1-\mu^2)\frac{\partial}{\partial \mu}\right]\right\}\begin{pmatrix}\varphi\\ \psi\end{pmatrix}=0 \tag{7.50}$$

We use the method of separation of variables to solve Eqs. (7.50). If the displacement functions φ and ψ can be expressed by

$$\begin{pmatrix}\varphi\\ \psi\end{pmatrix}=F(r)G(\mu) \tag{7.51}$$

Eq. (7.50) reduces to

$$\frac{1}{G(\mu)}\frac{d}{d\mu}\left[(1-\mu^2)\frac{dG(\mu)}{d\mu}\right]=-\frac{r^2}{F(r)}\left[\frac{d^2F(r)}{dr^2}+\frac{2}{r}\frac{dF(r)}{dr}\right]\equiv-\nu(1+\nu) \tag{7.52}$$

where $-\nu(1+\nu)$ means arbitrary constant. Equation (7.52) reduces to two separate equations

$$\frac{d}{d\mu}\left[(1-\mu^2)\frac{dG(\mu)}{d\mu}\right]+\nu(1+\nu)G(\mu)=0 \tag{7.53}$$

$$\frac{d^2F(r)}{dr^2}+\frac{2}{r}\frac{dF(r)}{dr}-\frac{\nu(1+\nu)}{r^2}F(r)=0 \tag{7.54}$$

Equation (7.53) is called Legendre's differential equation. It has two particular solutions

$$P_\nu(\mu) \quad \text{and} \quad Q_\nu(\mu)$$

where $P_\nu(\mu)$ is the Legendre function of the first kind of order ν and $Q_\nu(\mu)$ is the Legendre's function of the second kind of order ν. For more information, see Appendix C. The fundamental solution of $G(\mu)$ is then obtained as

$$G(\mu)=\begin{pmatrix}P_\nu(\mu)\\ Q_\nu(\mu)\end{pmatrix} \tag{7.55}$$

On the other hand, Eq. (7.54) has two particular solutions

$$r^\nu \quad \text{and} \quad r^{-(1+\nu)}$$

The fundamental solution for $F(r)$ is obtained as

$$F(r)=\begin{pmatrix}r^\nu\\ r^{-(1+\nu)}\end{pmatrix} \tag{7.56}$$

From Eqs. (7.55) and (7.56), the fundamental solutions for φ and ψ are

$$\begin{pmatrix}\varphi\\ \psi\end{pmatrix}=F(r)G(\mu)=\begin{pmatrix}r^\nu\\ r^{-(1+\nu)}\end{pmatrix}\begin{pmatrix}P_\nu(\cos\theta)\\ Q_\nu(\cos\theta)\end{pmatrix} \tag{7.57}$$

where ν is an arbitrary constant $(-\infty \leq \nu \leq \infty)$.

7.4 Illustrative Examples of Axisymmetric Problems in Spherical Bodies

We now consider practical problems of axisymmetric thermoelastic deformation in the spherical coordinate system such as a solid sphere, a hollow sphere, and an infinite body with a spherical cavity. In these spherical bodies, the coordinate variable θ is defined within the interval $0 \leq \theta \leq \pi$. The Legendre functions $P_\nu(\cos\theta)$ and $Q_\nu(\cos\theta)$, which are given as the fundamental solution of the Boussinesq harmonic functions φ and ψ for the axisymmetric case in the spherical coordinate system, reveal singular values at $\cos\theta = -1$ $(\theta = \pi)$ when an arbitrary constant ν is positive or negative real number except for zero and positive or negative integers. Moreover, it is shown that the Legendre functions satisfy the following relations:

$$P_{-n-1}(\cos\theta) = P_n(\cos\theta), \quad Q_{-n-1}(\cos\theta) = Q_n(\cos\theta) \qquad (n = 0, \pm 1, \pm 2, \ldots) \tag{7.58}$$

The Legendre function of the second kind $Q_n(\cos\theta)$ has a singular value when $\cos\theta = 1$ $(\theta = 0)$. From these conditions, the fundamental solutions φ and ψ for the axisymmetric case in the spherical coordinate system, which are adaptable to practical problems of spherical bodies, can be expressed as

$$\begin{pmatrix} \varphi \\ \psi \end{pmatrix} = \begin{pmatrix} r^n \\ r^{-n-1} \end{pmatrix} P_n(\cos\theta) \qquad (n = 0, 1, 2, \ldots) \tag{7.59}$$

7.4.1 Fundamental Solution of Temperature Field

We now consider heat conduction problems in the spherical coordinate system. Referring to Eq. (3.103), the heat conduction equation in a transient state without internal heat generation for the axisymmetric problem is given as

$$\frac{\partial T}{\partial t} = \kappa \left[\frac{\partial^2 T}{\partial r^2} + \frac{2}{r}\frac{\partial T}{\partial r} + \frac{1}{r^2 \sin\theta}\frac{\partial}{\partial\theta}\left(\sin\theta \frac{\partial T}{\partial\theta}\right)\right] \tag{7.60}$$

Furthermore, Eq. (7.60) for the steady state reduces to

$$\frac{\partial^2 T}{\partial r^2} + \frac{2}{r}\frac{\partial T}{\partial r} + \frac{1}{r^2 \sin\theta}\frac{\partial}{\partial\theta}\left(\sin\theta \frac{\partial T}{\partial\theta}\right) = 0 \tag{7.61}$$

Referring to Eq. (7.59), the fundamental solution of Eq. (7.61) is

$$T = \begin{pmatrix} r^n \\ r^{-n-1} \end{pmatrix} P_n(\cos\theta) \qquad (n = 0, 1, 2, \ldots) \tag{7.62}$$

Let us consider the general solution of Eq. (7.60) by use of the Laplace transform. Performing the Laplace transform Eq. (7.60) with respect to the variable t with consideration of the initial temperature $T_i(r, \theta)$, the heat conduction equation in a transient state in the transform domain reduces to

$$\frac{\partial^2 \overline{T}}{\partial r^2} + \frac{2}{r}\frac{\partial \overline{T}}{\partial r} + \frac{1}{r^2 \sin\theta}\frac{\partial}{\partial \theta}\left(\sin\theta \frac{\partial \overline{T}}{\partial \theta}\right) - \frac{p}{\kappa}\overline{T} = -\frac{1}{\kappa}T_i(r,\theta) \tag{a}$$

where p is the Laplace transform parameter.

The general solution of Eq. (a) may be expressed as the sum of the complementary solution $\overline{T}_c$ and the particular solution $\overline{T}_p$:

$$\overline{T} = \overline{T}_c + \overline{T}_p \tag{b}$$

where the complementary solution $\overline{T}_c$ and the particular solution $\overline{T}_p$ are governed by a linear homogeneous differential equation and a linear nonhomogeneous differential equation, respectively

$$\begin{aligned} &\frac{\partial^2 \overline{T}_c}{\partial r^2} + \frac{2}{r}\frac{\partial \overline{T}_c}{\partial r} + \frac{1}{r^2 \sin\theta}\frac{\partial}{\partial \theta}\left(\sin\theta \frac{\partial \overline{T}_c}{\partial \theta}\right) - \frac{p}{\kappa}\overline{T}_c = 0 \\ &\frac{\partial^2 \overline{T}_p}{\partial r^2} + \frac{2}{r}\frac{\partial \overline{T}_p}{\partial r} + \frac{1}{r^2 \sin\theta}\frac{\partial}{\partial \theta}\left(\sin\theta \frac{\partial \overline{T}_p}{\partial \theta}\right) - \frac{p}{\kappa}\overline{T}_p = -\frac{1}{\kappa}T_i(r,\theta) \end{aligned} \tag{c}$$

We assume that the complementary solution $\overline{T}_c$ takes a form

$$\overline{T}_c = F(r)P_n(\cos\theta) \qquad (n = 0, 1, 2, \ldots) \tag{d}$$

Substitution of Eq. (d) into the first of Eqs. (c) gives a differential equation for function $F(r)$

$$\frac{d^2 F}{dr^2} + \frac{2}{r}\frac{dF}{dr} + \left[\omega^2 - \frac{n(1+n)}{r^2}\right]F = 0 \tag{7.63}$$

where

$$\omega^2 = -\frac{p}{\kappa} \tag{7.64}$$

Equation (7.63) is called the spherical Bessel's differential equation. It has two particular solutions

$$j_n(\omega r) \quad \text{and} \quad y_n(\omega r)$$

where j_n and y_n are the spherical Bessel functions of the first and second kind of order n, respectively. For more information, see Appendix B. Thus, the fundamental solution $\overline{T}_c$ on the subsidiary domain is given as

$$\overline{T}_c = \begin{pmatrix} j_n(\omega r) \\ y_n(\omega r) \end{pmatrix} P_n(\cos\theta) \qquad (n = 0, 1, 2, \ldots) \tag{7.65}$$

7.4.2 Components of Displacement and Stress in Terms of Displacement Functions φ and ψ

Making use of Eq. (7.59), the displacement functions φ and ψ are represented by series forms

$$\begin{aligned}\varphi &= \sum_{n=0}^{\infty}(C'_{1,n}r^n + C'_{2,n}r^{-n-1})P_n(\cos\theta)\\ \psi &= \sum_{n=0}^{\infty}(D'_{1,n}r^n + D'_{2,n}r^{-n-1})P_n(\cos\theta)\end{aligned} \tag{7.66}$$

Substituting Eq. (7.66) into Eqs. (7.43), the radial displacement $\bar{\bar{u}}_r$ in terms of φ and ψ is obtained

$$\begin{aligned}\bar{\bar{u}}_r &= \frac{\partial\varphi}{\partial r} + \mu\left[r\frac{\partial\psi}{\partial r} - (3-4\nu)\psi\right]\\ &= \sum_{n=0}^{\infty}\left[nC'_{1,n}r^{n-1} - (n+1)C'_{2,n}r^{-n-2}\right]P_n(\mu)\\ &\quad + \sum_{n=0}^{\infty}\left[(n-3+4\nu)D'_{1,n}r^n - (n+4-4\nu)D'_{2,n}r^{-n-1}\right]\\ &\quad \times \frac{1}{2n+1}[(n+1)P_{n+1}(\mu) + nP_{n-1}(\mu)]\end{aligned} \tag{a}$$

in which $\mu = \cos\theta$. However, it can be seen that the expression given by Eq. (a) is not suitable for solving practical boundary value problems because it contains three kinds of Legendre functions with different orders $n-1, n, n+1$ under the summation signs. To solve this problem, we introduce new unknown constants given by

$$\begin{aligned}C'_{1,n} &= C_{1,n} - (n-4+4\nu)D_{1,n-2}\\ C'_{2,n} &= C_{2,n} - (n+5-4\nu)D_{2,n+2}\\ D'_{1,n} &= (2n+1)D_{1,n-1}, \quad D'_{2,n} = (2n+1)D_{2,n+1}\end{aligned} \tag{7.67}$$

The displacement functions φ and ψ are then reduced to

$$\begin{aligned}\varphi &= \sum_{n=0}^{\infty}\Big\{[C_{1,n} - (n-4+4\nu)D_{1,n-2}]r^n\\ &\qquad + [C_{2,n} - (n+5-4\nu)D_{2,n+2}]r^{-n-1}\Big\}P_n(\cos\theta)\\ \psi &= \sum_{n=0}^{\infty}\left[(2n+1)D_{1,n-1}r^n + (2n+1)D_{2,n+1}r^{-n-1}\right]P_n(\cos\theta)\end{aligned} \tag{7.68}$$

Using new unknown constants (7.67), the radial displacement $\bar{\bar{u}}_r$ can be expressed as

$$\bar{\bar{u}}_r = \sum_{n=0}^{\infty} \Big[nC_{1,n} r^{n-1} - (n+1) C_{2,n} r^{-n-2} + (n+1)(n-2+4\nu) D_{1,n} r^{n+1} - n(n+3-4\nu) D_{2,n} r^{-n} \Big] P_n(\mu) \tag{7.69a}$$

In a similar way, the hoop displacement and the thermal stress components are

$$\begin{aligned} \bar{\bar{u}}_\theta &= -(1-\mu^2)^{1/2} \left[\frac{1}{r}\frac{\partial \varphi}{\partial \mu} + \mu \frac{\partial \psi}{\partial \mu} - (3-4\nu)\psi \right] \\ &= -(1-\mu^2)^{1/2} \sum_{n=1}^{\infty} \Big[C_{1,n} r^{n-1} + C_{2,n} r^{-n-2} + (n+5-4\nu) D_{1,n} r^{n+1} \\ &\quad + (n-4+4\nu) D_{2,n} r^{-n} \Big] \frac{n+1}{1-\mu^2} [\mu P_n(\mu) - P_{n+1}(\mu)] \end{aligned} \tag{7.69b}$$

$$\begin{aligned} \bar{\bar{\sigma}}_{rr} &= 2G \left[\frac{\partial^2 \varphi}{\partial r^2} + \mu r \frac{\partial^2 \psi}{\partial r^2} - 2(1-\nu)\mu \frac{\partial \psi}{\partial r} - 2\nu \frac{1}{r}(1-\mu^2)\frac{\partial \psi}{\partial \mu} \right] \\ &= 2G \sum_{n=0}^{\infty} \Big[n(n-1) C_{1,n} r^{n-2} + (n+1)(n+2) C_{2,n} r^{-n-3} \\ &\quad + (n+1)(n^2-n-2-2\nu) D_{1,n} r^n \\ &\quad + n(n^2+3n-2\nu) D_{2,n} r^{-n-1} \Big] P_n(\mu) \end{aligned} \tag{7.70a}$$

$$\begin{aligned} \bar{\bar{\sigma}}_{\theta\theta} &= 2G \left\{ \frac{1}{r}\frac{\partial \varphi}{\partial r} + \frac{1}{r^2}\left[-\mu \frac{\partial \varphi}{\partial \mu} + (1-\mu^2)\frac{\partial^2 \varphi}{\partial \mu^2} \right] + (1-2\nu)\mu \frac{\partial \psi}{\partial r} \right. \\ &\quad \left. + \frac{\mu}{r}\left[-\mu \frac{\partial \psi}{\partial \mu} + (1-\mu^2)\frac{\partial^2 \psi}{\partial \mu^2} \right] - 2(1-\nu)\frac{1}{r}(1-\mu^2)\frac{\partial \psi}{\partial \mu} \right\} \\ &= 2G \left\{ - \sum_{n=0}^{\infty} \Big[n^2 C_{1,n} r^{n-2} + (n+1)^2 C_{2,n} r^{-n-3} \right. \\ &\quad + (n+1)(n^2+4n+2+2\nu) D_{1,n} r^n \\ &\quad + n(n^2-2n-1+2\nu) D_{2,n} r^{-n-1} \Big] P_n(\mu) \\ &\quad + \sum_{n=1}^{\infty} \Big[C_{1,n} r^{n-2} + C_{2,n} r^{-n-3} + (n+5-4\nu) D_{1,n} r^n + (n-4+4\nu) \\ &\quad \left. \times D_{2,n} r^{-n-1} \Big] (n+1) \frac{\mu}{1-\mu^2} [\mu P_n(\mu) - P_{n+1}(\mu)] \right\} \end{aligned} \tag{7.70b}$$

$$\bar{\bar{\sigma}}_{\phi\phi} = 2G\left\{\frac{1}{r}\frac{\partial\varphi}{\partial r} - \mu\frac{1}{r^2}\frac{\partial\varphi}{\partial\mu} + (1-2\nu)\mu\frac{\partial\psi}{\partial r} - \left[2\nu + (1-2\nu)\mu^2\right]\frac{1}{r}\frac{\partial\psi}{\partial\mu}\right\}$$

$$= 2G\left\{\sum_{n=0}^{\infty}\left\{nC_{1,n}r^{n-2} - (n+1)C_{2,n}r^{-n-3}\right.\right.$$

$$+ (n+1)[n-2-2\nu(2n+1)]D_{1,n}r^n$$

$$\left. - n[n+3-2\nu(2n+1)]D_{2,n}r^{-n-1}\right\} P_n(\mu)$$

$$-\sum_{n=1}^{\infty}\left[C_{1,n}r^{n-2} + C_{2.n}r^{-n-3} + (n+5-4\nu)D_{1,n}r^n\right.$$

$$\left. + (n-4+4\nu)\ D_{2,n}r^{-n-1}\right]$$

$$\left.\times(n+1)\frac{\mu}{1-\mu^2}\left[\mu P_n(\mu) - P_{n+1}(\mu)\right]\right\} \tag{7.70c}$$

$$\bar{\bar{\sigma}}_{r\theta} = 2G(1-\mu^2)^{1/2}\left[-\frac{\partial^2}{\partial r\partial\mu}\left(\frac{\varphi}{r}\right) + (1-2\nu)\frac{\partial\psi}{\partial r}\right.$$

$$\left. - \mu\frac{\partial^2\psi}{\partial r\partial\mu} + 2(1-\nu)\frac{1}{r}\mu\frac{\partial\psi}{\partial\mu}\right]$$

$$= -2G(1-\mu^2)^{1/2}\sum_{n=1}^{\infty}\left[(n-1)C_{1,n}r^{n-2} - (n+2)C_{2,n}r^{-n-3}\right.$$

$$\left. + (n^2+2n-1+2\nu)D_{1,n}r^n - (n^2-2+2\nu)D_{2,n}r^{-n-1}\right]$$

$$\times\frac{n+1}{1-\mu^2}\left[\mu P_n(\mu) - P_{n+1}(\mu)\right] \tag{7.70d}$$

Example 7.8 When the boundary surface ($r = a$) of a solid sphere at initial temperature T_i is heated symmetrically with respect to the coordinate axis z by the surrounding medium at a temperature $T_a F(\theta)$ with heat transfer coefficient h_a, find the thermoelastic displacements and stresses in the sphere under a steady temperature field.

Solution The steady state heat conduction equation and associated boundary condition are

$$\frac{\partial^2 T}{\partial r^2} + \frac{2}{r}\frac{\partial T}{\partial r} + \frac{1}{r^2\sin\theta}\frac{\partial}{\partial\theta}\left(\sin\theta\frac{\partial T}{\partial\theta}\right) = 0 \tag{a}$$

$$\lambda\frac{\partial T}{\partial r} = h_a\left[T_a F(\theta) - T\right] \qquad \text{on } r = a \tag{b}$$

Referring to the fundamental solution of the temperature distribution given by Eq. (7.62), the temperature can be expressed by the use of suitable form for the solid sphere

$$T = \sum_{n=0}^{\infty} A_n r^n P_n(\mu) \tag{7.71}$$

in which $\mu = \cos\theta$ $(-1 \leq \mu \leq 1)$ and $0 \leq \theta \leq \pi$. We expand the arbitrary temperature function $F(\theta)$ of the surrounding medium into a series of Legendre polynomials $P_n(\mu)$, see Appendix C

$$F(\theta) = \sum_{n=0}^{\infty} f_n P_n(\mu) \tag{c}$$

Multiplying both sides of Eq. (c) by $P_m(\mu)$, integrating from -1 to 1, and using the integration (see Eqs. (C.20) in Appendix C)

$$\begin{aligned} &\int_{-1}^{1} P_n(\mu) P_m(\mu)\, d\mu = 0 && m \neq n \\ &\int_{-1}^{1} P_n(\mu) P_m(\mu)\, d\mu = \frac{2}{2n+1} && m = n \end{aligned} \tag{d}$$

we obtain the coefficients f_n

$$f_n = \frac{2n+1}{2} \int_{-1}^{1} F(\theta) P_n(\mu) d\mu \tag{e}$$

Substituting Eqs. (7.71) and (c) into Eq. (b), the unknown constants A_n can be determined as

$$A_n = \frac{h_a T_a f_n}{n\lambda a^{n-1} + h_a a^n} \tag{7.72}$$

Therefore, the steady state temperature distribution is

$$T = T_a h_a \sum_{n=0}^{\infty} \frac{f_n}{n\lambda a^{n-1} + h_a a^n} r^n P_n(\mu) \tag{7.73}$$

As an illustrative example, we assume that the temperature function $F(\theta)$ of the surrounding medium is given by the partially distributed step-wise function

$$F(\theta) = H(\theta_0 - \theta) \tag{f}$$

in which H is the Heaviside step function. Using the relationships

$$P_0(\mu) = 1, \qquad \int P_n(\mu)\, d\mu = -\frac{1}{n}[\mu P_n(\mu) - P_{n+1}(\mu)] \tag{g}$$

the substitution of Eq. (f) into Eq. (e) gives the coefficients f_n

$$f_n = \begin{cases} \dfrac{2n+1}{2n}[\mu_0 P_n(\mu_0) - P_{n+1}(\mu_0)] & (n = 1, 2, \ldots) \\ \dfrac{1}{2}(1-\mu_0) & (n = 0) \end{cases} \tag{h}$$

in which $\mu_0 = \cos\theta_0$. Making use of Eq. (7.71), the temperature change τ is

$$\tau = T - T_i = -T_i + \sum_{n=0}^{\infty} A_n r^n P_n(\mu) \tag{7.74}$$

where A_n is given by Eq. (7.72). Then, the Goodier's thermoelastic displacement potential Φ satisfying the first of Eqs. (7.44) is given as

$$\Phi = K\left[-\frac{1}{6}T_i r^2 + \sum_{n=0}^{\infty}\frac{1}{2(2n+3)}A_n r^{n+2} P_n(\mu)\right] \tag{i}$$

Making use of Eqs. (7.43) and (7.47), the displacement and stress components corresponding to Φ and the temperature change τ are

$$\begin{aligned} \bar{u}_r &= K\left[-\frac{1}{3}T_i r + \sum_{n=0}^{\infty}\frac{n+2}{2(2n+3)}A_n r^{n+1} P_n(\mu)\right] \\ \bar{u}_\theta &= -K(1-\mu^2)^{1/2}\sum_{n=1}^{\infty}\frac{1}{2(2n+3)}A_n r^{n+1}\frac{n+1}{1-\mu^2}[\mu P_n(\mu) - P_{n+1}(\mu)] \end{aligned} \tag{7.75}$$

$$\begin{aligned} \bar{\sigma}_{rr} &= 2GK\left[\frac{2}{3}T_i + \sum_{n=0}^{\infty}\frac{n^2-n-4}{2(2n+3)}A_n r^n P_n(\mu)\right]. \\ \bar{\sigma}_{\theta\theta} &= 2GK\left(\frac{2}{3}T_i + \sum_{n=0}^{\infty}\frac{1}{2(2n+3)}A_n r^n\right. \\ &\quad \left.\times\left\{(n+1)\frac{\mu}{1-\mu^2}[\mu P_n(\mu) - P_{n+1}(\mu)] - (n+2)^2 P_n(\mu)\right\}\right) \\ \bar{\sigma}_{\phi\phi} &= 2GK\left(\frac{2}{3}T_i + \sum_{n=0}^{\infty}\frac{1}{2(2n+3)}A_n r^n\right. \\ &\quad \left.\times\left\{-(n+1)\frac{\mu}{1-\mu^2}[\mu P_n(\mu) - P_{n+1}(\mu)] - (3n+4)P_n(\mu)\right\}\right) \\ \bar{\sigma}_{r\theta} &= -2GK(1-\mu^2)^{1/2}\sum_{n=1}^{\infty}\frac{n+1}{2(2n+3)}A_n r^n\frac{n+1}{1-\mu^2}[\mu P_n(\mu) - P_{n+1}(\mu)] \end{aligned} \tag{7.76}$$

The bars indicate that these components correspond to Φ and temperature change τ. The displacements and stresses due to the displacement functions φ and ψ given by Eqs. (7.68) can be expressed as for the problems of the solid sphere

$$\bar{\bar{u}}_r = \sum_{n=0}^{\infty}\left[nC_{1,n}r^{n-1} + (n+1)(n-2+4\nu)D_{1,n}r^{n+1}\right]P_n(\mu)$$

$$\bar{\bar{u}}_\theta = -(1-\mu^2)^{1/2}\sum_{n=1}^{\infty}\left[C_{1,n}r^{n-1} + (n+5-4\nu)D_{1,n}r^{n+1}\right]\frac{n+1}{1-\mu^2}[\mu P_n(\mu) - P_{n+1}(\mu)] \tag{7.77}$$

$$\bar{\bar{\sigma}}_{rr} = 2G\sum_{n=0}^{\infty}\left[n(n-1)C_{1,n}r^{n-2} + (n+1)(n^2-n-2-2\nu)D_{1,n}r^n\right]P_n(\mu)$$

$$\bar{\bar{\sigma}}_{\theta\theta} = 2G\left\{-\sum_{n=0}^{\infty}\left[n^2C_{1,n}r^{n-2} + (n+1)(n^2+4n+2+2\nu)D_{1,n}r^n\right]P_n(\mu)\right.$$
$$+\sum_{n=1}^{\infty}\left[C_{1,n}r^{n-2} + (n+5-4\nu)D_{1,n}r^n\right]$$
$$\left.\times(n+1)\frac{\mu}{1-\mu^2}[\mu P_n(\mu) - P_{n+1}(\mu)]\right\}$$

$$\bar{\bar{\sigma}}_{\phi\phi} = 2G\left\{\sum_{n=0}^{\infty}\left\{nC_{1,n}r^{n-2} + (n+1)[n-2-2\nu(2n+1)]D_{1,n}r^n\right\}P_n(\mu)\right.$$
$$-\sum_{n=1}^{\infty}\left[C_{1,n}r^{n-2} + (n+5-4\nu)D_{1,n}r^n\right]$$
$$\left.\times(n+1)\frac{\mu}{1-\mu^2}[\mu P_n(\mu) - P_{n+1}(\mu)]\right\}$$

$$\bar{\bar{\sigma}}_{r\theta} = -2G(1-\mu^2)^{1/2}\sum_{n=1}^{\infty}\left[(n-1)C_{1,n}r^{n-2} + (n^2+2n-1+2\nu)D_{1,n}r^n\right]$$
$$\times\frac{n+1}{1-\mu^2}[\mu P_n(\mu) - P_{n+1}(\mu)] \tag{7.78}$$

The displacements are obtained by adding Eqs. (7.75) and (7.77), and stresses are obtained by adding Eqs. (7.76) and (7.78):

$$u_r = \bar{u}_r + \bar{\bar{u}}_r, \qquad u_\theta = \bar{u}_\theta + \bar{\bar{u}}_\theta$$
$$\sigma_{rr} = \bar{\sigma}_{rr} + \bar{\bar{\sigma}}_{rr}, \qquad \sigma_{\theta\theta} = \bar{\sigma}_{\theta\theta} + \bar{\bar{\sigma}}_{\theta\theta} \tag{j}$$
$$\sigma_{\phi\phi} = \bar{\sigma}_{\phi\phi} + \bar{\bar{\sigma}}_{\phi\phi}, \qquad \sigma_{r\theta} = \bar{\sigma}_{r\theta} + \bar{\bar{\sigma}}_{r\theta}$$

The mechanical boundary conditions on the traction free surface $r = a$ are

$$\sigma_{rr} = 0, \quad \sigma_{r\theta} = 0 \qquad \text{on } r = a \tag{7.79}$$

The unknown constants $C_{1,n}$ and $D_{1,n}$ are then obtained as

$$\begin{gathered}
C_{1,n} = K\frac{1-\nu}{\Delta_n}A_n a^2 \qquad (n=2,3,\ldots) \\
D_{1,0} = -\frac{K}{3(1+\nu)}(A_0 - T_i), \quad D_{1,n} = -K\frac{n+2}{(2n+3)\Delta_n}A_n \qquad (n=1,2,\ldots) \\
\Delta_n = 2\left[(n^2+n+1)+\nu(2n+1)\right]
\end{gathered} \tag{7.80}$$

in which the constants $C_{1,0}$ and $C_{1,1}$ are omitted here because $C_{1,0}$ is not included in the expressions for displacements and stresses, and $C_{1,1}$ corresponds to a rigid displacement in the axial direction z.

Example 7.9 When both the inner and the outer surfaces with radii a and b, respectively, of a hollow sphere at initial temperature T_i are heated symmetrically with respect to the coordinate axis z by the surrounding medium at temperatures $T_a F_a(\theta)$ and $T_b F_b(\theta)$, find the thermal displacements and thermal stresses due to a steady temperature field.

Solution Let the heat transfer coefficients on the inner and outer surfaces be h_a and h_b, respectively. The steady state heat conduction equation and associated boundary conditions are

$$\frac{\partial^2 T}{\partial r^2} + \frac{2}{r}\frac{\partial T}{\partial r} + \frac{1}{r^2 \sin\theta}\frac{\partial}{\partial\theta}\left(\sin\theta\frac{\partial T}{\partial\theta}\right) = 0 \tag{a}$$

$$\begin{aligned}
-\lambda\frac{\partial T}{\partial r} &= h_a[T_a F_a(\theta) - T] \qquad \text{on } r = a \\
\lambda\frac{\partial T}{\partial r} &= h_b[T_b F_b(\theta) - T] \qquad \text{on } r = b
\end{aligned} \tag{b}$$

Referring to Eq. (7.62), the temperature solution suitable for the hollow sphere is

$$T = \sum_{n=0}^{\infty}\left(A_n r^n + B_n r^{-n-1}\right)P_n(\mu) \tag{7.81}$$

We now expand the arbitrary temperature functions $F_a(\theta)$ and $F_b(\theta)$ into the series of Legendre polynomials

$$\begin{aligned}
\begin{pmatrix} F_a(\theta) \\ F_b(\theta) \end{pmatrix} &= \sum_{n=0}^{\infty}\begin{pmatrix} f_n \\ g_n \end{pmatrix}P_n(\mu) \\
\begin{pmatrix} f_n \\ g_n \end{pmatrix} &= \frac{2n+1}{2}\int_{-1}^{1}\begin{pmatrix} F_a(\theta) \\ F_b(\theta) \end{pmatrix}P_n(\mu)d\mu
\end{aligned} \tag{c}$$

Substituting Eqs. (7.81) and (c) into the boundary conditions (b), the unknown constants A_n and B_n are

$$
\begin{aligned}
A_n &= \frac{1}{\Delta_n}\left\{h_a T_a f_n\left[\lambda(n+1) - h_b b\right] b^{-n-2}\right.\\
&\qquad \left. + h_b T_b g_n\left[\lambda(n+1) + h_a a\right] a^{-n-2}\right\}\\
B_n &= \frac{1}{\Delta_n}\left[h_a T_a f_n (n\lambda + h_b b) b^{n-1} + h_b T_b g_n (n\lambda - h_a a) a^{n-1}\right]\\
\Delta_n &= (n\lambda + h_b b)\left[\lambda(n+1) + h_a a\right] a^{-n-2} b^{n-1}\\
&\qquad - (n\lambda - h_a a)\left[\lambda(n+1) - h_b b\right] a^{n-1} b^{-n-2}
\end{aligned}
\tag{7.82}
$$

Therefore, the temperature change τ is

$$
\begin{aligned}
\tau &= T - T_i\\
&= -T_i + \sum_{n=0}^{\infty}\left(A_n r^n + B_n r^{-n-1}\right) P_n(\mu)
\end{aligned}
\tag{d}
$$

Next, the Goodier's thermoelastic displacement potential Φ satisfying the first of Eqs. (7.44) is

$$
\Phi = K\left\{-\frac{1}{6}T_i r^2 + \sum_{n=0}^{\infty}\left[\frac{1}{2(2n+3)}A_n r^{n+2} - \frac{1}{2(2n-1)}B_n r^{-n+1}\right]P_n(\mu)\right\}
\tag{7.83}
$$

Substituting Eqs. (d) and (7.83) into Eqs. (7.43) and (7.47) and using Eqs. (7.69) and (7.70), the displacements and stresses corresponding to the Goodier's thermoelastic potential functions Φ and the displacement functions φ and ψ are

$$
\begin{aligned}
u_r &= K\left\{-\frac{1}{3}T_i r + \sum_{n=0}^{\infty}\left[\frac{n+2}{2(2n+3)}A_n r^{n+1} + \frac{n-1}{2(2n-1)}B_n r^{-n}\right]P_n(\mu)\right\}\\
&\quad + \sum_{n=0}^{\infty}\left[nC_{1,n} r^{n-1} - (n+1)C_{2,n} r^{-n-2} + (n+1)(n-2+4\nu)D_{1,n} r^{n+1}\right.\\
&\qquad \left. - n(n+3-4\nu)D_{2,n} r^{-n}\right]P_n(\mu)\\
u_\theta &= -K(1-\mu^2)^{1/2}\sum_{n=1}^{\infty}\left[\frac{1}{2(2n+3)}A_n r^{n+1} - \frac{1}{2(2n-1)}B_n r^{-n}\right]\\
&\quad \times \frac{n+1}{1-\mu^2}\left[\mu P_n(\mu) - P_{n+1}(\mu)\right]
\end{aligned}
$$

$$-(1-\mu^2)^{1/2}\sum_{n=1}^{\infty}\Big[C_{1,n}r^{n-1}+C_{2,n}r^{-n-2}+(n+5-4\nu)D_{1,n}r^{n+1}$$

$$+(n-4+4\nu)D_{2,n}r^{-n}\Big]\frac{n+1}{1-\mu^2}[\mu P_n(\mu)-P_{n+1}(\mu)] \tag{7.84}$$

$$\sigma_{rr}=2GK\left\{\frac{2}{3}T_i+\sum_{n=0}^{\infty}\left[\frac{n^2-n-4}{2(2n+3)}A_nr^n-\frac{n^2+3n-2}{2(2n-1)}B_nr^{-n-1}\right]P_n(\mu)\right\}$$

$$+2G\sum_{n=0}^{\infty}\Big[n(n-1)C_{1,n}r^{n-2}+(n+1)(n+2)C_{2,n}r^{-n-3}$$

$$+(n+1)(n^2-n-2-2\nu)D_{1,n}r^n+n(n^2+3n-2\nu)D_{2,n}r^{-n-1}\Big]P_n(\mu)$$

$$\sigma_{\theta\theta}=2GK\Bigg[\frac{2}{3}T_i+\sum_{n=0}^{\infty}\Bigg(\frac{1}{2(2n+3)}A_nr^n$$

$$\times\left\{(n+1)\frac{\mu}{1-\mu^2}[\mu P_n(\mu)-P_{n+1}(\mu)]-(n+1)^2P_n(\mu)\right\}$$

$$+\frac{1}{2(2n-1)}B_nr^{-n-1}$$

$$\times\left\{-(n+1)\frac{\mu}{1-\mu^2}[\mu P_n(\mu)-P_{n+1}(\mu)]+(n-1)^2P_n(\mu)\right\}\Bigg)\Bigg]$$

$$+2G\Bigg\{-\sum_{n=0}^{\infty}\Big[n^2C_{1,n}r^{n-2}+(n+1)^2C_{2,n}r^{-n-3}+(n+1)$$

$$\times(n^2+4n+2+2\nu)D_{1,n}r^n+n(n^2-2n-1+2\nu)D_{2,n}r^{-n-1}\Big]P_n(\mu)$$

$$+\sum_{n=1}^{\infty}\Big[C_{1,n}r^{n-2}+C_{2,n}r^{-n-3}+(n+5-4\nu)D_{1,n}r^n$$

$$+(n-4+4\nu)D_{2,n}r^{-n-1}\Big](n+1)\frac{\mu}{1-\mu^2}[\mu P_n(\mu)-P_{n+1}(\mu)]\Bigg\}$$

$$\sigma_{\phi\phi}=2GK\Bigg[\frac{2}{3}T_i+\sum_{n=0}^{\infty}\Bigg(\frac{1}{2(2n+3)}A_nr^n$$

$$\times\left\{-(n+1)\frac{\mu}{1-\mu^2}[\mu P_n(\mu)-P_{n+1}(\mu)]-(3n+4)P_n(\mu)\right\}$$

$$+\frac{1}{2(2n-1)}B_n r^{-n-1}$$

$$\times\left\{(n+1)\frac{\mu}{1-\mu^2}[\mu P_n(\mu)-P_{n+1}(\mu)]-(3n-1)P_n(\mu)\right\}\Bigg)\Bigg]$$

$$+2G\left\{\sum_{n=0}^{\infty}\Big[nC_{1,n}r^{n-2}-(n+1)C_{2,n}r^{-n-3}+(n+1)[n-2-2\nu(2n+1)]D_{1,n}r^n\right.$$

$$-n[n+3-2\nu(2n+1)]D_{2,n}r^{-n-1}\Big]P_n(\mu)$$

$$-\sum_{n=1}^{\infty}\Big[C_{1,n}r^{n-2}+C_{2,n}r^{-n-3}+(n+5-4\nu)D_{1,n}r^n$$

$$\left.+(n-4+4\nu)D_{2,n}r^{-n-1}\Big](n+1)\frac{\mu}{1-\mu^2}[\mu P_n(\mu)-P_{n+1}(\mu)]\right\}$$

$$\sigma_{r\theta}=-2GK(1-\mu^2)^{1/2}\sum_{n=1}^{\infty}\left[\frac{n+1}{2(2n+3)}A_n r^n+\frac{n}{2(2n-1)}B_n r^{-n-1}\right]$$

$$\times\frac{n+1}{1-\mu^2}[\mu P_n(\mu)-P_{n+1}(\mu)]$$

$$-2G(1-\mu^2)^{1/2}\sum_{n=1}^{\infty}\Big[(n-1)C_{1,n}r^{n-2}-(n+2)C_{2,n}r^{-n-3}$$

$$+(n^2+2n-1+2\nu)D_{1,n}r^n-(n^2-2+2\nu)D_{2,n}r^{-n-1}\Big]$$

$$\times\frac{n+1}{1-\mu^2}[\mu P_n(\mu)-P_{n+1}(\mu)] \tag{7.85}$$

When the mechanical boundary conditions on the traction free surfaces $r = a$ and $r = b$ are

$$\sigma_{rr}=0,\quad \sigma_{r\theta}=0 \qquad \text{on } r=a \text{ and } r=b \tag{7.86}$$

the unknown constants $C_{i,n}$ and $D_{i,n}$ $(i = 1, 2)$ are determined both for $n = 0$ and for $n = 1$,

$$C_{2,0}=\frac{1}{2}KB_0\frac{a^2b^2}{a^2+ab+b^2},\qquad D_{1,0}=-\frac{1}{2}\frac{K}{1+\nu}\left[\frac{2}{3}(A_0-T_i)+\frac{b+a}{b^2+ab+a^2}B_0\right]$$

$$C_{2,1}=\frac{1}{6}KB_1\frac{a^2b^2(b^3-a^3)}{b^5-a^5},\qquad D_{1,1}=-\frac{1}{2}\frac{K}{1+\nu}\left[\frac{1}{5}A_1+\frac{1}{2}\frac{b^2-a^2}{b^5-a^5}B_1\right] \tag{7.87}$$

$$D_{2,1}=0$$

in which the constants $C_{1,0}$, $D_{2,0}$ and $C_{1,1}$ are omitted here because $C_{1,0}$ and $D_{2,0}$ are not included in the expressions for displacements and stresses, and $C_{1,1}$ corresponds to a rigid displacement in the axial direction z. Furthermore, the unknown constants for $n \geq 2$ can be determined as a solution of the following simultaneous equations with 4 unknowns from 4 boundary conditions (7.86):

$$
\begin{aligned}
&n(n-1)C_{1,n}r^{n-2}+(n+1)(n+2)C_{2,n}r^{-n-3} \\
&\quad +(n+1)(n^2-n-2-2\nu)D_{1,n}r^n+n(n^2+3n-2\nu)D_{2,n}r^{-n-1} \\
&\quad =-K\left[\frac{n^2-n-4}{2(2n+3)}A_n r^n-\frac{n^2+3n-2}{2(2n-1)}B_n r^{-n-1}\right] \qquad \text{for } r=a \text{ and } r=b \\
&(n-1)C_{1,n}r^{n-2}-(n+2)C_{2,n}r^{-n-3} \\
&\quad +(n^2+2n-1+2\nu)D_{1,n}r^n-(n^2-2+2\nu)D_{2,n}r^{-n-1} \\
&\quad =-K\left[\frac{n+1}{2(2n+3)}A_n r^n+\frac{n}{2(2n-1)}B_n r^{-n-1}\right] \qquad \text{for } r=a \text{ and } r=b
\end{aligned}
\tag{7.88}
$$

Example 7.10 Let us consider an infinite body with a spherical cavity of radius a at initial temperature T_i. When the boundary surface of the spherical cavity is heated symmetrically with respect to the coordinate axis z by the surrounding medium at a temperature $T_a F(\theta)$ with heat transfer coefficient h_a, find the thermoelastic displacements and stresses in the infinite body under a steady temperature field.

Solution The steady state heat conduction equation and associated boundary condition are

$$
\frac{\partial^2 T}{\partial r^2}+\frac{2}{r}\frac{\partial T}{\partial r}+\frac{1}{r^2\sin\theta}\frac{\partial}{\partial\theta}\left(\sin\theta\frac{\partial T}{\partial\theta}\right)=0 \tag{a}
$$

$$
-\lambda\frac{\partial T}{\partial r}=h_a[T_a F(\theta)-T] \qquad \text{on } r=a \tag{b}
$$

Referring to Eq. (7.62), the temperature solution suitable for an infinite body with a spherical cavity is

$$
T=\sum_{n=0}^{\infty} B_n r^{-n-1} P_n(\mu) \tag{7.89}
$$

We now expand the arbitrary temperature function $F(\theta)$ into the series of Legendre polynomials

$$
\begin{aligned}
F(\theta) &= \sum_{n=0}^{\infty} f_n P_n(\mu) \\
f_n &= \frac{2n+1}{2}\int_{-1}^{1} F(\theta)P_n(\mu)\,d\mu
\end{aligned}
\tag{c}
$$

Substituting Eqs. (7.89) and (c) into Eq. (b), the unknown coefficients B_n are obtained

$$B_n = \frac{h_a T_a f_n}{\lambda(n+1)a^{-n-2} + h_a a^{-n-1}} \tag{7.90}$$

The temperature distribution in the steady state becomes

$$T = h_a T_a \sum_{n=0}^{\infty} \frac{f_n}{\lambda(n+1)a^{-n-2} + h_a a^{-n-1}} r^{-n-1} P_n(\mu) \tag{7.91}$$

Making use of Eq. (7.89), the temperature change τ is

$$\tau = T - T_i = -T_i + \sum_{n=0}^{\infty} B_n r^{-n-1} P_n(\mu) \tag{d}$$

and the Goodier's thermoelastic displacement potential Φ satisfying the first relation of Eqs. (7.44) is

$$\Phi = -K\left[\frac{1}{6}T_i r^2 + \sum_{n=0}^{\infty} \frac{1}{2(2n-1)} B_n r^{-n+1} P_n(\mu)\right] \tag{7.92}$$

Substituting the Goodier's thermoelastic displacement potential functions (7.92) and the displacement functions φ and ψ given by Eq. (7.68) into displacements Eqs. (7.43) and stresses (7.47), we have

$$\begin{aligned}
u_r &= K\left[-\frac{1}{3}T_i r + \sum_{n=0}^{\infty} \frac{n-1}{2(2n-1)} B_n r^{-n} P_n(\mu)\right] \\
&\quad - \sum_{n=0}^{\infty} \left[(n+1)C_{2,n} r^{-n-2} + n(n+3-4\nu)D_{2,n} r^{-n}\right] P_n(\mu) \\
u_\theta &= K(1-\mu^2)^{1/2} \sum_{n=1}^{\infty} \frac{1}{2(2n-1)} B_n r^{-n} \frac{n+1}{1-\mu^2}[\mu P_n(\mu) - P_{n+1}(\mu)] \\
&\quad - (1-\mu^2)^{1/2} \sum_{n=1}^{\infty} \left[C_{2,n} r^{-n-2} + (n-4+4\nu)D_{2,n} r^{-n}\right] \\
&\quad \times \frac{n+1}{1-\mu^2}[\mu P_n(\mu) - P_{n+1}(\mu)]
\end{aligned} \tag{7.93}$$

$$\begin{aligned}
\sigma_{rr} &= 2GK\left[\frac{2}{3}T_i - \sum_{n=0}^{\infty} \frac{n^2+3n-2}{2(2n-1)} B_n r^{-n-1} P_n(\mu)\right] \\
&\quad + 2G \sum_{n=0}^{\infty} \left[(n+1)(n+2)C_{2,n} r^{-n-3} + n(n^2+3n-2\nu)D_{2,n} r^{-n-1}\right] P_n(\mu)
\end{aligned}$$

$$\sigma_{\theta\theta} = 2GK\left(\frac{2}{3}T_i + \sum_{n=0}^{\infty}\frac{1}{2(2n-1)}B_n r^{-n-1}\right.$$

$$\left.\times\left\{-(n+1)\frac{\mu}{1-\mu^2}\left[\mu P_n(\mu) - P_{n+1}(\mu)\right] + (n-1)^2 P_n(\mu)\right\}\right)$$

$$+ 2G\left\{-\sum_{n=0}^{\infty}\left[(n+1)^2 C_{2,n} r^{-n-3} + n(n^2 - 2n - 1 + 2\nu)D_{2,n} r^{-n-1}\right]P_n(\mu)\right.$$

$$+\sum_{n=1}^{\infty}\left[C_{2,n} r^{-n-3} + (n - 4 + 4\nu)D_{2,n} r^{-n-1}\right]$$

$$\left.\times (n+1)\frac{\mu}{1-\mu^2}\left[\mu P_n(\mu) - P_{n+1}(\mu)\right]\right\}$$

$$\sigma_{\phi\phi} = 2GK\left(\frac{2}{3}T_i + \sum_{n=0}^{\infty}\frac{1}{2(2n-1)}B_n r^{-n-1}\right.$$

$$\left.\times\left\{(n+1)\frac{\mu}{1-\mu^2}\left[\mu P_n(\mu) - P_{n+1}(\mu)\right] - (3n-1)P_n(\mu)\right\}\right)$$

$$+ 2G\left(-\sum_{n=0}^{\infty}\left\{(n+1)C_{2,n} r^{-n-3} + n\left[n + 3 - 2\nu(2n+1)\right]D_{2,n} r^{-n-1}\right\}P_n(\mu)\right.$$

$$-\sum_{n=1}^{\infty}\left[C_{2,n} r^{-n-3} + (n - 4 + 4\nu)D_{2,n} r^{-n-1}\right]$$

$$\left.\times (n+1)\frac{\mu}{1-\mu^2}\left[\mu P_n(\mu) - P_{n+1}(\mu)\right]\right)$$

$$\sigma_{r\theta} = -2GK(1-\mu^2)^{1/2}\sum_{n=1}^{\infty}\frac{n}{2(2n-1)}B_n r^{-n-1}$$

$$\times \frac{n+1}{1-\mu^2}\left[\mu P_n(\mu) - P_{n+1}(\mu)\right]$$

$$+ 2G(1-\mu^2)^{1/2}\sum_{n=1}^{\infty}\left[(n+2)C_{2,n} r^{-n-3} + (n^2 - 2 + 2\nu)D_{2,n} r^{-n-1}\right]$$

$$\times \frac{n+1}{1-\mu^2}\left[\mu P_n(\mu) - P_{n+1}(\mu)\right] \qquad (7.94)$$

The mechanical boundary conditions on the traction free surface $r = a$ are

$$\sigma_{rr} = 0, \quad \sigma_{r\theta} = 0 \qquad \text{on } r = a \tag{e}$$

Then the unknown constants $C_{2,n}$ and $D_{2,n}$ $(n \geq 0)$ are determined as

$$\begin{aligned}
C_{2,0} &= -\frac{1}{2}Ka^3\left[\frac{2}{3}T_i - B_0a^{-1}\right] \\
C_{2,n} &= KB_na^2\frac{1-\nu}{\Delta_n} \\
D_{2,n} &= KB_n\frac{n-1}{2n-1}\frac{1}{\Delta_n} \qquad (n = 1, 2, \ldots) \\
\Delta_n &= 2\left[n^2 + n + 1 - \nu(2n+1)\right]
\end{aligned} \tag{7.95}$$

in which the coefficient $D_{2,0}$ is omitted here because $D_{2,0}$ is not included in the expressions for displacements and stresses.

Example 7.11 In Example 7.8, find the thermoelastic displacements and stresses in a solid sphere under a transient temperature field.

Solution Referring to Eq. (3.103), the heat conduction equation is given as

$$\frac{\partial T}{\partial t} = \kappa\left[\frac{\partial^2 T}{\partial r^2} + \frac{2}{r}\frac{\partial T}{\partial r} + \frac{1}{r^2\sin\theta}\frac{\partial}{\partial\theta}\left(\sin\theta\frac{\partial T}{\partial\theta}\right)\right] \tag{a}$$

The initial and boundary conditions are

$$T = T_i \qquad \text{at } t = 0 \tag{b}$$

$$-\lambda\frac{\partial T}{\partial r} = h_a[T - T_aF(\theta)] \qquad \text{on } r = a \tag{c}$$

Introducing the temperature change τ

$$\tau = T - T_i \tag{d}$$

the fundamental equation system (a) to (c) reduces to

$$\frac{\partial \tau}{\partial t} = \kappa\left[\frac{\partial^2 \tau}{\partial r^2} + \frac{2}{r}\frac{\partial \tau}{\partial r} + \frac{1}{r^2\sin\theta}\frac{\partial}{\partial\theta}\left(\sin\theta\frac{\partial \tau}{\partial\theta}\right)\right] \tag{a$'$}$$

$$\tau = 0 \qquad \text{at } t = 0 \tag{b$'$}$$

$$-\lambda\frac{\partial \tau}{\partial r} = h_a\left[\tau + T_i - T_aF_a(\theta)\right] \qquad \text{on } r = a \tag{c$'$}$$

Performing the Laplace transform with respect to the variable t on Eq. (a′), the fundamental solution $\bar{\tau}$ in the transform domain is given as

$$\bar{\tau} = \sum_{n=0}^{\infty} C_n j_n(\omega r) P_n(\mu) \tag{7.96}$$

where

$$\omega^2 = -\frac{p}{\kappa} \tag{7.97}$$

and p is the Laplace transform parameter. The unknown constants C_n are determined from the thermal boundary condition (c′)

$$\begin{aligned} C_0 &= -\frac{h_a(T_i - T_a f_0)}{p\,[h_a j_0(\omega a) - \lambda\omega j_1(\omega a)]} \\ C_n &= \frac{h_a a T_a f_n}{p\,[(n\lambda + h_a a) j_n(\omega a) - \lambda\omega a j_{n+1}(\omega a)]} \qquad (n = 1, 2, \ldots) \end{aligned} \tag{7.98}$$

in which f_n $(n \geq 0)$ is defined by Eq. (e) of Example 7.8. Applying the residual theorem for Laplace transform inversion, the temperature change τ is

$$\tau = -T_i + \sum_{n=0}^{\infty} A_n r^n P_n(\mu) + 2\sum_{n=0}^{\infty}\sum_{j=1}^{\infty} e^{-\kappa\omega_{nj}^2 t} A_{nj} j_n(\omega_{nj} r) P_n(\mu) \tag{7.99}$$

where

$$\begin{aligned} A_n &= \frac{T_a h_a f_n}{(n\lambda + h_a a)a^{n-1}} \qquad (n = 0, 1, \ldots) \\ A_{0j} &= -\frac{h_a a(T_i - T_a f_0)}{\omega_{0j} D_{0j}} \\ A_{nj} &= \frac{h_a a T_a f_n}{\omega_{nj} D_{nj}} \qquad (n = 1, 2, \ldots) \\ D_{nj} &= \left[\frac{n}{\omega_{nj}}(n\lambda + h_a a) - \lambda\omega_{nj} a^2\right] j_n(\omega_{nj} a) \\ &\quad + (\lambda - h_a a) a j_{n+1}(\omega_{nj} a) \qquad (n = 0, 1, \ldots) \end{aligned} \tag{7.100}$$

In the expressions (7.99) and (7.100), ω_{nj} $(n \geq 0)$ is the j-th eigenvalue for each value of n, which is given by the j-th positive root of the transcendental equation:

$$(n\lambda + h_a a) j_n(\omega_n a) - \lambda\omega_n a j_{n+1}(\omega_n a) = 0 \tag{7.101}$$

The Goodier's thermoelastic displacement potential Φ satisfying the first relation of Eqs. (7.44) is

$$\Phi = K\left[-\frac{1}{6}T_i r^2 + \sum_{n=0}^{\infty}\frac{1}{2(2n+3)}A_n r^{n+2}P_n(\mu) - 2\sum_{n=0}^{\infty}\sum_{j=1}^{\infty}e^{-\kappa\omega_{nj}^2 t}\frac{1}{\omega_{nj}^2}A_{nj}\,j_n(\omega_{nj}r)P_n(\mu)\right] \tag{7.102}$$

From Eqs. (7.68), the displacement functions φ and ψ, with $\mu = \cos\theta$, are given by

$$\varphi = \sum_{n=0}^{\infty}[C_{1,n} - (n-4+4\nu)D_{1,n-2}]r^n P_n(\mu)$$
$$\psi = \sum_{n=0}^{\infty}(2n+1)D_{1,n-1}r^n P_n(\mu) \tag{7.103}$$

Using Eqs. (7.43), (7.47), (7.102), and (7.103), the displacements and stresses due to the Goodier's thermoelastic displacement potential Φ and the displacement functions φ and ψ are given as

$$u_r = K\left\{-\frac{1}{3}T_i r + \sum_{n=0}^{\infty}\frac{n+2}{2(2n+3)}A_n r^{n+1}P_n(\mu) - 2\sum_{n=0}^{\infty}\sum_{j=1}^{\infty}e^{-\kappa\omega_{nj}^2 t}\frac{1}{\omega_{nj}^2}A_{nj}\right.$$
$$\left.\times\left[\frac{n}{r}j_n(\omega_{nj}r) - \omega_{nj}j_{n+1}(\omega_{nj}r)\right]P_n(\mu)\right\}$$
$$+\sum_{n=0}^{\infty}\left[nC_{1,n}r^{n-1} + (n+1)(n-2+4\nu)D_{1,n}r^{n+1}\right]P_n(\mu)$$

$$u_\theta = -K(1-\mu^2)^{1/2}\left\{\sum_{n=1}^{\infty}\frac{1}{2(2n+3)}A_n r^{n+1}\frac{n+1}{1-\mu^2}[\mu P_n(\mu) - P_{n+1}(\mu)]\right.$$
$$\left.-2\sum_{n=1}^{\infty}\sum_{j=1}^{\infty}e^{-\kappa\omega_{nj}^2 t}\frac{1}{\omega_{nj}^2}A_{nj}\frac{1}{r}j_n(\omega_{nj}r)\frac{n+1}{1-\mu^2}[\mu P_n(\mu) - P_{n+1}(\mu)]\right\}$$
$$-(1-\mu^2)^{1/2}\sum_{n=1}^{\infty}\left[C_{1,n}r^{n-1} + (n+5-4\nu)D_{1,n}r^{n+1}\right]$$
$$\times\frac{n+1}{1-\mu^2}[\mu P_n(\mu) - P_{n+1}(\mu)]$$

$$\sigma_{rr}=2GK\left\{\frac{2}{3}T_i+\sum_{n=0}^{\infty}\frac{n^2-n-4}{2(2n+3)}A_n r^n P_n(\mu)-2\sum_{n=0}^{\infty}\sum_{j=1}^{\infty}e^{-\kappa\omega_{nj}^2 t}\frac{1}{\omega_{nj}^2}A_{nj}\right.$$

$$\left.\times\left[\frac{n(n-1)}{r^2}j_n(\omega_{nj}r)+2\frac{\omega_{nj}}{r}j_{n+1}(\omega_{nj}r)\right]P_n(\mu)\right\}$$

$$+2G\sum_{n=0}^{\infty}\left[n(n-1)C_{1,n}r^{n-2}+(n+1)(n^2-n-2-2\nu)D_{1,n}r^n\right]P_n(\mu)$$

$$\sigma_{\theta\theta}=2GK\left(\frac{2}{3}T_i+\sum_{n=0}^{\infty}\frac{1}{2(2n+3)}A_n r^n\left\{(n+1)\frac{\mu}{1-\mu^2}[\mu P_n(\mu)-P_{n+1}(\mu)]\right.\right.$$

$$\left.-(n+2)^2P_n(\mu)\right\}-2\sum_{n=0}^{\infty}\sum_{j=1}^{\infty}e^{-\kappa\omega_{nj}^2 t}\frac{1}{\omega_{nj}^2}A_{nj}$$

$$\times\left\{\left[\left(\omega_{nj}^2-\frac{n^2}{r^2}\right)j_n(\omega_{nj}r)-\frac{\omega_{nj}}{r}j_{n+1}(\omega_{nj}r)\right]P_n(\mu)\right.$$

$$\left.\left.+\frac{1}{r^2}j_n(\omega_{nj}r)(n+1)\frac{\mu}{1-\mu^2}[\mu P_n(\mu)-P_{n+1}(\mu)]\right\}\right)$$

$$+2G\left\{-\sum_{n=0}^{\infty}\left[n^2C_{1,n}r^{n-2}+(n+1)(n^2+4n+2+2\nu)D_{1,n}r^n\right]P_n(\mu)\right.$$

$$\left.+\sum_{n=1}^{\infty}\left[C_{1,n}r^{n-2}+(n+5-4\nu)D_{1,n}r^n\right](n+1)\frac{\mu}{1-\mu^2}[\mu P_n(\mu)-P_{n+1}(\mu)]\right\}$$

$$\sigma_{\phi\phi}=2GK\left(\frac{2}{3}T_i+\sum_{n=0}^{\infty}\frac{1}{2(2n+3)}A_n r^n\left\{-(n+1)\frac{\mu}{1-\mu^2}[\mu P_n(\mu)-P_{n+1}(\mu)]\right.\right.$$

$$\left.-(3n+4)P_n(\mu)\right\}-2\sum_{n=0}^{\infty}\sum_{j=1}^{\infty}e^{-\kappa\omega_{nj}^2 t}\frac{1}{\omega_{nj}^2}A_{nj}\left\{\left[\left(\frac{n}{r^2}+\omega_{nj}^2\right)j_n(\omega_{nj}r)\right.\right.$$

$$\left.\left.\left.-\frac{\omega_{nj}}{r}j_{n+1}(\omega_{nj}r)\right]P_n(\mu)-\frac{1}{r^2}j_n(\omega_{nj}r)(n+1)\frac{\mu}{1-\mu^2}[\mu P_n(\mu)-P_{n+1}(\mu)]\right\}\right)$$

$$+2G\left(\sum_{n=0}^{\infty}\left\{nC_{1,n}r^{n-2}+(n+1)[n-2-2\nu(2n+1)]D_{1,n}r^n\right\}P_n(\mu)\right.$$

$$\left.-\sum_{n=1}^{\infty}\left[C_{1,n}r^{n-2}+(n+5-4\nu)D_{1,n}r^n\right](n+1)\frac{\mu}{1-\mu^2}[\mu P_n(\mu)-P_{n+1}(\mu)]\right)$$

$$P_n(\mu) - P_{n+1}(\mu)\big]$$

$$\frac{j}{\,} j_{n+1}(\omega_{nj} r)\Big]$$

$$^{1/2} \sum_{n=1}^{\infty} \big[(n-1) C_{1,n} r^{n-2}$$

$$) - P_{n+1}(\mu)\big] \tag{7.104}$$

...on free surface $r = a$ are given by Eq. ...own constants $C_{1,n}$ and $D_{1,n}$ $(n \geq 0)$

$$- 2 \sum_{j=1}^{\infty} e^{-\kappa \omega_{nj}^2 t} \frac{1}{\omega_{nj}^2} A_{nj}$$

$$\times \Big[\ldots a \ldots n + 4 - 2\nu) \frac{\omega_{nj}}{a} j_{n+1}(\omega_{nj} a) \Big] \Bigg\}$$

$$(n = 2, 3, \ldots)$$

$$D_{1,0} = -\frac{1}{1+\nu} K \left[\frac{1}{3}(A_0 - T_i) + 2 \sum_{j=1}^{\infty} e^{-\kappa \omega_{0j}^2 t} \frac{1}{\omega_{0j} a} A_{0j} j_1(\omega_{0j} a) \right]$$

$$D_{1,n} = -K \frac{n+2}{\Delta_n} a^{-n} \left[\frac{1}{2n+3} A_n a^n + 2 \sum_{j=1}^{\infty} e^{-\kappa \omega_{nj}^2 t} \frac{1}{\omega_{nj} a} A_{nj} j_{n+1}(\omega_{nj} a) \right]$$

$$(n = 1, 2, \ldots)$$

$$\Delta_n = 2\big[n^2 + n + 1 + \nu(2n+1)\big] \tag{7.105}$$

in which the constants $C_{1,0}$ and $C_{1,1}$ are omitted because $C_{1,0}$ is not included in the expressions for displacements and stresses, and $C_{1,1}$ corresponds to a rigid displacement in the axial direction z.

Example 7.12 In Example 7.9, find the thermoelastic displacements and stresses in a hollow sphere under a transient temperature field.

Solution Referring to Eq. (3.103), the heat conduction equation is given as

$$\frac{\partial T}{\partial t} = \kappa \left[\frac{\partial^2 T}{\partial r^2} + \frac{2}{r} \frac{\partial T}{\partial r} + \frac{1}{r^2 \sin\theta} \frac{\partial}{\partial \theta} \left(\sin\theta \frac{\partial T}{\partial \theta} \right) \right] \tag{a}$$

The initial and boundary conditions are

$$T = T_i \qquad \text{at } t = 0 \tag{b}$$

$$\begin{aligned} \lambda \frac{\partial T}{\partial r} &= h_a[T - T_a F_a(\theta)] \qquad \text{on } r = a \\ -\lambda \frac{\partial T}{\partial r} &= h_b[T - T_b F_b(\theta)] \qquad \text{on } r = b \end{aligned} \tag{c}$$

Introducing the temperature change τ

$$\tau = T - T_i \tag{d}$$

the fundamental equation system (a) to (c) reduces to

$$\frac{\partial \tau}{\partial t} = \kappa \left[\frac{\partial^2 \tau}{\partial r^2} + \frac{2}{r}\frac{\partial \tau}{\partial r} + \frac{1}{r^2 \sin\theta}\frac{\partial}{\partial \theta}\left(\sin\theta \frac{\partial \tau}{\partial \theta}\right)\right] \tag{a$'$}$$

$$\tau = 0 \qquad \text{at } t = 0 \tag{b$'$}$$

$$\begin{aligned} \lambda \frac{\partial \tau}{\partial r} &= h_a[\tau + T_i - T_a F_a(\theta)] \qquad \text{on } r = a \\ \lambda \frac{\partial \tau}{\partial r} &= -h_b[\tau + T_i - T_b F_b(\theta)] \qquad \text{on } r = b \end{aligned} \tag{c$'$}$$

Performing the Laplace transform with respect to the variable t on Eq. (a′), the fundamental solution $\bar{\tau}$ in the transform domain is given as

$$\bar{\tau} = \sum_{n=0}^{\infty} \left[A_n^* j_n(\omega r) + B_n^* y_n(\omega r) \right] P_n(\mu) \tag{7.106}$$

where

$$\omega^2 = -\frac{p}{\kappa} \tag{e}$$

p is the Laplace transform parameter, and A_n^* and B_n^* are unknown constants. From Eqs. (c′), the unknown constants A_n^* and B_n^* are determined as

$$\begin{aligned} A_0^* = -\frac{1}{p\Delta_0^*} &\{h_a(T_i - T_a f_0)\,[h_b y_0(\omega b) - \lambda\omega y_1(\omega b)] \\ &- h_b(T_i - T_b g_0)\,[h_a y_0(\omega a) + \lambda\omega y_1(\omega a)]\} \\ B_0^* = -\frac{1}{p\Delta_0^*} &\{h_b(T_i - T_b g_0)\,[h_a j_0(\omega a) + \lambda\omega j_1(\omega a)] \\ &- h_a(T_i - T_a f_0)\,[h_b j_0(\omega b) - \lambda\omega j_1(\omega b)]\} \\ \Delta_0^* = [h_a j_0(\omega a) &+ \lambda\omega j_1(\omega a)]\,[h_b y_0(\omega b) - \lambda\omega y_1(\omega b)] \\ &- [h_a y_0(\omega a) + \lambda\omega y_1(\omega a)]\,[h_b j_0(\omega b) - \lambda\omega j_1(\omega b)] \end{aligned}$$

$$A_n^* = -\frac{1}{p\Delta_n^*}\{h_a a T_a f_n [(n\lambda + h_b b) y_n(\omega b) - \lambda\omega b y_{n+1}(\omega b)]$$

$$+ h_b b T_b g_n [(n\lambda - h_a a) y_n(\omega a) - \lambda\omega a y_{n+1}(\omega a)]\}$$

$$B_n^* = \frac{1}{p\Delta_n^*}\{h_b b T_b g_n [(n\lambda - h_a a) j_n(\omega a) - \lambda\omega a j_{n+1}(\omega a)]$$

$$+ h_a a T_a f_n [(n\lambda + h_b b) j_n(\omega b) - \lambda\omega b j_{n+1}(\omega b)]\}$$

$$\Delta_n^* = [(n\lambda - h_a a) j_n(\omega a) - \lambda\omega a j_{n+1}(\omega a)] [(n\lambda + h_b b) y_n(\omega b)$$

$$-\lambda\omega b y_{n+1}(\omega b)] - [(n\lambda - h_a a) y_n(\omega a) - \lambda\omega a y_{n+1}(\omega a)]$$

$$\times [(n\lambda + h_b b) j_n(\omega b) - \lambda\omega b j_{n+1}(\omega b)] \qquad (n = 1, 2, \ldots) \tag{7.107}$$

In Eqs. (7.107), f_n and g_n $(n \geq 0)$ are defined by Eq. (c) of Example 7.9. Applying the residual theorem for the Laplace transform inversion, the temperature change τ is obtained

$$\tau = -T_i + \sum_{n=0}^{\infty}\left(A_n r^n + B_n r^{-n-1}\right) P_n(\mu)$$

$$+ 2\sum_{n=0}^{\infty}\sum_{j=1}^{\infty} e^{-\kappa\omega_{nj}^2 t}[A_{nj} j_n(\omega_{nj} r) + B_{nj} y_n(\omega_{nj} r)] P_n(\mu) \tag{7.108}$$

where

$$A_n = \frac{1}{\Delta_n}\left\{h_a T_a f_n [(n+1)\lambda - h_b b] b^{-n-2} + h_b T_b g_n [(n+1)\lambda + h_a a] a^{-n-2}\right\}$$

$$B_n = \frac{1}{\Delta_n}\left[h_a T_a f_n (n\lambda + h_b b) b^{n-1} + h_b T_b g_n (n\lambda - h_a a) a^{n-1}\right]$$

$$\Delta_n = (n\lambda + h_b b) [(n+1)\lambda + h_a a] a^{-n-2} b^{n-1}$$

$$- (n\lambda - h_a a) [(n+1)\lambda - h_b b] a^{n-1} b^{-n-2}$$

$$A_{0j} = -\frac{1}{\omega_{0j} D_{0j}}\left\{h_a (T_i - T_a f_0)\left[h_b y_0(\omega_{0j} b) - \lambda\omega_{0j} y_1(\omega_{0j} b)\right]\right.$$

$$\left. - h_b (T_i - T_b g_0)\left[h_a y_0(\omega_{0j} a) + \lambda\omega_{0j} y_1(\omega_{0j} a)\right]\right\}$$

$$B_{0j} = -\frac{1}{\omega_{0j} D_{0j}}\left\{h_b (T_i - T_b g_0)\left[h_a j_0(\omega_{0j} a) + \lambda\omega_{0j} j_1(\omega_{0j} a)\right]\right.$$

$$\left. - h_a (T_i - T_a f_0)\left[h_b j_0(\omega_{0j} b) - \lambda\omega_{0j} j_1(\omega_{0j} b)\right]\right\}$$

$$A_{nj} = -\frac{1}{\omega_{nj} D_{nj}}\left\{h_a a T_a f_n \left[(n\lambda + h_b b) y_n(\omega_{nj} b) - \lambda\omega_{nj} b y_{n+1}(\omega_{nj} b)\right]\right.$$

$$\left. + h_b b T_b g_n \left[(n\lambda - h_a a) y_n(\omega_{nj} a) - \lambda\omega_{nj} a y_{n+1}(\omega_{nj} a)\right]\right\}$$

$$
\begin{aligned}
B_{nj} &= \frac{1}{\omega_{nj} D_{nj}} \left\{ h_b b T_b g_n \left[(n\lambda - h_a a) j_n(\omega_{nj} a) - \lambda \omega_{nj} a j_{n+1}(\omega_{nj} a) \right] \right. \\
&\qquad \left. + h_a a T_a f_n \left[(n\lambda + h_b b) j_n(\omega_{nj} b) - \lambda \omega_{nj} b j_{n+1}(\omega_{nj} b) \right] \right\} \\
D_{0j} &= \left. \frac{d\Delta_0^*}{d\omega} \right|_{\omega \to \omega_{0j}} \\
&= \{ [\lambda \omega a j_0(\omega a) - (\lambda + h_a a) j_1(\omega a)] [h_b y_0(\omega b) - \lambda \omega y_1(\omega b)] \\
&\quad - [h_a j_0(\omega a) + \lambda \omega j_1(\omega a)] [\lambda \omega b y_0(\omega b) - (\lambda - h_b b) y_1(\omega b)] \\
&\quad - [\lambda \omega a y_0(\omega a) - (\lambda + h_a a) y_1(\omega a)] [h_b j_0(\omega b) - \lambda \omega j_1(\omega b)] \\
&\quad + [h_a y_0(\omega a) + \lambda \omega y_1(\omega a)] [\lambda \omega b j_0(\omega b) - (\lambda - h_b b) j_1(\omega b)] \} |_{\omega \to \omega_{0j}} \\
D_{nj} &= \left. \frac{d\Delta_n^*}{d\omega} \right|_{\omega \to \omega_{nj}} \\
&= \left\{ \left[\frac{n}{\omega} (n\lambda - h_a a) - \lambda \omega a^2 \right] j_n(\omega a) + a(\lambda + h_a a) j_{n+1}(\omega a) \right\} \\
&\quad \times [(n\lambda + h_b b) y_n(\omega b) - \lambda \omega b y_{n+1}(\omega b)] \\
&\quad + [(n\lambda - h_a a) j_n(\omega a) - \lambda \omega a j_{n+1}(\omega a)] \\
&\quad \times \left\{ \left[\frac{n}{\omega} (n\lambda + h_b b) - \lambda \omega b^2 \right] y_n(\omega b) + b(\lambda - h_b b) y_{n+1}(\omega b) \right\} \\
&\quad - \left\{ \left[\frac{n}{\omega} (n\lambda - h_a a) - \lambda \omega a^2 \right] y_n(\omega a) + a(\lambda + h_a a) y_{n+1}(\omega a) \right\} \\
&\quad \times [(n\lambda + h_b b) j_n(\omega b) - \lambda \omega b j_{n+1}(\omega b)] \\
&\quad - [(n\lambda - h_a a) y_n(\omega a) - \lambda \omega a y_{n+1}(\omega a)] \\
&\quad \times \left. \left\{ \left[\frac{n}{\omega} (n\lambda + h_b b) - \lambda \omega b^2 \right] j_n(\omega b) + b(\lambda - h_b b) j_{n+1}(\omega b) \right\} \right|_{\omega \to \omega_{nj}} \\
&\qquad\qquad (n = 1, 2, \ldots) \qquad (7.109)
\end{aligned}
$$

In expressions (7.108) to (7.109), ω_{nj} $(n \geq 0)$ is the j-th eigenvalue for each value of n, which is given by the j-th positive root of the transcendental equation:

$$
\begin{aligned}
&[(n\lambda - h_a a) j_n(\omega a) - \lambda \omega a j_{n+1}(\omega a)] \\
&\quad \times [(n\lambda + h_b b) y_n(\omega b) - \lambda \omega b y_{n+1}(\omega b)] \\
&\quad - [(n\lambda - h_a a) y_n(\omega a) - \lambda \omega a y_{n+1}(\omega a)] \\
&\quad \times [(n\lambda + h_b b) j_n(\omega b) - \lambda \omega b j_{n+1}(\omega b)] = 0 \qquad (n = 0, 1, \ldots) \qquad (7.110)
\end{aligned}
$$

Next, the Goodier's thermoelastic displacement potential Φ satisfying Eqs. (7.44) is

$$\Phi = K\left\{ -\frac{1}{6}T_i r^2 + \sum_{n=0}^{\infty}\left[\frac{1}{2(2n+3)}A_n r^{n+2} - \frac{1}{2(2n-1)}B_n r^{-n+1}\right]P_n(\mu) \right.$$

$$\left. -2\sum_{n=0}^{\infty}\sum_{j=1}^{\infty} e^{-\kappa\omega_{nj}^2 t}\frac{1}{\omega_{nj}^2}\left[A_{nj}j_n(\omega_{nj}r) + B_{nj}y_n(\omega_{nj}r)\right]P_n(\mu)\right\} \tag{7.111}$$

The displacements and stresses corresponding to the Goodier's thermoelastic displacement potential Φ and the displacement functions φ and ψ given by Eqs. (7.68) are

$$u_r = K\left(-\frac{1}{3}T_i r + \sum_{n=0}^{\infty}\left[\frac{n+2}{2(2n+3)}A_n r^{n+1} + \frac{n-1}{2(2n-1)}B_n r^{-n}\right]P_n(\mu) \right.$$

$$-2\sum_{n=0}^{\infty}\sum_{j=1}^{\infty} e^{-\kappa\omega_{nj}^2 t}\frac{1}{\omega_{nj}^2}\left\{A_{nj}\left[\frac{n}{r}j_n(\omega_{nj}r) - \omega_{nj}j_{n+1}(\omega_{nj}r)\right]\right.$$

$$\left.\left. + B_{nj}\left[\frac{n}{r}y_n(\omega_{nj}r) - \omega_{nj}y_{n+1}(\omega_{nj}r)\right]\right\}P_n(\mu)\right) + \bar{\bar{u}}_r$$

$$u_\theta = -K(1-\mu^2)^{1/2}\left\{\sum_{n=1}^{\infty}\left[\frac{1}{2(2n+3)}A_n r^{n+1} - \frac{1}{2(2n-1)}B_n r^{-n}\right]\right.$$

$$\times\frac{n+1}{1-\mu^2}\left[\mu P_n(\mu) - P_{n+1}(\mu)\right]$$

$$-2\sum_{n=1}^{\infty}\sum_{j=1}^{\infty} e^{-\kappa\omega_{nj}^2 t}\frac{1}{r\omega_{nj}^2}\left[A_{nj}j_n(\omega_{nj}r) + B_{nj}y_n(\omega_{nj}r)\right]$$

$$\left.\times\frac{n+1}{1-\mu^2}\left[\mu P_n(\mu) - P_{n+1}(\mu)\right]\right\} + \bar{\bar{u}}_\theta$$

$$\sigma_{rr} = 2GK\left(\frac{2}{3}T_i + \sum_{n=0}^{\infty}\left[\frac{n^2-n-4}{2(2n+3)}A_n r^n - \frac{n^2+3n-2}{2(2n-1)}B_n r^{-n-1}\right]P_n(\mu)\right.$$

$$-2\sum_{n=0}^{\infty}\sum_{j=1}^{\infty} e^{-\kappa\omega_{nj}^2 t}\frac{1}{\omega_{nj}^2}\left\{A_{nj}\left[\frac{n(n-1)}{r^2}j_n(\omega_{nj}r) + 2\frac{\omega_{nj}}{r}j_{n+1}(\omega_{nj}r)\right]\right.$$

$$\left.\left. + B_{nj}\left[\frac{n(n-1)}{r^2}y_n(\omega_{nj}r) + 2\frac{\omega_{nj}}{r}y_{n+1}(\omega_{nj}r)\right]\right\}P_n(\mu)\right) + \bar{\bar{\sigma}}_{rr}$$

$$
\begin{aligned}
\sigma_{\theta\theta} = 2GK\Bigg[&\frac{2}{3}T_i + \sum_{n=0}^{\infty}\Bigg(\frac{1}{2(2n+3)}A_n r^n \\
&\times\left\{(n+1)\frac{\mu}{1-\mu^2}\left[\mu P_n(\mu) - P_{n+1}(\mu)\right] - (n+2)^2 P_n(\mu)\right\} \\
&+ \frac{1}{2(2n-1)}B_n r^{-n-1} \\
&\times\left\{-(n+1)\frac{\mu}{1-\mu^2}\left[\mu P_n(\mu) - P_{n+1}(\mu)\right] + (n-1)^2 P_n(\mu)\right\}\Bigg) \\
&- 2\sum_{n=0}^{\infty}\sum_{j=1}^{\infty} e^{-\kappa\omega_{nj}^2 t}\frac{1}{\omega_{nj}^2}\left\{A_{nj}\left[\left(\omega_{nj}^2 - \frac{n^2}{r^2}\right) j_n(\omega_{nj} r) - \frac{\omega_{nj}}{r} j_{n+1}(\omega_{nj} r)\right]\right. \\
&\left. + B_{nj}\left[\left(\omega_{nj}^2 - \frac{n^2}{r^2}\right) y_n(\omega_{nj} r) - \frac{\omega_{nj}}{r} y_{n+1}(\omega_{nj} r)\right]\right\} P_n(\mu) \\
&- 2\sum_{n=0}^{\infty}\sum_{j=1}^{\infty} e^{-\kappa\omega_{nj}^2 t}\frac{1}{\omega_{nj}^2}\left[A_{nj}\frac{1}{r^2} j_n(\omega_{nj} r) + B_{nj}\frac{1}{r^2} y_n(\omega_{nj} r)\right] \\
&\times (n+1)\frac{\mu}{1-\mu^2}\left[\mu P_n(\mu) - P_{n+1}(\mu)\right]\Bigg] + \bar{\bar{\sigma}}_{\theta\theta}
\end{aligned}
$$

$$
\begin{aligned}
\sigma_{\phi\phi} = 2GK\Bigg[&\frac{2}{3}T_i + \sum_{n=0}^{\infty}\Bigg(\frac{1}{2(2n+3)}A_n r^n \\
&\times\left\{-(n+1)\frac{\mu}{1-\mu^2}\left[\mu P_n(\mu) - P_{n+1}(\mu)\right] - (3n+4) P_n(\mu)\right\} \\
&+ \frac{1}{2(2n-1)}B_n r^{-n-1} \\
&\times\left\{(n+1)\frac{\mu}{1-\mu^2}\left[\mu P_n(\mu) - P_{n+1}(\mu)\right] - (3n-1) P_n(\mu)\right\}\Bigg) \\
&- 2\sum_{n=0}^{\infty}\sum_{j=1}^{\infty} e^{-\kappa\omega_{nj}^2 t}\frac{1}{\omega_{nj}^2}\left\{A_{nj}\left[\left(\frac{n}{r^2} + \omega_{nj}^2\right) j_n(\omega_{nj} r) - \frac{\omega_{nj}}{r} j_{n+1}(\omega_{nj} r)\right]\right. \\
&\left. + B_{nj}\left[\left(\frac{n}{r^2} + \omega_{nj}^2\right) y_n(\omega_{nj} r) - \frac{\omega_{nj}}{r} y_{n+1}(\omega_{nj} r)\right]\right\} P_n(\mu) \\
&+ 2\sum_{n=0}^{\infty}\sum_{j=1}^{\infty} e^{-\kappa\omega_{nj}^2 t}\frac{1}{\omega_{nj}^2}\left[A_{nj}\frac{1}{r^2} j_n(\omega_{nj} r) + B_{nj}\frac{1}{r^2} y_n(\omega_{nj} r)\right] \\
&\times (n+1)\frac{\mu}{1-\mu^2}\left[\mu P_n(\mu) - P_{n+1}(\mu)\right]\Bigg] + \bar{\bar{\sigma}}_{\phi\phi}
\end{aligned}
$$

$$
\begin{aligned}
\sigma_{r\theta} = &-2GK(1-\mu^2)^{1/2}\left(\sum_{n=1}^{\infty}\left[\frac{n+1}{2(2n+3)}A_n r^n + \frac{n}{2(2n-1)}B_n r^{-n-1}\right]\right.\\
&\times \frac{n+1}{1-\mu^2}\left[\mu P_n(\mu) - P_{n+1}(\mu)\right] - 2\sum_{n=1}^{\infty}\sum_{j=1}^{\infty} e^{-\kappa\omega_{nj}^2 t}\frac{1}{\omega_{nj}^2}\\
&\times\left\{A_{nj}\left[\frac{n-1}{r^2}j_n(\omega_{nj}r) - \frac{\omega_{nj}}{r}j_{n+1}(\omega_{nj}r)\right]\right.\\
&\left.+ B_{nj}\left[\frac{n-1}{r^2}y_n(\omega_{nj}r) - \frac{\omega_{nj}}{r}y_{n+1}(\omega_{nj}r)\right]\right\}\\
&\left.\times\frac{n+1}{1-\mu^2}\left[\mu P_n(\mu) - P_{n+1}(\mu)\right]\right) + \bar{\bar{\sigma}}_{r\theta}
\end{aligned}
\tag{7.112}
$$

in which $\bar{\bar{u}}_r$, $\bar{\bar{u}}_\theta$, $\bar{\bar{\sigma}}_{rr}$, $\bar{\bar{\sigma}}_{\theta\theta}$, $\bar{\bar{\sigma}}_{\phi\phi}$, and $\bar{\bar{\sigma}}_{r\theta}$ are given by Eqs. (7.69) and (7.70). The mechanical boundary conditions on the traction free surfaces $r = a$ and $r = b$ are given by Eqs. (7.86). Using the boundary conditions, the unknown constants $C_{i,n}$ and $D_{i,n}$ $(i = 1, 2)$ are determined for $n = 0$ and $n = 1$

$$
\begin{aligned}
C_{2,0} = &\frac{1}{2}K\frac{a^2b^2}{b^3-a^3}\left(B_0(b-a) + 4\sum_{j=1}^{\infty}e^{-\kappa\omega_{0j}^2 t}\frac{1}{\omega_{0j}}\right.\\
&\left.\times\left\{A_{0j}\left[bj_1(\omega_{0j}a) - aj_1(\omega_{0j}b)\right] + B_{0j}\left[by_1(\omega_{0j}a) - ay_1(\omega_{0j}b)\right]\right\}\right)\\
D_{1,0} = &\frac{1}{2(1+\nu)}K\left(\frac{2}{3}(T_i - A_0) - B_0\frac{b^2-a^2}{b^3-a^3} - 4\sum_{j=1}^{\infty}e^{-\kappa\omega_{0j}^2 t}\frac{1}{\omega_{0j}}\frac{1}{b^3-a^3}\right.\\
&\left.\times\left\{A_{0j}\left[b^2 j_1(\omega_{0j}b) - a^2 j_1(\omega_{0j}a)\right] + B_{0j}\left[b^2 y_1(\omega_{0j}b) - a^2 y_1(\omega_{0j}a)\right]\right\}\right)\\
C_{2,1} = &\frac{1}{3}K\frac{a^4b^4}{b^5-a^5}\left(\frac{1}{2}B_1\frac{b^3-a^3}{a^2b^2} - 2\sum_{j=1}^{\infty}e^{-\kappa\omega_{1j}^2 t}\frac{1}{\omega_{1j}}\frac{1}{ab}\right.\\
&\left.\times\left\{A_{1j}\left[a^2 j_2(\omega_{1j}b) - b^2 j_2(\omega_{1j}a)\right] + B_{1j}\left[a^2 y_2(\omega_{1j}b) - b^2 y_2(\omega_{1j}a)\right]\right\}\right)\\
D_{1,1} = &-\frac{1}{2(1+\nu)}K\left(\frac{1}{5}A_1 + \frac{1}{2}B_1\frac{b^2-a^2}{b^5-a^5} + 2\sum_{j=1}^{\infty}e^{-\kappa\omega_{1j}^2 t}\frac{1}{\omega_{1j}}\frac{1}{b^5-a^5}\right.\\
&\left.\times\left\{A_{1j}\left[b^3 j_2(\omega_{1j}b) - a^3 j_2(\omega_{1j}a)\right] + B_{1j}\left[b^3 y_2(\omega_{1j}b) - a^3 y_2(\omega_{1j}a)\right]\right\}\right)\\
D_{2,1} = &\,0
\end{aligned}
\tag{7.113}
$$

in which the constants $C_{1,0}$, $D_{2,0}$, and $C_{1,1}$ are omitted because $C_{1,0}$ and $D_{2,0}$ are not included in the expressions for displacements and stresses, and $C_{1,1}$ corresponds to a rigid displacement in the axial direction z. Furthermore, the unknown constants for $n \geq 2$ can be determined as a solution of the following simultaneous equations with 4 unknowns from 4 boundary conditions (7.86):

$$
\begin{aligned}
& n(n-1)C_{1,n}r^{n-2} + (n+1)(n+2)C_{2,n}r^{-n-3} \\
& \quad + (n+1)(n^2 - n - 2 - 2\nu)D_{1,n}r^n + n(n^2 + 3n - 2\nu)D_{2,n}r^{-n-1} \\
& \quad = -K\left(\frac{n^2 - n - 4}{2(2n+3)}A_n r^n - \frac{n^2 + 3n - 2}{2(2n-1)}B_n r^{-n-1} - 2\sum_{j=1}^{\infty} e^{-\kappa\omega_{nj}^2 t}\frac{1}{\omega_{nj}^2}\right. \\
& \qquad \times \left\{A_{nj}\left[\frac{n(n-1)}{r^2} j_n(\omega_{nj}r) + 2\frac{\omega_{nj}}{r} j_{n+1}(\omega_{nj}r)\right]\right. \\
& \qquad \left.\left. + B_{nj}\left[\frac{n(n-1)}{r^2} y_n(\omega_{nj}r) + 2\frac{\omega_{nj}}{r} y_{n+1}(\omega_{nj}r)\right]\right\}\right) \\
& \qquad \text{for } r = a \text{ and } r = b \\
& (n-1)C_{1,n}r^{n-2} - (n+2)C_{2,n}r^{-n-3} \\
& \quad + (n^2 + 2n - 1 + 2\nu)D_{1,n}r^n - (n^2 - 2 + 2\nu)D_{2,n}r^{-n-1} \\
& \quad = -K\left(\frac{n+1}{2(2n+3)}A_n r^n + \frac{n}{2(2n-1)}B_n r^{-n-1} - 2\sum_{j=1}^{\infty} e^{-\kappa\omega_{nj}^2 t}\frac{1}{\omega_{nj}^2}\right. \\
& \qquad \times \left\{A_{nj}\left[\frac{n-1}{r^2} j_n(\omega_{nj}r) - \frac{\omega_{nj}}{r} j_{n+1}(\omega_{nj}r)\right]\right. \\
& \qquad \left.\left. + B_{nj}\left[\frac{n-1}{r^2} y_n(\omega_{nj}r) - \frac{\omega_{nj}}{r} y_{n+1}(\omega_{nj}r)\right]\right\}\right) \\
& \qquad \text{for } r = a \text{ and } r = b
\end{aligned}
\tag{7.114}
$$

Problems

7.1 Derive the formulas for the one-dimensional temperature change (7.35) in a transient state for an infinite medium with a spherical cavity, and the associated thermal displacement u_r and thermal stresses σ_{rr} and $\sigma_{\theta\theta}$ given by Eqs. (g) and (7.36) in Example 7.7. To solve this problem, the following relations are required:

$$L^{-1}\left\{\frac{e^{-\sqrt{p/\kappa}x}}{p\left(\sqrt{p/\kappa}+\alpha\right)}\right\} = \frac{1}{\alpha}\operatorname{erfc}\left(\frac{x}{\sqrt{2\kappa t}}\right) - \frac{1}{\alpha}e^{\alpha x+\kappa t\alpha^2}\operatorname{erfc}\left(\frac{x}{2\sqrt{\kappa t}}+\alpha\sqrt{\kappa t}\right)$$

$$\operatorname{erfc}(x) = \frac{2}{\sqrt{\pi}}\int_x^\infty e^{-u^2}\,du, \qquad \operatorname{erf}(x) = \frac{2}{\sqrt{\pi}}\int_0^x e^{-u^2}\,du$$

$$\operatorname{erf}(x) + \operatorname{erfc}(x) = 1$$

$$\int \operatorname{erf}(x)\,dx = x\operatorname{erf}(x) + \frac{1}{\sqrt{\pi}}e^{-x^2}$$

$$\int \operatorname{erfc}(x)\,dx = x\operatorname{erfc}(x) - \frac{1}{\sqrt{\pi}}e^{-x^2}$$

$$\int x\operatorname{erfc}(x)\,dx = \frac{1}{4}(2x^2-1)\operatorname{erfc}(x) - \frac{1}{2\sqrt{\pi}}xe^{-x^2}$$

$$\int e^{ax}\operatorname{erfc}(x)\,dx = \frac{1}{a}\left[e^{ax}\operatorname{erfc}(x) + e^{a^2/4}\operatorname{erf}\left(x-\frac{a}{2}\right)\right]$$

$$\int xe^{ax}\operatorname{erfc}(x)\,dx = \frac{1}{a^2}\left[(ax-1)e^{ax}\operatorname{erfc}(x) - \frac{a}{\sqrt{\pi}}e^{-x^2+ax} + \frac{1}{2}(a^2-2)e^{a^2/4}\operatorname{erf}\left(x-\frac{a}{2}\right)\right]$$

7.2 In Example 7.8, determine the coefficient f_n $(n \geq 0)$ when the band heat source $F(\theta)$ is supplied, which is symmetrically distributed with respect to x-y plane as shown in Fig. 7.1.

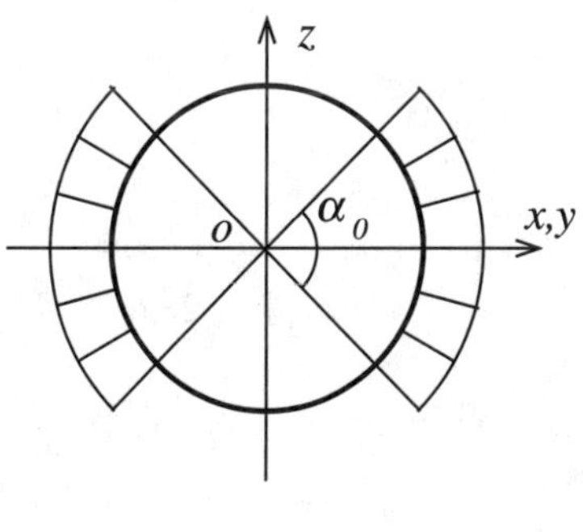

Figure 7.1 Band heat source.

Answer

$$f_n = \frac{2n+1}{2n}\{[\mu_1 P_n(\mu_1) - P_{n+1}(\mu_1)] - [\mu_2 P_n(\mu_2) - P_{n+1}(\mu_2)]\}; \qquad n \geq 1$$

$$f_0 = \frac{1}{2}(\mu_2 - \mu_1)$$

where $\mu_1 = -\sin(\alpha_0/2)$, $\mu_2 = \sin(\alpha_0/2)$.

7.3 Derive the relations for displacements and stresses given by Eqs. (7.69a) and (7.70a) due to the displacement functions φ and ψ given by Eqs. (7.68). For the problem, mathematical formulas for the Legendre functions given in Appendix C are required.

7.4 Derive the unknown constants $C_{1,n}$, $D_{1,n}$, $C_{2,n}$, and $D_{2,n}$ for $n = 0$ and $n = 1$ given by Eqs. (7.87) in Example 7.9.

7.5 Derive the particular solution Φ of the thermoelastic displacement potential given by Eq. (i) in Example 7.8.

7.6 Derive the particular solution Φ of the thermoelastic displacement potential given by Eq. (7.102) in Example 7.11, in which the following integral formulas are used:

$$\int r^{-n} j_{n+1}(\omega r)\, dr = -\frac{1}{\omega} r^{-n} j_n(\omega r)$$

$$\int r^{n+1} j_{n-1}(\omega r)\, dr = \frac{1}{\omega} r^{n+1} j_n(\omega r)$$

7.7 In Example 7.1, let $(a/b) = (1/2)$ and $\nu = 0.3$. Calculate the temperature change τ and extremal values of u_r, σ_{rr}, and $\sigma_{\theta\theta}$ for the two cases:

Case 1: Heating from the inner surface

$$T_a - T_i = 50\,\text{K}, \qquad T_b - T_i = 0\,\text{K}$$

Case 2: Heating from the outer surface

$$T_a - T_i = 0\,\text{K}, \qquad T_b - T_i = 50\,\text{K}$$

Answer

Case 1:

$$\tau = 50\left(2\frac{a}{r} - 1\right)\text{K}, \qquad u_r|_{r=a} = 14.3\alpha a, \qquad u_r|_{r=b} = 28.6\alpha a$$

$$\sigma_{rr}|_{\text{extremum}} = -11.5\alpha E, \qquad \sigma_{\theta\theta}|_{r=a} = -51.0\alpha E, \qquad \sigma_{\theta\theta}|_{r=b} = 20.4\alpha E$$

Case 2:

$$\tau = 100\left(1 - \frac{a}{r}\right)\text{K}, \qquad u_r|_{r=a} = 35.7\alpha a, \qquad u_r|_{r=b} = 71.4\alpha a$$

$$\sigma_{rr}|_{\text{extremum}} = 11.5\alpha E, \qquad \sigma_{\theta\theta}|_{r=a} = 51.0\alpha E, \qquad \sigma_{\theta\theta}|_{r=b} = -20.4\alpha E$$

7.8 In Example 7.5, let $\nu = 0.3$ and $T_a - T_i = 50\,\text{K}$. Calculate the temperature change τ and extremum values of σ_{rr} and $\sigma_{\theta\theta}$ when $h_a = \infty$.

Answer

$$\tau = 50\frac{a}{r}\,\text{K}, \qquad \sigma_{rr}|_{\text{extremum}} = \sigma_{rr}|_{r=\sqrt{3}\alpha} = -27.5\alpha E$$

$$\sigma_{\text{extremum}} = \sigma_{\theta\theta}|_{r=a} = -71.4\alpha E$$

(Stress distributions show large values compared with Case 1 of Problem 7.7)

7.9 In Example 7.8, calculate the temperature change and components of displacement and stress when $F(\theta)$ is given by $\cos\theta$ and $h_a = \infty$.

Answer

$$\tau = -T_i + T_a\frac{r}{a}\cos\theta$$

$$u_r = \frac{1}{2}T_a a\alpha\frac{r^2}{a^2}\cos\theta, \qquad u_\theta = \frac{1}{2}T_a a\alpha\frac{r^2}{a^2}\sin\theta$$

$$\sigma_{rr} = 0, \qquad \sigma_{\theta\theta} = 0, \qquad \sigma_{\phi\phi} = 0, \qquad \sigma_{r\theta} = 0$$

7.10 In Example 7.9, determine the coefficients A_n and B_n as given by Eq. (7.82) when $h_a = h_b = \infty$.

Answer

$$A_n = \frac{T_b g_n a^{-n-1} - T_a f_n b^{-n-1}}{a^{-n-1}b^n - a^n b^{-n-1}}, \qquad B_n = \frac{T_a f_n b^n - T_b g_n a^n}{a^{-n-1}b^n - a^n b^{-n-1}}$$

Chapter 8

THERMAL STRESSES IN PLATES

The problem of thermal stresses in thin plates is developed with the Kirchhoff-Love hypothesis. First, thermal stresses in thin plates due to the temperature change along the thickness only are discussed. Next, the thin plate theory based on the Kirchhoff-Love hypothesis is introduced in the Cartesian coordinate system, and also the boundary conditions are stated. Rectangular plates with various kinds of boundary conditions are considered. Finally, basic equations for circular plates are introduced and then a number of axisymmetric and non-axisymmetric problems are presented.

8.1 Thermal Stress in a Plate Due to Uniform Thermal Load

We now consider stress and strain in plates subjected to thermal loads. The coordinate system of the plate is shown in Fig. 8.1. Referring to Section 2.2 where the thermal stresses in beams were described, the strain components ϵ_x and ϵ_y are

$$\epsilon_x = \epsilon_y = \epsilon_0 + \frac{z}{\rho} \tag{8.1}$$

where ϵ_0 and ρ denote, respectively, the in-plane strain and the radius of curvature at the neutral plane of $z = 0$. Moreover, assuming that the temperature change τ changes in the thickness direction only, the strain components ϵ_x and ϵ_y at a distance z from the neutral plane are expressed as the sum of the free thermal strain and the strain due to

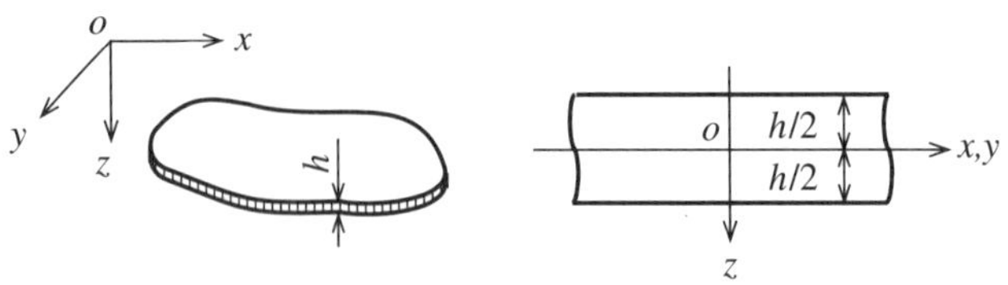

Figure 8.1 A plate.

stress σ_x and σ_y:

$$\begin{aligned} \epsilon_x &= \frac{1}{E}(\sigma_x - \nu\sigma_y) + \alpha\tau = \epsilon_0 + \frac{z}{\rho} \\ \epsilon_y &= \frac{1}{E}(\sigma_y - \nu\sigma_x) + \alpha\tau = \epsilon_0 + \frac{z}{\rho} \end{aligned} \tag{8.2}$$

Therefore, the stress components σ_x and σ_y are

$$\sigma_x(=\sigma_y) = \frac{E}{1-\nu}\left[\epsilon_0 + \frac{z}{\rho} - \alpha\tau\right] \tag{8.3}$$

When the plate is free from external forces, the in-plane forces and bending moments per unit length in the x and y directions must be zero:

$$\int_{-h/2}^{h/2} \sigma_x \, dz = 0 \tag{a}$$

$$\int_{-h/2}^{h/2} \sigma_x z \, dz = 0 \tag{b}$$

in which the thickness of the plate is denoted by h. Substitution of Eq. (8.3) into Eqs. (a) and (b) gives the in-plane strain ϵ_0 and the curvature $1/\rho$ at the neutral plane $z = 0$

$$\epsilon_0 = \frac{\alpha}{h}\int_{-h/2}^{h/2} \tau(z)\, dz \tag{c}$$

$$\frac{1}{\rho} = \frac{12\alpha}{h^3}\int_{-h/2}^{h/2} \tau(z) z \, dz \tag{d}$$

Then, from Eqs. (8.3), (c), and (d), the thermal stress component $\sigma_x(=\sigma_y)$ for the plate is

$$\sigma_x(=\sigma_y) = \frac{\alpha E}{1-\nu}\left[-\tau(z) + \frac{1}{h}\int_{-h/2}^{h/2} \tau(z)\, dz + \frac{12z}{h^3}\int_{-h/2}^{h/2} \tau(z) z \, dz\right] \tag{8.4}$$

Referring to Eqs. (2.16), it can be seen that the results given by Eq. (8.4) for the plate shows the value of $1/(1-\nu)$ times the value for the beam.

Example 8.1 When both the lower surface ($z = h/2$) and the upper surface ($z = -(h/2)$) of an infinite plate at initial temperature T_i are heated by the surrounding media at constant temperatures T_a and T_b, respectively, find the thermal stresses in the plate under a steady temperature field.

Solution Let the heat transfer coefficient on the lower and upper surfaces of the plate be h_a and h_b, respectively. The steady state heat conduction equation and associated boundary conditions are

$$\frac{d^2T}{dz^2} = 0 \tag{a}$$

$$-\lambda \frac{dT}{dz} = h_a(T - T_a) \qquad \text{on } z = \frac{h}{2} \tag{b}$$

$$\lambda \frac{dT}{dz} = h_b(T - T_b) \qquad \text{on } z = -\frac{h}{2} \tag{c}$$

where λ is the thermal conductivity. The temperature solution is

$$T(z) = C_1 + C_2 z \tag{d}$$

Substituting Eq. (d) into Eqs. (b) and (c), the one-dimensional steady temperature distribution in the thickness direction of the plate is

$$T(z) = \frac{1}{\lambda(h_a + h_b) + h_a h_b h} \times \left[h_a \left(\lambda + h_b \frac{h}{2} \right) T_a + h_b \left(\lambda + h_a \frac{h}{2} \right) T_b + h_a h_b (T_a - T_b) z \right] \tag{8.5}$$

Thus, the temperature change $\tau(z)$ is

$$\begin{aligned} \tau(z) &= T(z) - T_i \\ &= -T_i + \frac{1}{\lambda(h_a + h_b) + h_a h_b h} \\ &\quad \times \left[h_a \left(\lambda + h_b \frac{h}{2} \right) T_a + h_b \left(\lambda + h_a \frac{h}{2} \right) T_b + h_a h_b (T_a - T_b) z \right] \end{aligned} \tag{e}$$

Substituting Eq. (e) into Eq. (8.4), the thermal stress $\sigma_x (=\sigma_y)$ in the plate is

$$\sigma_x (=\sigma_y) = 0 \tag{8.6}$$

From the result of Eq. (8.6), it can be seen that thermal stress in the plate does not result from a one-dimensional steady temperature field irrespective of the variations of the temperature T_a and T_b of the surrounding media and the heat transfer coefficients h_a and h_b on the boundary surfaces.

Example 8.2 For the problem in Example 8.1, find the thermal stresses in an infinite plate under a transient temperature field.

Solution The transient heat conduction equation, the initial condition, and the boundary conditions are

$$\frac{\partial T}{\partial t} = \kappa \frac{\partial^2 T}{\partial z^2} \tag{a}$$

$$T = T_i \qquad \text{at } t = 0 \tag{b}$$

$$-\lambda \frac{dT}{dz} = h_a(T - T_a) \qquad \text{on } z = \frac{h}{2} \tag{c}$$

$$\lambda \frac{dT}{dz} = h_b(T - T_b) \qquad \text{on } z = -\frac{h}{2} \tag{d}$$

in which, λ and κ denote the thermal conductivity and the thermal diffusivity, respectively, and h_a and h_b denote the heat transfer coefficients on the lower surface ($z = h/2$) and the upper surface ($z = -(h/2)$), respectively.

Introducing the temperature change of τ

$$\tau = T - T_i \tag{e}$$

the fundamental equation system (a)–(d) reduces to

$$\frac{\partial \tau}{\partial t} = \kappa \frac{\partial^2 \tau}{\partial z^2} \tag{a$'$}$$

$$\tau = 0 \qquad \text{at } t = 0 \tag{b$'$}$$

$$-\lambda \frac{\partial \tau}{\partial z} = h_a[\tau - (T_a - T_i)] \qquad \text{on } z = \frac{h}{2} \tag{c$'$}$$

$$\lambda \frac{\partial \tau}{\partial z} = h_b[\tau - (T_b - T_i)] \qquad \text{on } z = -\frac{h}{2} \tag{d$'$}$$

Performing the Laplace transform with respect to the variable t on Eq. (a$'$), the fundamental solution in the transform domain is given as

$$\overline{\tau} = C_1 \cos \omega z + C_2 \sin \omega z \tag{f}$$

$$\omega^2 \equiv -p/\kappa \tag{g}$$

where p is Laplace transform parameter, and C_1 and C_2 are the unknown constants that must satisfy the thermal boundary conditions given by Eqs. (c$'$) and (d$'$). The constants C_1 and C_2 are determined as

$$C_1 = \frac{C_1^*(\omega)}{p\Delta(\omega)}, \qquad C_2 = \frac{C_2^*(\omega)}{p\Delta(\omega)} \tag{h}$$

$$
\begin{aligned}
C_1^*(\omega) &= h_a\,(T_a - T_i)\left(\lambda\omega\cos\frac{\omega h}{2} + h_b\sin\frac{\omega h}{2}\right)\\
&\quad + h_b\,(T_b - T_i)\left(\lambda\omega\cos\frac{\omega h}{2} + h_a\sin\frac{\omega h}{2}\right)\\
C_2^*(\omega) &= -h_a\,(T_a - T_i)\left(\lambda\omega\sin\frac{\omega h}{2} - h_b\cos\frac{\omega h}{2}\right)\\
&\quad - h_b\,(T_b - T_i)\left(-\lambda\omega\sin\frac{\omega h}{2} + h_a\cos\frac{\omega h}{2}\right)\\
\Delta(\omega) &= \left(h_a h_b - \lambda^2\omega^2\right)\sin\omega h + \lambda\omega(h_a + h_b)\cos\omega h
\end{aligned}
\tag{8.7}
$$

Applying the residue theorem for the Laplace inversion, the temperature change τ is obtained

$$
\begin{aligned}
\tau &= \frac{1}{\lambda\,(h_a + h_b) + h_a h_b h}\left[\left(\lambda + \frac{h_b h}{2}\right)h_a(T_a - T_i)\right.\\
&\quad \left. + \left(\lambda + \frac{h_a h}{2}\right)h_b(T_b - T_i) + h_a h_b(T_a - T_b)z\right]\\
&\quad + 2\sum_{j=1}^{\infty} e^{-\kappa\omega_j{}^2 t}\frac{1}{\omega_j D'(\omega_j)}\left[C_1^*(\omega_j)\cos\omega_j z + C_2^*(\omega_j)\sin\omega_j z\right]
\end{aligned}
\tag{8.8}
$$

$$
\begin{aligned}
D'(\omega_j) = \left.\frac{d\Delta(\omega)}{d\omega}\right|_{\omega\to\omega_j} &= \left[\lambda(h_a + h_b) + h\left(h_a h_b - \lambda^2\omega_j{}^2\right)\right]\cos\omega_j h\\
&\quad - \lambda\omega_j\,[h\,(h_a + h_b) + 2\lambda]\sin\omega_j h
\end{aligned}
\tag{8.9}
$$

In Eqs. (8.8) and (8.9), the eigenvalue ω_j is the j-th positive root of the transcendental equation

$$
\Delta\,(\omega) = 0 \tag{8.10}
$$

Substituting of Eq. (8.8) into Eq. (8.4), the thermal stress $\sigma_x(=\sigma_y)$ in the plate is obtained

$$
\begin{aligned}
\sigma_x(=\sigma_y) &= 2\frac{\alpha E}{1-\nu}\sum_{j=1}^{\infty} e^{-\kappa\omega_j{}^2 t}\frac{1}{\omega_j D'(\omega_j)}\\
&\quad \times\left[-C_1^*(\omega_j)\cos\omega_j z - C_2^*(\omega_j)\sin\omega_j z + C_1^*(\omega_j)\frac{2}{\omega_j h}\sin\frac{\omega_j h}{2}\right.\\
&\quad \left. + C_2^*(\omega_j)\frac{24z}{\omega_j{}^2 h^3}\left(\sin\frac{\omega_j h}{2} - \frac{\omega_j h}{2}\cos\frac{\omega_j h}{2}\right)\right]
\end{aligned}
\tag{8.11}
$$

There are some special cases of Eqs. (8.8)–(8.11).

[1] Prescribed surface temperatures T_a and T_b at the lower and upper surfaces, respectively.

$$\tau = \frac{1}{2}(T_a + T_b) - T_i + (T_a - T_b)\frac{z}{h} + 2\sum_{j=1}^{\infty} e^{-\kappa(j\pi/h)^2 t} \frac{1}{j\pi \cos j\pi}$$
$$\times \left[(T_a + T_b - 2T_i) \sin\frac{j\pi}{2} \cos\frac{j\pi z}{h} + (T_a - T_b) \cos\frac{j\pi}{2} \sin\frac{j\pi z}{h} \right]$$
$$= \frac{1}{2}(T_a + T_b) - T_i + (T_a - T_b)\frac{z}{h}$$
$$+ (T_a + T_b - 2T_i)\frac{2}{\pi}\sum_{n=1}^{\infty} e^{-\kappa(2n-1)^2\pi^2 t/h^2} \frac{(-1)^n}{2n-1} \cos\frac{(2n-1)\pi z}{h}$$
$$+ (T_a - T_b)\frac{2}{\pi}\sum_{n=1}^{\infty} e^{-\kappa(2n\pi)^2 t/h^2} \frac{(-1)^n}{2n} \sin\frac{2n\pi z}{h} \tag{8.12}$$

$$\sigma_x (=\sigma_y) = 2\frac{\alpha E}{1-\nu}\sum_{j=1}^{\infty} e^{-\kappa(j\pi/h)^2 t} \frac{1}{j\pi \cos j\pi}$$
$$\times \left[(T_a + T_b - 2T_i) \sin\frac{j\pi}{2} \left(\frac{2}{j\pi} \sin\frac{j\pi}{2} - \cos\frac{j\pi z}{h} \right) \right.$$
$$\left. -(T_a - T_b) \cos\frac{j\pi}{2} \left(\frac{12}{j\pi}\frac{z}{h} \cos\frac{j\pi}{2} + \sin\frac{j\pi z}{h} \right) \right]$$
$$= -\frac{\alpha E}{1-\nu} \left\{ (T_a + T_b - 2T_i)\frac{2}{\pi}\sum_{n=1}^{\infty} e^{-\kappa(2n-1)^2\pi^2 t/h^2} \right.$$
$$\times \frac{1}{2n-1} \left[\frac{2}{(2n-1)\pi} + (-1)^n \cos\frac{(2n-1)\pi z}{h} \right]$$
$$+ (T_a - T_b)\frac{2}{\pi}\sum_{n=1}^{\infty} e^{-\kappa(2n\pi)^2 t/h^2}$$
$$\left. \times \frac{1}{2n} \left[\frac{12}{2n\pi}\frac{z}{h} + (-1)^n \sin\frac{2n\pi z}{h} \right] \right\} \tag{8.13}$$

[2] Prescribed lower and upper surface temperatures T_a and T_b equal to a constant temperature T_0.

$$\tau = T_0 - T_i + (T_0 - T_i)\frac{4}{\pi}\sum_{n=1}^{\infty} e^{-\kappa(2n-1)^2\pi^2 t/h^2} \frac{(-1)^n}{2n-1} \cos\frac{(2n-1)\pi z}{h} \tag{8.14}$$

$$\sigma_x(=\sigma_y) = -\frac{\alpha E}{1-\nu}(T_0 - T_i)\frac{4}{\pi}\sum_{n=1}^{\infty} e^{-\kappa(2n-1)^2\pi^2 t/h^2} \times \frac{1}{2n-1}\left[\frac{2}{(2n-1)\pi} + (-1)^n \cos\frac{(2n-1)\pi z}{h}\right] \tag{8.15}$$

[3] Prescribed lower surface temperature T_a equals a constant temperature T_0, and prescribed upper surface temperature T_b equals the initial temperature T_i.

$$\tau = (T_0 - T_i)\left(\frac{1}{2} + \frac{z}{h}\right) + (T_0 - T_i)\frac{2}{\pi}\left[\sum_{n=1}^{\infty} e^{-\kappa(2n-1)^2\pi^2 t/h^2}\frac{(-1)^n}{2n-1}\cos\frac{(2n-1)\pi z}{h} + \sum_{n=1}^{\infty} e^{-\kappa(2n\pi)^2 t/h^2}\frac{(-1)^n}{2n}\sin\frac{2n\pi z}{h}\right] \tag{8.16}$$

$$\sigma_x(=\sigma_y) = -\frac{\alpha E}{1-\nu}(T_0 - T_i)\frac{2}{\pi}\left\{\sum_{n=1}^{\infty} e^{-\kappa(2n-1)^2\pi^2 t/h^2} \times \frac{1}{2n-1}\left[\frac{2}{(2n-1)\pi} + (-1)^n\cos\frac{(2n-1)\pi z}{h}\right] + \sum_{n=1}^{\infty} e^{-\kappa(2n\pi)^2 t/h^2}\frac{1}{2n}\left[\frac{12}{2n\pi}\frac{z}{h} + (-1)^n \sin\frac{2n\pi z}{h}\right]\right\} \tag{8.17}$$

[4] Temperatures T_a and T_b of the lower and upper surrounding media equal to a constant temperature T_0, and the heat transfer coefficients h_a and h_b at the lower and upper surfaces equal to h_0.

$$\tau = T_0 - T_i + 4h_0(T_0 - T_i)\sum_{j=1}^{\infty} e^{-\kappa\omega_j{}^2 t}\frac{1}{\omega_j D'(\omega_j)} \times \left(\lambda\omega_j \cos\frac{\omega_j h}{2} + h_0 \sin\frac{\omega_j h}{2}\right)\cos\omega_j z \tag{8.18}$$

$$\sigma_x(=\sigma_y) = \frac{\alpha E}{1-\nu}4h_0(T_0 - T_i)\sum_{j=1}^{\infty} e^{-\kappa\omega_j{}^2 t}\frac{1}{\omega_j D'(\omega_j)} \times \left(\lambda\omega_j \cos\frac{\omega_j h}{2} + h_0 \sin\frac{\omega_j h}{2}\right)\left(\frac{2}{\omega_j h}\sin\frac{\omega_j h}{2} - \cos\omega_j z\right) \tag{8.19}$$

where

$$D'(\omega_j) = \left[2\lambda h_0 + h\left(h_0^2 - \lambda^2 {\omega_j}^2\right)\right] \cos \omega_j h - 2\lambda \omega_j (h h_0 + \lambda) \sin \omega_j h \tag{8.20}$$

The eigenvalue ω_j is the j-th positive root of the transcendental equation

$$\left(h_0^2 - \lambda^2 \omega^2\right) \sin \omega h + 2\lambda \omega h_0 \cos \omega h = 0 \tag{8.21}$$

[5] Temperature T_a of the lower surrounding medium equals a constant temperature T_0, and temperature T_b of the upper surrounding medium equals the initial temperature T_i of the plate, and the heat transfer coefficients h_a and h_b at the lower and upper surfaces equal h_0.

$$\begin{aligned}
\tau &= (T_0 - T_i) \frac{1}{2\lambda + h h_0} \left[\lambda + h_0 \left(\frac{h}{2} + z\right)\right] \\
&\quad + 2h_0 (T_0 - T_i) \sum_{j=1}^{\infty} e^{-\kappa {\omega_j}^2 t} \frac{1}{\omega_j D'(\omega_j)} \\
&\quad \times \left\{\lambda \omega_j \cos\left[\omega_j \left(\frac{h}{2} + z\right)\right] + h_0 \sin\left[\omega_j \left(\frac{h}{2} + z\right)\right]\right\}
\end{aligned} \tag{8.22}$$

$$\begin{aligned}
\sigma_x (=\sigma_y) &= \frac{\alpha E}{1 - \nu} 2h_0 (T_0 - T_i) \sum_{j=1}^{\infty} e^{-\kappa {\omega_j}^2 t} \frac{1}{\omega_j D'(\omega_j)} \\
&\quad \times \left[-\lambda \omega_j \cos \omega_j \left(\frac{h}{2} + z\right) - h_0 \sin \omega_j \left(\frac{h}{2} + z\right)\right. \\
&\quad + \frac{\lambda}{h} \sin \omega_j h + \frac{2h_0}{\omega_j h} \sin^2 \frac{\omega_j h}{2} \\
&\quad - \frac{24z}{{\omega_j}^2 h^3} \left(\lambda \omega_j \sin \frac{\omega_j h}{2} - h_0 \cos \frac{\omega_j h}{2}\right) \\
&\quad \left. \times \left(\sin \frac{\omega_j h}{2} - \frac{\omega_j h}{2} \cos \frac{\omega_j h}{2}\right)\right]
\end{aligned} \tag{8.23}$$

In Eqs. (8.22) and (8.23), $D'(\omega_j)$ is defined by Eq. (8.20) and the eigenvalue ω_j is the j-th positive root of the transcendental equation (8.21).

Until now we have considered plates of isotropic materials. We now consider the stresses and strains in anisotropic plates subjected to thermal loads. If the plate has orthotropic material properties in the in-plane direction of the z plane, and the principal axes of orthotropy coincide with the coordinate axes of x and y, the strain components ϵ_x and ϵ_y are

$$\epsilon_x = \epsilon_{0x} + \frac{z}{\rho_x}, \qquad \epsilon_y = \epsilon_{0y} + \frac{z}{\rho_y} \tag{8.24}$$

where ϵ_{0x} and ϵ_{0y} denote the strain components in the x and y directions at the neutral plane, respectively, and ρ_x and ρ_y denote the radii of curvature of the neutral plane along the orthogonal coordinate axes x and y, respectively.

Assuming that the temperature change along the z axis is $\tau(z)$, the strains ϵ_x and ϵ_y at a distance z from the neutral plane are expressed as the sum of the free thermal strain and the strain due to stresses σ_x and σ_y

$$\begin{aligned}\epsilon_x &= \frac{1}{E_x}(\sigma_x - \nu_{xy}\sigma_y) + \alpha_x\tau \\ \epsilon_y &= \frac{1}{E_y}(\sigma_y - \nu_{yx}\sigma_x) + \alpha_y\tau\end{aligned} \tag{8.25}$$

where E_x and E_y denote the Young's moduli in the x and y directions, α_x and α_y are the coefficients of linear thermal expansion in the x and y directions, respectively, ν_{xy} denotes Poisson's ratio concerned with the strain component in the x direction due to the load in the y direction, and ν_{yx} means the opposite. Furthermore, from the reciprocal theorem in Chapter 10, the following relation of the elastic constants between ν_{xy} and ν_{yx} must be satisfied

$$\frac{\nu_{xy}}{E_x} = \frac{\nu_{yx}}{E_y} \tag{8.26}$$

Substituting Eqs. (8.24) into Eqs. (8.25), the stress components σ_x and σ_y are written as

$$\sigma_x = \frac{1}{1-\nu_{xy}\nu_{yx}}\left[E_x\epsilon_{0x} + \nu_{xy}E_y\epsilon_{0y} + E_x\frac{z}{\rho_x} + \nu_{xy}E_y\frac{z}{\rho_y} - (\alpha_x E_x + \alpha_y\nu_{xy}E_y)\tau\right] \tag{8.27}$$

$$\sigma_y = \frac{1}{1-\nu_{xy}\nu_{yx}}\left[E_y\epsilon_{0y} + \nu_{yx}E_x\epsilon_{0x} + E_y\frac{z}{\rho_y} + \nu_{yx}E_x\frac{z}{\rho_x} - (\alpha_y E_y + \alpha_x\nu_{yx}E_x)\tau\right] \tag{8.28}$$

When the plate is free from external forces, the in-plane forces and bending moments per unit length in the x and y directions must be zero

$$\int_{-h/2}^{h/2} \sigma_x \, dz = 0 \tag{a}$$

$$\int_{-h/2}^{h/2} \sigma_x z \, dz = 0 \tag{b}$$

$$\int_{-h/2}^{h/2} \sigma_y \, dz = 0 \tag{c}$$

$$\int_{-h/2}^{h/2} \sigma_y z \, dz = 0 \tag{d}$$

Substitution of Eqs. (8.27) and (8.28) into Eqs. (a)–(d) gives the in-plane strains ϵ_{0x} and ϵ_{0y}, and the curvatures $1/\rho_x$ and $1/\rho_y$ at the neutral plane $z = 0$

$$\epsilon_{0x} = \alpha_x \frac{1}{h} \int_{-h/2}^{h/2} \tau(z)\, dz, \qquad \epsilon_{0y} = \alpha_y \frac{1}{h} \int_{-h/2}^{h/2} \tau(z)\, dz \tag{8.29}$$

$$\frac{1}{\rho_x} = \alpha_x \frac{12}{h^3} \int_{-h/2}^{h/2} \tau(z) z\, dz, \qquad \frac{1}{\rho_y} = \alpha_y \frac{12}{h^3} \int_{-h/2}^{h/2} \tau(z) z\, dz \tag{8.30}$$

Substitution of Eqs. (8.29) and (8.30) into Eqs. (8.27) and (8.28) gives the stress components σ_x and σ_y

$$\sigma_x = \frac{\alpha_x E_x + \alpha_y \nu_{xy} E_y}{1 - \nu_{xy}\nu_{yx}} \left[-\tau(z) + \frac{1}{h} \int_{-h/2}^{h/2} \tau(z)\, dz + \frac{12z}{h^3} \int_{-h/2}^{h/2} \tau(z) z\, dz \right] \tag{8.31}$$

$$\begin{aligned} \sigma_y &= \frac{\alpha_y E_y + \alpha_x \nu_{yx} E_x}{1 - \nu_{xy}\nu_{yx}} \left[-\tau(z) + \frac{1}{h} \int_{-h/2}^{h/2} \tau(z)\, dz + \frac{12z}{h^3} \int_{-h/2}^{h/2} \tau(z) z\, dz \right] \\ &= \frac{\alpha_y E_y + \alpha_x \nu_{yx} E_x}{\alpha_x E_x + \alpha_y \nu_{xy} E_y} \sigma_x \end{aligned} \tag{8.32}$$

Comparing the results given by Eqs. (8.31) and (8.32) with the result of Eq. (8.4) for the isotropic plate, it can be found that the stress distributions σ_x and σ_y for the orthotropic plate show the analogous distribution with the one for the isotropic plate, however, the values of σ_x and σ_y for the orthotropic plate are different from the ones for the isotropic plate due to the variations of the material properties E_x, E_y, α_x, α_y, and ν_{xy}, ν_{yx}. In other words, the results of $\sigma_x (=\sigma_y)$ for the isotropic plate, which are derived in Examples 8.1 and 8.2 are easily extended to the stress components σ_x and σ_y for the orthotropic plate by exchanging the term containing material constants in the following manner:

$$\begin{aligned} \frac{\alpha E}{1-\nu} &\rightarrow \frac{\alpha_x E_x + \alpha_y \nu_{xy} E_y}{1 - \nu_{xy}\nu_{yx}} \quad \text{for } \sigma_x \\ \frac{\alpha E}{1-\nu} &\rightarrow \frac{\alpha_y E_y + \alpha_x \nu_{yx} E_x}{1 - \nu_{xy}\nu_{yx}} \quad \text{for } \sigma_y \end{aligned} \tag{8.33}$$

When the temperature change in the plate is given by Eq. (8.5) in a steady state, it can easily be found that the thermal stress distributions σ_x and σ_y in an orthotropic plate result in zero stress because of zero thermal stress in the isotropic plate. Furthermore, as for a transient temperature distribution given by Eq. (8.8), the thermal stress distributions σ_x and σ_y are derived by making use of the relations of Eqs. (8.33) and (8.11), which are given as

$$\sigma_x = 2\frac{\alpha_x E_x + \alpha_y \nu_{xy} E_y}{1 - \nu_{xy}\nu_{yx}} \sum_{j=1}^{\infty} e^{-\kappa \omega_j{}^2 t} \frac{1}{\omega_j D'(\omega_j)}$$
$$\times \left[-C_1^*(\omega_j)\cos\omega_j z - C_2^*(\omega_j)\sin\omega_j z + C_1^*(\omega_j)\frac{2}{\omega_j h}\sin\frac{\omega_j h}{2} \right.$$
$$\left. + C_2^*(\omega_j)\frac{24z}{\omega_j{}^2 h^3}\left(\sin\frac{\omega_j h}{2} - \frac{\omega_j h}{2}\cos\frac{\omega_j h}{2}\right)\right] \tag{8.34}$$

$$\sigma_y = \frac{\alpha_y E_y + \alpha_x \nu_{yx} E_x}{\alpha_x E_x + \alpha_y \nu_{xy} E_y}\sigma_x \tag{8.35}$$

in which $C_1^*(\omega_j)$, $C_2^*(\omega_j)$, and $D'(\omega_j)$ are given by Eqs. (8.7) and (8.9), and ω_j is the eigenvalue, which is given by the j-th positive root of $\Delta(\omega) = 0$, and $\Delta(\omega)$ is given by Eq. (8.7).

8.2 Basic Equations for a Rectangular Plate

8.2.1 Fundamental Relations of Thermal Bending

In discussing the thermal bending problems of a thin plate with a thickness h, it will be assumed that the deflection, which means a deformation in the out-of-plane direction of the plate, is small. We introduce the Kirchhoff-Love hypothesis that the plane initially perpendicular to the neutral plane of the plate remains a plane after deformation and is perpendicular to the deformed neutral plane. The coordinate system of the plate is taken as shown in Fig. 8.1.

Let u, v, and w be displacement components in the x, y, and z direction at the neutral plane ($z = 0$). Then, referring to Fig. 8.2, the displacement components u' and v' in the in-plane direction x and y at the arbitrary point of the plate are

$$u' = u - z\frac{\partial w}{\partial x}$$
$$v' = v - z\frac{\partial w}{\partial y} \tag{a}$$

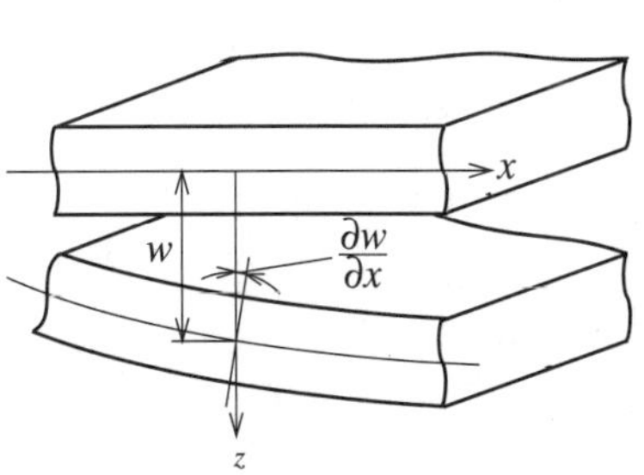

Figure 8.2 Displacement.

Therefore, the strain components in the in-plane direction are

$$\begin{aligned}
\epsilon_{xx} &= \frac{\partial u'}{\partial x} = \frac{\partial u}{\partial x} - z\frac{\partial^2 w}{\partial x^2} \\
\epsilon_{yy} &= \frac{\partial v'}{\partial y} = \frac{\partial v}{\partial y} - z\frac{\partial^2 w}{\partial y^2} \\
\epsilon_{xy} &= \frac{1}{2}\left(\frac{\partial u'}{\partial y} + \frac{\partial v'}{\partial x}\right) = \frac{1}{2}\left(\frac{\partial u}{\partial y} + \frac{\partial v}{\partial x}\right) - z\frac{\partial^2 w}{\partial x \partial y}
\end{aligned} \tag{8.36}$$

Two-dimensional stress–strain relations in the in-plane directions for an isotropic plate are

$$\begin{aligned}
\epsilon_{xx} &= \frac{1}{E}(\sigma_{xx} - \nu\sigma_{yy}) + \alpha\tau \\
\epsilon_{yy} &= \frac{1}{E}(\sigma_{yy} - \nu\sigma_{xx}) + \alpha\tau \\
\epsilon_{xy} &= \frac{1}{2G}\sigma_{xy} = \frac{1+\nu}{E}\sigma_{xy}
\end{aligned} \tag{b}$$

These relations are written in the following form:

$$\begin{aligned}
\sigma_{xx} &= \frac{E}{1-\nu^2}[\epsilon_{xx} + \nu\epsilon_{yy} - (1+\nu)\alpha\tau] \\
\sigma_{yy} &= \frac{E}{1-\nu^2}[\epsilon_{yy} + \nu\epsilon_{xx} - (1+\nu)\alpha\tau] \\
\sigma_{xy} &= 2G\epsilon_{xy} = \frac{E}{1+\nu}\epsilon_{xy}
\end{aligned} \tag{c}$$

Substitution of Eqs. (8.36) into Eqs. (c) gives the relations

$$\begin{aligned}
\sigma_{xx} &= \frac{E}{1-\nu^2}\left[\frac{\partial u}{\partial x} + \nu\frac{\partial v}{\partial y} - z\left(\frac{\partial^2 w}{\partial x^2} + \nu\frac{\partial^2 w}{\partial y^2}\right) - (1+\nu)\alpha\tau\right] \\
\sigma_{yy} &= \frac{E}{1-\nu^2}\left[\frac{\partial v}{\partial y} + \nu\frac{\partial u}{\partial x} - z\left(\frac{\partial^2 w}{\partial y^2} + \nu\frac{\partial^2 w}{\partial x^2}\right) - (1+\nu)\alpha\tau\right] \\
\sigma_{xy} &= \frac{E}{2(1+\nu)}\left(\frac{\partial u}{\partial y} + \frac{\partial v}{\partial x} - 2z\frac{\partial^2 w}{\partial x \partial y}\right)
\end{aligned} \tag{8.37}$$

In order to analyze the thermoelastic behavior of the plate, we now introduce the concept of resultant forces N_x, N_y, N_{xy} and resultant moments M_x, M_y, M_{xy} per unit length of the plate, which are defined as

$$\begin{aligned}
N_x &= \int_{-h/2}^{h/2} \sigma_{xx}\,dz, \qquad N_y = \int_{-h/2}^{h/2} \sigma_{yy}\,dz, \qquad N_{xy} = \int_{-h/2}^{h/2} \sigma_{xy}\,dz \\
M_x &= \int_{-h/2}^{h/2} \sigma_{xx} z\,dz, \qquad M_y = \int_{-h/2}^{h/2} \sigma_{yy} z\,dz, \qquad M_{xy} = -\int_{-h/2}^{h/2} \sigma_{xy} z\,dz
\end{aligned} \tag{8.38}$$

Making use of Eqs. (8.37), these resultant forces and resultant moments are expressed by the displacement components u, v, w in the following form:

$$
\begin{aligned}
N_x &= \frac{Eh}{1-\nu^2}\left(\frac{\partial u}{\partial x} + \nu\frac{\partial v}{\partial y}\right) - \frac{1}{1-\nu}N_T \\
N_y &= \frac{Eh}{1-\nu^2}\left(\frac{\partial v}{\partial y} + \nu\frac{\partial u}{\partial x}\right) - \frac{1}{1-\nu}N_T \\
N_{xy} &= \frac{Eh}{2(1+\nu)}\left(\frac{\partial u}{\partial y} + \frac{\partial v}{\partial x}\right)
\end{aligned}
\tag{8.39}
$$

$$
\begin{aligned}
M_x &= -D\left(\frac{\partial^2 w}{\partial x^2} + \nu\frac{\partial^2 w}{\partial y^2}\right) - \frac{1}{1-\nu}M_T \\
M_y &= -D\left(\frac{\partial^2 w}{\partial y^2} + \nu\frac{\partial^2 w}{\partial x^2}\right) - \frac{1}{1-\nu}M_T \\
M_{xy} &= (1-\nu)D\frac{\partial^2 w}{\partial x \partial y}
\end{aligned}
\tag{8.40}
$$

In Eqs. (8.39) and (8.40), D is the bending rigidity of the plate, N_T and M_T are the so-called thermally induced resultant force and resultant moment, respectively, which are defined by

$$
D = \frac{Eh^3}{12(1-\nu^2)}
\tag{8.41}
$$

$$
N_T = \alpha E \int_{-h/2}^{h/2} \tau \, dz, \qquad M_T = \alpha E \int_{-h/2}^{h/2} \tau z \, dz
\tag{8.42}
$$

Substituting Eqs. (8.39) and (8.40) into Eqs. (8.37), the thermal stress components are shown in terms of the resultant forces and resultant moments, which are given as

$$
\begin{aligned}
\sigma_{xx} &= \frac{1}{h}N_x + \frac{12z}{h^3}M_x + \frac{1}{1-\nu}\left(\frac{1}{h}N_T + \frac{12z}{h^3}M_T - \alpha E \tau\right) \\
\sigma_{yy} &= \frac{1}{h}N_y + \frac{12z}{h^3}M_y + \frac{1}{1-\nu}\left(\frac{1}{h}N_T + \frac{12z}{h^3}M_T - \alpha E \tau\right) \\
\sigma_{xy} &= \frac{1}{h}N_{xy} - \frac{12z}{h^3}M_{xy}
\end{aligned}
\tag{8.43}
$$

We now consider the equilibrium state in the in-plane directions of x and y. When body forces are absent, the equilibrium equations of the plate are given by Eqs. (5.17).

$$
\frac{\partial \sigma_{xx}}{\partial x} + \frac{\partial \sigma_{yx}}{\partial y} = 0, \qquad \frac{\partial \sigma_{xy}}{\partial x} + \frac{\partial \sigma_{yy}}{\partial y} = 0
\tag{d}
$$

Performing the integration for Eqs. (d) with respect to the thickness direction of z, we obtain

$$\frac{\partial N_x}{\partial x} + \frac{\partial N_{xy}}{\partial y} = 0, \qquad \frac{\partial N_{xy}}{\partial x} + \frac{\partial N_y}{\partial y} = 0 \tag{8.44}$$

We now introduce a thermal stress function F, which is analogous to the thermal stress function χ available for the plane thermoelastic problems, defined by

$$N_x = \frac{\partial^2 F}{\partial y^2}, \qquad N_y = \frac{\partial^2 F}{\partial x^2}, \qquad N_{xy} = -\frac{\partial^2 F}{\partial x \partial y} \tag{8.45}$$

Then the equilibrium equations (8.44) are satisfied automatically. Moreover, the stress function F should satisfy the compatibility equation in terms of the strain components with respect to the in-plane directions, which is given by Eq. (5.18)

$$\frac{\partial^2 \epsilon_{xx}}{\partial y^2} + \frac{\partial^2 \epsilon_{yy}}{\partial x^2} = 2\frac{\partial^2 \epsilon_{xy}}{\partial x \partial y} \tag{e}$$

Substituting Eq. (b) into Eq. (e), and performing the integration with respect to the thickness direction of z, we have

$$\begin{aligned}\left(\frac{\partial^2}{\partial y^2} - \nu\frac{\partial^2}{\partial x^2}\right) N_x + \left(\frac{\partial^2}{\partial x^2} - \nu\frac{\partial^2}{\partial y^2}\right) N_y - 2(1+\nu)\frac{\partial^2 N_{xy}}{\partial x \partial y} \\ = -\left(\frac{\partial^2}{\partial x^2} + \frac{\partial^2}{\partial y^2}\right) N_T\end{aligned} \tag{8.46}$$

Substituting Eqs. (8.45) into Eq. (8.46), the governing equation for F is derived as

$$\nabla^2\nabla^2 F = -\nabla^2 N_T \tag{8.47}$$

where

$$\nabla^2 = \frac{\partial^2}{\partial x^2} + \frac{\partial^2}{\partial y^2} \tag{8.48}$$

Applying the analytical procedure for the thermal stress function χ discussed in Section 5.2, we can find the stress function F governed by Eq. (8.47) under the appropriate mechanical boundary conditions, therefore, the resultant forces N_x, N_y, and N_{xy} are obtained from Eqs. (8.45).

In addition to the bending moments M_x, M_y, and the twisting moment M_{xy} in Eqs. (8.38), we define the twisting moment M_{yx}, and the shearing forces Q_x and Q_y

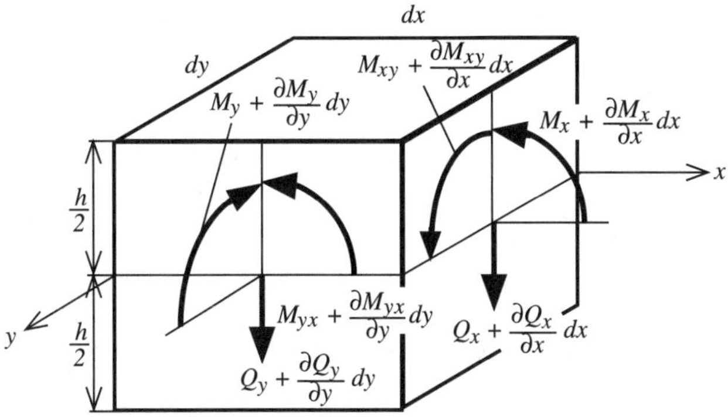

Figure 8.3 Moments and shearing forces.

per unit length parallel to axes x and y in Fig. 8.3

$$Q_x = \int_{-h/2}^{h/2} \sigma_{xz}\, dz, \qquad Q_y = \int_{-h/2}^{h/2} \sigma_{yz}\, dz$$
$$M_{yx} = \int_{-h/2}^{h/2} \sigma_{yx} z\, dz \tag{8.49}$$

The equilibrium equations of moments about y and x axes are

$$\frac{\partial M_x}{\partial x} + \frac{\partial M_{yx}}{\partial y} - Q_x = 0$$
$$\frac{\partial M_y}{\partial y} - \frac{\partial M_{xy}}{\partial x} - Q_y = 0 \tag{8.50}$$

Comparing the definition of M_{xy} given by Eqs. (8.38) with M_{yx} in Eqs. (8.49), it follows that

$$M_{yx} = -M_{xy} \tag{8.51}$$

Furthermore, we consider the equilibrium in the out-of-plane direction of z, which gives us the following relation:

$$\frac{\partial Q_x}{\partial x} + \frac{\partial Q_y}{\partial y} + p = 0 \tag{8.52}$$

where p denotes an arbitrarily distributed external load acting on the surface of the plate. Substitution of Eqs. (8.50) into Eq. (8.52) gives the fundamental relation for thermal bending problems of a plate

$$\frac{\partial^2 M_x}{\partial x^2} + \frac{\partial^2 M_y}{\partial y^2} - 2\frac{\partial^2 M_{xy}}{\partial x \partial y} + p = 0 \tag{8.53}$$

Taking into account the relations of Eqs. (8.40), this relation can be rewritten in terms of deflection w alone

$$\frac{\partial^2}{\partial x^2}\left[D\left(\frac{\partial^2 w}{\partial x^2}+\nu\frac{\partial^2 w}{\partial y^2}\right)\right]+\frac{\partial^2}{\partial y^2}\left[D\left(\frac{\partial^2 w}{\partial y^2}+\nu\frac{\partial^2 w}{\partial x^2}\right)\right]$$
$$+2(1-\nu)\frac{\partial^2}{\partial x\partial y}\left(D\frac{\partial^2 w}{\partial x\partial y}\right)=p-\frac{1}{1-\nu}\nabla^2 M_T \tag{8.54}$$

This result is the fundamental equation of the bending problems of the plate due to the thermal load. If the bending rigidity of the plate D is not changed over the entire region of the plate, the fundamental equation (8.54) becomes

$$\nabla^2\nabla^2 w=\frac{1}{D}\left(p-\frac{1}{1-\nu}\nabla^2 M_T\right) \tag{8.55}$$

Furthermore, when the external load p is absent, the fundamental equation (8.55) results in

$$\nabla^2\nabla^2 w=-\frac{1}{(1-\nu)D}\nabla^2 M_T \tag{8.56}$$

8.2.2 Boundary Conditions of the Plate

We now consider boundary conditions for plate bending problems due to thermal loads. Let n and s be the normal and tangential directions of the boundary surface of the plate shown in Fig. 8.4.

[1] For a built-in edge or fixed end, the deflection w and its first derivative with respect to the normal n on the boundary surface must be zero

$$w=0, \qquad \frac{\partial w}{\partial n}=0 \tag{8.57}$$

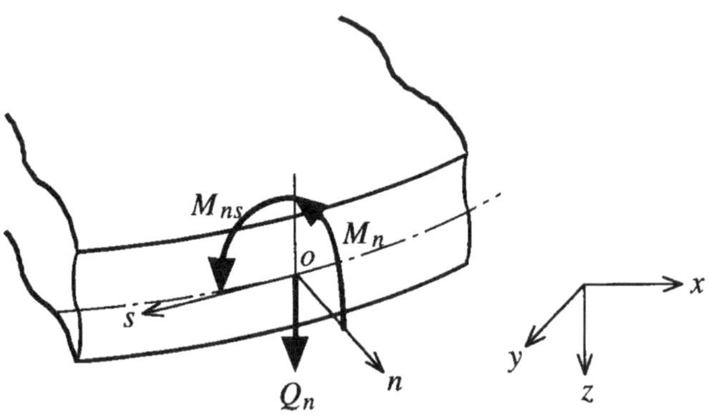

Figure 8.4 Boundary conditions.

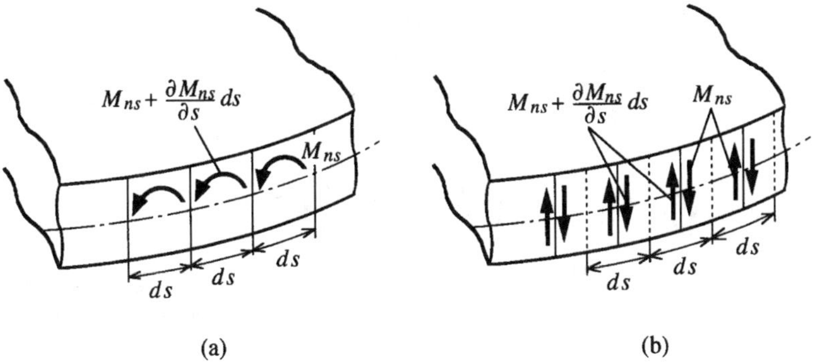

Figure 8.5 Kirchhoff's boundary condition.

[2] For a simply supported edge, the deflection w and bending moment M_n along the tangential axis s on the boundary surface must be zero

$$w = 0, \qquad M_n = 0 \tag{8.58}$$

[3] For a free edge, the bending moment M_n, twisting moment M_{ns} along the normal axis n, and the shearing force Q_n on the boundary surface must be zero

$$M_n = 0, \qquad M_{ns} = 0, \qquad Q_n = 0 \tag{a}$$

However, three boundary conditions in Eqs. (a) are too many. Kirchhoff* showed that the second and third terms in Eqs. (a) can be replaced by one boundary condition. Referring to Fig. 8.5, it can be found that the twisting moment M_{ns} is replaced by the shearing force Q'_n as

$$Q'_n = -\frac{\partial M_{ns}}{\partial s} \tag{b}$$

Thus, the latter two relations of Eqs. (a) are combined into the following form

$$V_n \equiv Q_n + Q'_n = Q_n - \frac{\partial M_{ns}}{\partial s} = 0 \tag{c}$$

in which V_n is called the equivalent shearing force per unit length. Therefore, the boundary conditions for a free edge are

$$M_n = 0, \qquad Q_n - \frac{\partial M_{ns}}{\partial s} = 0 \tag{8.59}$$

and are called the Kirchhoff's boundary conditions.

---

*See: G. Kirchhoff, *J. f. Math.* (Crelle), Vol. 40, p. 51 (1850).

Let α be an angle between x axis and the normal direction n on the boundary surface. Then, the derivatives with respect to n and s are

$$\begin{aligned} \frac{\partial}{\partial n} &= \cos\alpha \frac{\partial}{\partial x} + \sin\alpha \frac{\partial}{\partial y} \\ \frac{\partial}{\partial s} &= -\sin\alpha \frac{\partial}{\partial x} + \cos\alpha \frac{\partial}{\partial y} \end{aligned} \tag{d}$$

On the other hand, the moments M_n, M_s, and M_{ns}, and the shearing force Q_n are related to the components in the Cartesian coordinate system (x, y) in the following manner:

$$\begin{aligned} M_n &= M_x \cos^2\alpha + M_y \sin^2\alpha - 2M_{xy} \sin\alpha\cos\alpha \\ M_s &= M_x \sin^2\alpha + M_y \cos^2\alpha + 2M_{xy} \sin\alpha\cos\alpha \\ M_{ns} &= (M_x - M_y)\sin\alpha\cos\alpha + M_{xy}(\cos^2\alpha - \sin^2\alpha) \\ Q_n &= Q_x \cos\alpha + Q_y \sin\alpha \\ Q_s &= -Q_x \sin\alpha + Q_y \cos\alpha \end{aligned} \tag{e}$$

Substituting Eqs. (8.40) into Eqs. (8.50), shearing forces Q_x and Q_y are represented in terms of deflection w

$$\begin{aligned} Q_x &= -\frac{\partial}{\partial x}\left(D\nabla^2 w + \frac{1}{1-\nu}M_T\right) \\ Q_y &= -\frac{\partial}{\partial y}\left(D\nabla^2 w + \frac{1}{1-\nu}M_T\right) \end{aligned} \tag{8.60}$$

In the similar manner, we have following relations in terms of w for moments and shearing forces in the coordinate system (n, s):

$$\begin{aligned} M_n &= -D\left[\cos^2\alpha\left(\frac{\partial^2 w}{\partial x^2} + \nu\frac{\partial^2 w}{\partial y^2}\right) + \sin^2\alpha\left(\frac{\partial^2 w}{\partial y^2} + \nu\frac{\partial^2 w}{\partial x^2}\right)\right. \\ &\qquad \left. + 2(1-\nu)\sin\alpha\cos\alpha\frac{\partial^2 w}{\partial x\partial y}\right] - \frac{1}{1-\nu}M_T \\ M_s &= -D\left[\sin^2\alpha\left(\frac{\partial^2 w}{\partial x^2} + \nu\frac{\partial^2 w}{\partial y^2}\right) + \cos^2\alpha\left(\frac{\partial^2 w}{\partial y^2} + \nu\frac{\partial^2 w}{\partial x^2}\right)\right. \\ &\qquad \left. - 2(1-\nu)\sin\alpha\cos\alpha\frac{\partial^2 w}{\partial x\partial y}\right] - \frac{1}{1-\nu}M_T \\ M_{ns} &= (1-\nu)D\left[\sin\alpha\cos\alpha\left(\frac{\partial^2 w}{\partial y^2} - \frac{\partial^2 w}{\partial x^2}\right) + (\cos^2\alpha - \sin^2\alpha)\frac{\partial^2 w}{\partial x\partial y}\right] \\ Q_n &= -\left(\cos\alpha\frac{\partial}{\partial x} + \sin\alpha\frac{\partial}{\partial y}\right)\left(D\nabla^2 w + \frac{1}{1-\nu}M_T\right) \end{aligned}$$

$$Q_s = \left(\sin\alpha\frac{\partial}{\partial x} - \cos\alpha\frac{\partial}{\partial y}\right)\left(D\nabla^2 w + \frac{1}{1-\nu}M_T\right)$$

$$\begin{aligned}
V_n &= Q_n - \frac{\partial M_{ns}}{\partial s} \\
&= -D\left(\cos\alpha\frac{\partial}{\partial x}\left\{\left[1+(1-\nu)\sin^2\alpha\right]\frac{\partial^2 w}{\partial x^2}\right.\right. \\
&\qquad \left. + \left[2-\nu-3(1-\nu)\sin^2\alpha\right]\frac{\partial^2 w}{\partial y^2}\right\} \\
&\qquad + \sin\alpha\frac{\partial}{\partial y}\left\{\left[1+(1-\nu)\cos^2\alpha\right]\frac{\partial^2 w}{\partial y^2}\right. \\
&\qquad \left.\left. + \left[2-\nu-3(1-\nu)\cos^2\alpha\right]\frac{\partial^2 w}{\partial x^2}\right\}\right) \\
&\quad - \frac{1}{1-\nu}\left(\cos\alpha\frac{\partial}{\partial x} + \sin\alpha\frac{\partial}{\partial y}\right)M_T
\end{aligned} \tag{8.61}$$

The second derivatives with respect to n and s are

$$\begin{aligned}
\frac{\partial^2}{\partial n^2} &= \cos^2\alpha\frac{\partial^2}{\partial x^2} + 2\cos\alpha\sin\alpha\frac{\partial^2}{\partial x\partial y} + \sin^2\alpha\frac{\partial^2}{\partial y^2} \\
\frac{\partial^2}{\partial s^2} &= \sin^2\alpha\frac{\partial^2}{\partial x^2} - 2\cos\alpha\sin\alpha\frac{\partial^2}{\partial x\partial y} + \cos^2\alpha\frac{\partial^2}{\partial y^2} \\
\frac{\partial^2}{\partial n\partial s} &= -\cos\alpha\sin\alpha\frac{\partial^2}{\partial x^2} + (\cos^2\alpha - \sin^2\alpha)\frac{\partial^2}{\partial x\partial y} + \cos\alpha\sin\alpha\frac{\partial^2}{\partial y^2}
\end{aligned} \tag{8.62}$$

and alternative forms of Eqs. (d) and (8.62) are

$$\begin{aligned}
\frac{\partial}{\partial x} &= \cos\alpha\frac{\partial}{\partial n} - \sin\alpha\frac{\partial}{\partial s} \\
\frac{\partial}{\partial y} &= \sin\alpha\frac{\partial}{\partial n} + \cos\alpha\frac{\partial}{\partial s} \\
\frac{\partial^2}{\partial x^2} &= \cos^2\alpha\frac{\partial^2}{\partial n^2} - 2\cos\alpha\sin\alpha\frac{\partial^2}{\partial n\partial s} + \sin^2\alpha\frac{\partial^2}{\partial s^2} \\
\frac{\partial^2}{\partial y^2} &= \sin^2\alpha\frac{\partial^2}{\partial n^2} + 2\cos\alpha\sin\alpha\frac{\partial^2}{\partial n\partial s} + \cos^2\alpha\frac{\partial^2}{\partial s^2} \\
\frac{\partial^2}{\partial x\partial y} &= \cos\alpha\sin\alpha\frac{\partial^2}{\partial n^2} + (\cos^2\alpha - \sin^2\alpha)\frac{\partial^2}{\partial n\partial s} - \cos\alpha\sin\alpha\frac{\partial^2}{\partial s^2}
\end{aligned} \tag{f}$$

The moments M_n, M_s, M_{ns}, and the shearing force Q_n in Eqs. (8.61) can be rewritten by use of the coordinate system (n, s)

$$
\begin{aligned}
M_n &= -D\left(\frac{\partial^2 w}{\partial n^2} + \nu\frac{\partial^2 w}{\partial s^2}\right) - \frac{1}{1-\nu}M_T \\
M_s &= -D\left(\frac{\partial^2 w}{\partial s^2} + \nu\frac{\partial^2 w}{\partial n^2}\right) - \frac{1}{1-\nu}M_T \\
M_{ns} &= (1-\nu)D\frac{\partial^2 w}{\partial n \partial s} \\
Q_n &= -\frac{\partial}{\partial n}\left(D\nabla^2 w + \frac{1}{1-\nu}M_T\right) \\
Q_s &= -\frac{\partial}{\partial s}\left(D\nabla^2 w + \frac{1}{1-\nu}M_T\right) \\
V_n &= Q_n - \frac{\partial M_{ns}}{\partial s} = -\frac{\partial}{\partial n}\left\{D\left[\frac{\partial^2 w}{\partial n^2} + (2-\nu)\frac{\partial^2 w}{\partial s^2}\right] + \frac{1}{1-\nu}M_T\right\}
\end{aligned}
\tag{8.63}
$$

where

$$
\nabla^2 = \frac{\partial^2}{\partial n^2} + \frac{\partial^2}{\partial s^2} \tag{g}
$$

The boundary conditions are from Eqs. (8.57) to (8.59) in terms of w as follows:

[1] For a built-in edge

$$
w = 0, \qquad \frac{\partial w}{\partial n} = 0 \tag{8.64}
$$

[2] For a simply supported edge

$$
w = 0, \qquad \frac{\partial^2 w}{\partial n^2} + \nu\frac{\partial^2 w}{\partial s^2} = -\frac{1}{(1-\nu)D}M_T \tag{h}
$$

However, it can be found for the simply supported edge that the curvature $(\partial^2 w/\partial s^2)$ equals zero on the boundary surface of $n = \text{const}$ because of a prescribed condition of $w = 0$ on $n = \text{const}$. Thus, the boundary condition for a simply supported edge becomes

$$
w = 0, \qquad \frac{\partial^2 w}{\partial n^2} = -\frac{1}{(1-\nu)D}M_T \tag{8.65}
$$

[3] For a free edge

$$\frac{\partial^2 w}{\partial n^2} + \nu \frac{\partial^2 w}{\partial s^2} = -\frac{1}{(1-\nu)D} M_T$$
$$\frac{\partial}{\partial n}\left[\frac{\partial^2 w}{\partial n^2} + (2-\nu)\frac{\partial^2 w}{\partial s^2}\right] = -\frac{1}{(1-\nu)D}\frac{\partial M_T}{\partial n} \tag{8.66}$$

8.3 Fundamental Solutions for Rectangular Plates

8.3.1 Simply Supported Rectangular Plate

Consider a simply supported rectangular plate with its dimensions $a \times b$ subjected to thermal load, as shown in Fig. 8.6. The fundamental equation and the associated boundary conditions in the Cartesian coordinate system are given from Eqs. (8.56) and (8.65) as

$$\nabla^2 \nabla^2 w = -\frac{1}{(1-\nu)D} \nabla^2 M_T \tag{a}$$

$$w = 0, \quad \frac{\partial^2 w}{\partial x^2} = -\frac{1}{(1-\nu)D} M_T \quad \text{on } x = 0 \text{ and } x = a$$
$$w = 0, \quad \frac{\partial^2 w}{\partial y^2} = -\frac{1}{(1-\nu)D} M_T \quad \text{on } y = 0 \text{ and } y = b \tag{b}$$

Now, the fundamental equation (a) is resolved into the following two equation system

$$\nabla^2 w + \frac{1}{(1-\nu)D} M_T = F \tag{c}$$

$$\nabla^2 F = 0 \tag{d}$$

Thus, Eq. (d) is the governing equation for an unknown function F, and Eq. (c) is recognized as a boundary condition for F

$$F = \frac{\partial^2 w}{\partial x^2} + \frac{\partial^2 w}{\partial y^2} + \frac{1}{(1-\nu)D} M_T \tag{e}$$

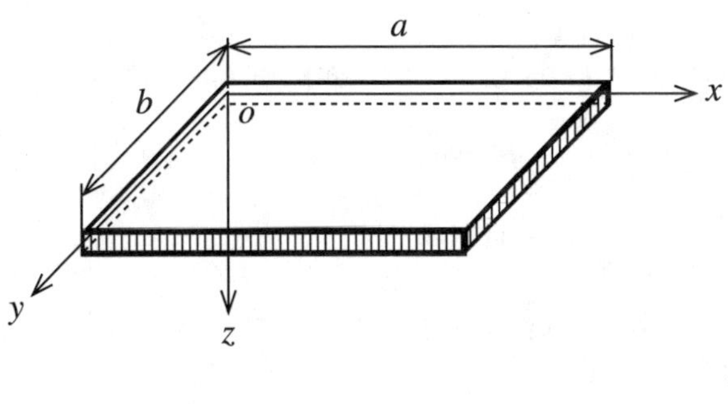

Figure 8.6 A simply supported rectangular plate.

We find that the curvature along the boundary surface becomes equal to zero,

$$\begin{aligned} \frac{\partial^2 w}{\partial y^2} &= 0 \qquad \text{on } x = 0 \text{ and } x = a \\ \frac{\partial^2 w}{\partial x^2} &= 0 \qquad \text{on } y = 0 \text{ and } y = b \end{aligned} \tag{f}$$

Using Eqs. (f) and (e), the second terms of boundary conditions (b) are reduced to the boundary conditions for F

$$\begin{aligned} F &= 0 \qquad \text{on } x = 0 \text{ and } x = a \\ F &= 0 \qquad \text{on } y = 0 \text{ and } y = b \end{aligned} \tag{g}$$

Therefore, the solution of F, the fundamental equation system given by Eqs. (d) and (g), is equal to zero

$$F = 0 \tag{h}$$

Substituting Eq. (h) into Eq. (c), the governing equation for a simply supported rectangular plate is

$$\nabla^2 w = -\frac{1}{(1-\nu)D} M_T \tag{8.67}$$

and the corresponding boundary conditions are

$$\begin{aligned} w &= 0 \qquad \text{on } x = 0 \text{ and } x = a \\ w &= 0 \qquad \text{on } y = 0 \text{ and } y = b \end{aligned} \tag{8.68}$$

We assume that the deflection w and the resultant moment M_T can be expanded in a double trigonometric series

$$w = \sum_{m=1}^{\infty} \sum_{n=1}^{\infty} w_{mn} \sin \alpha_m x \sin \beta_n y \tag{8.69}$$

$$M_T = \sum_{m=1}^{\infty} \sum_{n=1}^{\infty} a_{mn} \sin \alpha_m x \sin \beta_n y \tag{8.70}$$

in which α_m and β_n are determined so as to satisfy the boundary conditions (8.68)

$$\alpha_m = \frac{m\pi}{a}, \qquad \beta_n = \frac{n\pi}{b} \tag{8.71}$$

and the coefficient of the Fourier series expansion of M_T is

$$a_{mn} = \frac{4}{ab} \int_0^a \int_0^b M_T(x, y) \sin \alpha_m x \sin \beta_n y \, dx \, dy \tag{8.72}$$

Substitution of Eqs. (8.69) and (8.70) into Eqs. (8.67) gives

$$w_{mn} = \frac{a_{mn}}{(1-\nu)D} \frac{1}{\alpha_m^2 + \beta_n^2} \tag{8.73}$$

Thus, we have obtained the deflection w for a simply supported rectangular plate. Substituting Eqs. (8.69) and (8.70) into Eqs. (8.40) and (8.50), we obtain the following expressions for the resultant moments and shearing forces

$$\begin{aligned}
M_x &= -\sum_{m=1}^{\infty}\sum_{n=1}^{\infty} a_{mn} \frac{\beta_n^2}{\alpha_m^2 + \beta_n^2} \sin \alpha_m x \sin \beta_n y \\
M_y &= -\sum_{m=1}^{\infty}\sum_{n=1}^{\infty} a_{mn} \frac{\alpha_m^2}{\alpha_m^2 + \beta_n^2} \sin \alpha_m x \sin \beta_n y \\
M_{xy} &= \sum_{m=1}^{\infty}\sum_{n=1}^{\infty} a_{mn} \frac{\alpha_m \beta_n}{\alpha_m^2 + \beta_n^2} \cos \alpha_m x \cos \beta_n y \\
Q_x &= 0 \\
Q_y &= 0
\end{aligned} \tag{8.74}$$

Furthermore, if we are required to evaluate the thermal stress components due to thermal bending, the stress components σ_{xx}, σ_{yy}, and σ_{xy} can be calculated from Eqs. (8.43), in which the in-plane resultant forces N_x, N_y, and N_{xy} are omitted.

8.3.2 A Rectangular Plate with Two Opposite Edges Simply Supported and the Other Two Edges Clamped

We assume that the boundary surfaces $y = \pm(b/2)$ are built-in, and the other two boundary surfaces $x = \pm(a/2)$ are simply supported (see Fig. 8.7). Then the governing equation of the plate bending problem under thermal loading and the associated boundary

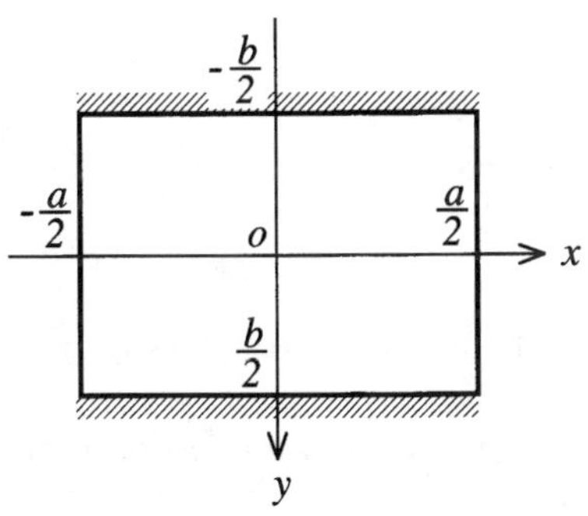

Figure 8.7 A rectangular plate with two built-in edges.

conditions are given from Eqs. (8.56), (8.64), and (8.65)

$$\nabla^2\nabla^2 w = -\frac{1}{(1-\nu)D}\nabla^2 M_T \tag{a}$$

and

$$\begin{aligned} w &= 0, \quad \frac{\partial^2 w}{\partial x^2} = -\frac{M_T}{(1-\nu)D} \qquad \text{on } x = \pm\frac{a}{2} \\ w &= 0, \quad \frac{\partial w}{\partial y} = 0 \qquad \text{on } y = \pm\frac{b}{2} \end{aligned} \tag{b}$$

Thermoelastic problems with built-in edge conditions can successfully be solved by making use of the method of superposition of the thermoelastic solution with simply supported edges as mentioned in Subsection 8.3.1 and the isothermal elastic solution with simply supported edges subjected to a bending moment along the edges. To simplify the analytical process, we assume that the thermal loads are applied symmetrically with respect to the coordinate axes x and y. Considering the parallel translation of the origin of the coordinate system between this problem and the one in Subsection 8.3.1, the thermoelastic deflection w_1 and thermally induced resultant moment M_T for the simply supported plate are reduced from Eqs. (8.69) to (8.73) to

$$\begin{aligned} w_1 &= \frac{1}{(1-\nu)D}\sum_{m=1,3,5,\ldots}^{\infty}\sum_{n=1,3,5,\ldots}^{\infty}\frac{a_{mn}}{\alpha_m^2+\beta_n^2}\cos\alpha_m x\cos\beta_n y \\ M_T &= \sum_{m=1,3,5,\ldots}^{\infty}\sum_{n=1,3,5,\ldots}^{\infty} a_{mn}\cos\alpha_m x\cos\beta_n y \end{aligned} \tag{8.75}$$

where

$$a_{mn} = \frac{4}{ab}\int_{-a/2}^{a/2}\int_{-b/2}^{b/2} M_T(x,y)\cos\alpha_m x\cos\beta_n y\,dx\,dy \tag{8.76}$$

and $\alpha_m = m\pi/a$ and $\beta_n = n\pi/b$. The gradients of w_1 on the boundary surfaces are

$$\begin{aligned} \left.\frac{\partial w_1}{\partial x}\right|_{x=a/2} &= -\frac{1}{(1-\nu)D}\sum_{m=1,3,5,\ldots}^{\infty}\sum_{n=1,3,5,\ldots}^{\infty}\frac{a_{mn}\alpha_m}{\alpha_m^2+\beta_n^2}(-1)^{(m-1)/2}\cos\beta_n y \\ \left.\frac{\partial w_1}{\partial y}\right|_{y=b/2} &= -\frac{1}{(1-\nu)D}\sum_{m=1,3,5,\ldots}^{\infty}\sum_{n=1,3,5,\ldots}^{\infty}\frac{a_{mn}\beta_n}{\alpha_m^2+\beta_n^2}(-1)^{(n-1)/2}\cos\alpha_m x \end{aligned} \tag{8.77}$$

The second equation in Eqs. (8.77) does not satisfy the boundary conditions on $y = \pm(b/2)$ in Eqs. (b).

Next, in order to satisfy all the boundary conditions, we consider an isothermal elastic problem of a simply supported rectangular plate where a bending moment $f(x)$

is applied on the edge surfaces $y = \pm(b/2)$. In this case, the bending moment $f(x)$ is symmetrically distributed with respect to y axis and can be expanded in the following series

$$f(x) = \sum_{m=1,3,5,\ldots}^{\infty} E_m \cos \alpha_m x \tag{8.78}$$

in which E_m are unknown constants to be determined from the built-in edge boundary condition. The governing equation of the deflection w_2 of the isothermal elastic problem is reduced from Eq. (8.56) to

$$\nabla^2 \nabla^2 w_2 = 0 \tag{c}$$

and the boundary conditions for a simply supported edge are

$$\begin{aligned} w_2 &= 0, \quad \frac{\partial^2 w_2}{\partial x^2} = 0 \qquad \text{on } x = \pm\frac{a}{2} \\ w_2 &= 0, \quad \frac{\partial^2 w_2}{\partial y^2} = -\frac{1}{D} f(x) \qquad \text{on } y = \pm\frac{b}{2} \end{aligned} \tag{d}$$

The suitable solution of Eq. (c) can be constructed from Eqs. (5.77)

$$w_2 = \sum_{m=1,3,5,\ldots}^{\infty} [C_{1m} \cosh \alpha_m y + C_{2m} \alpha_m y \sinh \alpha_m y] \cos \alpha_m x \tag{e}$$

As the solution (e) automatically satisfies the boundary condition on $x = \pm(a/2)$, we get from the boundary condition on $y = \pm(b/2)$

$$\begin{aligned} C_{1m} \cosh \gamma_m + C_{2m} \gamma_m \sinh \gamma_m &= 0 \\ C_{1m} \alpha_m^2 \cosh \gamma_m + C_{2m} \alpha_m^2 (2 \cosh \gamma_m + \gamma_m \sinh \gamma_m) &= -\frac{E_m}{D} \end{aligned} \tag{f}$$

Solving Eqs. (f) for C_{1m} and C_{2m}, and substituting C_{1m} and C_{2m} into Eq. (e) gives

$$\begin{aligned} w_2 = -\frac{1}{2D} \sum_{m=1,3,5,\ldots}^{\infty} \frac{E_m}{\alpha_m^2 \cosh \gamma_m} \\ \times (\alpha_m y \sinh \alpha_m y - \gamma_m \tanh \gamma_m \cosh \alpha_m y) \cos \alpha_m x \end{aligned} \tag{8.79}$$

where

$$\gamma_m = \frac{m\pi}{2} \frac{b}{a} \tag{8.80}$$

Therefore, the gradients of w_2 on the boundary surfaces are

$$
\begin{aligned}
\left.\frac{\partial w_2}{\partial x}\right|_{x=a/2} &= \frac{1}{2D}\sum_{m=1,3,5,\ldots}^{\infty}\frac{E_m(-1)^{(m-1)/2}}{\alpha_m\cosh\gamma_m} \\
&\quad\times(\alpha_m y\sinh\alpha_m y-\gamma_m\tanh\gamma_m\cosh\alpha_m y) \\
\left.\frac{\partial w_2}{\partial y}\right|_{y=b/2} &= -\frac{1}{2D}\sum_{m=1,3,5,\ldots}^{\infty}\frac{E_m}{\alpha_m\cosh^2\gamma_m} \\
&\quad\times(\gamma_m+\sinh\gamma_m\cosh\gamma_m)\cos\alpha_m x
\end{aligned}
\tag{8.81}
$$

Now we need to satisfy the second relation of the boundary condition (b) by

$$
\frac{\partial w_1}{\partial y}+\frac{\partial w_2}{\partial y}=0 \qquad \text{on } y=\pm\frac{b}{2} \tag{g}
$$

Substitution of Eqs. (8.77) and (8.81) into Eq. (g) gives the unknown constants E_m as

$$
E_m=-\frac{2}{1-\nu}\left(\frac{\alpha_m\cosh^2\gamma_m}{\gamma_m+\sinh\gamma_m\cosh\gamma_m}\right)\sum_{n=1,3,5,\ldots}^{\infty}\frac{a_{mn}\beta_n(-1)^{(n-1)/2}}{\alpha_m^2+\beta_n^2} \tag{8.82}
$$

Thus, we have obtained the deflection w by the superposition of w_1 and w_2. The resultant moments and shearing forces due to the deflection w_2 are calculated from Eqs. (8.40) and (8.50) as

$$
\begin{aligned}
M_x &= -\frac{1}{2}\sum_{m=1,3,5,\ldots}^{\infty}\frac{E_m}{\cosh\gamma_m}\{(1-\nu)\alpha_m y\sinh\alpha_m y \\
&\quad-[2\nu+(1-\nu)\gamma_m\tanh\gamma_m]\cosh\alpha_m y\}\cos\alpha_m x \\
M_y &= \frac{1}{2}\sum_{m=1,3,5,\ldots}^{\infty}\frac{E_m}{\cosh\gamma_m}\{(1-\nu)\alpha_m y\sinh\alpha_m y \\
&\quad+[2-(1-\nu)\gamma_m\tanh\gamma_m]\cosh\alpha_m y\}\cos\alpha_m x \\
M_{xy} &= \frac{1}{2}(1-\nu)\sum_{m=1,3,5,\ldots}^{\infty}\frac{E_m}{\cosh\gamma_m}[\alpha_m y\cosh\alpha_m y \\
&\quad+(1-\gamma_m\tanh\gamma_m)\sinh\alpha_m y]\sin\alpha_m x \\
Q_x &= -\sum_{m=1,3,5,\ldots}^{\infty}\frac{E_m\alpha_m}{\cosh\gamma_m}\cosh\alpha_m y\sin\alpha_m x \\
Q_y &= \sum_{m=1,3,5,\ldots}^{\infty}\frac{E_m\alpha_m}{\cosh\gamma_m}\sinh\alpha_m y\cos\alpha_m x
\end{aligned}
\tag{h}
$$

For the case where the boundary surface $y = \pm(b/2)$ is supported with a built-in edge, the deflection w, the resultant moments and shearing forces are obtained from Eqs. (8.75), (8.79), (8.40) and (8.50)

$$
\begin{aligned}
w = & \frac{1}{(1-\nu)D} \sum_{m=1,3,5,\dots}^{\infty} \sum_{n=1,3,5,\dots}^{\infty} \frac{a_{mn}}{\alpha_m^2 + \beta_n^2} \cos \alpha_m x \cos \beta_n y \\
& - \frac{1}{2D} \sum_{m=1,3,5,\dots}^{\infty} \frac{E_m}{\alpha_m^2 \cosh \gamma_m} (\alpha_m y \sinh \alpha_m y - \gamma_m \tanh \gamma_m \cosh \alpha_m y) \cos \alpha_m x \\
M_x = & - \sum_{m=1,3,5,\dots}^{\infty} \sum_{n=1,3,5,\dots}^{\infty} \frac{a_{mn} \beta_n^2}{\alpha_m^2 + \beta_n^2} \cos \alpha_m x \cos \beta_n y - \frac{1}{2} \sum_{m=1,3,5,\dots}^{\infty} \frac{E_m}{\cosh \gamma_m} \\
& \times \{(1-\nu)\alpha_m y \sinh \alpha_m y - [2\nu + (1-\nu)\gamma_m \tanh \gamma_m] \cosh \alpha_m y\} \cos \alpha_m x \\
M_y = & - \sum_{m=1,3,5,\dots}^{\infty} \sum_{n=1,3,5,\dots}^{\infty} \frac{a_{mn} \alpha_m^2}{\alpha_m^2 + \beta_n^2} \cos \alpha_m x \cos \beta_n y + \frac{1}{2} \sum_{m=1,3,5,\dots}^{\infty} \frac{E_m}{\cosh \gamma_m} \\
& \times \{(1-\nu)\alpha_m y \sinh \alpha_m y + [2 - (1-\nu)\gamma_m \tanh \gamma_m] \cosh \alpha_m y\} \cos \alpha_m x \\
M_{xy} = & \sum_{m=1,3,5,\dots}^{\infty} \sum_{n=1,3,5,\dots}^{\infty} \frac{a_{mn} \alpha_m \beta_n}{\alpha_m^2 + \beta_n^2} \sin \alpha_m x \sin \beta_n y + \frac{1}{2}(1-\nu) \sum_{m=1,3,5,\dots}^{\infty} \frac{E_m}{\cosh \gamma_m} \\
& \times [\alpha_m y \cosh \alpha_m y + (1 - \gamma_m \tanh \gamma_m) \sinh \alpha_m y] \sin \alpha_m x \\
Q_x = & - \sum_{m=1,3,5,\dots}^{\infty} \frac{E_m \alpha_m}{\cosh \gamma_m} \cosh \alpha_m y \sin \alpha_m x \\
Q_y = & \sum_{m=1,3,5,\dots}^{\infty} \frac{E_m \alpha_m}{\cosh \gamma_m} \sinh \alpha_m y \cos \alpha_m x
\end{aligned}
\tag{8.83}
$$

8.3.3 A Rectangular Plate with Built-in All Edges

We assume the boundary surfaces $x = \pm(a/2)$ and $y = \pm(b/2)$ are built-in edges in Fig. 8.8. In this case the governing equation of the plate bending problem under thermal load and the associated boundary conditions are given from Eqs. (8.56) and (8.64)

$$
\nabla^2 \nabla^2 w = -\frac{1}{(1-\nu)D} \nabla^2 M_T \tag{a}
$$

$$
\begin{aligned}
& w = 0, \quad \frac{\partial w}{\partial x} = 0 \quad \text{on } x = \pm \frac{a}{2} \\
& w = 0, \quad \frac{\partial w}{\partial y} = 0 \quad \text{on } y = \pm \frac{b}{2}
\end{aligned}
\tag{b}
$$

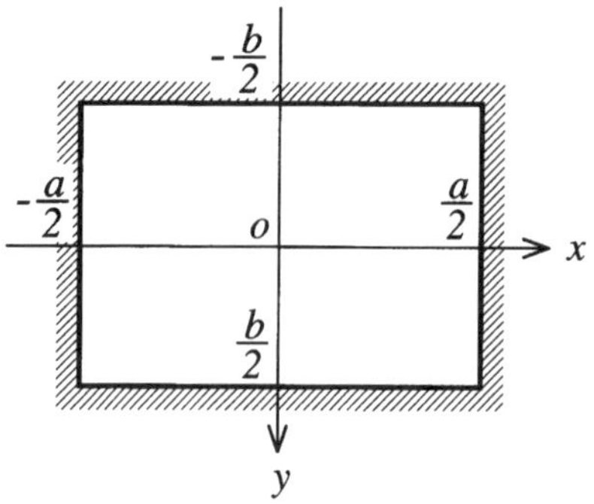

Figure 8.8 A rectangular plate with all edges built-in.

It is difficult to solve this problem directly. Therefore, we divide the problem into three problems. The first is a thermal deflection problem for a simply supported plate subject to a thermal load

$$\nabla^2\nabla^2 w_1 = -\frac{1}{(1-\nu)D}\nabla^2 M_T \tag{c}$$

$$\begin{aligned} w_1 &= 0, \quad \frac{\partial^2 w_1}{\partial x^2} = -\frac{1}{(1-\nu)D}M_T \qquad \text{on } x = \pm\frac{a}{2} \\ w_1 &= 0, \quad \frac{\partial^2 w_1}{\partial y^2} = -\frac{1}{(1-\nu)D}M_T \qquad \text{on } y = \pm\frac{b}{2} \end{aligned} \tag{d}$$

The second is an isothermal problem for a rectangular plate with two opposite edges $[x = \pm(a/2)]$ simply supported and the other two edges $[y = \pm(b/2)]$ with bending moment $f(x)$

$$\nabla^2\nabla^2 w_2 = 0 \tag{e}$$

$$\begin{aligned} w_2 &= 0, \quad \frac{\partial^2 w_2}{\partial x^2} = 0 \qquad \text{on } x = \pm\frac{a}{2} \\ w_2 &= 0, \quad \frac{\partial^2 w_2}{\partial y^2} = -\frac{1}{D}f(x) \qquad \text{on } y = \pm\frac{b}{2} \end{aligned} \tag{f}$$

The third is an isothermal problem for a rectangular plate with two opposite edges $[y = \pm(b/2)]$ simply supported and the other two edges $[x = \pm(a/2)]$ with bending moment $g(y)$

$$\nabla^2\nabla^2 w_3 = 0 \tag{g}$$

$$\begin{aligned} w_3 &= 0, \quad \frac{\partial^2 w_3}{\partial x^2} = -\frac{1}{D}g(y) \qquad \text{on } x = \pm\frac{a}{2} \\ w_3 &= 0, \quad \frac{\partial^2 w_3}{\partial y^2} = 0 \qquad \text{on } y = \pm\frac{b}{2} \end{aligned} \tag{h}$$

The first and second problems are discussed in Subsection 8.3.2. Now we consider the third problem. The bending moment $g(y)$ is symmetrically distributed with respect to the x axis and can be expanded into the series

$$g(y) = \sum_{n=1,3,5,\dots}^{\infty} F_n \cos \beta_n y \tag{8.84}$$

in which F_n are unknown constants to be determined from the boundary condition of the built-in edge and $\beta_n = (n\pi/b)$. Thus, the deflection w_3 and its derivatives on the edge surfaces $x = \pm(a/2)$ or $y = \pm(b/2)$ are

$$w_3 = -\frac{1}{2D} \sum_{n=1,3,5,\dots}^{\infty} \frac{F_n}{\beta_n^2 \cosh \delta_n} \times (\beta_n x \sinh \beta_n x - \delta_n \tanh \delta_n \cosh \beta_n x) \cos \beta_n y \tag{8.85}$$

$$\left.\frac{\partial w_3}{\partial x}\right|_{x=a/2} = -\frac{1}{2D} \sum_{n=1,3,5,\dots}^{\infty} \frac{F_n}{\beta_n \cosh^2 \delta_n} (\delta_n + \sinh \delta_n \cosh \delta_n) \cos \beta_n y$$

$$\left.\frac{\partial w_3}{\partial y}\right|_{y=b/2} = \frac{1}{2D} \sum_{n=1,3,5,\dots}^{\infty} \frac{F_n(-1)^{(n-1)/2}}{\beta_n \cosh \delta_n} (\beta_n x \sinh \beta_n x - \delta_n \tanh \delta_n \cosh \beta_n x) \tag{8.86}$$

where

$$\delta_n = \frac{n\pi}{2} \frac{a}{b} \tag{8.87}$$

Now, assuming that the two opposite edges $x = \pm(a/2)$ are built-in, the relation for the gradient of deflection is rewritten as

$$\frac{\partial w_1}{\partial x} + \frac{\partial w_3}{\partial x} = 0 \qquad \text{on } x = \pm\frac{a}{2} \tag{i}$$

Substituting Eqs. (8.77) and (8.86) into Eq. (i), the unknown constants F_n can be determined as

$$F_n = -\frac{2}{1-\nu} \left(\frac{\beta_n \cosh^2 \delta_n}{\delta_n + \sinh \delta_n \cosh \delta_n} \right) \sum_{m=1,3,5,\dots}^{\infty} \frac{a_{mn} \alpha_m}{\alpha_m^2 + \beta_n^2} (-1)^{(m-1)/2} \tag{8.88}$$

Thus, we have obtained the deflection w by the superposition of w_1 and w_3. The resultant moments and shearing forces corresponding to w_1 are given in terms of the double series form of Eq. (8.83), and the resultant moments and the shearing forces corresponding to

w_3 are calculated from w_3 given by Eq. (8.85). They are

$$
\begin{aligned}
M_x &= \frac{1}{2} \sum_{n=1,3,5,\ldots}^{\infty} \frac{F_n}{\cosh \delta_n} \{(1-\nu)\beta_n x \sinh \beta_n x \\
&\quad + [2 - (1-\nu)\delta_n \tanh \delta_n] \cosh \beta_n x\} \cos \beta_n y \\
M_y &= -\frac{1}{2} \sum_{n=1,3,5,\ldots}^{\infty} \frac{F_n}{\cosh \delta_n} \{(1-\nu)\beta_n x \sinh \beta_n x \\
&\quad - [2\nu + (1-\nu)\delta_n \tanh \delta_n] \cosh \beta_n x\} \cos \beta_n y \\
M_{xy} &= \frac{1}{2}(1-\nu) \sum_{n=1,3,5,\ldots}^{\infty} \frac{F_n}{\cosh \delta_n} [\beta_n x \cosh \beta_n x \\
&\quad + (1 - \delta_n \tanh \delta_n) \sinh \beta_n x] \sin \beta_n y \\
Q_x &= \sum_{n=1,3,5,\ldots}^{\infty} \frac{F_n \beta_n}{\cosh \delta_n} \sinh \beta_n x \cos \beta_n y \\
Q_y &= -\sum_{n=1,3,5,\ldots}^{\infty} \frac{F_n \beta_n}{\cosh \delta_n} \cosh \beta_n x \sin \beta_n y
\end{aligned}
\tag{8.89}
$$

We finally consider the rectangular plate with all edges built-in subject to thermal loading. The method of superposition of the solutions w_1, w_2, and w_3 will now be utilized. In other words, we assume that the bending moments $f(x)$ and $g(y)$, which are defined by Eqs. (8.78) and (8.84), act on the edge surface $y = \pm(b/2)$ and $x = \pm(a/2)$ simultaneously. Then the boundary conditions for the deformation gradient are written as

$$
\begin{aligned}
&\frac{\partial w_1}{\partial x} + \frac{\partial w_2}{\partial x} + \frac{\partial w_3}{\partial x} = 0 \qquad \text{on } x = \pm\frac{a}{2} \\
&\frac{\partial w_1}{\partial y} + \frac{\partial w_2}{\partial y} + \frac{\partial w_3}{\partial y} = 0 \qquad \text{on } y = \pm\frac{b}{2}
\end{aligned}
\tag{j}
$$

Substituting Eqs. (8.77), (8.81), and (8.86) into Eqs. (j), the following system of simultaneous equations for the unknown constants E_m and F_n is obtained:

$$
\begin{aligned}
&-\frac{1}{(1-\nu)D} \sum_{m=1,3,5,\ldots}^{\infty} \sum_{n=1,3,5,\ldots}^{\infty} \frac{a_{mn}\alpha_m}{\alpha_m^2 + \beta_n^2} (-1)^{(m-1)/2} \cos \beta_n y \\
&+ \frac{1}{2D} \sum_{m=1,3,5,\ldots}^{\infty} \frac{E_m (-1)^{(m-1)/2}}{\alpha_m \cosh \gamma_m} (\alpha_m y \sinh \alpha_m y - \gamma_m \tanh \gamma_m \cosh \alpha_m y)
\end{aligned}
$$

$$-\frac{1}{2D}\sum_{n=1,3,5,\ldots}^{\infty}\frac{F_n}{\beta_n\cosh^2\delta_n}(\delta_n+\sinh\delta_n\cosh\delta_n)\cos\beta_n y=0 \tag{k}$$

$$-\frac{1}{(1-\nu)D}\sum_{m=1,3,5,\ldots}^{\infty}\sum_{n=1,3,5,\ldots}^{\infty}\frac{a_{mn}\beta_n}{\alpha_m^2+\beta_n^2}(-1)^{(n-1)/2}\cos\alpha_m x$$

$$-\frac{1}{2D}\sum_{m=1,3,5,\ldots}^{\infty}\frac{E_m}{\alpha_m\cosh^2\gamma_m}(\gamma_m+\sinh\gamma_m\cosh\gamma_m)\cos\alpha_m x$$

$$+\frac{1}{2D}\sum_{n=1,3,5,\ldots}^{\infty}\frac{F_n(-1)^{(n-1)/2}}{\beta_n\cosh\delta_n}(\beta_n x\sinh\beta_n x-\delta_n\tanh\delta_n\cosh\beta_n x)=0 \tag{l}$$

We now introduce trigonometric series expansions

$$\begin{aligned}
\cosh\alpha_m y &= \frac{4}{b}\cosh\gamma_m\sum_{n=1,3,5,\ldots}^{\infty}\frac{\beta_n(-1)^{(n-1)/2}}{\alpha_m^2+\beta_n^2}\cos\beta_n y\\
\alpha_m y\sinh\alpha_m y &= \frac{4}{b}\sum_{n=1,3,5,\ldots}^{\infty}\frac{\beta_n(-1)^{(n-1)/2}}{(\alpha_m^2+\beta_n^2)^2}\\
&\quad\times\left[(\alpha_m^2+\beta_n^2)\gamma_m\sinh\gamma_m-2\alpha_m^2\cosh\gamma_m\right]\cos\beta_n y\\
\cosh\beta_n x &= \frac{4}{a}\cosh\delta_n\sum_{m=1,3,5,\ldots}^{\infty}\frac{\alpha_m(-1)^{(m-1)/2}}{\alpha_m^2+\beta_n^2}\cos\alpha_m x\\
\beta_n x\sinh\beta_n x &= \frac{4}{a}\sum_{m=1,3,5,\ldots}^{\infty}\frac{\alpha_m(-1)^{(m-1)/2}}{(\alpha_m^2+\beta_n^2)^2}\\
&\quad\times\left[(\alpha_m^2+\beta_n^2)\delta_n\sinh\delta_n-2\beta_n^2\cosh\delta_n\right]\cos\alpha_m x\\
&\quad\text{for } m,n=1,3,5,\ldots
\end{aligned} \tag{8.90}$$

Substituting Eqs. (8.90) into Eqs. (k) and (l), the unknown constants E_m and F_n are determined from the system of simultaneous equations

$$\left(\frac{\delta_n+\sinh\delta_n\cosh\delta_n}{2\beta_n\cosh^2\delta_n}\right)F_n+\frac{4}{b}\beta_n(-1)^{(n-1)/2}\sum_{m=1,3,5,\ldots}^{\infty}\frac{\alpha_m(-1)^{(m-1)/2}}{\left(\alpha_m^2+\beta_n^2\right)^2}E_m$$

$$=-\frac{1}{1-\nu}\sum_{m=1,3,5,\ldots}^{\infty}\frac{a_{mn}\alpha_m}{\alpha_m^2+\beta_n^2}(-1)^{(m-1)/2}\qquad\text{for } n=1,3,5,\ldots \tag{8.91}$$

$$\left(\frac{\gamma_m + \sinh\gamma_m \cosh\gamma_m}{2\alpha_m \cosh^2\gamma_m}\right) E_m + \frac{4}{a}\alpha_m(-1)^{(m-1)/2} \sum_{n=1,3,5,\ldots}^{\infty} \frac{\beta_n(-1)^{(n-1)/2}}{(\alpha_m^2+\beta_n^2)^2} F_n$$

$$= -\frac{1}{1-\nu} \sum_{n=1,3,5,\ldots}^{\infty} \frac{a_{mn}\beta_n}{\alpha_m^2+\beta_n^2}(-1)^{(n-1)/2} \qquad \text{for } m = 1, 3, 5, \ldots$$

Therefore, the deflection w of the plate with built-in edge conditions for all the boundary surfaces is evaluated by superposing the solutions of w_1, w_2, and w_3 represented by Eqs. (8.75), (8.79), and (8.85), and the components of the resultant moments and shearing forces are obtained by superposing the results given by Eqs. (8.83) and (8.89).

8.4 Basic Equations for a Circular Plate

8.4.1 Fundamental Relations of Thermal Bending

We now consider thermal bending problems of a circular plate of thickness h and radius a as shown in Fig. 8.9. The fundamental equation for deflection w and the components of the resultant moment and shearing force are derived by introducing a coordinate transformation between the Cartesian coordinate system (x, y) and the cylindrical coordinate system (r, θ). The resultant moments M_{rr}, $M_{\theta\theta}$, $M_{r\theta}$, and $M_{\theta r}$, and the shearing forces Q_r and Q_θ acting on a small element of the circular plate are shown in Fig. 8.10.

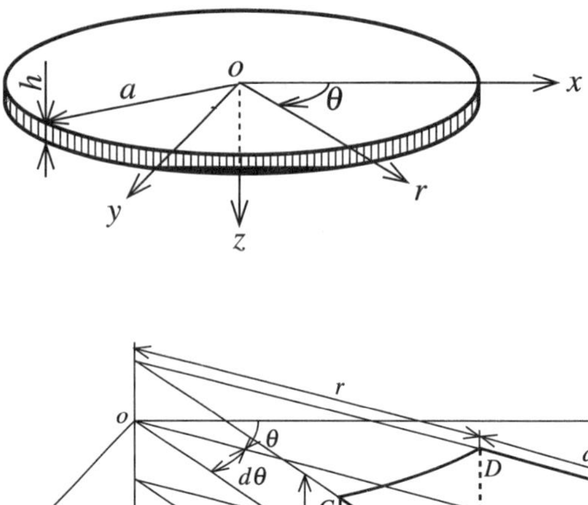

Figure 8.9 A circular plate.

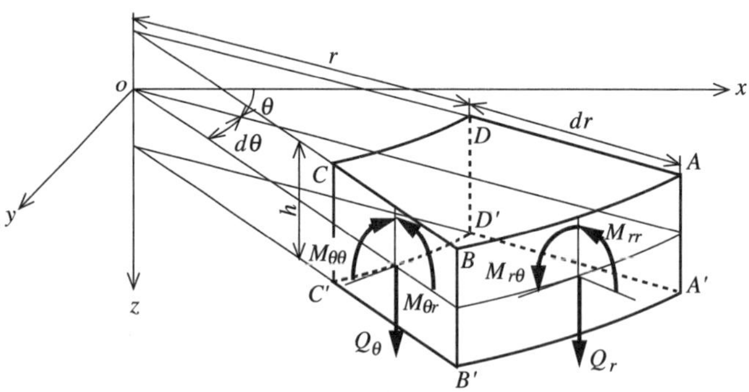

Figure 8.10 Moments and shearing forces.

Referring to Eqs. (8.56) and (8.61), and identifying the angle α with angle θ, the fundamental equation for w and the resultant moments and shearing forces in the cylindrical coordinate system are represented by the following expressions in terms of variables x and y:

$$\left(\frac{\partial^2}{\partial x^2}+\frac{\partial^2}{\partial y^2}\right)^2 w(r,\theta) = -\frac{1}{(1-\nu)D}\left(\frac{\partial^2}{\partial x^2}+\frac{\partial^2}{\partial y^2}\right) M_T(r,\theta) \tag{a}$$

$$\begin{aligned}
M_{rr} &= -D\left[\cos^2\theta\left(\frac{\partial^2 w}{\partial x^2}+\nu\frac{\partial^2 w}{\partial y^2}\right)+\sin^2\theta\left(\frac{\partial^2 w}{\partial y^2}+\nu\frac{\partial^2 w}{\partial x^2}\right)\right.\\
&\qquad \left.+\,2(1-\nu)\sin\theta\cos\theta\frac{\partial^2 w}{\partial x\partial y}\right]-\frac{1}{1-\nu}M_T\\
M_{\theta\theta} &= -D\left[\sin^2\theta\left(\frac{\partial^2 w}{\partial x^2}+\nu\frac{\partial^2 w}{\partial y^2}\right)+\cos^2\theta\left(\frac{\partial^2 w}{\partial y^2}+\nu\frac{\partial^2 w}{\partial x^2}\right)\right.\\
&\qquad \left.-\,2(1-\nu)\sin\theta\cos\theta\frac{\partial^2 w}{\partial x\partial y}\right]-\frac{1}{1-\nu}M_T
\end{aligned} \tag{b}$$

$$\begin{aligned}
M_{r\theta} &= (1-\nu)D\left[\sin\theta\cos\theta\left(\frac{\partial^2 w}{\partial y^2}-\frac{\partial^2 w}{\partial x^2}\right)+(\cos^2\theta-\sin^2\theta)\frac{\partial^2 w}{\partial x\partial y}\right]\\
Q_r &= -\left(\cos\theta\frac{\partial}{\partial x}+\sin\theta\frac{\partial}{\partial y}\right)\left[D\left(\frac{\partial^2 w}{\partial x^2}+\frac{\partial^2 w}{\partial y^2}\right)+\frac{1}{1-\nu}M_T\right]\\
Q_\theta &= \left(\sin\theta\frac{\partial}{\partial x}-\cos\theta\frac{\partial}{\partial y}\right)\left[D\left(\frac{\partial^2 w}{\partial x^2}+\frac{\partial^2 w}{\partial y^2}\right)+\frac{1}{1-\nu}M_T\right]
\end{aligned}$$

Using the relations between the coordinate systems (x, y) and (r, θ)

$$\begin{aligned}
x &= r\cos\theta, \qquad y = r\sin\theta\\
r^2 &= x^2+y^2, \qquad \theta = \tan^{-1}\frac{y}{x}
\end{aligned} \tag{c}$$

the relations for partial derivatives with respect to x or y are obtained

$$\begin{aligned}
\frac{\partial w}{\partial x} &= \cos\theta\frac{\partial w}{\partial r}-\sin\theta\frac{1}{r}\frac{\partial w}{\partial \theta}\\
\frac{\partial w}{\partial y} &= \sin\theta\frac{\partial w}{\partial r}+\cos\theta\frac{1}{r}\frac{\partial w}{\partial \theta}\\
\frac{\partial^2 w}{\partial x^2} &= \cos^2\theta\frac{\partial^2 w}{\partial r^2}+\sin^2\theta\left(\frac{1}{r}\frac{\partial w}{\partial r}+\frac{1}{r^2}\frac{\partial^2 w}{\partial \theta^2}\right)\\
&\qquad -2\sin\theta\cos\theta\left(\frac{1}{r}\frac{\partial^2 w}{\partial r\partial\theta}-\frac{1}{r^2}\frac{\partial w}{\partial \theta}\right)
\end{aligned}$$

$$\frac{\partial^2 w}{\partial y^2} = \sin^2\theta \frac{\partial^2 w}{\partial r^2} + \cos^2\theta \left(\frac{1}{r}\frac{\partial w}{\partial r} + \frac{1}{r^2}\frac{\partial^2 w}{\partial \theta^2} \right)$$
$$+ 2\sin\theta\cos\theta \left(\frac{1}{r}\frac{\partial^2 w}{\partial r \partial \theta} - \frac{1}{r^2}\frac{\partial w}{\partial \theta} \right)$$
$$\frac{\partial^2 w}{\partial x \partial y} = \sin\theta\cos\theta \left(\frac{\partial^2 w}{\partial r^2} - \frac{1}{r}\frac{\partial w}{\partial r} - \frac{1}{r^2}\frac{\partial^2 w}{\partial \theta^2} \right)$$
$$+ (\cos^2\theta - \sin^2\theta) \left(\frac{1}{r}\frac{\partial^2 w}{\partial r \partial \theta} - \frac{1}{r^2}\frac{\partial w}{\partial \theta} \right) \qquad \text{(d)}$$

Substituting Eqs. (d) into Eqs. (a) and (b), the governing equation of plate bending for a circular plate is

$$\nabla^2\nabla^2 w(r,\theta) = -\frac{1}{(1-\nu)D}\nabla^2 M_T(r,\theta) \qquad (8.92)$$

where

$$\nabla^2 = \frac{\partial^2}{\partial r^2} + \frac{1}{r}\frac{\partial}{\partial r} + \frac{1}{r^2}\frac{\partial^2}{\partial \theta^2} \qquad (8.93)$$

and the components of the resultant moments and shearing forces are represented as

$$M_{rr} = -D\left[\frac{\partial^2 w}{\partial r^2} + \nu \left(\frac{1}{r}\frac{\partial w}{\partial r} + \frac{1}{r^2}\frac{\partial^2 w}{\partial \theta^2} \right) \right] - \frac{1}{1-\nu} M_T$$
$$M_{\theta\theta} = -D\left[\nu\frac{\partial^2 w}{\partial r^2} + \left(\frac{1}{r}\frac{\partial w}{\partial r} + \frac{1}{r^2}\frac{\partial^2 w}{\partial \theta^2} \right) \right] - \frac{1}{1-\nu} M_T$$
$$M_{r\theta} = (1-\nu)D\frac{\partial}{\partial r}\left(\frac{1}{r}\frac{\partial w}{\partial \theta} \right) \qquad (8.94)$$
$$Q_r = -D\frac{\partial}{\partial r}(\nabla^2 w) - \frac{1}{1-\nu}\frac{\partial M_T}{\partial r}$$
$$Q_\theta = -D\frac{1}{r}\frac{\partial}{\partial \theta}(\nabla^2 w) - \frac{1}{1-\nu}\frac{1}{r}\frac{\partial M_T}{\partial \theta}$$

Furthermore, the thermal stress components σ_{rr}, $\sigma_{\theta\theta}$, and $\sigma_{r\theta}$ are evaluated from Eqs. (8.43) by adopting the coordinate transformation rule

$$\sigma_{rr} = \frac{1}{h}N_r + \frac{12z}{h^3}M_{rr} + \frac{1}{1-\nu}\left(\frac{1}{h}N_T + \frac{12z}{h^3}M_T - \alpha E \tau \right)$$
$$\sigma_{\theta\theta} = \frac{1}{h}N_\theta + \frac{12z}{h^3}M_{\theta\theta} + \frac{1}{1-\nu}\left(\frac{1}{h}N_T + \frac{12z}{h^3}M_T - \alpha E \tau \right) \qquad (8.95)$$
$$\sigma_{r\theta} = \frac{1}{h}N_{r\theta} - \frac{12z}{h^3}M_{r\theta}$$

in which the resultant forces and resultant moments in the cylindrical coordinate system are defined as

$$
\begin{aligned}
N_r &= \int_{-h/2}^{h/2} \sigma_{rr}\, dz, \qquad N_\theta = \int_{-h/2}^{h/2} \sigma_{\theta\theta}\, dz, \qquad N_{r\theta} = \int_{-h/2}^{h/2} \sigma_{r\theta}\, dz, \\
M_{rr} &= \int_{-h/2}^{h/2} \sigma_{rr} z\, dz, \qquad M_{\theta\theta} = \int_{-h/2}^{h/2} \sigma_{\theta\theta} z\, dz, \qquad M_{r\theta} = -\int_{-h/2}^{h/2} \sigma_{r\theta} z\, dz
\end{aligned}
\tag{8.96}
$$

8.4.2 Boundary Conditions for a Circular Plate

Let us consider the boundary conditions for a circular plate subjected to thermal loading. Referring to Eqs. (8.57) to (8.59), three different kinds of boundary conditions on the edge surface $r = \text{const}$ can be identified:

[1] For a built-in edge

$$
w = 0, \qquad \frac{\partial w}{\partial r} = 0 \tag{a}
$$

[2] For a simply supported edge

$$
w = 0, \qquad M_{rr} = 0 \tag{b}
$$

[3] For a free edge

$$
M_{rr} = 0, \qquad Q_r - \frac{1}{r}\frac{\partial M_{r\theta}}{\partial \theta} = 0 \tag{c}
$$

Making use of the relations of Eqs. (8.94), these boundary conditions are rewritten in terms of deflection w as

[1] For the built-in edge

$$
w = 0, \qquad \frac{\partial w}{\partial r} = 0 \tag{8.97}
$$

[2] For the simply supported edge

$$
w = 0, \qquad \frac{\partial^2 w}{\partial r^2} + \nu\left(\frac{1}{r}\frac{\partial w}{\partial r} + \frac{1}{r^2}\frac{\partial^2 w}{\partial \theta^2}\right) = -\frac{1}{(1-\nu)D} M_T \tag{8.98}
$$

[3] For the free edge

$$
\begin{aligned}
&\frac{\partial^2 w}{\partial r^2} + \nu\left(\frac{1}{r}\frac{\partial w}{\partial r} + \frac{1}{r^2}\frac{\partial^2 w}{\partial \theta^2}\right) = -\frac{1}{(1-\nu)D} M_T \\
&\frac{\partial}{\partial r}(\nabla^2 w) + (1-\nu)\frac{1}{r}\frac{\partial}{\partial r}\left(\frac{1}{r}\frac{\partial^2 w}{\partial \theta^2}\right) = -\frac{1}{(1-\nu)D}\frac{\partial M_T}{\partial r}
\end{aligned}
\tag{8.99}
$$

8.4.3 Axisymmetric Thermal Bending Problems

We now consider an axisymmetric thermal bending problem where the circular plate is subjected to axisymmetric thermal loads and its end surfaces are subjected to axisymmetric boundary conditions.

Referring to the results of the Subsections 8.4.1 and 8.4.2, the governing equation (8.92) is changed to the form

$$\nabla^2\nabla^2 w(r) = -\frac{1}{(1-\nu)D}\nabla^2 M_T(r) \tag{8.100}$$

where

$$\nabla^2 = \frac{\partial^2}{\partial r^2} + \frac{1}{r}\frac{\partial}{\partial r} = \frac{1}{r}\frac{\partial}{\partial r}\left(r\frac{\partial}{\partial r}\right) \tag{8.101}$$

and the boundary conditions given by Eqs. (8.97) to (8.99) are rewritten as

[1] For the built-in edge

$$w = 0, \qquad \frac{\partial w}{\partial r} = 0 \tag{8.102}$$

[2] For the simply supported edge

$$w = 0, \qquad \frac{\partial^2 w}{\partial r^2} + \nu\frac{1}{r}\frac{\partial w}{\partial r} = -\frac{1}{(1-\nu)D}M_T \tag{8.103}$$

[3] For the free edge

$$\begin{aligned} \frac{\partial^2 w}{\partial r^2} + \nu\frac{1}{r}\frac{\partial w}{\partial r} &= -\frac{1}{(1-\nu)D}M_T \\ \frac{\partial}{\partial r}(\nabla^2 w) &= -\frac{1}{(1-\nu)D}\frac{\partial M_T}{\partial r} \end{aligned} \tag{8.104}$$

The resultant moments and shearing forces are (from Eqs. (8.94))

$$\begin{aligned} M_{rr} &= -D\left(\frac{\partial^2 w}{\partial r^2} + \nu\frac{1}{r}\frac{\partial w}{\partial r}\right) - \frac{1}{1-\nu}M_T \\ M_{\theta\theta} &= -D\left(\nu\frac{\partial^2 w}{\partial r^2} + \frac{1}{r}\frac{\partial w}{\partial r}\right) - \frac{1}{1-\nu}M_T \\ Q_r &= -D\frac{\partial}{\partial r}(\nabla^2 w) - \frac{1}{1-\nu}\frac{\partial M_T}{\partial r} \\ M_{r\theta} &= 0 \\ Q_\theta &= 0 \end{aligned} \tag{8.105}$$

8.5 Fundamental Solutions for Circular Plates

8.5.1 Axisymmetric Problems

Let us consider axisymmetric deformations of circular plates. The governing equation of deflection w is given by Eq. (8.100)

$$\frac{1}{r}\frac{\partial}{\partial r}\left\{r\frac{\partial}{\partial r}\left[\frac{1}{r}\frac{\partial}{\partial r}\left(r\frac{\partial w}{\partial r}\right)+\frac{1}{(1-\nu)D}M_T\right]\right\}=0 \tag{a}$$

Performing successive integrations on Eq. (a), the general solution of deflection w is

$$w(r)=\frac{1}{D}\left[C_1r^2+C_2+C_3r^2\ln r+C_4\ln r-\frac{1}{1-\nu}\int_r\frac{1}{r}\int_r rM_T(r)\,dr\,dr\right] \tag{8.106}$$

in which the unknown constants C_1 to C_4 are determined from the boundary conditions. Substituting Eq. (8.106) into Eqs. (8.105), the resultant moments M_{rr}, $M_{\theta\theta}$, and shearing force Q_r are represented as

$$\begin{aligned}
M_{rr}&=-\Big\{2(1+\nu)C_1+[2(1+\nu)\ln r+3+\nu]\,C_3\\
&\qquad-(1-\nu)C_4\frac{1}{r^2}+\frac{1}{r^2}\int_r rM_T(r)\,dr\Big\}\\
M_{\theta\theta}&=-\Big\{2(1+\nu)C_1+[2(1+\nu)\ln r+1+3\nu]\,C_3\\
&\qquad+(1-\nu)C_4\frac{1}{r^2}+M_T-\frac{1}{r^2}\int_r rM_T(r)\,dr\Big\}\\
Q_r&=-4C_3\frac{1}{r}
\end{aligned} \tag{8.107}$$

Furthermore, the boundary conditions (8.102) to (8.104) on the edge surface are given as follows:

[1] For the built-in edge

$$\begin{aligned}
C_1r^2+C_2+C_3r^2\ln r+C_4\ln r-\frac{1}{1-\nu}\int_r\frac{1}{r}\int_r rM_T(r)\,dr\,dr&=0\\
2C_1r+C_3r(2\ln r+1)+C_4\frac{1}{r}-\frac{1}{1-\nu}\frac{1}{r}\int_r rM_T(r)\,dr&=0
\end{aligned} \tag{8.108}$$

[2] For the simply supported edge

$$
\begin{aligned}
&C_1 r^2 + C_2 + C_3 r^2 \ln r + C_4 \ln r \\
&\quad - \frac{1}{1-\nu} \int_r \frac{1}{r} \int_r r M_T(r)\, dr\, dr = 0 \\
&2(1+\nu)C_1 + [2(1+\nu)\ln r + 3 + \nu]C_3 \\
&\quad - (1-\nu)C_4 \frac{1}{r^2} + \frac{1}{r^2} \int_r r M_T(r)\, dr = 0
\end{aligned}
\tag{8.109}
$$

[3] For the free edge

$$
\begin{aligned}
&2(1+\nu)C_1 + [2(1+\nu)\ln r + 3 + \nu]C_3 \\
&\quad -(1-\nu)C_4 \frac{1}{r^2} + \frac{1}{r^2} \int_r r M_T(r)\, dr = 0 \\
&C_3 = 0
\end{aligned}
\tag{8.110}
$$

Now let us consider a solid circular plate with radius a. It can be found from Eqs. (8.106) and (8.107) that the solutions for such a plate are easily derived by neglecting the terms containing C_3 and C_4. Changing the indefinite integrals in Eqs. (8.106) and (8.107) into definite integrals with integral path from 0 to r, the expressions for w, M_{rr}, $M_{\theta\theta}$, and Q_r are given as

$$
\begin{aligned}
w &= \frac{1}{D}\left[C_1 r^2 + C_2 - \frac{1}{1-\nu} \int_0^r \frac{1}{r} \int_0^r r M_T(r)\, dr\, dr\right] \\
M_{rr} &= -\left[2(1+\nu)C_1 + \frac{1}{r^2} \int_0^r r M_T(r)\, dr\right] \\
M_{\theta\theta} &= -\left[2(1+\nu)C_1 + M_T - \frac{1}{r^2} \int_0^r r M_T(r)\, dr\right] \\
Q_r &= 0
\end{aligned}
\tag{b}
$$

The unknown constants C_1 and C_2 are determined from Eqs. (8.108) to (8.110), which are given as

[1] For the built-in edge condition
The unknown constants C_1 and C_2 are

$$
\begin{aligned}
C_1 &= \frac{1}{1-\nu} \frac{1}{2a^2} \int_0^a r M_T(r)\, dr \\
C_2 &= \frac{1}{1-\nu}\left[\int_0^a \frac{1}{r} \int_0^r r M_T(r)\, dr\, dr - \frac{1}{2} \int_0^a r M_T(r)\, dr\right]
\end{aligned}
\tag{8.111}
$$

w, M_{rr}, $M_{\theta\theta}$, and Q_r are

$$w = \frac{1}{(1-\nu)D}\left[\int_r^a \frac{1}{r}\int_0^r rM_T(r)\,dr\,dr - \frac{1}{2}\left(1-\frac{r^2}{a^2}\right)\int_0^a rM_T(r)\,dr\right]$$
$$M_{rr} = -\frac{1+\nu}{1-\nu}\frac{1}{a^2}\int_0^a rM_T(r)\,dr - \frac{1}{r^2}\int_0^r rM_T(r)\,dr$$
$$M_{\theta\theta} = -\frac{1+\nu}{1-\nu}\frac{1}{a^2}\int_0^a rM_T(r)\,dr - M_T(r) + \frac{1}{r^2}\int_0^r rM_T(r)\,dr$$
$$Q_r = 0 \tag{8.112}$$

[2] For the simply supported edge condition
The unknown constants C_1 and C_2 are

$$C_1 = -\frac{1}{2(1+\nu)}\frac{1}{a^2}\int_0^a rM_T(r)\,dr$$
$$C_2 = \frac{1}{1-\nu}\int_0^a \frac{1}{r}\int_0^r rM_T(r)\,dr\,dr + \frac{1}{2(1+\nu)}\int_0^a rM_T(r)\,dr \tag{8.113}$$

w, M_{rr}, $M_{\theta\theta}$, and Q_r are

$$w=\frac{1}{D}\left[\frac{1}{1-\nu}\int_r^a \frac{1}{r}\int_0^r rM_T(r)\,dr\,dr+\frac{1}{2(1+\nu)}\left(1-\frac{r^2}{a^2}\right)\int_0^a rM_T(r)\,dr\right]$$
$$M_{rr}=\frac{1}{a^2}\int_0^a rM_T(r)\,dr-\frac{1}{r^2}\int_0^r rM_T(r)\,dr \tag{8.114}$$
$$M_{\theta\theta}=\frac{1}{a^2}\int_0^a rM_T(r)\,dr+\frac{1}{r^2}\int_0^r rM_T(r)\,dr-M_T(r)$$
$$Q_r=0$$

[3] For the free edge condition
The unknown constant C_1 is

$$C_1 = -\frac{1}{2(1+\nu)}\frac{1}{a^2}\int_0^a rM_T(r)\,dr \tag{8.115}$$

w, M_{rr}, $M_{\theta\theta}$, and Q_r are

$$w=\frac{1}{D}\left[C_2-\frac{1}{1-\nu}\int_0^r \frac{1}{r}\int_0^r rM_T(r)\,dr\,dr-\frac{1}{2(1+\nu)}\frac{r^2}{a^2}\int_0^a rM_T(r)\,dr\right]$$
$$M_{rr}=\frac{1}{a^2}\int_0^a rM_T(r)\,dr-\frac{1}{r^2}\int_0^r rM_T(r)\,dr \tag{8.116}$$
$$M_{\theta\theta}=\frac{1}{a^2}\int_0^a rM_T(r)\,dr+\frac{1}{r^2}\int_0^r rM_T(r)\,dr-M_T(r)$$
$$Q_r=0$$

The deflection w in Eqs. (8.116) has an unknown constant C_2. In order to determine C_2, a condition for the deflection w on the circular edge $r = a$ should be added. When the additional condition is zero, C_2 is

$$C_2 = \frac{1}{1-\nu}\int_0^a \frac{1}{r}\int_0^r rM_T(r)\,dr\,dr + \frac{1}{2(1+\nu)}\int_0^a rM_T(r)\,dr \tag{8.117}$$

It can be found that these results are identical with Eqs. (8.113) for the simply supported edge condition. Therefore, the solution for a solid circular plate with free edge condition is identical with the one for the simply supported edge condition given by Eqs. (8.114).

Let us consider axisymmetric problems of a hollow circular plate with inner radius a and outer radius b. In such cases, the deflection w, resultant moments M_{rr} and $M_{\theta\theta}$, and shearing force Q_r are formulated by Eqs. (8.106) to (8.110). Many combinations of boundary conditions can be considered for a hollow circular plate, and the solutions for a few typical cases will be given here.

[1] Both edges $r = a$ and $r = b$ are built-in

$$\begin{aligned}
w = \frac{1}{(1-\nu)D}\Bigg\{&\frac{1}{B}\left[-\frac{1}{ab}(b^2-a^2)(r^2-a^2) - 2\frac{b}{a}r^2\ln\frac{b}{r}\right.\\
&\left.-2\frac{a}{b}r^2\ln\frac{r}{a} + 2ab\ln\frac{b}{a}\left(1+2\ln\frac{r}{a}\right)\right]\int_a^b \frac{1}{r}\int_a^r rM_T(r)\,dr\,dr\\
&+\frac{1}{B}\left[-2\frac{a}{b}(b^2-r^2)\ln\frac{b}{a}\ln\frac{r}{a} + \frac{1}{ab}(b^2-a^2)(r^2-a^2)\ln\frac{b}{r}\right]\\
&\times\int_a^b rM_T(r)\,dr - \int_a^r \frac{1}{r}\int_a^r rM_T(r)\,dr\,dr\Bigg\}
\end{aligned} \tag{8.118}$$

$$\begin{aligned}
M_{rr} = -\frac{4}{(1-\nu)B}&\left[-(1+\nu)\left(\frac{b}{a}\ln\frac{b}{r} + \frac{a}{b}\ln\frac{r}{a}\right)\right.\\
&\left.-(1-\nu)\frac{ab}{r^2}\ln\frac{b}{a} + \frac{b^2-a^2}{ab}\right]\int_a^b \frac{1}{r}\int_a^r rM_T(r)\,dr\,dr\\
-\frac{1}{(1-\nu)B}&\left\{4(1+\nu)\frac{a}{b}\ln\frac{b}{a}\ln\frac{r}{a} + 2(1+\nu)\frac{b^2-a^2}{ab}\ln\frac{b}{r}\right.\\
&\left.+2\frac{a}{b}\left[3+\nu+(1-\nu)\frac{b^2}{r^2}\right]\ln\frac{b}{a} - \frac{b^2-a^2}{ab}\left[3+\nu+(1-\nu)\frac{a^2}{r^2}\right]\right\}\\
&\times\int_a^b rM_T(r)\,dr - \frac{1}{r^2}\int_a^r rM_T(r)\,dr
\end{aligned}$$

$$\begin{aligned}
M_{\theta\theta} = -\frac{4}{(1-\nu)B}&\left[-(1+\nu)\left(\frac{b}{a}\ln\frac{b}{r} + \frac{a}{b}\ln\frac{r}{a}\right)\right.\\
&\left.+(1-\nu)\frac{ab}{r^2}\ln\frac{b}{a} + \nu\frac{b^2-a^2}{ab}\right]\int_a^b \frac{1}{r}\int_a^r rM_T(r)\,dr\,dr
\end{aligned}$$

$$- \frac{1}{(1-\nu)B}\left\{4(1+\nu)\frac{a}{b}\ln\frac{b}{a}\ln\frac{r}{a} + 2(1+\nu)\frac{b^2-a^2}{ab}\ln\frac{b}{r}\right.$$

$$\left. + 2\frac{a}{b}\left[1+3\nu-(1-\nu)\frac{b^2}{r^2}\right]\ln\frac{b}{a} - \frac{b^2-a^2}{ab}\left[1+3\nu-(1-\nu)\frac{a^2}{r^2}\right]\right\}$$

$$\times \int_a^b rM_T(r)\,dr - M_T(r) + \frac{1}{r^2}\int_a^r rM_T(r)\,dr \tag{8.119}$$

$$Q_r = -\frac{4}{(1-\nu)B}\frac{1}{r}\left[2\frac{b^2-a^2}{ab}\int_a^b \frac{1}{r}\int_a^r rM_T(r)\,dr\,dr\right.$$

$$\left. + \left(2\frac{a}{b}\ln\frac{b}{a} - \frac{b^2-a^2}{ab}\right)\int_a^b rM_T(r)\,dr\right] \tag{8.120}$$

where

$$B = 4ab\left(\ln\frac{b}{a}\right)^2 - \frac{1}{ab}(b^2-a^2)^2 \tag{8.121}$$

[2] Both edges $r = a$ and $r = b$ are simply supported

$$w = \frac{1}{(1-\nu)DB}\left[4(1+\nu)^2\ln\frac{r}{a}\ln\frac{b}{a}\right.$$

$$+ 2(1-\nu^2)\left(\frac{r^2}{a^2}\ln\frac{b}{r} + \frac{r^2}{b^2}\ln\frac{r}{a} - \ln\frac{b}{a}\right)$$

$$\left. +(3+\nu)(1-\nu)\frac{1}{a^2b^2}(b^2-a^2)(r^2-a^2)\right]\int_a^b \frac{1}{r}\int_a^r rM_T(r)\,dr\,dr$$

$$- \frac{1}{DB}\frac{1}{b^2}\left\{-2(1+\nu)(b^2-r^2)\ln\frac{r}{a}\ln\frac{b}{a} - 4(b^2-r^2)\ln a\right.$$

$$- \left[3+\nu+(1-\nu)\frac{b^2}{a^2}\right](r^2-a^2)\ln b$$

$$\left. + \left[3+\nu+(1-\nu)\frac{r^2}{a^2}\right](b^2-a^2)\ln r\right\}\int_a^b rM_T(r)\,dr$$

$$- \frac{1}{(1-\nu)D}\int_a^r \frac{1}{r}\int_a^r rM_T(r)\,dr\,dr \tag{8.122}$$

$$M_{rr} = \frac{4(1+\nu)^2}{a^2b^2}\frac{1}{B}\left(-b^2\ln\frac{b}{r} - a^2\ln\frac{r}{a} + \frac{a^2b^2}{r^2}\ln\frac{b}{a}\right)\int_a^b \frac{1}{r}\int_a^r rM_T(r)\,dr\,dr$$

$$- \frac{1}{b^2}\frac{1}{B}\left[-4(1+\nu)^2\ln\frac{r}{a}\ln\frac{b}{a} - (3+\nu)(1-\nu)\frac{(b^2-a^2)(r^2-a^2)}{a^2r^2}\right.$$

$$+2(1-\nu^2)\left(\frac{b^2}{a^2}\ln\frac{b}{r}+\ln\frac{r}{a}-\frac{b^2}{r^2}\ln\frac{b}{a}\right)\Bigg]\int_a^b rM_T(r)\,dr$$

$$-\frac{1}{r^2}\int_a^r rM_T(r)\,dr \tag{8.123}$$

$$M_{\theta\theta}=-\frac{4(1+\nu)}{a^2b^2}\frac{1}{B}\left[(1+\nu)\left(b^2\ln\frac{b}{r}+a^2\ln\frac{r}{a}+\frac{a^2b^2}{r^2}\ln\frac{b}{a}\right)\right.$$

$$\left.+(1-\nu)(b^2-a^2)\right]\int_a^b\frac{1}{r}\int_a^r rM_T(r)\,dr\,dr$$

$$-\frac{1}{b^2}\frac{1}{B}\left\{-4(1+\nu)^2\ln\frac{r}{a}\ln\frac{b}{a}-(1-\nu)(b^2-a^2)\right.$$

$$\times\left[(1+3\nu)\frac{1}{a^2}+(3+\nu)\frac{1}{r^2}\right]+2(1-\nu^2)\left[\frac{b^2}{a^2}\ln\frac{b}{r}+\ln\frac{r}{a}\right.$$

$$\left.\left.+\left(2+\frac{b^2}{r^2}\right)\ln\frac{b}{a}\right]\right\}\int_a^b rM_T(r)\,dr-M_T+\frac{1}{r^2}\int_a^r rM_T(r)\,dr \tag{8.124}$$

$$Q_r=\frac{4}{r}\frac{1}{B}\left\{2(1+\nu)\frac{b^2-a^2}{a^2b^2}\int_a^b\frac{1}{r}\int_a^r rM_T(r)\,dr\,dr\right.$$

$$\left.+\frac{1}{b^2}\left[2(1+\nu)\ln\frac{b}{a}+(1-\nu)\frac{b^2-a^2}{a^2}\right]\int_a^b rM_T(r)\,dr\right\} \tag{8.125}$$

where

$$B=4(1+\nu)^2\left(\ln\frac{b}{a}\right)^2+(1-\nu)(3+\nu)\frac{(b^2-a^2)^2}{a^2b^2} \tag{8.126}$$

[3] The inner edge $r=a$ is built-in and the outer edge $r=b$ is free

$$w=-\frac{1}{D}\frac{1}{B}\left(\frac{r^2-a^2}{2b^2}-\frac{a^2}{b^2}\ln\frac{r}{a}\right)\int_a^b rM_T(r)\,dr$$

$$-\frac{1}{(1-\nu)D}\int_a^r\frac{1}{r}\int_a^r rM_T(r)\,dr\,dr \tag{8.127}$$

$$M_{rr}=\frac{1}{Bb^2}\left[(1+\nu)+(1-\nu)\frac{a^2}{r^2}\right]\int_a^b rM_T(r)\,dr$$

$$-\frac{1}{r^2}\int_a^r rM_T(r)\,dr \tag{8.128}$$

$$M_{\theta\theta}=\frac{1}{Bb^2}\left[(1+\nu)-(1-\nu)\frac{a^2}{r^2}\right]\int_a^b rM_T(r)\,dr$$

$$-M_T(r)+\frac{1}{r^2}\int_a^r rM_T(r)\,dr \tag{8.129}$$

$$Q_r=0 \tag{8.130}$$

where

$$B = 1 + \nu + (1 - \nu)\frac{a^2}{b^2} \tag{8.131}$$

8.5.2 Non-axisymmetric Problems

Let us consider non-axisymmetric thermal bending problems of a circular plate. The governing equation for such problems is given by Eq. (8.92). We assume that the deflection $w(r, \theta)$ and the thermally induced resultant moment $M_T(r, \theta)$ can be expanded into the following trigonometric series:

$$\begin{aligned} w(r, \theta) &= \sum_{n=0}^{\infty} W_{nc}(r) \cos n\theta + \sum_{n=1}^{\infty} W_{ns}(r) \sin n\theta \\ M_T(r, \theta) &= \sum_{n=0}^{\infty} M_{nc}(r) \cos n\theta + \sum_{n=1}^{\infty} M_{ns}(r) \sin n\theta \end{aligned} \tag{8.132}$$

Substituting Eqs. (8.132) into Eq. (8.92), the fundamental equations for the unknown functions $W_{nc}(r)$ and $W_{ns}(r)$ are

$$\left(\frac{d^2}{dr^2}+\frac{1}{r}\frac{d}{dr}-\frac{n^2}{r^2}\right)^2 \begin{pmatrix} W_{nc} \\ W_{ns} \end{pmatrix} = -\frac{1}{(1-\nu)D}\left(\frac{d^2}{dr^2}+\frac{1}{r}\frac{d}{dr}-\frac{n^2}{r^2}\right)\begin{pmatrix} M_{nc} \\ M_{ns} \end{pmatrix} \tag{8.133}$$

After some algebraic manipulations, the solutions $W_{nc}(r)$ and $W_{ns}(r)$ for $n \geq 0$ are obtained in the form

$$\begin{aligned}
W_{0c}(r) &= \frac{1}{D}\left[C_{01}r^2 + C_{02} + C_{03}r^2 \ln r + C_{04} \ln r - \frac{1}{1-\nu}\int_r \frac{1}{r}\int_r r M_{0c}(r)\,dr\,dr\right] \\
W_{1c}(r) &= \frac{1}{D}\left[C_{11}r^3 + C_{12}r + C_{13}r \ln r + C_{14}\frac{1}{r} - \frac{1}{1-\nu}\frac{1}{r}\int_r r \int_r M_{1c}(r)\,dr\,dr\right] \\
W_{1s}(r) &= \frac{1}{D}\left[D_{11}r^3 + D_{12}r + D_{13}r \ln r + D_{14}\frac{1}{r} - \frac{1}{1-\nu}\frac{1}{r}\int_r r \int_r M_{1s}(r)\,dr\,dr\right] \\
W_{nc}(r) &= \frac{1}{D}\Bigg\{C_{n1}r^{n+2} + C_{n2}r^n + C_{n3}r^{-n+2} + C_{n4}r^{-n} \\
&\quad - \frac{1}{1-\nu}\frac{1}{2n}\left[r^n \int_r r^{1-n} M_{nc}(r)\,dr - r^{-n}\int_r r^{1+n} M_{nc}(r)\,dr\right]\Bigg\} \\
W_{ns}(r) &= \frac{1}{D}\Bigg\{D_{n1}r^{n+2} + D_{n2}r^n + D_{n3}r^{-n+2} + D_{n4}r^{-n} \\
&\quad - \frac{1}{1-\nu}\frac{1}{2n}\left[r^n \int_r r^{1-n} M_{ns}(r)\,dr - r^{-n}\int_r r^{1+n} M_{ns}(r)\,dr\right]\Bigg\} \\
&\quad \text{for } n \geq 2
\end{aligned} \tag{8.134}$$

where C_{n1} to C_{n4} $(n \geq 0)$ and D_{n1} to D_{n4} $(n \geq 1)$ are the unknown constants determined from the boundary conditions. In Eqs. (8.134), the terms containing the unknown constants C_{n3} and C_{n4} $(n \geq 0)$, D_{n3} and D_{n4} $(n \geq 1)$ can be neglected for the solid circular plate, while all the terms concerned with the unknown constants are required for the hollow circular plate with a concentric hole. Substituting Eqs. (8.132) and (8.134) into Eqs. (8.94), the resultant moments and shearing forces for the non-axisymmetric cases of a circular plate are given as

$$
\begin{aligned}
M_{rr} &= -D\sum_{n=0}^{\infty}\left\{\left[\frac{d^2}{dr^2}+\nu\left(\frac{1}{r}\frac{d}{dr}-\frac{n^2}{r^2}\right)\right]W_{nc}(r)+\frac{1}{(1-\nu)D}M_{nc}(r)\right\}\cos n\theta \\
&\quad - D\sum_{n=1}^{\infty}\left\{\left[\frac{d^2}{dr^2}+\nu\left(\frac{1}{r}\frac{d}{dr}-\frac{n^2}{r^2}\right)\right]W_{ns}(r)+\frac{1}{(1-\nu)D}M_{ns}(r)\right\}\sin n\theta \\
M_{\theta\theta} &= -D\sum_{n=0}^{\infty}\left[\left(\nu\frac{d^2}{dr^2}+\frac{1}{r}\frac{d}{dr}-\frac{n^2}{r^2}\right)W_{nc}(r)+\frac{1}{(1-\nu)D}M_{nc}(r)\right]\cos n\theta \\
&\quad - D\sum_{n=1}^{\infty}\left[\left(\nu\frac{d^2}{dr^2}+\frac{1}{r}\frac{d}{dr}-\frac{n^2}{r^2}\right)W_{ns}(r)+\frac{1}{(1-\nu)D}M_{ns}(r)\right]\sin n\theta \\
M_{r\theta} &= -(1-\nu)D\sum_{n=1}^{\infty}n\left(\frac{1}{r}\frac{d}{dr}-\frac{1}{r^2}\right)W_{nc}(r)\sin n\theta \\
&\quad + (1-\nu)D\sum_{n=1}^{\infty}n\left(\frac{1}{r}\frac{d}{dr}-\frac{1}{r^2}\right)W_{ns}(r)\cos n\theta \\
Q_r &= -D\sum_{n=0}^{\infty}\frac{d}{dr}\left[\left(\frac{d^2}{dr^2}+\frac{1}{r}\frac{d}{dr}-\frac{n^2}{r^2}\right)W_{nc}(r)+\frac{1}{(1-\nu)D}M_{nc}(r)\right]\cos n\theta \\
&\quad - D\sum_{n=1}^{\infty}\frac{d}{dr}\left[\left(\frac{d^2}{dr^2}+\frac{1}{r}\frac{d}{dr}-\frac{n^2}{r^2}\right)W_{ns}(r)+\frac{1}{(1-\nu)D}M_{ns}(r)\right]\sin n\theta \\
Q_\theta &= D\sum_{n=1}^{\infty}\frac{n}{r}\left[\left(\frac{d^2}{dr^2}+\frac{1}{r}\frac{d}{dr}-\frac{n^2}{r^2}\right)W_{nc}(r)+\frac{1}{(1-\nu)D}M_{nc}(r)\right]\sin n\theta \\
&\quad - D\sum_{n=1}^{\infty}\frac{n}{r}\left[\left(\frac{d^2}{dr^2}+\frac{1}{r}\frac{d}{dr}-\frac{n^2}{r^2}\right)W_{ns}(r)+\frac{1}{(1-\nu)D}M_{ns}(r)\right]\cos n\theta
\end{aligned}
\tag{8.135}
$$

Let us consider a solid circular plate of radius a. For simplicity, and without loss of generality, we assume that the deflection $w(r,\theta)$ is symmetric with respect to the coordinate axis $\theta = 0$. Then, the deflection w is given from Eqs. (8.132) and (8.134) as

$$
\begin{aligned}
w(r,\theta) &= \sum_{n=0}^{\infty} W_{nc}(r)\cos n\theta \\
&= \frac{1}{D}\left[C_{01}r^2 + C_{02} - \frac{1}{1-\nu}\int_0^r \frac{1}{r}\int_0^r rM_{0c}(r)\,dr\,dr\right] \\
&\quad + \frac{1}{D}\left[C_{11}r^3 + C_{12}r - \frac{1}{1-\nu}\frac{1}{r}\int_0^r r\int_0^r M_{1c}(r)\,dr\,dr\right]\cos\theta \\
&\quad + \frac{1}{D}\sum_{n=2}^{\infty}\left\{C_{n1}r^{n+2} + C_{n2}r^n - \frac{1}{1-\nu}\frac{1}{2n}\right. \\
&\quad \times \left.\left[r^n\int_0^r r^{-n+1}M_{nc}(r)\,dr - r^{-n}\int_0^r r^{n+1}M_{nc}(r)\,dr\right]\right\}\cos n\theta \qquad (8.136)
\end{aligned}
$$

Substituting Eq. (8.136) into Eqs. (8.135), the resultant moments and shearing forces are

$$
\begin{aligned}
M_{rr} &= -\left[2(1+\nu)C_{01} + \frac{1}{r^2}\int_0^r rM_{0c}(r)\,dr\right] \\
&\quad - \left[2(3+\nu)C_{11}r + \frac{1}{r}\int_0^r M_{1c}(r)\,dr - \frac{2}{r^3}\int_0^r r\int_0^r M_{1c}(r)\,dr\,dr\right]\cos\theta \\
&\quad - \sum_{n=2}^{\infty}\left\{(n+1)\left[(n+2)-\nu(n-2)\right]C_{n1}r^n + n(n-1)(1-\nu)C_{n2}r^{n-2}\right. \\
&\qquad - \frac{1}{2}\left[(n-1)r^{n-2}\int_0^r r^{-n+1}M_{nc}(r)\,dr\right. \\
&\qquad\qquad \left.\left. -(n+1)r^{-n-2}\int_0^r r^{n+1}M_{nc}(r)\,dr\right]\right\}\cos n\theta \\
M_{\theta\theta} &= -\left[2(1+\nu)C_{01} + M_{0c}(r) - \frac{1}{r^2}\int_0^r rM_{0c}(r)\,dr\right] \\
&\quad - \left[2(1+3\nu)C_{11}r + M_{1c}(r) - \frac{1}{r}\int_0^r M_{1c}(r)\,dr\right. \\
&\qquad \left. + \frac{2}{r^3}\int_0^r r\int_0^r M_{1c}(r)\,dr\,dr\right]\cos\theta \\
&\quad - \sum_{n=2}^{\infty}\left\{(n+1)\left[\nu(n+2)-(n-2)\right]C_{n1}r^n - n(n-1)(1-\nu)C_{n2}r^{n-2}\right. \\
&\qquad + \frac{1}{2}\left[2M_{nc}(r) + (n-1)r^{n-2}\int_0^r r^{-n+1}M_{nc}(r)\,dr\right. \\
&\qquad\qquad \left.\left. -(n+1)r^{-n-2}\int_0^r r^{n+1}M_{nc}(r)\,dr\right]\right\}\cos n\theta
\end{aligned}
$$

$$M_{r\theta} = -\left[2(1-\nu)C_{11}r - \frac{1}{r}\int_0^r M_{1c}(r)\,dr + \frac{2}{r^3}\int_0^r r\int_0^r M_{1c}(r)\,dr\,dr\right]\sin\theta$$

$$- \sum_{n=2}^{\infty}\left\{n(n+1)(1-\nu)C_{n1}r^n + n(n-1)(1-\nu)C_{n2}r^{n-2}\right.$$

$$-\frac{1}{2}\left[(n-1)r^{n-2}\int_0^r r^{-n+1}M_{nc}(r)\,dr\right.$$

$$\left.\left.+(n+1)r^{-n-2}\int_0^r r^{n+1}M_{nc}(r)\,dr\right]\right\}\sin n\theta$$

$$Q_r = -\sum_{n=1}^{\infty} 4n(n+1)C_{n1}r^{n-1}\cos n\theta$$

$$Q_\theta = \sum_{n=1}^{\infty} 4n(n+1)C_{n1}r^{n-1}\sin n\theta \tag{8.137}$$

There are some special cases of Eqs. (8.137) for three kinds of boundary conditions.

[1] For a built-in edge condition
Making use of the boundary condition (8.97), the unknown constants C_{n1} and C_{n2} $(n \geq 0)$ are

$$C_{01} = \frac{1}{1-\nu}\frac{1}{2a^2}\int_0^a rM_{0c}(r)\,dr$$

$$C_{02} = \frac{1}{1-\nu}\left[\int_0^a \frac{1}{r}\int_0^r rM_{0c}(r)\,dr\,dr - \frac{1}{2}\int_0^a rM_{0c}(r)\,dr\right]$$

$$C_{11} = \frac{1}{1-\nu}\frac{1}{2a^2}\left[\int_0^a M_{1c}(r)\,dr - \frac{2}{a^2}\int_0^a r\int_0^r M_{1c}(r)\,dr\,dr\right]$$

$$C_{12} = \frac{1}{1-\nu}\frac{1}{2}\left[\frac{4}{a^2}\int_0^a r\int_0^r M_{1c}(r)\,dr\,dr - \int_0^a M_{1c}(r)\,dr\right] \tag{8.138}$$

$$C_{n1} = \frac{1}{1-\nu}\frac{1}{2a^{2n+2}}\int_0^a r^{n+1}M_{nc}(r)\,dr$$

$$C_{n2} = \frac{1}{1-\nu}\frac{1}{2n}\left[\int_0^a r^{-n+1}M_{nc}(r)\,dr - (n+1)a^{-2n}\int_0^a r^{n+1}M_{nc}(r)\,dr\right]$$

for $n \geq 2$

[2] For a simply supported edge condition
Making use of the boundary condition (8.98), the unknown constants C_{n1} and C_{n2} $(n \geq 0)$ are

$$C_{01} = -\frac{1}{2(1+\nu)}\frac{1}{a^2}\int_0^a rM_{0c}(r)\,dr$$

$$C_{02} = \frac{1}{1-\nu}\int_0^a \frac{1}{r}\int_0^r rM_{0c}(r)\,dr\,dr + \frac{1}{2(1+\nu)}\int_0^a rM_{0c}(r)\,dr$$

$$C_{11} = \frac{1}{2(3+\nu)}\left[-\frac{1}{a^2}\int_0^a M_{1c}(r)\,dr + \frac{2}{a^4}\int_0^a r\int_0^r M_{1c}(r)\,dr\,dr\right]$$

$$C_{12} = \frac{1}{2(3+\nu)}\left[\int_0^a M_{1c}(r)\,dr + \frac{4(1+\nu)}{1-\nu}\frac{1}{a^2}\int_0^a r\int_0^r M_{1c}(r)\,dr\,dr\right]$$

$$C_{n1} = -\frac{1}{2(2n+1+\nu)a^{2n+2}}\int_0^a r^{n+1}M_{nc}(r)\,dr$$

$$C_{n2} = -\frac{1}{2(2n+1+\nu)n(1-\nu)}\left[-(2n+1+\nu)\int_0^a r^{-n+1}M_{nc}(r)\,dr + (n+1)(1+\nu)a^{-2n}\int_0^a r^{n+1}M_{nc}(r)\,dr\right] \quad \text{for } n \geq 2 \tag{8.139}$$

[3] For a free edge condition
Referring to the relation (8.137), the equivalent shearing force $Q_r - (1/r)\times(\partial M_{r\theta}/\partial\theta)$ is calculated as

$$\begin{aligned} Q_r - \frac{1}{r}\frac{\partial M_{r\theta}}{\partial\theta} &= \left[-2(3+\nu)C_{11} - \frac{1}{r^2}\int_0^r M_{1c}(r)\,dr + \frac{2}{r^4}\int_0^r r\int_0^r M_{1c}(r)\,dr\,dr\right]\cos\theta \\ &\quad + \sum_{n=2}^{\infty} n\Bigg\{(n+1)(n-4-n\nu)C_{n1}r^{n-1} + n(n-1)(1-\nu)C_{n2}r^{n-3} \\ &\quad - \frac{1}{2}\left[(n-1)r^{n-3}\int_0^r r^{-n+1}M_{nc}(r)\,dr + (n+1)r^{-n-3}\int_0^r r^{n+1}M_{nc}(r)\,dr\right]\Bigg\}\cos n\theta \end{aligned} \tag{8.140}$$

Making use of the boundary condition (8.99) and Eq. (8.140), the unknown constants C_{n1} and C_{n2} $(n \geq 0)$ are

$$C_{01} = -\frac{1}{2(1+\nu)}\frac{1}{a^2}\int_0^a rM_{0c}(r)\,dr$$

$$C_{11} = -\frac{1}{2(3+\nu)}\left[\frac{1}{a^2}\int_0^a M_{1c}(r)\,dr - \frac{2}{a^4}\int_0^a r\int_0^r M_{1c}(r)\,dr\,dr\right]$$

$$C_{n1} = -\frac{1}{2(3+\nu)a^{2n+2}}\int_0^a r^{n+1}M_{nc}(r)\,dr$$

$$C_{n2} = \frac{1}{2(3+\nu)n(1-\nu)}\left[(3+\nu)\int_0^a r^{-n+1}M_{nc}(r)\,dr\right.$$
$$\left. + (n+1)(1-\nu)a^{-2n}\int_0^a r^{n+1}M_{nc}(r)\,dr\right] \tag{8.141}$$

The unknown constants C_{02} and C_{12} are not determinated for the free edge condition of a solid circular plate. It may be recognized, however, that the terms of C_{02} and C_{12} represent a rigid translation in the z direction and a rigid rotation around y axis, respectively. When the rigid translation in the z direction and the rigid rotation around the y axis are omitted, these unknown constants are zero

$$C_{02} = 0, \qquad C_{12} = 0 \tag{8.142}$$

Problems

8.1 Derive thermal stress components σ_x and σ_y for a two-layered thin plate due to temperature change $\tau(z)$ as shown in Fig. 8.11.

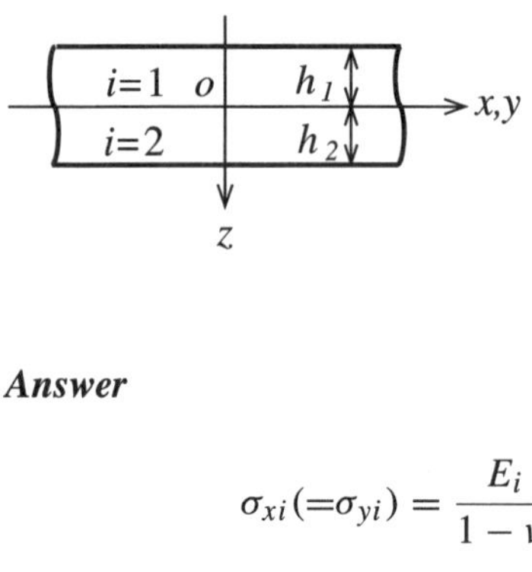

Figure 8.11 A two-layered thin plate.

Answer

$$\sigma_{xi}(=\sigma_{yi}) = \frac{E_i}{1-\nu_i}\left[\epsilon_0 + \frac{z}{\rho} - \alpha_i\tau(z)\right], \qquad (i = 1, 2)$$

where

$$\epsilon_0 = \frac{1}{D}\left\{\frac{1}{3}\left[(1-\nu_2)E_1h_1^3 + (1-\nu_1)E_2h_2^3\right]\right.$$
$$\times\left[(1-\nu_2)\alpha_1E_1\int_{-h_1}^0 \tau\,dz + (1-\nu_1)\alpha_2E_2\int_0^{h_2}\tau\,dz\right]$$

$$
\begin{aligned}
&-\frac{1}{2}\left[-(1-\nu_2)E_1h_1^2+(1-\nu_1)E_2h_2^2\right]\\
&\times\left[(1-\nu_2)\alpha_1E_1\int_{-h_1}^{0}\tau z\,dz+(1-\nu_1)\alpha_2E_2\int_{0}^{h_2}\tau z\,dz\right]\Bigg\}
\end{aligned}
$$

$$
\begin{aligned}
\frac{1}{\rho}=\frac{1}{D}\Bigg\{&[(1-\nu_2)E_1h_1+(1-\nu_1)E_2h_2]\\
&\times\left[(1-\nu_2)\alpha_1E_1\int_{-h_1}^{0}\tau z\,dz+(1-\nu_1)\alpha_2E_2\int_{0}^{h_2}\tau z\,dz\right]\\
&-\frac{1}{2}\left[-(1-\nu_2)E_1h_1^2+(1-\nu_1)E_2h_2^2\right]\\
&\times\left[(1-\nu_2)\alpha_1E_1\int_{-h_1}^{0}\tau\,dz+(1-\nu_1)\alpha_2E_2\int_{0}^{h_2}\tau\,dz\right]\Bigg\}
\end{aligned}
$$

in which

$$
\begin{aligned}
D=\frac{1}{12}\Bigg\{&\left[(1-\nu_2)E_1h_1^2+(1-\nu_1)E_2h_2^2\right]^2\\
&+4(1-\nu_1)(1-\nu_2)E_1E_2h_1h_2(h_1^2+h_1h_2+h_2^2)\Bigg\}
\end{aligned}
$$

In a special case $h_1=h_2=h$,

$$
\begin{aligned}
\epsilon_0=\frac{h^2}{D}\Bigg\{&\frac{h}{3}[(1-\nu_2)E_1+(1-\nu_1)E_2]\\
&\times\left[(1-\nu_2)\alpha_1E_1\int_{-h}^{0}\tau\,dz+(1-\nu_1)\alpha_2E_2\int_{0}^{h}\tau\,dz\right]\\
&-\frac{1}{2}[-(1-\nu_2)E_1+(1-\nu_1)E_2]\\
&\times\left[(1-\nu_2)\alpha_1E_1\int_{-h}^{0}\tau z\,dz+(1-\nu_1)\alpha_2E_2\int_{0}^{h}\tau z\,dz\right]\Bigg\}
\end{aligned}
$$

$$
\begin{aligned}
\frac{1}{\rho}=\frac{h}{D}\Bigg\{&[(1-\nu_2)E_1+(1-\nu_1)E_2]\\
&\times\left[(1-\nu_2)\alpha_1E_1\int_{-h}^{0}\tau z\,dz+(1-\nu_1)\alpha_2E_2\int_{0}^{h}\tau z\,dz\right]\\
&-\frac{h}{2}[-(1-\nu_2)E_1+(1-\nu_1)E_2]\\
&\times\left[(1-\nu_2)\alpha_1E_1\int_{-h}^{0}\tau\,dz+(1-\nu_1)\alpha_2E_2\int_{0}^{h}\tau\,dz\right]\Bigg\}
\end{aligned}
$$

in which

$$D = \frac{h^4}{12}\{[(1-\nu_2)E_1 + (1-\nu_1)E_2]^2 + 12(1-\nu_1)(1-\nu_2)E_1E_2\}$$

8.2 Derive thermal stress components σ_x and σ_y for a nonhomogeneous plate due to temperature change $\tau(z)$ as shown in Fig. 8.1.

Answer

$$\sigma_x(=\sigma_y) = \frac{E(z)}{1-\nu(z)}\left[\epsilon_0 + \frac{z}{\rho} - \alpha(z)\tau(z)\right]$$

where

$$\epsilon_0 = \frac{1}{D}\left[\int_{-h/2}^{h/2} \frac{E(z)}{1-\nu(z)} z^2\, dz \int_{-h/2}^{h/2} \frac{\alpha(z)E(z)}{1-\nu(z)}\tau(z)\, dz - \int_{-h/2}^{h/2} \frac{E(z)}{1-\nu(z)} z\, dz \int_{-h/2}^{h/2} \frac{\alpha(z)E(z)}{1-\nu(z)}\tau(z) z\, dz\right]$$

$$\frac{1}{\rho} = \frac{1}{D}\left[\int_{-h/2}^{h/2} \frac{E(z)}{1-\nu(z)}\, dz \int_{-h/2}^{h/2} \frac{\alpha(z)E(z)}{1-\nu(z)}\tau(z) z\, dz - \int_{-h/2}^{h/2} \frac{E(z)}{1-\nu(z)} z\, dz \int_{-h/2}^{h/2} \frac{\alpha(z)E(z)}{1-\nu(z)}\tau(z)\, dz\right]$$

$$D = \int_{-h/2}^{h/2} \frac{E(z)}{1-\nu(z)}\, dz \int_{-h/2}^{h/2} \frac{E(z)}{1-\nu(z)} z^2\, dz - \left[\int_{-h/2}^{h/2} \frac{E(z)}{1-\nu(z)} z\, dz\right]^2$$

8.3 Derive the formulas for thermal stress components σ_x and σ_y for a multi-layered plate due to temperature change $\tau(z)$ as shown in Fig. 8.12.

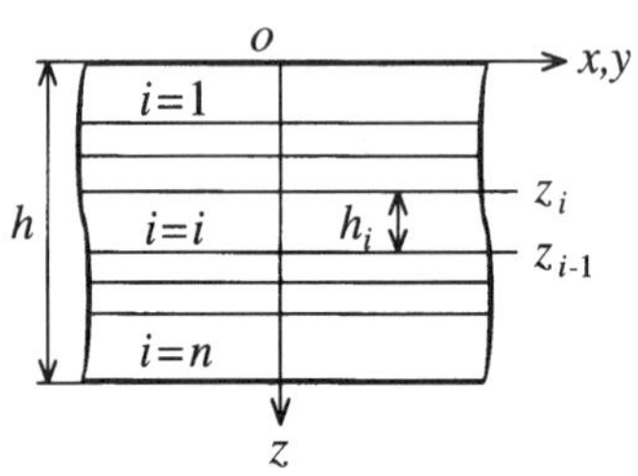

Figure 8.12 A multi-layered plate.

Answer

$$\sigma_{xi}(=\sigma_{yi}) = \frac{E_i}{1-\nu_i}\left[\epsilon_0 + \frac{z}{\rho} - \alpha_i\tau(z)\right], \qquad (i = 1, \ldots, n)$$

where

$$\epsilon_0 = \frac{1}{D}\left[\sum_{i=1}^{n}\frac{E_i h_i}{1-\nu_i}\frac{1}{3}\left(z_i^2 + z_i z_{i-1} + z_{i-1}^2\right)\sum_{j=1}^{n}\frac{\alpha_j E_j}{1-\nu_j}\int_{z_{j-1}}^{z_j}\tau(z)\,dz\right.$$

$$\left.-\sum_{i=1}^{n}\frac{E_i h_i}{1-\nu_i}\frac{1}{2}(z_i + z_{i-1})\sum_{j=1}^{n}\frac{\alpha_j E_j}{1-\nu_j}\int_{z_{j-1}}^{z_j}\tau(z) z\,dz\right]$$

$$\frac{1}{\rho} = \frac{1}{D}\left[\sum_{i=1}^{n}\frac{E_i h_i}{1-\nu_i}\sum_{j=1}^{n}\frac{\alpha_j E_j}{1-\nu_j}\int_{z_{j-1}}^{z_j}\tau(z) z\,dz\right.$$

$$\left.-\sum_{i=1}^{n}\frac{E_i h_i}{1-\nu_i}\frac{1}{2}(z_i + z_{i-1})\sum_{j=1}^{n}\frac{\alpha_j E_j}{1-\nu_j}\int_{z_{j-1}}^{z_j}\tau(z)\,dz\right]$$

$$D = \sum_{i=1}^{n}\frac{E_i h_i}{1-\nu_i}\sum_{i=1}^{n}\frac{E_i h_i}{1-\nu_i}\frac{1}{3}\left(z_i^2 + z_i z_{i-1} + z_{i-1}^2\right) - \left[\sum_{i=1}^{n}\frac{E_i h_i}{1-\nu_i}\frac{1}{2}(z_i + z_{i-1})\right]^2$$

in which

$$h_i = z_i - z_{i-1} \quad (i = 1, \ldots, n), \qquad z_n = h, \qquad z_0 = 0$$

8.4 For a simply supported rectangular plate described in Subsection 8.3.1, derive the coefficients a_{mn} given by Eq. (8.72) when the temperature solution is given by

$$T = \frac{1}{2}(T_b + T_a) + (T_b - T_a)\frac{z}{h}$$

and the initial temperature is $T = T_i$. (See Fig. 8.6 and Eq. (3.35).)

Answer

$$a_{mn} = \begin{cases} \dfrac{16}{\pi^2} M_T \dfrac{1}{mn}; & m = 1, 3, 5, \ldots, n = 1, 3, 5, \ldots \\ 0; & m = 2, 4, 6, \ldots, n = 2, 4, 6, \ldots \end{cases}$$

where

$$M_T = \frac{h^2}{12}\alpha E(T_b - T_a) = \text{const}$$

8.5 Making use of Eqs. (8.112), (8.114), and (8.116), derive the thermal deflection w and bending moments M_{rr} and $M_{\theta\theta}$ for a solid circular plate when the temperature distribution T is given in the same form as in Problem 8.4.

Answer

[1] For built-in edge condition

$$w = 0, \qquad M_{rr} = M_{\theta\theta} = -\frac{1}{1-\nu} M_T$$

[2] For simply supported edge condition and for free edge condition

$$w = \frac{1}{2(1-\nu^2)D}(a^2 - r^2)M_T, \qquad M_{rr} = M_{\theta\theta} = 0$$

where

$$M_T = \frac{h^2}{12}\alpha E(T_b - T_a)$$

8.6 Derive the relations Eq. (8.27) through Eqs. (8.30) for an anisotropic plate.

Hint Substituting Eqs. (8.24) into Eqs. (8.25) and solving them for σ_x and σ_y, we obtain Eqs. (8.27) and (8.28).

Equations (8.29) are obtained by substituting Eqs. (8.27) and (8.28) into Eqs. (a) and (c) on p. 353 and considering formulas

$$\int_{-h/2}^{h/2} dz = h, \qquad \int_{-h/2}^{h/2} z\, dz = 0$$

Equations (8.30) are obtained by substituting Eqs. (8.27) and (8.28) into Eqs. (b) and (d) on p. 353 and considering formula

$$\int_{-h/2}^{h/2} z^2\, dz = \frac{h^3}{12}$$

8.7 Derive the equilibrium equations (8.50) and (8.52) for a plate.

8.8 Derive the relation of the coefficients a_{mn} given by Eq. (8.72) for a simply supported plate.

8.9 Derive the fundamental solution w given by Eq. (e) in Subsection 8.3.2.

8.10 Derive the formula on the trigonometric series expansion given by Eqs. (8.90).

Chapter 9

THERMALLY INDUCED INSTABILITY

Thermoelastic buckling of beam-columns is considered for various boundary conditions. The thermally-induced deflection of a beam in the pre-buckling state is assumed. The buckling loads are derived for various boundary conditions. Furthermore, the thermoelastic buckling behavior of rectangular and circular plates is discussed, and the fundamental equation systems are derived. Illustrative examples for thin plates are presented when the in-plane deformations are prescribed. Considered are: a simply supported rectangular plate subjected to a uniform temperature change, and an axisymmetrical problem of a circular plate with simply supported or clamped edge conditions.

9.1 Instability of Beam-Column

We consider beam-columns as bodies which undergo buckling. Beams, which are treated in Chapter 2, are assumed to be subjected to lateral loads under some support conditions. However, beam-columns treated here are assumed to be subjected to both axial and lateral loads.

9.1.1 Introduction

Let x be a coordinate variable in the axial direction of a beam, y and z be the orthogonal coordinate axes perpendicular to the x axis, as shown in Fig. 9.1. We now consider the thermal stress σ_{xx} in the axial direction induced by arbitrarily distributed temperature change. Assuming that the beam deforms according to the Bernoulli-Euler hypothesis

Figure 9.1 Coordinate system.

(see Chapter 2) and is a homogeneous body, the thermal stress σ_{xx} in the axial direction, from Subsection 2.5.1, is given as

$$\sigma_x = -\alpha E\tau(x, y, z) + \epsilon_0 E + \frac{Ey}{\rho_y} + \frac{Ez}{\rho_z} \tag{9.1}$$

where ϵ_0, ρ_y, and ρ_z are

$$\begin{aligned} \epsilon_0 &= \frac{P_T}{EA} \\ \frac{1}{\rho_y} &= \frac{I_y M_{Tz} - I_{yz} M_{Ty}}{E(I_y I_z - I_{yz}^2)} \\ \frac{1}{\rho_z} &= \frac{I_z M_{Ty} - I_{yz} M_{Tz}}{E(I_y I_z - I_{yz}^2)} \end{aligned} \tag{9.2}$$

and

$$\begin{aligned} &I_z = \int_A y^2\, dA, \quad I_y = \int_A z^2\, dA, \quad I_{yz} = \int_A yz\, dA, \quad A = \int_A dA \\ &P_T = \int_A \alpha E\tau\, dA, \quad M_{Tz} = \int_A \alpha E\tau y\, dA, \quad M_{Ty} = \int_A \alpha E\tau z\, dA \end{aligned} \tag{9.3}$$

The relations between the deflections v, w, and radii of curvature ρ_y, ρ_z are given as

$$\frac{d^2 v}{dx^2} = -\frac{1}{\rho_y}, \qquad \frac{d^2 w}{dx^2} = -\frac{1}{\rho_z} \tag{9.4}$$

Combining Eqs. (9.2) and (9.4), the fundamental equations for thermal bending problems of a beam are

$$\frac{d^2 v}{dx^2} = -\frac{1}{E}\left[\frac{M_{Tz} I_y - M_{Ty} I_{yz}}{I_y I_z - I_{yz}^2}\right] \tag{9.5}$$

$$\frac{d^2 w}{dx^2} = -\frac{1}{E}\left[\frac{M_{Ty} I_z - M_{Tz} I_{yz}}{I_y I_z - I_{yz}^2}\right] \tag{9.6}$$

For the special case where the y and z are principal axes, which means that $I_{yz} = 0$, the fundamental equations reduce to the following simplified forms

$$\frac{d^2 v}{dx^2} = -\frac{M_{Tz}}{EI_z}$$
$$\frac{d^2 w}{dx^2} = -\frac{M_{Ty}}{EI_y} \tag{9.7}$$

9.1.2 Beam-Column Problems by Axially Imposed External Load

In order to simplify the analytical procedure, we assume that the coordinates y and z of the cross section of the beam are principal axes. Then, the fundamental equations for thermal bending problems of the beam are given by Eqs. (9.7). To analyze the thermal buckling behavior of the beam, we now consider the beam-column problem with axially imposed external load P, in which P is assumed positive in compression. Denoting the bending moments due to the compressive load P by M_y and M_z about the y and z axes respectively, the relations between M_y, M_z and v, w are given as

$$\frac{d^2 v}{dx^2} = -\frac{M_z}{EI_z}$$
$$\frac{d^2 w}{dx^2} = -\frac{M_y}{EI_y} \tag{9.8}$$

Therefore, the fundamental equations for thermal buckling problems for the beam-column due to both thermal and compressive loads are derived by superposing Eqs. (9.7) with Eqs. (9.8)

$$EI_z \frac{d^2 v}{dx^2} = -M_z - M_{Tz} \qquad \text{for } y \text{ direction} \tag{9.9}$$

$$EI_y \frac{d^2 w}{dx^2} = -M_y - M_{Ty} \qquad \text{for } z \text{ direction} \tag{9.10}$$

In other words, the thermal buckling behavior for the y and z directions can be evaluated from Eqs. (9.9) and (9.10), respectively, with appropriate boundary conditions.

Now, we consider the typical cases of thermal buckling behavior in the z direction. The boundary conditions in the z direction are expressed as follows:

[1] For a built-in edge

$$w = 0, \qquad \frac{dw}{dx} = 0 \tag{9.11}$$

[2] For a simply supported edge

$$w = 0, \qquad M_y = 0 \tag{9.12}$$

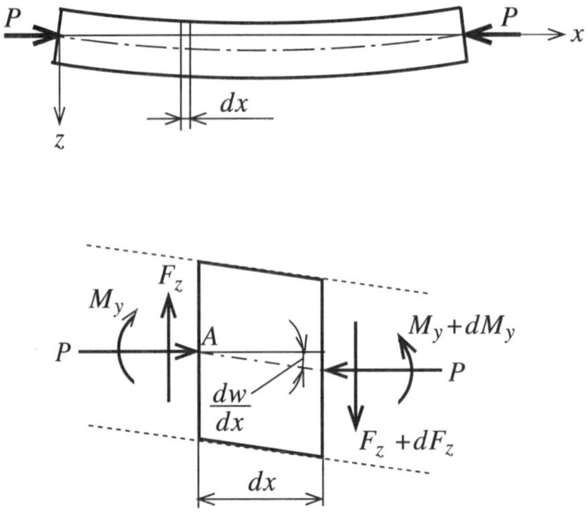

Figure 9.2 Equilibrium of moments.

[3] For a free edge

$$M_y = 0, \qquad F_z = 0 \tag{9.13}$$

in which F_z is the shearing force in the z direction.

Referring to Fig. 9.2, and taking into account the equilibrium of moments about a point A in that figure, we obtain

$$F_z = \frac{dM_y}{dx} - P\frac{dw}{dx} \tag{9.14}$$

Considering the fundamental equation (9.10), the boundary conditions (9.12) and (9.13) are represented in terms of deflection w as

[2] For the simply supported edge

$$w = 0, \qquad EI_y\frac{d^2w}{dx^2} + M_{Ty} = 0 \tag{9.12$'$}$$

[3] For the free edge

$$EI_y\frac{d^2w}{dx^2} + M_{Ty} = 0, \qquad EI_y\frac{d^3w}{dx^3} + P\frac{dw}{dx} + \frac{dM_{Ty}}{dx} = 0 \tag{9.13$'$}$$

Now, let us evaluate the deflection w of a beam-column and buckling load P_{cr}, when $M_{Ty}(x)$ is represented by a linear function of x

$$M_{Ty}(x) = D_1 + D_2x \tag{9.15}$$

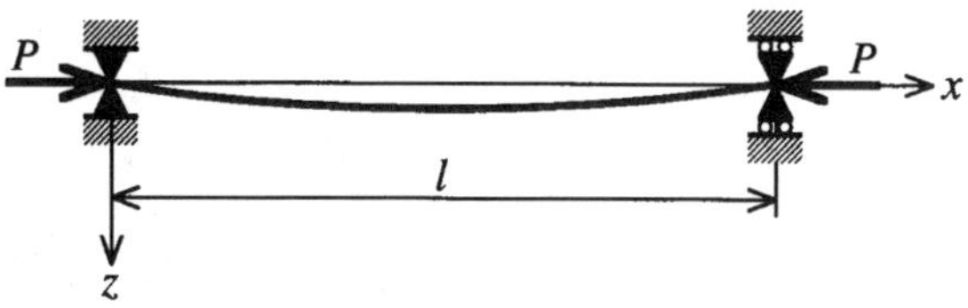

Figure 9.3 Simply supported beam-column.

The cross section of the beam-column is uniform and the length of the beam-column is assumed to be l.

[1] For a beam-column with edges $x = 0$ and $x = l$ simply supported
When both edges are simply supported, the boundary conditions are

$$w = 0, \quad EI_y \frac{d^2 w}{dx^2} + (D_1 + D_2 x) = 0 \qquad \text{at } x = 0 \text{ and } x = l \tag{9.16}$$

Referring to Fig. 9.3, the bending moment M_y is

$$M_y = Pw \tag{9.17}$$

Substituting Eq. (9.17) into Eq. (9.10), the fundamental equation for w is rewritten as

$$EI_y \frac{d^2 w}{dx^2} + Pw = -M_{Ty}(x) \tag{9.18}$$

Furthermore, we have a following fundamental equation through substitution of Eq. (9.15)

$$EI_y \frac{d^2 w}{dx^2} + Pw = -(D_1 + D_2 x) \tag{9.19}$$

Then, we have the fundamental solution of Eq. (9.19)

$$w = C_1 \cos kx + C_2 \sin kx - \frac{1}{P}(D_1 + D_2 x) \tag{9.20}$$

where

$$k^2 = \frac{P}{EI_y} \tag{9.21}$$

and C_1 and C_2 are unknown constants. The solution w satisfying the boundary conditions (9.16) is thus

$$w = \frac{1}{P}\left[D_1\left(\cos kx - 1 + \frac{1 - \cos kl}{\sin kl}\sin kx\right) + \frac{D_2}{k}\left(\frac{kl \sin kx}{\sin kl} - kx\right)\right] \tag{9.22}$$

The deflection w tends to infinity when kl satisfies the condition

$$\sin kl = 0 \tag{9.23}$$

that is, the deflection w becomes infinite for $kl = n\pi$ $(n = 1, 2, 3, \ldots)$. For $kl = \pi$, we find from Eqs. (9.21) and (9.23)

$$P_{cr} = \frac{\pi^2 E I_y}{l^2} \tag{9.24}$$

When the axial compressive load approaches the critical value given by Eq. (9.24), a very large deflection w is produced. The critical value of the axial compressive load is called the buckling load and is denoted as P_{cr}.

[2] For a cantilever beam-column
We assume that the one edge $x = 0$ is built-in and the other edge $x = l$ is free. The boundary conditions are

$$\begin{aligned} &w = 0, \quad \frac{dw}{dx} = 0 \quad \text{at } x = 0 \\ &EI_y \frac{d^2 w}{dx^2} + (D_1 + D_2 x) = 0, \quad EI_y \frac{d^3 w}{dx^3} + P\frac{dw}{dx} + D_2 = 0 \quad \text{at } x = l \end{aligned} \tag{9.25}$$

Referring to Fig. 9.4, the bending moment M_y is

$$M_y = -P\,(w_0 - w) \tag{9.26}$$

where w_0 is the displacement at the free edge. Substituting Eq. (9.26) into Eq. (9.10), the fundamental equation for w is rewritten as

$$EI_y \frac{d^2 w}{dx^2} + Pw = Pw_0 - M_{Ty} \tag{9.27}$$

or

$$EI_y \frac{d^2 w}{dx^2} + Pw = Pw_0 - (D_1 + D_2 x) \tag{9.28}$$

Thus, we have the fundamental solution w in the form

$$w = C_1 \cos kx + C_2 \sin kx + w_0 - \frac{1}{P}(D_1 + D_2 x) \tag{9.29}$$

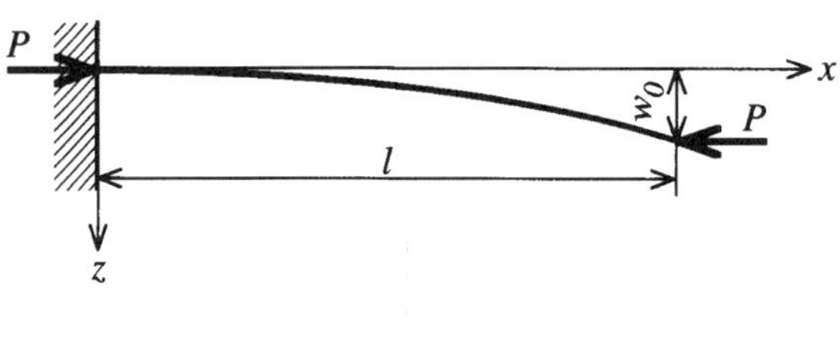

Figure 9.4 Cantilever beam-column.

For the prebuckling state, the solution w satisfying the boundary conditions (9.25) is represented by

$$w = \frac{1}{P \cos kl} \left\{ D_1 (\cos kx - 1) + \frac{D_2}{k} [(kl - \sin kl)(\cos kx - 1) + \cos kl (\sin kx - kx)] \right\} \tag{9.30}$$

Therefore, the buckling load P_{cr} is derived from the condition

$$\cos kl = 0 \tag{9.31}$$

We then obtain

$$P_{cr} = \frac{\pi^2 E I_y}{4l^2} \tag{9.32}$$

[3] For the beam-column with one edge simply supported and the other one built-in We assume that the edge $x = 0$ is simply supported and the edge $x = l$ is built-in. Referring to Fig. 9.5, the bending moment M_y is given by

$$M_y = Pw + R_0 x \tag{9.33}$$

in which R_0 is the reaction force at the edge $x = 0$. Substituting Eq. (9.33) into Eq. (9.10), the fundamental equation for w is rewritten as

$$E I_y \frac{d^2 w}{dx^2} + Pw = -R_0 x - M_{T_y}(x) \tag{9.34}$$

or

$$E I_y \frac{d^2 w}{dx^2} + Pw = -R_0 x - (D_1 + D_2 x) \tag{9.35}$$

We write the fundamental solution w as

$$w = C_1 \cos kx + C_2 \sin kx - \frac{1}{P}[D_1 + (D_2 + R_0)x] \tag{9.36}$$

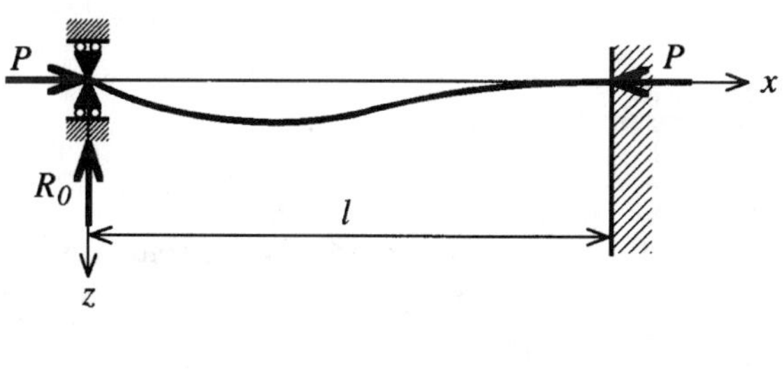

Figure 9.5 Beam-column with simply supported edge and built-in edge.

The boundary conditions are

$$\begin{aligned} &w = 0, \quad EI_y \frac{d^2 w}{dx^2} + (D_1 + D_2 x) = 0 \qquad \text{at } x = 0 \\ &w = 0, \quad \frac{dw}{dx} = 0 \qquad \text{at } x = l \end{aligned} \tag{9.37}$$

The solution w satisfying the boundary conditions (9.37) is thus

$$w = \frac{D_1}{P}\left\{\cos kx - 1 + \frac{1}{kl\cos kl - \sin kl} \times [(\cos kl - 1)(kx + \sin kx) + kl \sin kl \sin kx]\right\} \tag{9.38}$$

Therefore, the buckling load P_{cr} is derived from the condition

$$kl\cos kl - \sin kl = 0 \quad \text{i.e., } \tan kl = kl \tag{9.39}$$

We then have

$$P_{cr} = \frac{4.4934^2 EI_y}{l^2} \simeq \frac{2.046\pi^2 EI_y}{l^2} \tag{9.40}$$

[4] For the beam-column with both edges built-in
Referring to Fig. 9.6, the bending moment M_y is given by

$$M_y = Pw + R_0 x + M_0 \tag{9.41}$$

in which R_0 means the reaction force at the edge $x = 0$, M_0 is the bending moment at the edge $x = 0$ caused by the thermally induced bending moment M_{Ty}. Substituting Eq. (9.41) into Eq. (9.10), the fundamental equation for w is rewritten as

$$EI_y \frac{d^2 w}{dx^2} + Pw = -R_0 x - M_0 - M_{T_y}(x) \tag{9.42}$$

or

$$EI_y \frac{d^2 w}{dx^2} + Pw = -[D_1 + M_0 + (D_2 + R_0)x] \tag{9.43}$$

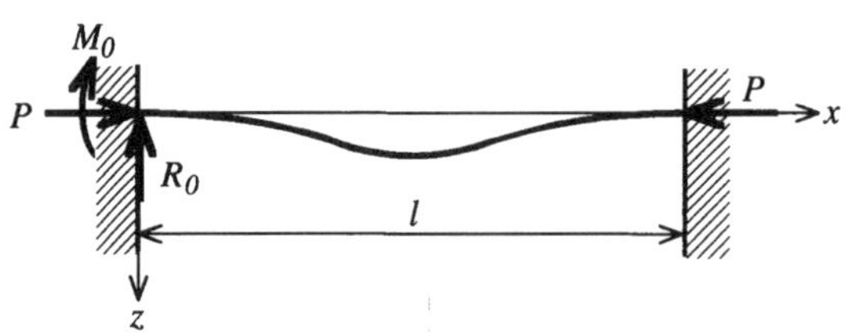

Figure 9.6 Beam-column with built-in edges.

We write the fundamental solution w as

$$w = C_1 \cos kx + C_2 \sin kx - \frac{1}{P}[D_1 + M_0 + (D_2 + R_0)x] \tag{9.44}$$

The boundary conditions are

$$w = 0, \quad \frac{dw}{dx} = 0 \qquad \text{at } x = 0 \text{ and } x = l \tag{9.45}$$

From the boundary conditions, we have

$$\begin{aligned} D_1 + M_0 = C_1 P, \qquad & D_2 + R_0 = PkC_2 \\ C_1 \cos kl + C_2 \sin kl &= \frac{1}{P}[D_1 + M_0 + (D_2 + R_0)l] \\ -kC_1 \sin kl + kC_2 \cos kl &= \frac{1}{P}(D_2 + R_0) \end{aligned} \tag{9.46}$$

Elimination of $(D_1 + M_0)$ and $(D_2 + R_0)$ from Eqs. (9.46) gives

$$\begin{bmatrix} \cos kl - 1 & \sin kl - kl \\ -\sin kl & \cos kl - 1 \end{bmatrix} \begin{bmatrix} C_1 \\ C_2 \end{bmatrix} = \begin{bmatrix} 0 \\ 0 \end{bmatrix} \tag{9.47}$$

In order to have a non-trivial solution for Eq. (9.44), we find

$$\begin{vmatrix} \cos kl - 1 & \sin kl - kl \\ -\sin kl & \cos kl - 1 \end{vmatrix} = 0 \tag{9.48}$$

Eq. (9.48) can be rewritten as

$$\sin \frac{kl}{2} \cos \frac{kl}{2} \left(\tan \frac{kl}{2} - \frac{kl}{2} \right) = 0 \tag{9.49}$$

The smallest value of kl at which the buckling shape can exist is $kl/2 = \pi$. We find the corresponding buckling load P_{cr}

$$P_{cr} = \frac{4\pi^2 E I_y}{l^2} \tag{9.50}$$

Example 9.1 Assuming that the thermally induced bending moment $M_{Ty}(x)$ is given as an arbitrary function of x, find the deflection w of a beam-column for several kinds of boundary conditions.

Solution

[1] For both edges simply supported
We have the fundamental solution of w from Eq. (9.18)

$$w = C_1 \cos kx + C_2 \sin kx + \frac{k}{P}\left[\cos kx \int_0^x M_{T_y}(x') \sin kx'\, dx' - \sin kx \int_0^x M_{T_y}(x') \cos kx'\, dx'\right] \tag{9.51}$$

The boundary conditions at $x = 0$ and $x = l$ are given from Eqs. (9.12′). Therefore, the form of w that satisfies the boundary conditions is

$$w = \frac{k}{P}\left[\sin kx \int_x^l M_{T_y}(x') \cos kx'\, dx' + \cos kx \int_0^x M_{T_y}(x') \sin kx'\, dx' - \cot kl \sin kx \int_0^l M_{T_y}(x') \sin kx'\, dx'\right] \tag{9.52}$$

[2] For a cantilever beam-column shown in Fig. 9.4
We have the fundamental solution w from Eq. (9.27)

$$w = C_1 \cos kx + C_2 \sin kx + w_0 + \frac{k}{P}\left[\cos kx \int_0^x M_{T_y}(x') \sin kx'\, dx' - \sin kx \int_0^x M_{T_y}(x') \cos kx'\, dx'\right] \tag{9.53}$$

Referring to Eqs. (9.11) and (9.13′), the boundary conditions are

$$\begin{aligned} &w = 0, \quad \frac{dw}{dx} = 0 \quad \text{at } x = 0 \\ &EI_y \frac{d^2w}{dx^2} + M_{T_y}(x) = 0, \quad EI_y \frac{d^3w}{dx^3} + P\frac{dw}{dx} + \frac{dM_{T_y}}{dx} = 0 \quad \text{at } x = l \end{aligned} \tag{a}$$

Therefore, the form of w that satisfies the boundary conditions is

$$w = \frac{k}{P}\left\{(\cos kx - 1)\left[-\int_0^l M_{T_y}(x') \sin kx'\, dx' + \tan kl \int_0^l M_{T_y}(x') \cos kx'\, dx'\right] + \cos kx \int_0^x M_{T_y}(x') \sin kx'\, dx' - \sin kx \int_0^x M_{T_y}(x') \cos kx'\, dx'\right\} \tag{9.54}$$

[3] For one edge simply supported and the other one built-in as shown in Fig. 9.5
We have the fundamental solution of w from Eq. (9.34) as

$$w = C_1 \cos kx + C_2 \sin kx - \frac{R_0}{P}x + \frac{k}{P}\left[\cos kx \int_0^x M_{T_y}(x') \sin kx'\, dx' - \sin kx \int_0^x M_{T_y}(x') \cos kx'\, dx'\right] \tag{9.55}$$

The boundary conditions are given by Eqs. (9.11) and (9.12′). Therefore, we have the solution w for the prebuckling state as

$$w = \frac{k}{P}\left[\left(\frac{\sin kx(kl \sin kl + \cos kl) - kx}{kl \cos kl - \sin kl}\right)\int_0^l M_{T_y}(x') \sin kx'\, dx' + \sin kx \int_x^l M_{T_y}(x') \cos kx'\, dx' + \cos kx \int_0^x M_{T_y}(x') \sin kx'\, dx'\right] \tag{9.56}$$

[4] For both edges built-in
We have the fundamental solution of w from Eq. (9.42) as

$$w = C_1 \cos kx + C_2 \sin kx - \frac{1}{P}(R_0 x + M_0) + \frac{k}{P}\left[\cos kx \int_0^x M_{T_y}(x') \sin kx'\, dx' - \sin kx \int_0^x M_{T_y}(x') \cos kx'\, dx'\right] \tag{9.57}$$

The boundary conditions are given by Eqs. (9.45). Then, we have the solution w for the prebuckling state as

$$\begin{aligned} w = \frac{k}{P}\Bigg\{&\left(\frac{1 - \cos kx}{2(1 - \cos kl) - kl \sin kl}\right) \\ &\times \left[(1 - \cos kl - kl \sin kl) \int_0^l M_{T_y}(x') \sin kx'\, dx' \right. \\ &\left. +(\sin kl - kl \cos kl) \int_0^l M_{T_y}(x') \cos kx'\, dx'\right] \\ &-\left(\frac{\sin kx - kx}{2(1 - \cos kl) - kl \sin kl}\right) \\ &\times \left[\sin kl \int_0^l M_{T_y}(x) \sin kx'\, dx' + (\cos kl - 1) \int_0^l M_{T_y}(x') \cos kx'\, dx'\right] \\ &+ \cos kx \int_0^x M_{T_y}(x') \sin kx'\, dx' - \sin kx \int_0^x M_{T_y}(x') \cos kx'\, dx'\Bigg\} \end{aligned} \tag{9.58}$$

Now, consider the case when the arbitrary function M_{Ty} is given by the parabolic function

$$M_{T_y}(x) = D_0 x(l - x) \tag{9.59}$$

Then, we have the deflection w and the buckling load P_{cr} for four different typical cases.

[1] For both edges simply supported
Substitution of Eq. (9.59) into Eq. (9.52) gives

$$w = \frac{D_0}{Pk^2}\left[2(\cos kx - 1) + 2\left(\frac{1-\cos kl}{\sin kl}\right)\sin kx - k^2x(l-x)\right] \tag{9.60}$$

The buckling load P_{cr} is derived from $\sin kl = 0$

$$P_{cr} = \frac{\pi^2 EI_y}{l^2} \tag{9.61}$$

[2] For a cantilever beam-column
From Eqs. (9.54) and (9.59), the deflection is

$$w = \frac{D_0}{Pk^2}\left[kl \sin kx + (\cos kx - 1)\left(\frac{2 - kl\sin kl}{\cos kl}\right) - k^2x(l-x)\right] \tag{9.62}$$

The buckling load P_{cr} is determined from $\cos kl = 0$

$$P_{cr} = \frac{\pi^2 EI_y}{4l^2} \tag{9.63}$$

[3] For one edge simply supported and the other edge built-in
Introduction of Eq. (9.59) into Eq. (9.56) yields

$$\begin{aligned} w = \frac{D_0}{Pk^2}\bigg[&2(\cos kx - 1) - k^2x(l-x) + kx\left(\frac{2(\cos kl - 1) + kl\sin kl}{kl\cos kl - \sin kl}\right) \\ &+ \sin kx\left(\frac{2(\cos kl - 1) + kl(2\sin kl - kl)}{kl\cos kl - \sin bl}\right)\bigg] \end{aligned} \tag{9.64}$$

The buckling load P_{cr} is given from $kl\cos kl = \sin kl$

$$P_{cr} \simeq \frac{2.046\pi^2 EI_y}{l^2} \tag{9.65}$$

[4] For both edges built-in
We have the fundamental solution of w from Eqs. (9.58) and (9.59) as

$$\begin{aligned} w = \frac{D_0}{Pk^2}\bigg\{&kl\sin kx - k^2x(l-x) \\ &+ \frac{1-\cos kx}{2(1-\cos kl) - kl\sin kl}[k^2l^2(1+\cos kl) - 2kl\sin kl]\bigg\} \end{aligned} \tag{9.66}$$

The buckling load P_{cr} is given from $2(1 - \cos kl) = kl \sin kl$

$$P_{cr} = \frac{4\pi^2 E I_y}{l^2} \tag{9.67}$$

9.2 Instability of Plate

We consider a thin plate as a body that undergoes buckling. Rectangular plates or circular plates, which are treated in Chapter 8, are assumed to be subject to thermal load under some support conditions. However, these plates treated here are assumed to be subject to both in-plane and lateral loads.

9.2.1 Basic Equations for a Rectangular Plate

Let u, v, and w be displacement components in the x, y, and z directions at the neutral plane ($z = 0$) of a thin plate with its thickness h, as shown in Fig. 8.1. Then, the stress components can be represented in terms of displacement components and temperature change τ, which are already given by Eqs. (8.37).

$$\begin{aligned}
\sigma_{xx} &= \frac{E}{1-\nu^2}\left[\frac{\partial u}{\partial x} + \nu\frac{\partial v}{\partial y} - z\left(\frac{\partial^2 w}{\partial x^2} + \nu\frac{\partial^2 w}{\partial y^2}\right) - (1+\nu)\alpha\tau\right] \\
\sigma_{yy} &= \frac{E}{1-\nu^2}\left[\frac{\partial v}{\partial y} + \nu\frac{\partial u}{\partial x} - z\left(\frac{\partial^2 w}{\partial y^2} + \nu\frac{\partial^2 w}{\partial x^2}\right) - (1+\nu)\alpha\tau\right] \\
\sigma_{xy} &= \frac{E}{2(1+\nu)}\left(\frac{\partial u}{\partial y} + \frac{\partial v}{\partial x} - 2z\frac{\partial^2 w}{\partial x \partial y}\right)
\end{aligned} \tag{9.68}$$

We now introduce the resultant forces N_x, N_y, N_{xy} and resultant moments M_x, M_y, M_{xy} per unit length of the plate, which are defined by Eqs. (8.38) and written down as

$$\begin{aligned}
N_x &= \int_{-h/2}^{h/2} \sigma_{xx}\, dz, \quad N_y = \int_{-h/2}^{h/2} \sigma_{yy}\, dz, \quad N_{xy} = \int_{-h/2}^{h/2} \sigma_{xy}\, dz \\
M_x &= \int_{-h/2}^{h/2} \sigma_{xx} z\, dz, \quad M_y = \int_{-h/2}^{h/2} \sigma_{yy} z\, dz, \quad M_{xy} = -\int_{-h/2}^{h/2} \sigma_{xy} z\, dz
\end{aligned} \tag{9.69}$$

By the substitution of Eqs. (9.68) into Eqs. (9.69), the resultant forces and the resultant moments are expressed in terms of the displacement components, which are given by Eqs. (8.39) and (8.40), and which are written down as

$$\begin{aligned}
N_x &= \frac{Eh}{1-\nu^2}\left(\frac{\partial u}{\partial x} + \nu\frac{\partial v}{\partial y}\right) - \frac{1}{1-\nu}N_T \\
N_y &= \frac{Eh}{1-\nu^2}\left(\frac{\partial v}{\partial y} + \nu\frac{\partial u}{\partial x}\right) - \frac{1}{1-\nu}N_T
\end{aligned} \tag{9.70}$$

$$N_{xy} = \frac{Eh}{2(1+\nu)}\left(\frac{\partial u}{\partial y} + \frac{\partial v}{\partial x}\right)$$

$$\begin{aligned} M_x &= -D\left(\frac{\partial^2 w}{\partial x^2} + \nu\frac{\partial^2 w}{\partial y^2}\right) - \frac{1}{1-\nu}M_T \\ M_y &= -D\left(\frac{\partial^2 w}{\partial y^2} + \nu\frac{\partial^2 w}{\partial x^2}\right) - \frac{1}{1-\nu}M_T \\ M_{xy} &= (1-\nu)D\frac{\partial^2 w}{\partial x \partial y} \end{aligned} \tag{9.71}$$

where the bending rigidity of the plate D, thermally induced resultant force N_T, and thermally induced resultant moment M_T are defined by Eqs. (8.41) and (8.42). They are written down as

$$D = \frac{Eh^3}{12(1-\nu^2)} \tag{9.72}$$

$$N_T = \alpha E \int_{-h/2}^{h/2} \tau(x, y, z)\, dz, \qquad M_T = \alpha E \int_{-h/2}^{h/2} \tau(x, y, z) z\, dz \tag{9.73}$$

Therefore, the thermal stress components given by Eqs. (9.68) are rewritten in terms of resultant forces and resultant moments in the following form as given by Eqs. (8.43):

$$\begin{aligned} \sigma_{xx} &= \frac{1}{h}N_x + \frac{12z}{h^3}M_x + \frac{1}{1-\nu}\left(\frac{1}{h}N_T + \frac{12z}{h^3}M_T - \alpha E\tau\right) \\ \sigma_{yy} &= \frac{1}{h}N_y + \frac{12z}{h^3}M_y + \frac{1}{1-\nu}\left(\frac{1}{h}N_T + \frac{12z}{h^3}M_T - \alpha E\tau\right) \\ \sigma_{xy} &= \frac{1}{h}N_{xy} - \frac{12z}{h^3}M_{xy} \end{aligned} \tag{9.74}$$

We now consider the equilibrium state in the in-plane direction x and y. Then, we have following relations, which are already given by Eqs. (8.44):

$$\frac{\partial N_x}{\partial x} + \frac{\partial N_{xy}}{\partial y} = 0, \qquad \frac{\partial N_{xy}}{\partial x} + \frac{\partial N_y}{\partial y} = 0 \tag{9.75}$$

In order to evaluate the resultant forces N_x, N_y, N_{xy}, we now introduce a thermal stress function F, which is analogous to the Airy's thermal stress function χ available to the plane problems, defined by Eqs. (8.45)

$$N_x = \frac{\partial^2 F}{\partial y^2}, \qquad N_y = \frac{\partial^2 F}{\partial x^2}, \qquad N_{xy} = -\frac{\partial^2 F}{\partial x \partial y} \tag{9.76}$$

Taking into account the compatibility condition of the strain components, the governing equation for F can be derived, which is already given by Eq. (8.47) and written as

$$\left(\frac{\partial^2}{\partial x^2}+\frac{\partial^2}{\partial y^2}\right)^2 F=-\left(\frac{\partial^2}{\partial x^2}+\frac{\partial^2}{\partial y^2}\right) N_T \tag{9.77}$$

or

$$\nabla^2\nabla^2 F=-\nabla^2 N_T, \qquad \nabla^2=\frac{\partial^2}{\partial x^2}+\frac{\partial^2}{\partial y^2} \tag{9.77$'$}$$

Applying the analytical procedure for the thermal stress function χ discussed in Section 5.2, we can find the stress function F governed by Eq. (9.77) under the appropriate mechanical boundary conditions, therefore, the resultant forces N_x, N_y, N_{xy} are obtained from Eqs. (9.76).

Two different kinds of boundary condition on the edge of the rectangular plate for the in-plane deformation can be identified:

[1] For a built-in edge

$$u_n=0, \quad N_{ns}=0 \qquad \text{or} \qquad u_n=0, \quad u_s=0 \tag{9.78}$$

[2] For a free edge

$$N_n=0, \qquad N_{ns}=0 \tag{9.79}$$

where u_i $(i=n,s)$ means an in-plane displacement component, n and s are the normal and tangential directions of the boundary surface of the plate.

Let us now consider the bending behavior of the plate applicable to the thermally induced buckling. We define the shearing forces Q_x and Q_y, and the twisting moment M_{yx} per unit length as is given by Eqs. (8.49)

$$\begin{gathered} Q_x=\int_{-h/2}^{h/2}\sigma_{xz}\,dz, \qquad Q_y=\int_{-h/2}^{h/2}\sigma_{yz}\,dz \\ M_{yx}=\int_{-h/2}^{h/2}\sigma_{yx}z\,dz=-M_{xy} \end{gathered} \tag{9.80}$$

Next, we consider the equilibrium of moments about y and x axes, which is given by Eqs. (8.50) as

$$\begin{gathered} \frac{\partial M_x}{\partial x}+\frac{\partial M_{yx}}{\partial y}-Q_x=0 \\ \frac{\partial M_y}{\partial y}-\frac{\partial M_{xy}}{\partial x}-Q_y=0 \end{gathered} \tag{9.81}$$

Furthermore, we consider the equilibrium of forces in the out-of-plane direction of z. In this case, we have to take into consideration the effect of resultant forces N_x, N_y, and N_{xy} due to the large deflection.

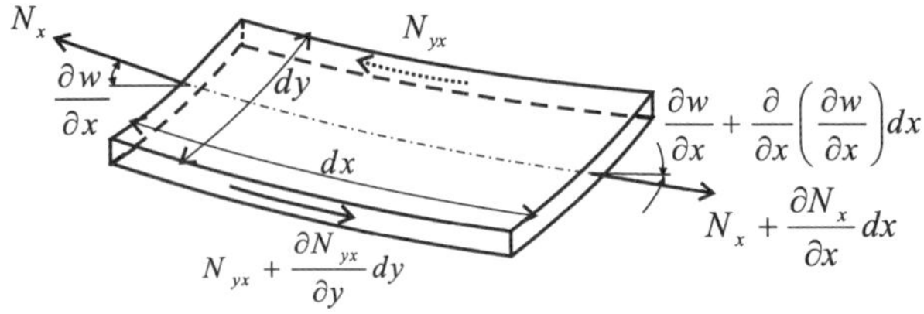

Figure 9.7 Equilibrium of forces in the out-of-plane direction.

Referring to Fig. 9.7, the resultant force of R_1 in the out-of-plane direction due to N_x and N_{yx} is given as

$$\begin{aligned}
R_1 &= \left(N_x + \frac{\partial N_x}{\partial x}dx\right) dy \sin\left(\frac{\partial w}{\partial x} + \frac{\partial^2 w}{\partial x^2}dx\right) - N_x\, dy \sin\left(\frac{\partial w}{\partial x}\right) \\
&\quad + \left(N_{yx} + \frac{\partial N_{yx}}{\partial y}dy\right) dx \sin\left(\frac{\partial w}{\partial x} + \frac{\partial^2 w}{\partial x \partial y}dy\right) - N_{yx} dx \sin\left(\frac{\partial w}{\partial x}\right) \\
&\cong \left(N_x + \frac{\partial N_x}{\partial x}dx\right) dy \left(\frac{\partial w}{\partial x} + \frac{\partial^2 w}{\partial x^2}dx\right) - N_x dy \frac{\partial w}{\partial x} \\
&\quad + \left(N_{yx} + \frac{\partial N_{yx}}{\partial y}dy\right) dx \left(\frac{\partial w}{\partial x} + \frac{\partial^2 w}{\partial x \partial y}dy\right) - N_{yx} dx \frac{\partial w}{\partial x} \\
&= \left(N_x \frac{\partial^2 w}{\partial x^2} + \frac{\partial N_x}{\partial x}\frac{\partial w}{\partial x}\right) dx\, dy + \left(N_{yx}\frac{\partial^2 w}{\partial x \partial y} + \frac{\partial N_{yx}}{\partial y}\frac{\partial w}{\partial x}\right) dx\, dy \\
&= \left[N_x \frac{\partial^2 w}{\partial x^2} + N_{yx}\frac{\partial^2 w}{\partial x \partial y} + \left(\frac{\partial N_x}{\partial x} + \frac{\partial N_{yx}}{\partial y}\right)\frac{\partial w}{\partial x}\right] dx\, dy \qquad \text{(a)}
\end{aligned}$$

By the substitution of Eqs. (9.75) into Eq. (a), it follows that

$$R_1 = \left(N_x \frac{\partial^2 w}{\partial x^2} + N_{yx}\frac{\partial^2 w}{\partial x \partial y}\right) dx\, dy \qquad \text{(b)}$$

Similarly, the resultant force R_2 in the out-of-plane direction due to N_y and N_{xy} is given as

$$R_2 = \left(N_y \frac{\partial^2 w}{\partial y^2} + N_{xy}\frac{\partial^2 w}{\partial x \partial y}\right) dx\, dy \qquad \text{(c)}$$

Finally, we can evaluate the resultant forces in the out-of-plane direction due to Q_x, Q_y and the transverse load p, which are given by the relations

$$\begin{gathered}
\left(\frac{\partial Q_x}{\partial x} + \frac{\partial Q_y}{\partial y} + p\right) dx\, dy + R_1 + R_2 = 0 \\
\frac{\partial Q_x}{\partial x} + \frac{\partial Q_y}{\partial y} + p + N_x \frac{\partial^2 w}{\partial x^2} + N_y \frac{\partial^2 w}{\partial y^2} + 2N_{xy}\frac{\partial^2 w}{\partial x \partial y} = 0
\end{gathered} \qquad (9.82)$$

Substitution of Eqs. (9.81) into Eq. (9.82) gives the fundamental relation for the thermal buckling problems of a plate

$$\frac{\partial}{\partial x}\left(\frac{\partial M_x}{\partial x}+\frac{\partial M_{yx}}{\partial y}\right)+\frac{\partial}{\partial y}\left(\frac{\partial M_y}{\partial y}-\frac{\partial M_{xy}}{\partial x}\right)+p$$
$$+N_x\frac{\partial^2 w}{\partial x^2}+N_y\frac{\partial^2 w}{\partial y^2}+2N_{xy}\frac{\partial^2 w}{\partial x\partial y}=0 \tag{d}$$

Taking into account Eqs. (9.71), this relation can be rewritten in terms of the deflection w alone

$$\nabla^2\nabla^2 w=\frac{1}{D}\left(p-\frac{1}{1-\nu}\nabla^2 M_T+N_x\frac{\partial^2 w}{\partial x^2}+N_y\frac{\partial^2 w}{\partial y^2}+2N_{xy}\frac{\partial^2 w}{\partial x\partial y}\right) \tag{9.83}$$

Furthermore, when the external transverse load p is absent, the fundamental equation (9.83) results in

$$\nabla^2\nabla^2 w=\frac{1}{D}\left(-\frac{1}{1-\nu}\nabla^2 M_T+N_x\frac{\partial^2 w}{\partial x^2}+N_y\frac{\partial^2 w}{\partial y^2}+2N_{xy}\frac{\partial^2 w}{\partial x\partial y}\right) \tag{9.84}$$

On the other hand, the mechanical boundary conditions for the thermal buckling problems governed by Eq. (9.83) or (9.84) are stated by Eqs. (8.57) to (8.59) and are represented as

[1] For a built-in edge or fixed end

$$w=0,\qquad \frac{\partial w}{\partial n}=0 \tag{9.85}$$

[2] For a simply supported edge

$$w=0,\qquad M_n=0 \tag{9.86}$$

[3] For a free edge

$$M_n=0,\qquad Q_n-\frac{\partial M_{ns}}{\partial s}=0 \tag{9.87}$$

where n and s are the normal and tangential directions of the boundary surface of the plate.

9.2.2 Illustrative Examples for a Rectangular Plate

Let us consider a rectangular plate subjected to a uniform temperature change τ_0 with its length $a \times b$ as shown in Fig. 8.6, and assume that the edges are restrained, so that the displacement u or v in the in-plane direction is zero. Then, the boundary conditions for the in-plane deformation are given as

$$\begin{aligned} u &= 0 \qquad \text{on } x = 0, a \\ v &= 0 \qquad \text{on } y = 0, b \end{aligned} \tag{9.88}$$

The strain components are then identically zero over the whole domain under consideration. Thus,

$$\epsilon_{xx} = 0, \qquad \epsilon_{yy} = 0, \qquad \epsilon_{xy} = 0 \tag{9.89}$$

For such case, making use of the stress–strain relations (5.16), the stress components for the in-plane deformation are obtained as

$$\sigma_{xx} = \sigma_{yy} = -\frac{1}{1-\nu}\alpha E \tau_0, \qquad \sigma_{xy} = 0 \tag{9.90}$$

By the substitution of Eqs. (9.90) into Eqs. (9.69) and (9.73), it follows that

$$N_x = N_y = -\frac{1}{1-\nu}\alpha E \tau_0 h \equiv N_0, \qquad N_{xy} = 0, \qquad M_T = 0 \tag{9.91}$$

Substituting Eqs. (9.91) into Eq. (9.84), the governing equation for the thermal buckling problem of the plate is

$$\nabla^2\left(\nabla^2 w - \frac{N_0}{D} w\right) = 0 \tag{9.92}$$

We assume that the edges of the plate are simply supported. Then, the boundary conditions for the thermal buckling problem are given by Eqs. (9.86) and (9.71), which are represented by

$$\begin{aligned} w = 0, \quad M_x &= -D\left(\frac{\partial^2 w}{\partial x^2} + \nu\frac{\partial^2 w}{\partial y^2}\right) = 0 \qquad \text{on } x = 0, a \\ w = 0, \quad M_y &= -D\left(\frac{\partial^2 w}{\partial y^2} + \nu\frac{\partial^2 w}{\partial x^2}\right) = 0 \qquad \text{on } y = 0, b \end{aligned} \tag{9.93}$$

The fundamental equation (9.92) may be resolved into the following two equations system:

$$\left(\frac{\partial^2 w}{\partial x^2} + \frac{\partial^2 w}{\partial y^2}\right) F = 0 \tag{9.94}$$

$$\left(\frac{\partial^2 w}{\partial x^2} + \frac{\partial^2 w}{\partial y^2}\right) w - \frac{N_0}{D} w = F \tag{9.95}$$

Now, it can be seen for a simply supported edge that the following conditions are satisfied on the boundaries:

$$\begin{aligned} w = 0, \quad \frac{\partial^2 w}{\partial y^2} = 0 \qquad \text{on } x = 0, a \\ w = 0, \quad \frac{\partial^2 w}{\partial x^2} = 0 \qquad \text{on } y = 0, b \end{aligned} \tag{9.96}$$

Therefore, making use of Eqs. (9.93) and (9.95), the mechanical boundary conditions for the unknown function F lead to

$$\begin{aligned} F = 0 \qquad \text{on } x = 0, a \\ F = 0 \qquad \text{on } y = 0, b \end{aligned} \tag{9.97}$$

Thus, the unknown function F governed by Eq. (9.94) becomes equal to zero, or

$$F = 0 \tag{9.98}$$

By the substitution of Eq. (9.98) into Eq. (9.95), the fundamental equation for the deflection w is reduced to

$$\left(\frac{\partial^2}{\partial x^2} + \frac{\partial^2}{\partial y^2}\right) w - \frac{N_0}{D} w = 0 \tag{9.99}$$

The deflection w satisfying the first relation of the boundary condition (9.93) may now be expanded into a double trigonometric series

$$w = \sum_{m=1}^{\infty} \sum_{n=1}^{\infty} w_{mn} \sin \alpha_m x \sin \beta_n y \tag{9.100}$$

where the coefficient w_{mn} is indeterminate, while α_m and β_n are determined so as to satisfy the boundary conditions $w = 0$, namely

$$\alpha_m = \frac{m\pi}{a}, \qquad \beta_n = \frac{n\pi}{b} \tag{9.101}$$

Substitution of Eq. (9.100) into Eq. (9.99) gives

$$\alpha_m^2 + \beta_n^2 + \frac{N_0}{D} = 0 \tag{9.102}$$

Furthermore, by the substitution of Eqs. (9.72) and (9.91) into Eq. (9.102), we have

$$\alpha\tau_0 = \frac{\pi^2}{12(1+\nu)} \left(\frac{h}{a}\right)^2 \left[m^2 + n^2 \left(\frac{a}{b}\right)^2\right]; \qquad m, n = 1, 3, 5, \ldots \tag{9.103}$$

from which it can be seen that the actual thermal buckling is caused when $m = n = 1$, and the critical temperature rise τ_{0cr} is given by

$$\alpha \tau_{0cr} = \frac{\pi^2}{12(1+\nu)} \left(\frac{h}{a}\right)^2 \left[1 + \left(\frac{a}{b}\right)^2\right] \tag{9.104}$$

The mode of the deflection w for this thermal buckling is given by

$$w = w_{11} \sin\frac{\pi x}{a} \sin\frac{\pi y}{b} \tag{9.105}$$

Now, assuming that $a = b$ (square plate), $\nu = 0.3$, and $\alpha = 1.0 \times 10^{-5}$ 1/K, then the critical temperature rise τ_{0cr} is given as

$$\tau_{0cr} = \begin{cases} 316\ [\mathrm{K}]; & a/h = 20 \\ 50.6\ [\mathrm{K}]; & a/h = 50 \\ 12.6\ [\mathrm{K}]; & a/h = 100 \end{cases} \tag{9.106}$$

Example 9.2 Assuming that the upper surface ($z = -h/2$) of the rectangular thin plate is held at zero temperature ($\tau = 0$) and the lower surface ($z = h/2$) of the plate is at constant temperature ($\tau = \tau_0$), derive the critical temperature rise τ_{0cr} for thermal buckling and the deflection w for a simply supported edge condition, and for the restricted deflection condition in the in-plane direction.

Solution The one-dimensional temperature change τ varying in the thickness direction z can be represented as

$$\tau = \tau_0 \left(\frac{1}{2} + \frac{z}{h}\right) \tag{9.107}$$

The strain components are identically zero because the in-plane deformation is prevented, so that

$$\epsilon_{xx} = 0, \qquad \epsilon_{yy} = 0, \qquad \epsilon_{xy} = 0 \tag{a}$$

Then, the stress components can be obtained from Eq. (5.16) as

$$\sigma_{xx} = \sigma_{yy} = -\frac{1}{1-\nu}\alpha E \tau(z) = -\frac{\alpha E \tau_0}{1-\nu}\left(\frac{1}{2} + \frac{z}{h}\right), \qquad \sigma_{xy} = 0 \tag{9.108}$$

By the substitution of Eqs. (9.108) into Eqs. (9.69) and (9.73), we have

$$N_x = N_y = -\frac{\alpha E \tau_0 h}{2(1-\nu)} \equiv N_0, \quad N_{xy} = 0, \quad M_T = \frac{1}{12}\alpha E \tau_0 h^2 \equiv M_{T0} \tag{9.109}$$

Substituting Eqs. (9.109) into Eq. (9.84), the governing equation for w is

$$\nabla^2\left(\nabla^2 w - \frac{N_0}{D}w\right) = 0 \tag{b}$$

The boundary conditions for simply supported edges are given from Eqs. (9.86) and (9.71) as

$$\begin{aligned} &w = 0, \quad M_x = -D\left(\frac{\partial^2 w}{\partial x^2} + \nu\frac{\partial^2 w}{\partial y^2}\right) - \frac{M_{T0}}{1-\nu} = 0 \qquad \text{on } x = 0, a \\ &w = 0, \quad M_y = -D\left(\frac{\partial^2 w}{\partial y^2} + \nu\frac{\partial^2 w}{\partial x^2}\right) - \frac{M_{T0}}{1-\nu} = 0 \qquad \text{on } y = 0, b \end{aligned} \tag{9.110}$$

Now, the fundamental equation (b) can be resolved into the following system of two equations:

$$\left(\frac{\partial^2}{\partial x^2} + \frac{\partial^2}{\partial y^2}\right) f = 0 \tag{c}$$

$$\left(\frac{\partial^2}{\partial x^2} + \frac{\partial^2}{\partial y^2}\right) w - \frac{N_0}{D}w = f \tag{d}$$

On the other hand, it can be seen from the boundary conditions of a simply supported edge that

$$\begin{aligned} &w = 0, \quad \frac{\partial^2 w}{\partial y^2} = 0 \qquad \text{on } x = 0, a \\ &w = 0, \quad \frac{\partial^2 w}{\partial x^2} = 0 \qquad \text{on } y = 0, b \end{aligned} \tag{e}$$

Therefore, making use of the boundary condition (9.110), the boundary conditions for unknown function f result in

$$\begin{aligned} &f + \frac{M_{T0}}{(1-\nu)D} = 0 \qquad \text{on } x = 0, a \\ &f + \frac{M_{T0}}{(1-\nu)D} = 0 \qquad \text{on } y = 0, b \end{aligned} \tag{f}$$

Therefore, the unknown function f governed by Eq. (c) can be evaluated as

$$f = -\frac{M_{T0}}{(1-\nu)D} \tag{9.111}$$

The fundamental equation for deflection w governed by Eq. (d) is reduced to

$$\left(\frac{\partial^2}{\partial x^2} + \frac{\partial^2}{\partial y^2}\right) w - \frac{N_0}{D}w = -\frac{M_{T0}}{(1-\nu)D} \tag{9.112}$$

The deflection w satisfying the boundary condition of $w = 0$ may be given by a double trigonometric series

$$w = \sum_{m=1}^{\infty}\sum_{n=1}^{\infty} w_{mn} \sin \alpha_m x \sin \beta_n y \tag{g}$$

where α_m and β_n are

$$\alpha_m = \frac{m\pi}{a}, \qquad \beta_n = \frac{n\pi}{b} \tag{h}$$

Similarly, the constant term M_{T0} in Eq. (9.112) can be represented by

$$M_{T0} = \sum_{m=1}^{\infty}\sum_{n=1}^{\infty} a_{mn} \sin \alpha_m x \sin \beta_n y \tag{i}$$

Now, the coefficient a_{mn} is evaluated from Eq. (8.72) as

$$a_{mn} = \begin{cases} \dfrac{4}{ab} M_{T0} \dfrac{4}{\alpha_m \beta_n}, & m, n = 1, 3, 5, \ldots \\[2ex] 0, & m, n = 2, 4, 6, \ldots \end{cases} \tag{9.113}$$

By the substitution of Eqs. (g) and (i) into Eq. (9.112), it follows that

$$w_{mn} = \begin{cases} \dfrac{1}{(1-\nu)D} \dfrac{a_{mn}}{\alpha_m^2 + \beta_n^2 + N_0/D}, & m, n = 1, 3, 5, \ldots \\[2ex] 0, & m, n = 2, 4, 6, \ldots \end{cases} \tag{j}$$

Thus, the deflection w takes the form

$$w = \frac{1}{(1-\nu)D} \sum_{m=1,3,5,\ldots}^{\infty} \sum_{n=1,3,5,\ldots}^{\infty} \frac{a_{mn}}{\alpha_m^2 + \beta_n^2 + N_0/D} \sin \alpha_m x \sin \beta_n y \tag{9.114}$$

The thermal buckling condition is then given by

$$\alpha_m^2 + \beta_n^2 + \frac{N_0}{D} = 0 \tag{k}$$

from which it can be seen that the actual thermal buckling appears when $m = n = 1$, and the critical temperature rise τ_{0cr} is given by

$$\alpha \tau_{0cr} = \frac{\pi^2}{6(1+\nu)} \left(\frac{h}{a}\right)^2 \left[1 + \left(\frac{a}{b}\right)^2\right] \tag{9.115}$$

Furthermore, the mode of the deflection w for this thermal buckling is given by

$$w = w_{11} \sin \frac{\pi x}{a} \sin \frac{\pi y}{b} \tag{l}$$

Assuming that $a = b$ (square plate), $\nu = 0.3$, and $\alpha = 1.0 \times 10^{-5}$ 1/K, then the critical temperature rise τ_{0cr} is then given as

$$\tau_{0cr} = \begin{cases} 632\ [\mathrm{K}]; & a/h = 20 \\ 101\ [\mathrm{K}]; & a/h = 50 \\ 25.3\ [\mathrm{K}]; & a/h = 100 \end{cases} \tag{m}$$

9.2.3 Basic Equations for a Circular Plate

Let u_r, u_θ, and w be displacement components in the r, θ, and z directions at the neutral plane ($z = 0$) of a thin circular plate with its thickness h, as shown in Fig. 8.9. The stress components may be derived by the method of coordinate transformation applied to relation (9.68)

$$\begin{aligned}
\sigma_{rr} &= \frac{E}{1-\nu^2}\left\{\frac{\partial u_r}{\partial r} + \nu\left(\frac{u_r}{r} + \frac{1}{r}\frac{\partial u_\theta}{\partial \theta}\right)\right. \\
&\qquad \left. - z\left[\frac{\partial^2 w}{\partial r^2} + \nu\left(\frac{1}{r}\frac{\partial w}{\partial r} + \frac{1}{r^2}\frac{\partial^2 w}{\partial \theta^2}\right)\right] - (1+\nu)\alpha\tau\right\} \\
\sigma_{\theta\theta} &= \frac{E}{1-\nu^2}\left\{\nu\frac{\partial u_r}{\partial r} + \left(\frac{u_r}{r} + \frac{1}{r}\frac{\partial u_\theta}{\partial \theta}\right)\right. \\
&\qquad \left. - z\left[\nu\frac{\partial^2 w}{\partial r^2} + \left(\frac{1}{r}\frac{\partial w}{\partial r} + \frac{1}{r^2}\frac{\partial^2 w}{\partial \theta^2}\right)\right] - (1+\nu)\alpha\tau\right\} \\
\sigma_{r\theta} &= \frac{E}{2(1+\nu)}\left[\frac{1}{r}\frac{\partial u_r}{\partial \theta} + r\frac{\partial}{\partial r}\left(\frac{u_\theta}{r}\right) - 2z\frac{\partial}{\partial r}\left(\frac{1}{r}\frac{\partial w}{\partial \theta}\right)\right]
\end{aligned} \tag{9.116}$$

We now introduce the resultant forces N_r, N_θ, $N_{r\theta}$ and the resultant moments M_r, M_θ, $M_{r\theta}$ per unit length of the plate, which are already defined by Eqs. (8.96) and written down as

$$\begin{aligned}
N_r &= \int_{-h/2}^{h/2} \sigma_{rr}\,dz, \quad & N_\theta &= \int_{-h/2}^{h/2} \sigma_{\theta\theta}\,dz, \quad & N_{r\theta} &= \int_{-h/2}^{h/2} \sigma_{r\theta}\,dz, \\
M_r &= \int_{-h/2}^{h/2} \sigma_{rr} z\,dz, \quad & M_\theta &= \int_{-h/2}^{h/2} \sigma_{\theta\theta} z\,dz, \quad & M_{r\theta} &= -\int_{-h/2}^{h/2} \sigma_{r\theta} z\,dz
\end{aligned} \tag{9.117}$$

By the substitution of Eqs. (9.116) into Eqs. (9.117), the resultant forces and the resultant moments are expressed by the displacement components, which are written down as

$$\begin{aligned}
N_r &= \frac{Eh}{1-\nu^2}\left[\frac{\partial u_r}{\partial r} + \nu\left(\frac{u_r}{r} + \frac{1}{r}\frac{\partial u_\theta}{\partial \theta}\right)\right] - \frac{1}{1-\nu}N_T \\
N_\theta &= \frac{Eh}{1-\nu^2}\left[\nu\frac{\partial u_r}{\partial r} + \left(\frac{u_r}{r} + \frac{1}{r}\frac{\partial u_\theta}{\partial \theta}\right)\right] - \frac{1}{1-\nu}N_T
\end{aligned} \tag{9.118}$$

$$
\begin{aligned}
N_{r\theta} &= \frac{Eh}{2(1+\nu)}\left[\frac{1}{r}\frac{\partial u_r}{\partial \theta} + r\frac{\partial}{\partial r}\left(\frac{u_\theta}{r}\right)\right] \\
M_r &= -D\left[\frac{\partial^2 w}{\partial r^2} + \nu\left(\frac{1}{r}\frac{\partial w}{\partial r} + \frac{1}{r^2}\frac{\partial^2 w}{\partial \theta^2}\right)\right] - \frac{1}{1-\nu}M_T \\
M_\theta &= -D\left[\nu\frac{\partial^2 w}{\partial r^2} + \left(\frac{1}{r}\frac{\partial w}{\partial r} + \frac{1}{r^2}\frac{\partial^2 w}{\partial \theta^2}\right)\right] - \frac{1}{1-\nu}M_T \\
N_{r\theta} &= (1-\nu)D\frac{\partial}{\partial r}\left(\frac{1}{r}\frac{\partial w}{\partial \theta}\right)
\end{aligned}
\tag{9.119}
$$

where the thermally induced resultant force N_T and the thermally induced resultant moment M_T are already defined by Eqs. (9.73). In the polar-cylindrical coordinates they are:

$$
N_T = \alpha E \int_{-h/2}^{h/2} \tau(r,\theta,z)\,dz, \qquad M_T = \alpha E \int_{-h/2}^{h/2} \tau(r,\theta,z) z\,dz \tag{9.120}
$$

The thermal stress components given by Eqs. (9.116) are rewritten in terms of resultant forces and resultant moments as given by Eqs. (8.95):

$$
\begin{aligned}
\sigma_{rr} &= \frac{1}{h}N_r + \frac{12z}{h^3}M_r + \frac{1}{1-\nu}\left(\frac{1}{h}N_T + \frac{12z}{h^3}M_T - \alpha E\tau\right) \\
\sigma_{\theta\theta} &= \frac{1}{h}N_\theta + \frac{12z}{h^3}M_\theta + \frac{1}{1-\nu}\left(\frac{1}{h}N_T + \frac{12z}{h^3}M_T - \alpha E\tau\right) \\
\sigma_{r\theta} &= \frac{1}{h}N_{r\theta} - \frac{12z}{h^3}M_{r\theta}
\end{aligned}
\tag{9.121}
$$

Now, the governing equation determining the resultant forces N_r, N_θ, $N_{r\theta}$ is given by Eqs. (9.77) or (9.77′), which is written down as

$$
\nabla^2\nabla^2 F = -\nabla^2 N_T \tag{9.122}
$$

where

$$
\nabla^2 = \frac{\partial^2}{\partial r^2} + \frac{1}{r}\frac{\partial}{\partial r} + \frac{1}{r^2}\frac{\partial^2}{\partial \theta^2} \tag{9.123}
$$

and F means the thermal stress function applicable to the plane problems. The relations between F and the resultant forces N_r, N_θ, $N_{r\theta}$ can be derived from Eqs. (9.76) by the method of coordinate transformation. They become

$$
N_r = \frac{1}{r}\frac{\partial F}{\partial r} + \frac{1}{r^2}\frac{\partial^2 F}{\partial \theta^2}, \qquad N_\theta = \frac{\partial^2 F}{\partial r^2}, \qquad N_{r\theta} = -\frac{\partial}{\partial r}\left(\frac{1}{r}\frac{\partial F}{\partial \theta}\right) \tag{9.124}
$$

Applying the analytical procedure for the thermal stress function χ discussed in Section 5.2, we can find the stress function F governed by Eq. (9.122) under the appropriate mechanical boundary conditions, and subsequently, the resultant forces N_r, N_θ, $N_{r\theta}$ are obtained from Eqs. (9.124).

Now, two different kinds of boundary condition on the edge ($r = a$) of the circular plate for the in-plane deformation can be identified:

[1] For a built-in edge

$$u_r = 0, \quad N_{r\theta} = 0 \qquad \text{or} \qquad u_r = 0, \quad u_\theta = 0 \tag{9.125}$$

[2] For a free edge

$$N_r = 0, \qquad N_{r\theta} = 0 \tag{9.126}$$

We now consider the bending behavior of the plate applicable to the thermally induced buckling. The governing equation for the thermal bending problem may be derived from Eq. (9.83), written in terms of Cartesian coordinates, by using the method of coordinate transformation. Thus, the governing equation for thermal bucking problems of the circular plate can be represented as

$$\nabla^2\nabla^2 w = \frac{1}{D}\left[p - \frac{1}{1-\nu}\nabla^2 M_T \right.$$
$$\left. + N_r\frac{\partial^2 w}{\partial r^2} + N_\theta\left(\frac{1}{r}\frac{\partial w}{\partial r} + \frac{1}{r^2}\frac{\partial^2 w}{\partial \theta^2}\right) + 2N_{r\theta}\frac{\partial}{\partial r}\left(\frac{1}{r}\frac{\partial w}{\partial \theta}\right)\right] \tag{9.127}$$

Furthermore, when the external transverse load p is absent, the fundamental equation (9.127) results in

$$\nabla^2\nabla^2 w = \frac{1}{D}\left[-\frac{1}{1-\nu}\nabla^2 M_T \right.$$
$$\left. + N_r\frac{\partial^2 w}{\partial r^2} + N_\theta\left(\frac{1}{r}\frac{\partial w}{\partial r} + \frac{1}{r^2}\frac{\partial^2 w}{\partial \theta^2}\right) + 2N_{r\theta}\frac{\partial}{\partial r}\left(\frac{1}{r}\frac{\partial w}{\partial \theta}\right)\right] \tag{9.128}$$

The mechanical boundary conditions for the thermally induced bending problems of the circular plate with its radius a are given by Eqs. (8.97) and (8.98), which are:

[1] For a built-in edge

$$w = 0, \quad \frac{\partial w}{\partial r} = 0 \qquad \text{on } r = a \tag{9.129}$$

[2] For a simply supported edge

$$w = 0, \quad \frac{\partial^2 w}{\partial r^2} + \nu\left(\frac{1}{r}\frac{\partial w}{\partial r} + \frac{1}{r^2}\frac{\partial^2 w}{\partial \theta^2}\right) = -\frac{1}{(1-\nu)D}M_T \qquad \text{on } r = a \tag{9.130}$$

Consider now the axisymmetric behavior of a circular plate. The non-zero stress components in terms of the displacement components are derived from Eqs. (9.116), which are represented as

$$
\begin{aligned}
\sigma_{rr} &= \frac{E}{1-\nu^2}\left[\frac{du_r}{dr} + \nu\frac{u_r}{r} - z\left(\frac{d^2w}{dr^2} + \nu\frac{1}{r}\frac{dw}{dr}\right) - (1+\nu)\alpha\tau\right] \\
\sigma_{\theta\theta} &= \frac{E}{1-\nu^2}\left[\nu\frac{du_r}{dr} + \frac{u_r}{r} - z\left(\nu\frac{d^2w}{dr^2} + \frac{1}{r}\frac{dw}{dr}\right) - (1+\nu)\alpha\tau\right]
\end{aligned}
\tag{9.131}
$$

The resultant forces N_r, N_θ and the resultant moments M_r, M_θ per unit length of the plate are defined by Eq. (9.117) and written down as

$$
\begin{aligned}
N_r &= \int_{-h/2}^{h/2} \sigma_{rr}\,dz, \qquad N_\theta = \int_{-h/2}^{h/2} \sigma_{\theta\theta}\,dz, \\
M_r &= \int_{-h/2}^{h/2} \sigma_{rr}z\,dz, \qquad M_\theta = \int_{-h/2}^{h/2} \sigma_{\theta\theta}z\,dz
\end{aligned}
\tag{9.132}
$$

The resultant forces and the resultant moments are expressed in terms of the displacement components, which are rewritten from Eqs. (9.118) and (9.119) as

$$
\begin{aligned}
N_r &= \frac{Eh}{1-\nu^2}\left(\frac{du_r}{dr} + \nu\frac{u_r}{r}\right) - \frac{1}{1-\nu}N_T, \\
N_\theta &= \frac{Eh}{1-\nu^2}\left(\nu\frac{du_r}{dr} + \frac{u_r}{r}\right) - \frac{1}{1-\nu}N_T, \\
M_r &= -D\left(\frac{d^2w}{dr^2} + \nu\frac{1}{r}\frac{dw}{dr}\right) - \frac{1}{1-\nu}M_T, \\
M_\theta &= -D\left(\nu\frac{d^2w}{dr^2} + \frac{1}{r}\frac{dw}{dr}\right) - \frac{1}{1-\nu}M_T
\end{aligned}
\tag{9.133}
$$

where the thermally induced resultant force N_T and the thermally induced resultant moment M_T are rewritten from Eqs. (9.120) as

$$
N_T = \alpha E\int_{-h/2}^{h/2} \tau(r,z)\,dz, \qquad M_T = \alpha E\int_{-h/2}^{h/2} \tau(r,z)z\,dz \tag{9.134}
$$

Therefore, the thermal stress components given by Eqs. (9.131) may be rewritten in terms of the resultant forces and resultant moments in the following form:

$$
\begin{aligned}
\sigma_{rr} &= \frac{1}{h}N_r + \frac{12z}{h^3}M_r + \frac{1}{1-\nu}\left(\frac{1}{h}N_T + \frac{12z}{h^3}M_T - \alpha E\tau\right) \\
\sigma_{\theta\theta} &= \frac{1}{h}N_\theta + \frac{12z}{h^3}M_\theta + \frac{1}{1-\nu}\left(\frac{1}{h}N_T + \frac{12z}{h^3}M_T - \alpha E\tau\right)
\end{aligned}
\tag{9.135}
$$

Now, the governing equation (9.122) determining the resultant forces N_r, N_θ is reduced to

$$\nabla^2\nabla^2 F = -\nabla^2 N_T, \tag{9.136}$$

where

$$\nabla^2 = \frac{d^2}{dr^2} + \frac{1}{r}\frac{d}{dr} = \frac{1}{r}\frac{d}{dr}\left(r\frac{d}{dr}\frac{d}{dr}\right) \tag{9.137}$$

and the relations between F and the resultant forces N_r, N_θ, which are given by Eqs. (9.124), are reduced to the form.

$$N_r = \frac{1}{r}\frac{dF}{dr}, \qquad N_\theta = \frac{d^2F}{dr^2} \tag{9.138}$$

For such case, two different kinds of boundary condition on the edge ($r = a$) of the circular plate for the in-plane deformation can be identified:

[1] For a built-in edge

$$u_r = 0 \tag{9.139}$$

[2] For a free edge

$$N_r = 0 \tag{9.140}$$

On the other sides, the governing equation (9.127) for the thermal bending problem is reduced to

$$\nabla^2\nabla^2 w = \frac{1}{D}\left(p - \frac{1}{1-\nu}\nabla^2 M_T + N_r\frac{d^2w}{dr^2} + N_\theta\frac{1}{r}\frac{dw}{dr}\right) \tag{9.141}$$

If the external transverse load p is absent, the fundamental equation (9.141) results in

$$\nabla^2\nabla^2 w = \frac{1}{D}\left(-\frac{1}{1-\nu}\nabla^2 M_T + N_r\frac{d^2w}{dr^2} + N_\theta\frac{1}{r}\frac{dw}{dr}\right) \tag{9.142}$$

Now, the mechanical boundary conditions (9.129) and (9.130) for the thermal buckling problems of the circular plate with radius a are rewritten as

[1] For the built-in edge

$$w = 0, \quad \frac{dw}{dr} = 0 \qquad \text{on } r = a \tag{9.143}$$

[2] For the simply supported edge

$$w = 0, \quad \frac{d^2w}{dr^2} + \nu\frac{1}{r}\frac{dw}{dr} = -\frac{1}{(1-\nu)D}M_T \qquad \text{on } r = a \tag{9.144}$$

9.2.4 Illustrative Examples for a Circular Plate

We now consider a circular plate of radius a subjected to a constant temperature rise τ_0. Let the displacement of the edge of the plate be restricted in the radial direction. Then, all the strain components become zero:

$$\epsilon_{rr} = 0, \qquad \epsilon_{\theta\theta} = 0, \qquad \epsilon_{r\theta} = 0 \tag{9.145}$$

Substituting relations (9.145) into relations (5.16), which are rewritten in the polar coordinate system, the stress components are given as

$$\sigma_{rr} = \sigma_{\theta\theta} = -\frac{1}{1-\nu}\alpha E \tau_0, \qquad \sigma_{r\theta} = 0 \tag{9.146}$$

By the substitution of Eqs. (9.146) into Eqs. (9.134), we have

$$N_T = \alpha E \tau_0 h \equiv N_{T0}, \qquad M_T = 0 \tag{9.147}$$

Furthermore, considering the relations (9.145), we find that

$$u_r = 0 \tag{9.148}$$

By the substitution of Eqs. (9.147) and (9.148) into Eqs. (9.133), we have

$$N_r = N_\theta = -\frac{1}{1-\nu} N_{T0} \tag{9.149}$$

Substituting Eqs. (9.147) and (9.149) into Eq. (9.142), the governing equation for the out-of-plane deflection w is given as

$$\nabla^2 \left[\nabla^2 w + \frac{N_{T0}}{(1-\nu)D} w \right] = 0 \tag{9.150}$$

where

$$\nabla^2 = \frac{1}{r}\frac{d}{dr}\left(r \frac{d}{dr}\frac{d}{dr} \right) \tag{9.151}$$

Performing the successive integration for Eq. (9.150), we have

$$\frac{d^2 w}{dr^2} + \frac{1}{r}\frac{dw}{dr} + \beta^2 w = C_1 \ln r + C_2 \tag{9.152}$$

where

$$\beta^2 = \frac{N_{T0}}{(1-\nu)D} = 12(1+\nu)\alpha\tau_0 \frac{1}{h^2} \tag{9.153}$$

The fundamental solution for Eq. (9.152) may be obtained by the method of variation of parameters (see the Hint of Problem 3.5), and it is given by

$$w = D_1 J_0(\beta r) + D_2 Y_0(\beta r) + \frac{1}{\beta^2} C_2 + C_1 \frac{\pi}{2}\left[-J_0(\beta r)\int r\ln r Y_0(\beta r)\,dr + Y_0(\beta r)\int r\ln r J_0(\beta r)\,dr\right] \tag{9.154}$$

By the substitution of Eqs. (9.154) and (9.147) into Eqs. (9.133), the resultant moment M_r is evaluated as

$$\begin{aligned} M_r = -D\Bigg(&-D_1\left[\beta^2 J_0(\beta r)-(1-\nu)\frac{\beta}{r}J_1(\beta r)\right]-D_2\left[\beta^2 Y_0(\beta r)-(1-\nu)\frac{\beta}{r}Y_1(\beta r)\right] \\ &+C_1\frac{\pi}{2}\left\{\frac{2}{\pi}\ln r+\left[\beta^2 J_0(\beta r)-(1-\nu)\frac{\beta}{r}J_1(\beta r)\right]\int r\ln r Y_0(\beta r)\,dr \right.\\ &\left.-\left[\beta^2 Y_0(\beta r)-(1-\nu)\frac{\beta}{r}Y_1(\beta r)\right]\int r\ln r J_0(\beta r)\,dr\right\}\Bigg) \end{aligned} \tag{9.155}$$

Taking into account the value at the origin $r=0$ of w and M_r, the terms involving $Y_0(\beta r)$ and $\ln r$ should be excluded. Then the fundamental solution of w suitable for the circular plate results in the form

$$w = D_1 J_0(\beta r) + \frac{1}{\beta^2} C_2 \tag{9.156}$$

in which, the unknown constants D_1, C_2 should be determined from the appropriate boundary conditions.

[1] For a built-in edge
Substituting Eq. (9.156) into the boundary condition (9.129), we have a following solution for w:

$$w = D_1[J_0(\beta r) - J_0(\beta a)] \tag{9.157}$$

Furthermore, the thermal buckling will be caused by the condition

$$J_1(\beta a) = 0 \tag{9.158}$$

The eigenvalues βa satisfying Eq. (9.158) are given as

$$\beta a = 3.8317,\ 7.0516,\ldots \tag{9.159}$$

from which it can be seen that the actual thermal buckling appears when $\beta a = 3.8317$, and the critical temperature rise τ_{0cr} is given by

$$\alpha\tau_{0cr} = \frac{1}{12(1+\nu)}\left(\frac{h}{a}\right)^2 (3.8317)^2 \tag{9.160}$$

Now, assuming that $\nu = 0.3$, and $\alpha = 1.0 \times 10^{-5}$ 1/K, the critical temperature rise τ_{0cr} is given as

$$\tau_{0cr} = \begin{cases} 235 \text{ [K]}; & a/h = 20 \\ 37.6 \text{ [K]}; & a/h = 50 \\ 9.4 \text{ [K]}; & a/h = 100 \end{cases} \tag{9.161}$$

[2] For a simply supported edge
Substituting Eq. (9.156) into the boundary condition (9.130), we have the following solution for w:

$$w = D_1[J_0(\beta r) - J_0(\beta a)] \tag{9.162}$$

Furthermore, the thermal buckling will be caused by the condition

$$\beta a J_0(\beta a) - (1-\nu) J_1(\beta a) = 0 \tag{9.163}$$

The eigenvalues βa satisfying Eq. (9.163) are given as

$$\beta a = 2.05, 5.39, \ldots \tag{9.164}$$

from which it can be seen that the actual thermal buckling appears when $\beta a = 2.05$, and the critical temperature rise τ_{0cr} is given by

$$\alpha \tau_{0cr} = \frac{1}{12(1+\nu)} \left(\frac{h}{a}\right)^2 (2.05)^2 \tag{9.165}$$

Now, assuming that $\nu = 0.3$, and $\alpha = 1.0 \times 10^{-5}$ 1/K, the critical temperature rise τ_{0cr} is given as

$$\tau_{0cr} = \begin{cases} 67.3 \text{ [K]}; & a/h = 20 \\ 10.7 \text{ [K]}; & a/h = 50 \end{cases} \tag{9.166}$$

Example 9.3 Assuming that the upper surface ($z = -h/2$) of a circular thin plate is held at zero temperature ($\tau = 0$) and the lower surface ($z = +h/2$) of the plate is at constant temperature ($\tau = \tau_0$), derive the critical temperature rise τ_{0cr} for thermal buckling and the deflection w when the in-plane deformations are restrained.

Solution The temperature change τ varying in the thickness direction z is taken as in Example 9.2.

$$\tau = \tau_0 \left(\frac{1}{2} + \frac{z}{h}\right) \tag{a}$$

The strain components are identically zero, so that

$$\epsilon_{rr} = 0, \qquad \epsilon_{\theta\theta} = 0, \qquad \epsilon_{r\theta} = 0 \tag{b}$$

The stress components may then be obtained from Eq. (5.16) as

$$\sigma_{rr} = \sigma_{\theta\theta} = -\frac{\alpha E \tau_0}{1-\nu}\left(\frac{1}{2} + \frac{z}{h}\right), \qquad \sigma_{r\theta} = 0 \tag{9.167}$$

By the substitution of Eqs. (9.167) into Eqs. (9.132) and (9.134), we have

$$N_r = N_\theta = -\frac{\alpha E \tau_0 h}{2(1-\nu)} \equiv N_0, \qquad N_{r\theta} = 0, \qquad M_T = \frac{1}{12}\alpha E \tau_0 h^2 \equiv M_{T0} \tag{9.168}$$

Substituting Eqs. (9.168) into Eq. (9.142), the governing equation for w is

$$\nabla^2\left(\nabla^2 w - \frac{N_0}{D} w\right) = 0 \tag{c}$$

in which

$$\nabla^2 = \frac{1}{r}\frac{d}{dr}\left(r\frac{d}{dr}\frac{d}{dr}\right) \tag{d}$$

Referring to Eq. (9.156), the fundamental solution of the deflection w is

$$w = D_1 J_0(\beta r) + \frac{1}{\beta^2} C_2 \tag{e}$$

where

$$\beta^2 = 6(1+\nu)\alpha \tau_0 \frac{1}{h^2} \tag{9.169}$$

[1] For a built-in edge
Substituting Eq. (e) into the boundary condition (9.129) results in a following solution for w:

$$w = D_1[J_0(\beta r) - J_0(\beta a)] \tag{f}$$

in which the coefficient D_1 is indeterminate. Furthermore, the thermal buckling will be caused by the condition

$$J_1(\beta a) = 0 \tag{g}$$

The eigenvalues βa satisfying Eq. (g) are given as

$$\beta a = 3.8317,\ 7.0516, \ldots \tag{h}$$

from which, it can be seen that the actual thermal buckling appears when $\beta a = 3.8317$, and the critical temperature rise τ_{0cr} is given by

$$\alpha \tau_{0cr} = \frac{1}{6(1+\nu)}\left(\frac{h}{a}\right)^2 (3.8317)^2 \tag{9.170}$$

Now, assuming that $\nu = 0.3$, and $\alpha = 1.0 \times 10^{-5}$ 1/K, the critical temperature rise τ_{0cr} is given as

$$\tau_{0cr} = \begin{cases} 471\ [\text{K}]; & a/h = 20 \\ 75.3\ [\text{K}]; & a/h = 50 \\ 18.8\ [\text{K}]; & a/h = 100 \end{cases} \tag{9.171}$$

[2] For a simply supported edge

In this case, the boundary condition is given by Eqs. (9.130), and they are

$$w = 0, \quad \frac{\partial^2 w}{\partial r^2} + \nu \frac{1}{r}\frac{\partial w}{\partial r} = -\frac{M_{T0}}{(1-\nu)D} \qquad \text{on } r = a \tag{9.172}$$

Substituting Eq. (e) into the first relation of Eqs. (9.172), we have

$$w = D_1[J_0(\beta r) - J_0(\beta a)] \tag{i}$$

Substituting now Eq. (i) into the second relation of Eqs. (9.172) gives

$$D_1 = \frac{M_{T0}}{(1-\nu)D}\frac{a}{\beta[\beta a J_0(\beta a) - (1-\nu)J_1(\beta a)]} \tag{j}$$

Therefore, the deflection w is finally determined as

$$w = \frac{M_{T0}}{(1-\nu)D}\frac{a}{\beta[\beta a J_0(\beta a) - (1-\nu)J_1(\beta a)]}[J_0(\beta r) - J_0(\beta a)] \tag{9.173}$$

from which it follows that the thermal buckling is caused by the condition

$$\beta a J_0(\beta a) - (1-\nu)J_1(\beta a) = 0 \tag{k}$$

The eigenvalues βa satisfying Eq. (k) are given as

$$\beta a = 2.05, 5.39, \ldots \tag{l}$$

Thus, it can be seen that the actual thermal buckling appears when $\beta a = 2.05$, and the critical temperature rise τ_{0cr} is given by

$$\alpha \tau_{0cr} = \frac{1}{6(1+\nu)}\left(\frac{h}{a}\right)^2 (2.05)^2 \tag{9.174}$$

Now, assuming that $\nu = 0.3$, and $\alpha = 1.0 \times 10^{-5}$ 1/K, the critical temperature rise τ_{0cr} is given as

$$\tau_{0cr} = \begin{cases} 134.6\ [\text{K}]; & a/h = 20 \\ 21.4\ [\text{K}]; & a/h = 50 \end{cases} \tag{9.175}$$

In this section, we have treated illustrative examples of the thermal buckling problem, with plates subjected to a uniform temperature change under the condition of prescribed displacement components in the in-plane direction. However, it may be easily predicted that the thermal buckling phenomena will also be caused by a non-uniform temperature change under the free edge condition for the in-plane deformation. The analytical procedures for such problems are complicated and difficult, and we omit here their description.

Problems

9.1 Derive the thermal stress distribution σ_x for a beam-column when M_{Ty} is given by an arbitrary function $f(x)$ in which x and y axes are principal axes, and temperature change $\tau(x,y,z)$ is given by $\tau_0(x)g(y,z)$.

Answer

$$\sigma_x = \frac{f(x)}{\int_A g(y,z)z\,dA}\left[-g(y,z) + \frac{1}{A}\int_A g(y,z)\,dA \right.$$
$$\left. + \frac{y}{I_z}\int_A g(y,z)y\,dA + \frac{z}{I_y}\int_A g(y,z)z\,dA\right]$$

$$f(x) = \alpha E\tau_0(x)\int_A g(y,z)z\,dA$$

9.2 Show that the deflection w equals zero in a pre-buckling state for a beam-column with both edge built-in condition when M_{Ty} is given by Eq. (9.15).

9.3 Derive the relation of shearing force F_z given by Eq. (9.14) for a beam-column.

9.4 Derive the relations for deflection w given by Eqs. (9.22), (9.30), and (9.38) of a beam-column for several kinds of boundary conditions when $M_{Ty} = D_1 + D_2 x$.

9.5 Derive the relations for deflection w given by Eqs. (9.52), (9.54), (9.56), and (9.58) of a beam-column for several kinds of boundary conditions when M_{Ty} is given by an arbitrary function of x.

9.6 Derive the relations for deflection w given by Eqs. (9.60), (9.62), (9.64), and (9.66) of a beam-column for several kinds of boundary conditions when $M_{Ty} = D_0 x(l - x)$.

9.7 Derive the stress–displacement relation given by Eqs. (9.116) for a circular plate.

9.8 Derive the relation between resultant forces N_r, N_θ, $N_{r\theta}$ with stress function F given by Eqs. (9.124) for a circular plate.

9.9 Derive the fundamental relation of thermal buckling given by Eqs. (9.127) or (9.128) for a circular plate.

9.10 Derive the fundamental solution for a deflection w of thermal buckling given by Eq. (9.154) for a circular plate.

Chapter 10

THERMODYNAMICS OF THERMOELASTICITY

The basic principles of classical thermodynamics are stated. The principle of energy conservation and the second law of thermodynamics are introduced. The Helmholtz free energy and the Gibbs thermodynamic potential are derived to discuss the thermodynamics of deformation of the body. The heat conduction equation and the equation of motion are considered from the viewpoint of the thermodynamics of deformation of the elastic body. The variational equations of the coupled dynamical thermoelasticity of 3D bodies are formulated from the basic principles. These equations are used to establish the uniqueness theorem and the reciprocal theorem.

10.1 The Principle of Energy Conservation

Thermoelasticity is concerned with equilibrium of the bodies subjected to mechanical and thermal loads, and which can be treated as thermodynamic systems. Thermodynamics is a very important subject in the study of the theory of thermoelasticity. The state of a thermodynamic system is defined by a number of independent variables. One of the independent variables is the temperature.

At first, we consider the principle of energy conservation which is called the first law of thermodynamics and which is one of the most important laws of physical science. When a system is subjected to an arbitrary process, the initial state of the system changes to the final state. In this case, the principle of energy conservation has the form

$$\frac{d}{dt}(U+K)=\frac{dW}{dt}+\frac{dQ}{dt} \tag{10.1}$$

where U is the internal energy, K denotes the kinetic energy, W stands for the work of external forces, and Q represents the total heat supplied to the system during the process. It states that the increment in time of the sum of the internal and kinetic energies is equal to the sum of the increment in time of the work of external forces and the increment in time of the total heat supplied to the system during the process.

A homogeneous continuous body occupies an inner region D with a boundary surface B. The functions $u_i(P,t)$ and $v_i(P,t) = du_i(P,t)/dt$ denote, respectively, the displacement and velocity components of the particle at a position P at time t, and $\rho(P,t)$ and $\epsilon(P,t)$ are the density and the internal energy per unit volume, respectively, of the particle located at P at time t. The kinetic energy K of the body at time t is defined as

$$K = \frac{1}{2}\int_D \rho v_i v_i \, dV \tag{10.2}$$

The internal energy U of the body at time t is given by

$$U = \int_D \epsilon \, dV \tag{10.3}$$

The power (time rate of doing work) of the external forces is

$$\frac{dW}{dt} = \int_D F_i v_i \, dV + \int_B p_i v_i \, dA \tag{10.4}$$

where F_i denotes the component of the body force and p_i the component of the surface traction applied at the boundary surface B. Using Cauchy's fundamental relation (4.5′) and Gauss' divergence theorem, the second term on the right-hand side of Eq. (10.4) yields

$$\int_B p_i v_i \, dA = \int_B \sigma_{ji} n_j v_i \, dA = \int_D (\sigma_{ji} v_i)_{,j} \, dV = \int_D (\sigma_{ji,j} v_i + \sigma_{ji} v_{i,j}) \, dV$$

Then, Eq. (10.4) reduces to

$$\frac{dW}{dt} = \int_D (\sigma_{ji,j} + F_i) v_i \, dV + \int_D \sigma_{ji} v_{i,j} \, dV \tag{10.5}$$

The total rate of energy transferred into D and due to internal heating is

$$\frac{dQ}{dt} = -\int_B q_i n_i dA + \int_D Q^* dV \tag{10.6}$$

where q_i denotes the component of heat flux per unit time and unit area transferred from D across B, n_i is the outer unit normal of the surface element, and Q^* stands for the quantity of heat generation in unit time and unit volume in D. Applying Gauss' divergence theorem

$$\int_B q_i n_i dA = \int_D q_{i,i} \, dV \tag{a}$$

to Eq. (10.6), reduces Eq. (10.6) to

$$\frac{dQ}{dt} = -\int_D (q_{i,i} - Q^*)\,dV \tag{10.7}$$

The substitution of Eqs. (10.2), (10.3), (10.5), and (10.7) into Eq. (10.1) yields

$$\frac{d}{dt}\left(\frac{1}{2}\int_D \rho v_i v_i\,dV + \int_D \epsilon\,dV\right) = \int_D (\sigma_{ji,j} + F_i)v_i\,dV + \int_D \sigma_{ji} v_{i,j}\,dV - \int_D (q_{i,i} - Q^*)\,dV \tag{10.8}$$

When the body is subjected to motion and deformation, the mass of the body does not change due to the principle of conservation of mass:

$$\frac{d}{dt}\int_D \rho\,dV = 0 \tag{10.9}$$

If it is assumed to be valid for Eq. (10.9) for any element of the body, we obtain

$$\frac{d}{dt}(\rho\,dV) = 0 \tag{10.10}$$

Equation (10.10) is called the equation of local conservation of mass or the equation of continuity. The first term on the left-hand side of Eq. (10.8) reduces to

$$\frac{d}{dt}\int_D \frac{1}{2}\rho v_i v_i\,dV = \int_D \rho v_i \dot{v}_i\,dV + \frac{1}{2}\int_D v_i v_i \frac{d}{dt}(\rho\,dV) \tag{b}$$

Using the equation of local conservation of mass (10.10), Eq. (b) yields

$$\frac{d}{dt}\int_D \frac{1}{2}\rho v_i v_i\,dV = \int_D \rho v_i \dot{v}_i\,dV \tag{c}$$

Therefore, Eq. (10.8) yields

$$\int_D \rho v_i \dot{v}_i\,dV + \int_D \dot{\epsilon}\,dV = \int_D (\sigma_{ji,j} + F_i)v_i\,dV + \int_D \sigma_{ji} v_{i,j}\,dV - \int_D (q_{i,i} - Q^*)\,dV$$

or

$$\int_D (\sigma_{ji,j} + F_i - \rho\dot{v}_i)v_i\,dV + \int_D [\sigma_{ji} v_{i,j} - (q_{i,i} - Q^*) - \dot{\epsilon}]\,dV = 0 \tag{10.11}$$

The integrand in the first integral denotes the equation of motion which will be discussed later, so the integrand is zero. Therefore, Eq. (10.11) reduces to

$$\int_D [\sigma_{ji} v_{i,j} - (q_{i,i} - Q^*) - \dot{\epsilon}]\,dV = 0 \tag{10.12}$$

Equation (10.12) should be valid for every part of the body, so the integrand in Eq. (10.12) is identically zero at every point of the body. Then we obtain the principle of energy conservation

$$\dot{\epsilon} = \sigma_{ji} v_{i,j} - (q_{i,i} - Q^*) \tag{10.13}$$

We may write $v_{i,j}$ in terms of its symmetric and antisymmetric portions:

$$v_{i,j} = \frac{1}{2}(v_{i,j} + v_{j,i}) + \frac{1}{2}(v_{i,j} - v_{j,i}) = \dot{\epsilon}_{ij} + \dot{\omega}_{ij} \tag{d}$$

where

$$\dot{\epsilon}_{ij} = \frac{1}{2}(v_{i,j} + v_{j,i}), \qquad \dot{\omega}_{ij} = \frac{1}{2}(v_{i,j} - v_{j,i}) \tag{e}$$

By use of the symmetry of the stress components $\sigma_{ij} = \sigma_{ji}$, the first term on the right-hand side of Eq. (10.13) reduces to

$$\sigma_{ji} v_{i,j} = \frac{1}{2}(\sigma_{ji} v_{i,j} + \sigma_{ij} v_{j,i}) = \frac{1}{2}\sigma_{ji}(v_{i,j} + v_{j,i}) = \sigma_{ji}\dot{\epsilon}_{ij} \tag{f}$$

so the principle of energy conservation is obtained as:

$$\dot{\epsilon} = \sigma_{ij}\dot{\epsilon}_{ij} - (q_{i,i} - Q^*) \tag{10.14}$$

10.2 The Second Law of Thermodynamics

10.2.1 Reversible and Irreversible Processes

The second law of thermodynamics is the result of a generalization of experimental data. It has two typical formulations:

Postulate of Clausius A transformation whose only final result is to transfer heat from a body at a given temperature to a body at a higher temperature is impossible.

Postulate of Lord Kelvin A transformation whose only final result is to transform into work heat extracted from a source which is at the same temperature throughout is impossible.

To prepare the way for a suitable formulation of the second law of thermodynamics for a deformable body, it is necessary to examine some concepts such as those of state parameter and process. When an elastic body is subjected to external forces and/or heat supply, a thermodynamic system of the body is determined by the values of certain physical quantities known as state parameters. In thermodynamics a complete set of state parameters includes four parameters such as the absolute temperature and spatial coordinates. The absolute temperature and the three displacements are often selected as the parameters in linear thermoelasticity. When some parameters are uniquely determined by the principal state parameters, the parameters are called functions of state.

There are two categories of the thermodynamic processes, reversible and irreversible processes. If a process can be carried out in the opposite direction exactly through the same intermediate state as the actual process, returning the system to its initial state, the process is called a reversible process. On the other hand, a process which does not satisfy these conditions is called an irreversible process.

10.2.2 Cycle of Carnot

N. L. Sadi Carnot proposed in 1824 a cycle consisting of reversible processes as a highly idealized reversible cycle. This process is called a Carnot cycle. The process is assumed to be carried under idealized conditions. The Carnot cycle consists of four processes as shown in Fig. 10.1:

First process (reversible isothermal expansion process) The expansion process is performed at a constant temperature from the state A to the state B and is called an isothermal process. During the expansion process, the system absorbs the quantity Q_1 of heat from the heat reservoir with constant temperature T_1.

Second process (reversible adiabatic expansion process) The expansion process is performed under the adiabatic condition from the state B to the state C. As the expansion process is performed under the adiabatic condition, heat exchange does not occur with the surrounding medium and the system's temperature drops from T_1 to T_2.

Third process (reversible isothermal compression process) The compression process is performed from the state C to the state D as an isothermal process. During the compression process, the system transfers the quantity Q_2 of heat to another heat reservoir with constant temperature T_2.

Fourth process (reversible adiabatic compression process) The compression process is performed under the adiabatic condition from the state D to the state A which is the initial state.

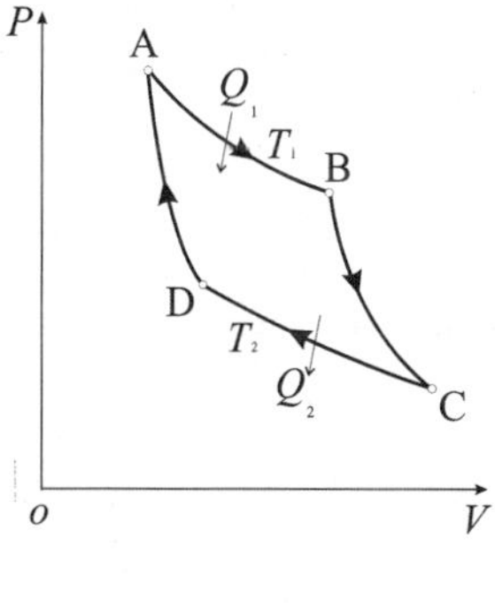

Figure 10.1 Carnot cycle.

We assume that an operational material is a perfect gas during Carnot cycle and discuss each process.

First process: The equation of state for the perfect gas is given by Boyle-Charles law as:

$$PV = GRT \tag{10.15}$$

where P denotes the pressure of the gas, V is the volume of the gas, G stands for the mass of the gas, R means the gas constant, and T represents the absolute temperature. The internal energy, which is the state parameter, does not change during the isothermal process, and the kinetic energy is zero because the process is sufficiently slow, so the principle of energy conservation (10.1) reduces to

$$\frac{dW}{dt} + \frac{dQ}{dt} = 0 \tag{10.16}$$

The heat supplied to the system during the reversible isothermal expansion process is

$$Q_1 = \int_A^B dQ = -\int_A^B \frac{dW}{dt}dt = -\int_A^B P\,dV$$

Using Eq. (10.15) in the above equation, we get

$$Q_1 = -\int_A^B \frac{GRT_1}{V}\,dV = -GRT_1 \ln \frac{V_B}{V_A} \tag{10.17}$$

Since the first process is the reversible isothermal expansion process, the Boyle-Charles law gives

$$P_A V_A = P_B V_B \tag{10.18}$$

Second and fourth processes: When the process is adiabatic, it is easily shown that the principle of energy conservation (10.1) and Boyle-Charles law (10.15) lead to the relation for the perfect gas

$$PV^{\kappa} = \text{constant} \tag{10.19}$$

where $\kappa = c_p/c_v$, and c_p and c_v denote the specific heats at constant pressure and constant volume, respectively. The two adiabatic processes give

$$P_B V_B^{\kappa} = P_C V_C^{\kappa}, \qquad P_D V_D^{\kappa} = P_A V_A^{\kappa} \tag{10.20}$$

Third process: The third process is a reversible isothermal compression process, so we can get

$$Q_2 = -\int_C^D \frac{GRT_2}{V} dV = -GRT_2 \ln \frac{V_D}{V_C}, \qquad P_C V_C = P_D V_D \tag{10.21}$$

Using Eqs. (10.17), (10.18), (10.20), and (10.21), we finally get

$$\frac{Q_1}{T_1} + \frac{Q_2}{T_2} = 0 \tag{10.22}$$

where it is assumed that the quantity Q_1 of heat absorbed in the system is positive but the quantity Q_2 of heat transferred from the system is negative.

Equation (10.22) has been proved for the ideal gas. However, it is valid for the Carnot cycle of any operational material.

10.2.3 Entropy

Now we apply the result of the Carnot cycle to any reversible cycle. Any reversible cycle can be replaced by a sum of infinitesimal Carnot cycles. The relation (10.22) is valid for each infinitesimal Carnot cycle so that we can get for any reversible cycle as the sum of the infinitesimal Carnot cycles

$$\left(\frac{\Delta Q_1}{T_1} + \frac{\Delta Q_2}{T_2}\right) + \left(\frac{\Delta Q_3}{T_3} + \frac{\Delta Q_4}{T_4}\right) + \left(\frac{\Delta Q_5}{T_5} + \frac{\Delta Q_6}{T_6}\right) + \cdots = 0 \tag{a}$$

The integral expression of the above equation for any reversible cycle reduces to

$$\oint \frac{dQ}{T} = 0 \tag{10.23}$$

This contour integral is called Clausius integral and the relation (10.23) is called the Clausius theorem. Equation (10.23) means that Clausius's integral becomes zero for a reversible process. Therefore, the value of $\int dQ/T$ along any path from a state A to a state B in the system is the same. This result indicates that the ratio dQ/T is a total differential. We define the total differential of the entropy S as

$$dS = \frac{dQ}{T} \tag{10.24}$$

for any reversible process. By definition, the entropy is a function of state and its dimension is [kJ/K]. The entropy per unit volume s is given by

$$ds = \frac{dq}{T} \tag{10.25}$$

where q is the heat per unit volume. A difference of the entropy between two states A and B of the system is

$$S(B) - S(A) = \int_A^B \frac{dQ}{T} \tag{10.26}$$

When a process is irreversible, the Clausius integral* relation holds

$$\oint \frac{dQ}{T} < 0 \tag{10.27}$$

Therefore, the Clausius integral can never be larger than zero for any process

$$\oint \frac{dQ}{T} \le 0 \tag{10.28}$$

Relation (10.28) is called the Clausius inequality. We take the system from a state A to a state B along an irreversible process 1 and back to the state A again along a reversible process 2. From the Clausius inequality

$$0 > \oint_{A1B2A} \frac{dQ}{T} = \left(\int_A^B \frac{dQ}{T}\right)_1 + \left(\int_B^A \frac{dQ}{T}\right)_2$$

The process 2 from the state B to the state A is reversible so the last term of the above equation can be expressed in terms of entropy using Eq. (10.26)

$$\left(\int_B^A \frac{dQ}{T}\right)_2 = S(A) - S(B)$$

Substitution of this into the preceding inequality, we get

$$S(B) - S(A) > \left(\int_A^B \frac{dQ}{T}\right)_1 \tag{10.29}$$

so that entropy increases for an irreversible process. For the general case of any type of process from the state A to the state B, we obtain

$$S(B) - S(A) \ge \int_A^B \frac{dQ}{T} \tag{10.30}$$

If the system is completely isolated, $dQ = 0$, and Eq. (10.30) reduces to

$$S(B) \ge S(A) \tag{10.31}$$

that is, for any process occurring in an isolated system, the entropy of the final state $S(B)$ can never be less than that of the initial state $S(A)$.

*See: Enrico Fermi, Thermodynamics, Dover Publications, New York (1936), p. 46.

The definition of entropy (10.25) and Eq. (10.7) give the local entropy balance

$$T\dot{s} = -q_{i,i} + Q^* \tag{10.32}$$

Equation (10.32) can be rewritten in the form

$$\frac{ds}{dt} = -\left(\frac{q_i}{T}\right)_{,i} - \frac{q_i T_{,i}}{T^2} + \frac{Q^*}{T} \tag{10.33}$$

Integration of Eq. (10.32) over the volume of the body yields

$$\dot{S} = -\int_D \left(\frac{q_i}{T}\right)_{,i} dV - \int_D \frac{q_i T_{,i}}{T^2} dV + \int_D \frac{Q^*}{T} dV \tag{10.34}$$

or

$$\dot{S} = -\int_B \frac{q_i n_i}{T} dA - \int_D \frac{q_i T_{,i}}{T^2} dV + \int_D \frac{Q^*}{T} dV \tag{10.34$'$}$$

The first term is the increase of the entropy due to heat exchange with the surrounding medium, while the second term is the entropy production in the volume by heat exchange, and the third term is the entropy production by volumetric heat sources (like embedded heating elements).

The second law of thermodynamics of the irreversible process leads to the Clausius-Duhem inequality

$$-\frac{q_i T_{,i}}{T^2} \geq 0 \tag{10.35}$$

or

$$\frac{ds}{dt} + \left(\frac{q_i}{T}\right)_{,i} - \frac{Q^*}{T} \geq 0 \tag{10.35$'$}$$

or

$$\frac{ds}{dt} + \frac{q_{i,i}}{T} - \frac{Q^*}{T} - \frac{q_i T_{,i}}{T^2} \geq 0 \tag{10.35$''$}$$

The first three terms in Eq. (10.35″) mean the local entropy production and the last term denotes the entropy production by heat conduction.

10.3 Thermodynamic Functions

When the thermodynamics of deformation of the body is studied, it is more convenient to introduce, in addition to the internal energy per unit volume ϵ and the entropy per unit volume s, the following two state functions:

the Helmholtz free energy

$$F = \epsilon - Ts \tag{10.36}$$

the Gibbs thermodynamic potential

$$G = F - \sigma_{ij}\epsilon_{ij} \tag{10.37}$$

The increments of the state parameters during a change of state of the elastic body are total differentials so that the Helmholtz free energy and the Gibbs thermodynamic potential are state functions. The material derivative of the Helmholtz free energy is

$$\dot{F} = \dot{\epsilon} - T\dot{s} - s\dot{T} = \sigma_{ij}\dot{\epsilon}_{ij} - T\dot{s} - s\dot{T} - q_{i,i} + Q^* \tag{a}$$

where Eq. (10.14) has been used. Elimination of the heat supply $(q_{i,i} - Q^*)$ between the inequality (10.35″) and Eq. (a) gives

$$\dot{F} + s\dot{T} - \sigma_{ij}\dot{\epsilon}_{ij} + \frac{q_i T_{,i}}{T} \leq 0 \tag{10.38}$$

When the Helmholtz free energy F can be expressed by

$$F = F(\epsilon_{ij}, T, T_{,i}) \tag{10.39}$$

the material derivative of F is

$$\dot{F} = \frac{\partial F}{\partial \epsilon_{ij}}\dot{\epsilon}_{ij} + \frac{\partial F}{\partial T}\dot{T} + \frac{\partial F}{\partial T_{,i}}\dot{T}_{,i} \tag{b}$$

Introducing the Helmholtz free energy F to the principle of energy conservation (10.14) and the inequality (10.38) we get

$$\left(\frac{\partial F}{\partial \epsilon_{ij}} - \sigma_{ij}\right)\dot{\epsilon}_{ij} + \left(\frac{\partial F}{\partial T} + s\right)\dot{T} + \frac{\partial F}{\partial T_{,i}}\dot{T}_{,i} + T\dot{s} + q_{i,i} - Q^* = 0 \tag{10.40}$$

and

$$\left(\frac{\partial F}{\partial \epsilon_{ij}} - \sigma_{ij}\right)\dot{\epsilon}_{ij} + \left(\frac{\partial F}{\partial T} + s\right)\dot{T} + \frac{\partial F}{\partial T_{,i}}\dot{T}_{,i} + \frac{q_i T_{,i}}{T} \leq 0 \tag{10.41}$$

The equation (10.40) and the inequality (10.41) should be satisfied for all permissible values of the time derivatives $\dot{\epsilon}_{ij}$, $\dot{T}$, and $\dot{T}_{,i}$. Hence the coefficients of the time derivatives $\dot{\epsilon}_{ij}$, $\dot{T}$, and $\dot{T}_{,i}$ must vanish:

$$\sigma_{ij} = \frac{\partial F}{\partial \epsilon_{ij}} \tag{10.42}$$

$$s = -\frac{\partial F}{\partial T} \tag{10.43}$$

$$\frac{\partial F}{\partial T_{,i}} = 0 \tag{10.44}$$

The equation (10.44) means that the Helmholtz free energy is independent of the spatial temperature gradient

$$F = F(\epsilon_{ij}, T) \tag{10.45}$$

From the remaining three terms in Eq. (10.40) and Eq. (10.41), we find

$$T\dot{s} = -(q_{i,i} - Q^*) \tag{10.46}$$

and

$$-\frac{q_i T_{,i}}{T} \geq 0 \tag{10.47}$$

Let us consider the irreversible process of heat conduction. The Fourier law of heat conduction for an anisotropic body is

$$q_i = -\lambda_{ij} T_{,j} \tag{10.48}$$

where λ_{ij} denote the components of the thermal conductivity tensor. Substitution of Eq. (10.48) into Eq. (10.47) yields

$$\lambda_{ij} T_{,i} T_{,j} \geq 0 \tag{10.49}$$

The heat conductivity λ_{ij} must be symmetric to satisfy a positive definite quadratic form Eq. (10.49). For an isotropic body we have

$$q_i = -\lambda T_{,i}, \qquad \lambda \geq 0 \tag{10.50}$$

Let us expand the Helmholtz free energy $F = F(\epsilon_{ij}, T)$ in a Taylor series in the vicinity of the natural state ($\epsilon_{ij} = 0,\ T = T_0$)

$$\begin{aligned} F(\epsilon_{ij}, T) = F(0, T_0) &+ \frac{\partial F(0, T_0)}{\partial \epsilon_{ij}} \epsilon_{ij} + \frac{\partial F(0, T_0)}{\partial T}(T - T_0) \\ &+ \frac{1}{2}\left[\frac{\partial^2 F(0, T_0)}{\partial \epsilon_{ij} \partial \epsilon_{kl}} \epsilon_{ij}\epsilon_{kl} + 2\frac{\partial^2 F(0, T_0)}{\partial \epsilon_{ij} \partial T} \epsilon_{ij}(T - T_0) \right. \\ &\left. + \frac{\partial^2 F(0, T_0)}{\partial T^2}(T - T_0)^2\right] + \cdots \end{aligned} \tag{10.51}$$

Here $F(0, T_0)$ is the energy of the natural state and is assumed to be zero. Moreover, we assume that the stress and the entropy are free in the elastic body for the natural state:

$$\frac{\partial F(0, T_0)}{\partial \epsilon_{ij}} = \sigma_{ij} = 0, \qquad \frac{\partial F(0, T_0)}{\partial T} = -s = 0$$

Thus, Eq. (10.51) reduces to

$$F(\epsilon_{ij}, T) = \frac{1}{2} c_{ijkl} \epsilon_{ij} \epsilon_{kl} - \beta_{ij} \epsilon_{ij} (T - T_0) - \frac{1}{2} m (T - T_0)^2 + \cdots \tag{10.52}$$

where

$$c_{ijkl} = \frac{\partial^2 F(0, T_0)}{\partial \epsilon_{ij} \partial \epsilon_{kl}}, \qquad \beta_{ij} = -\frac{\partial^2 F(0, T_0)}{\partial \epsilon_{ij} \partial T}, \qquad m = -\frac{\partial^2 F(0, T_0)}{\partial T^2} \tag{10.53}$$

Taking into account that the strains and the temperature change are small, we will truncate the expansion of F in ϵ_{ij} and $T - T_0$ after the second order terms, i.e., we will retain only quadratic terms in ϵ_{ij} and $T - T_0$

$$F(\epsilon_{ij}, T) = \frac{1}{2} c_{ijkl} \epsilon_{ij} \epsilon_{kl} - \beta_{ij} \epsilon_{ij} \tau - \frac{1}{2} m \tau^2 \tag{10.54}$$

where τ denotes temperature change from the natural state:

$$\tau = T - T_0 \tag{10.55}$$

Introduction of Eq. (10.54) into Eqs. (10.42) and (10.43) gives

$$\sigma_{ij} = \frac{\partial F}{\partial \epsilon_{ij}} = c_{ijkl} \epsilon_{kl} - \beta_{ij} \tau \tag{10.56}$$

$$s = -\frac{\partial F}{\partial T} = \beta_{ij} \epsilon_{ij} + m\tau \tag{10.57}$$

Equation (10.56) is the constitutive equation of an anisotropic thermoelastic body and is called the Duhamel-Neumann relations for the anisotropic body. The material constants c_{ijkl} and β_{ij} refer to the natural state of the body. For an isotropic body, the material constants c_{ijkl} and β_{ij} can be expressed by

$$c_{ijkl} = \mu(\delta_{ik}\delta_{jl} + \delta_{il}\delta_{jk}) + \lambda \delta_{ij} \delta_{kl}, \qquad \beta_{ij} = \beta \delta_{ij} \tag{10.58}$$

where δ_{ij} denotes the Kronecker's symbol. Substitution of Eq. (10.58) into Eq. (10.54) yields

$$F(\epsilon_{ij}, T) = \mu \epsilon_{ij} \epsilon_{ij} + \frac{1}{2} \lambda \epsilon_{kk} \epsilon_{mm} - \beta \epsilon_{kk} \tau - \frac{1}{2} m \tau^2 \tag{10.59}$$

where the material constants λ and μ are Lamé constants.

The Helmholtz free energy for the isotropic body can be expressed by the linear invariant of strain ϵ_{kk}, the two second order invariants $\epsilon_{ij}\epsilon_{ij}$ and $\epsilon_{kk}\epsilon_{mm}$, and the first and second orders of the temperature change τ. The constitutive equation for the isotropic thermoelastic body is

$$\sigma_{ij} = \frac{\partial F}{\partial \epsilon_{ij}} = 2\mu \epsilon_{ij} + \lambda \epsilon_{kk} \delta_{ij} - \beta \tau \delta_{ij} \tag{10.60}$$

$$s = -\frac{\partial F}{\partial T} = \beta \epsilon_{kk} + m\tau \tag{10.61}$$

Equation (10.60) is the Duhamel-Neumann relation for the isotropic body, cf. Eq. (4.18′). Performing the contraction of the stress tensor† we get

$$\sigma_{kk} = 3(K\epsilon_{kk} - \beta\tau), \qquad K = \lambda + \frac{2}{3}\mu \tag{10.62}$$

where K is the bulk modulus. Solving Eq. (10.60) with respect to ϵ_{ij} we obtain, cf. Eq. (4.15′)

$$\epsilon_{ij} = \frac{1}{2\mu}\left(\sigma_{ij} - \frac{\lambda}{3\lambda + 2\mu}\sigma_{kk}\delta_{ij}\right) + \frac{\beta}{3\lambda + 2\mu}\tau\delta_{ij} \tag{10.63}$$

and

$$\epsilon_{kk} = \frac{\sigma_{kk}}{3K} + \frac{\beta}{K}\tau \tag{10.64}$$

When a free thermal expansion is generated in the body, where no thermal stresses are produced in the body, Eq. (10.63) leads to the relation

$$\epsilon_{ij} = \frac{\beta}{3\lambda + 2\mu}\tau\delta_{ij} \equiv \alpha\tau\delta_{ij} \tag{10.65}$$

A change of the normal strain is proportional to the temperature change τ and the coefficient α is called the coefficient of linear thermal expansion:

$$\alpha = \frac{\beta}{3\lambda + 2\mu} \tag{10.66}$$

Equation (10.63) is frequently written in terms of other material constants

$$\begin{aligned}\epsilon_{ij} &= \frac{1}{2G}\left(\sigma_{ij} - \frac{\nu}{1+\nu}\sigma_{kk}\delta_{ij}\right) + \alpha\tau\delta_{ij} \\ &= \frac{1+\nu}{E}\left(\sigma_{ij} - \frac{\nu}{1+\nu}\sigma_{kk}\delta_{ij}\right) + \alpha\tau\delta_{ij}\end{aligned} \tag{10.67}$$

where G is the shear modulus, ν is the Poisson's ratio, and E is the Young's modulus. The following relations can be obtained between these material constants and Lamé constants λ and μ:

$$\begin{aligned}&G = \frac{E}{2(1+\nu)}, \qquad G = \mu, \qquad E = \frac{\mu(3\lambda + 2\mu)}{\lambda + \mu}, \qquad \nu = \frac{\lambda}{2(\lambda + \mu)} \\ &\lambda = \frac{\nu E}{(1+\nu)(1-2\nu)}, \qquad \mu = \frac{E}{2(1+\nu)}, \qquad \beta = (3\lambda + 2\mu)\alpha\end{aligned} \tag{10.68}$$

† See: H. Jeffreys, Cartesian Tensors, Cambridge University Press (1931).

Let us return to the Helmholtz free energy (10.52) and neglect here all terms higher than the second order in ϵ_{ij}, and collect all terms depending on the temperature in one term $f(\tau)$

$$F(\epsilon_{ij}, T) = \frac{1}{2} c_{ijkl}\epsilon_{ij}\epsilon_{kl} - \beta_{ij}\epsilon_{ij}\tau + f(\tau) \tag{10.69}$$

For the isotropic body, Eq. (10.69) reduces to

$$F(\epsilon_{ij}, T) = \mu\epsilon_{ij}\epsilon_{ij} + \frac{1}{2}\lambda\epsilon_{kk}\epsilon_{mm} - \beta\epsilon_{kk}\tau + f(\tau) \tag{10.70}$$

Since the entropy per unit volume s is a function of ϵ_{ij} and T, we have

$$ds = \left(\frac{\partial s}{\partial \epsilon_{ij}}\right)_T d\epsilon_{ij} + \left(\frac{\partial s}{\partial T}\right)_\epsilon dT \tag{c}$$

The specific heat at constant strain c_ϵ is defined by

$$c_\epsilon = T\left(\frac{\partial s}{\partial T}\right)_\epsilon \tag{10.71}$$

Substitution of Eq. (10.61) into Eq. (10.71) yields

$$c_\epsilon = T\left(\frac{\partial s}{\partial T}\right)_\epsilon = -T\left(\frac{\partial^2 F}{\partial T^2}\right)_\epsilon = -T\left(\frac{\partial^2 f(\tau)}{\partial T^2}\right) \tag{d}$$

Integration of Eq. (d) with respect to T gives

$$\frac{\partial f(\tau)}{\partial T} = -\int_{T_0}^{T} \frac{c_\epsilon}{T} dT = -c_\epsilon \ln\left(\frac{T}{T_0}\right) \tag{e}$$

Introduction of Eq. (10.70) to Eq. (10.43) gives

$$s = -\frac{\partial F}{\partial T} = \beta\epsilon_{kk} - \frac{\partial f(\tau)}{\partial T} \tag{f}$$

Substitution of Eq. (e) into Eq. (f) yields the constitutive equation for the isotropic thermoelastic body

$$\begin{aligned} s &= \beta\epsilon_{kk} + c_\epsilon \ln\left(\frac{T}{T_0}\right) \\ &= \beta\epsilon_{kk} + c_\epsilon \ln\left(1 + \frac{\tau}{T_0}\right) \end{aligned} \tag{10.72}$$

When $|\tau/T_0| \ll 1$, the constitutive equation (10.72) and the Helmholtz free energy F reduce, respectively, to

$$s = \beta\epsilon_{kk} + \frac{c_\epsilon}{T_0}\tau \tag{10.73}$$

and

$$F(\epsilon_{ij}, T) = \mu\epsilon_{ij}\epsilon_{ij} + \frac{1}{2}\lambda\epsilon_{kk}\epsilon_{mm} - \beta\epsilon_{kk}\tau - \frac{c_\epsilon}{2T_0}\tau^2 \tag{10.74}$$

Comparing Eqs. (10.59) and (10.74) we find $m = c_\epsilon/T_0$.

Let us define the specific heat at constant stress c_σ by

$$c_\sigma = T\left(\frac{\partial s}{\partial T}\right)_\sigma \tag{10.75}$$

Since the strain is a function of stress and temperature, we have

$$d\epsilon_{ij} = \left(\frac{\partial \epsilon_{ij}}{\partial \sigma_{kl}}\right)_T d\sigma_{kl} + \left(\frac{\partial \epsilon_{ij}}{\partial T}\right)_\sigma dT \tag{g}$$

Introducing Eq. (g) into Eq. (c) we have

$$\begin{aligned} ds &= \left(\frac{\partial s}{\partial \epsilon_{ij}}\right)_T d\epsilon_{ij} + \left(\frac{\partial s}{\partial T}\right)_\epsilon dT \\ &= \left(\frac{\partial s}{\partial \epsilon_{ij}}\right)_T \left(\frac{\partial \epsilon_{ij}}{\partial \sigma_{kl}} d\sigma_{kl} + \frac{\partial \epsilon_{ij}}{\partial T} dT\right) + \left(\frac{\partial s}{\partial T}\right)_\epsilon dT \\ &= \left(\frac{\partial s}{\partial \sigma_{ij}}\right)_T d\sigma_{ij} + \left[-\frac{\partial^2 F}{\partial \epsilon_{ij}\partial T}\left(\frac{\partial \epsilon_{ij}}{\partial T}\right)_\sigma + \left(\frac{\partial s}{\partial T}\right)_\epsilon\right] dT \end{aligned} \tag{h}$$

Since the entropy s is a function of strain and the temperature and the strain is a function of stress and temperature, the entropy per unit volume s can be expressed by the stress and the temperature

$$ds = \left(\frac{\partial s}{\partial \sigma_{ij}}\right)_T d\sigma_{ij} + \left(\frac{\partial s}{\partial T}\right)_\sigma dT \tag{i}$$

Comparing Eqs. (h) and (i), the specific heat at constant stress c_σ is given by

$$c_\sigma = T\left(\frac{\partial s}{\partial T}\right)_\sigma = T\left[-\frac{\partial^2 F}{\partial \epsilon_{ij}\partial T}\left(\frac{\partial \epsilon_{ij}}{\partial T}\right)_\sigma + \left(\frac{\partial s}{\partial T}\right)_\epsilon\right] \tag{10.76}$$

Then the difference between the specific heat at constant stress c_σ and the specific heat at constant strain c_ϵ is expressed by

$$c_\sigma - c_\epsilon = -T\frac{\partial^2 F}{\partial \epsilon_{ij}\partial T}\left(\frac{\partial \epsilon_{ij}}{\partial T}\right)_\sigma \tag{10.77}$$

10.4 Fundamental Differential Equations of Thermoelasticity

10.4.1 Heat Conduction Equation

We consider the heat conduction equation which is treated in Chapter 3. The Fourier law of heat conduction for an isotropic body is from Eq. (10.50)

$$q_i = -\lambda T_{,i} = -\lambda \tau_{,i} \qquad \lambda \geq 0 \tag{10.78}$$

where $\tau = T - T_0$, and T_0 is constant temperature at the natural state. The constitutive equation of the entropy is from Eq. (10.73)

$$s = \beta \epsilon_{kk} + \frac{c_\epsilon}{T_0}\tau \tag{10.79}$$

Substitution of Eqs. (10.78) and (10.79) into the local entropy balance (10.32) yields a non-linear equation:

$$T\left(\beta \dot{\epsilon}_{kk} + \frac{c_\epsilon}{T_0}\dot{\tau}\right) = \lambda \tau_{,ii} + Q^* \tag{10.80}$$

As Eq. (10.79) is derived under assumption $|\tau/T_0| \ll 1$, we obtain the linearized heat conduction equation by replacing T on the left-hand side of Eq. (10.80) by T_0

$$c_\epsilon \dot{\tau} + T_0 \beta \dot{\epsilon}_{kk} = \lambda \tau_{,ii} + Q^* \tag{10.81}$$

or

$$c_\epsilon \dot{\tau} = \lambda \tau_{,ii} + Q^* - (3\lambda + 2\mu)\alpha T_0 \dot{\epsilon}_{kk} \tag{10.81$'$}$$

The third term on the right-hand side means coupling term between deformation and heating. So, if we rapidly bend a flexible bar up and down, the bar will start to heat up. This is the coupling of deformation and heating. Using new coefficients

$$\kappa = \frac{\lambda}{c_\epsilon}, \qquad \eta = \frac{\beta T_0}{\lambda} \tag{10.82}$$

we receive the heat conduction equation which takes the mechanical coupling term into account

$$\dot{\tau} = \kappa \tau_{,ii} + \frac{Q^*}{c_\epsilon} - \kappa \eta \dot{\epsilon}_{kk} \tag{10.83}$$

where κ denotes the thermal diffusivity and η denotes the coupling parameter. When $\kappa\eta \ll 1$, the heat conduction equation with mechanical coupling term reduces to the ordinary heat conduction equation (3.5)

$$\frac{\partial \tau}{\partial t} = \kappa \tau_{,ii} + \frac{Q^*}{c_\epsilon} \tag{10.84}$$

10.4.2 Equation of Motion

Consider a homogeneous continuous body occupying an inner region D with a boundary surface B. Let each point P on the boundary surface B be subjected to traction p_i (force per unit area), and each small element of the body be subjected to a body force per unit volume F_i. Let $u_i(P,t)$ and $v_i(P,t) = (\partial/\partial t)u_i(P,t)$ denote, respectively, the displacement and velocity components of the particle at the point P at time t. The laws of Newtonian mechanics (Newton's second law) for the continuous body require for linear motion

$$\int_D \rho \dot{v}_i \, dV = \int_B p_i \, dA + \int_D F_i \, dV \tag{10.85}$$

and for rotational motion

$$\int_D \rho \frac{\partial}{\partial t}(\gamma_{ijk} v_j x_k) dV = \int_B \gamma_{ijk} p_j x_k \, dA + \int_D \gamma_{ijk} F_j x_k \, dV \tag{10.86}$$

where γ_{ijk} is permutation tensor defined by Eq. (4.25)

$$\gamma_{ijk} = \begin{cases} +1 & \text{if } ijk \text{ represents an even permutation of 123} \\ 0 & \text{if any two of the } ijk \text{ indices are equal} \\ -1 & \text{if } ijk \text{ represents an odd permutation of 123} \end{cases} \tag{10.87}$$

Introducing the Cauchy's fundamental relations (4.5′) to the first integral on the right-hand side in Eq. (10.85) we obtain

$$\int_D \rho \dot{v}_i \, dV = \int_B \sigma_{ji} n_j \, dA + \int_D F_i \, dV \tag{a}$$

Introducing the Gauss' divergence theorem:

$$\int_B \sigma_{ji} n_j \, dA = \int_D \sigma_{ji,j} \, dV \tag{b}$$

to Eq. (a), we obtain

$$\int_D (\sigma_{ji,j} + F_i - \rho \dot{v}_i) \, dV = 0 \tag{10.88}$$

Equation (10.88) should be valid for every part of the body, so the integrand in Eq. (10.88) is identically zero at every point in the body. We obtain the equation of motion for the deformable body

$$\sigma_{ji,j} + F_i = \rho \ddot{u}_i \tag{10.89}$$

Introducing the Cauchy's fundamental relations (4.5′) to the first integral on the right-hand side in Eq. (10.86) we have

$$\int_D \rho(\gamma_{ijk}\dot{v}_j x_k + \gamma_{ijk} v_j \dot{x}_k)\,dV = \int_B \gamma_{ijk}\sigma_{mj} n_m x_k\,dA + \int_D \gamma_{ijk} F_j x_k\,dV \tag{c}$$

By applying the Gauss' divergence theorem to Eq. (c) we get

$$\int_D \rho(\gamma_{ijk}\dot{v}_j x_k + \gamma_{ijk} v_j v_k)\,dV = \int_D \gamma_{ijk}(\sigma_{mj} x_k)_{,m}\,dV + \int_D \gamma_{ijk} F_j x_k\,dV$$

After some arrangements we find

$$\int_D \rho\gamma_{ijk} v_j v_k\,dV = \int_D \gamma_{ijk}(\sigma_{mj,m} + F_j - \rho\dot{v}_j) x_k\,dV + \int_D \gamma_{ijk}\sigma_{mj} x_{k,m}\,dV \tag{d}$$

Substitution of $\gamma_{ijk} v_j v_k = 0$, $x_{k,m} = \delta_{km}$ and of Eq. (10.89) into Eq. (d) gives

$$\int_D \gamma_{ijk}\sigma_{kj}\,dV = 0 \tag{10.90}$$

Equation (10.90) should be valid for every part of the body. Therefore,

$$\begin{aligned} \gamma_{ijk}\sigma_{kj} &= \frac{1}{2}(\gamma_{ijk}\sigma_{kj} + \gamma_{ijk}\sigma_{kj}) \\ &= \frac{1}{2}(\gamma_{ijk}\sigma_{kj} + \gamma_{ikj}\sigma_{jk}) \\ &= \frac{1}{2}\gamma_{ijk}(\sigma_{kj} - \sigma_{jk}) = 0 \end{aligned} \tag{e}$$

Thus, the stress tensor is symmetric

$$\sigma_{ij} = \sigma_{ji} \tag{10.91}$$

Introducing the constitutive equation (4.18′) of the isotropic thermoelastic body

$$\sigma_{ij} = 2\mu\epsilon_{ij} + \lambda\epsilon_{kk}\delta_{ij} - \beta\tau\delta_{ij}$$

into the equation of motion (10.89), we obtain the equation of motion expressed in terms of the strain components

$$2\mu\epsilon_{ji,j} + \lambda\epsilon_{kk,i} - \beta\tau_{,i} + F_i = \rho\ddot{u}_i \tag{10.92}$$

and an alternative form expressed in terms of the displacement components

$$\mu u_{i,kk} + (\lambda + \mu) u_{k,ki} - \beta\tau_{,i} + F_i = \rho\ddot{u}_i \tag{10.93}$$

10.4.3 Solution of Equation of Motion

Consider a homogeneous continuous linear thermoelastic body occupying an inner region D with a boundary surface B. When the body is subjected to mechanical and thermal loads, the system of governing equations is

governing differential equations of motion

$$\mu u_{i,kk} + (\lambda + \mu) u_{k,ki} - \beta \tau_{,i} + F_i = \rho \ddot{u}_i \tag{10.94}$$

governing differential equation of heat conduction

$$\dot{\tau} = \kappa \tau_{,ii} + \frac{Q^*}{c_\epsilon} - \kappa \eta \dot{\epsilon}_{kk} \tag{10.95}$$

and the mechanical boundary conditions

$$\sigma_{ji} n_j = p_i \qquad \text{on the part of the boundary } B_1 \tag{10.96}$$

$$u_i = u_{bi} \qquad \text{on the rest of the boundary } B_2 \tag{10.97}$$

and the thermal boundary condition

$$-\lambda \frac{\partial \tau}{\partial n} + q_b = h(\tau - \Theta) \qquad \text{on the boundary } B \tag{10.98}$$

in which Θ denotes the temperature change of the surrounding medium, and the initial mechanical conditions

$$u_i(\boldsymbol{x}, 0) = u_{0i}(\boldsymbol{x}), \quad \dot{u}_i(\boldsymbol{x}, 0) = \dot{u}_{0i}(\boldsymbol{x}) \qquad \text{at the initial time } t = 0 \tag{10.99}$$

and the initial thermal condition

$$\tau(\boldsymbol{x}, 0) = \tau_0(\boldsymbol{x}) \qquad \text{at the initial time } t = 0 \tag{10.100}$$

The differential equations (10.94) and (10.95) of thermoelasticity may be rewritten in the vector form

$$\mu \nabla^2 \boldsymbol{u} + (\lambda + \mu) \operatorname{grad} \operatorname{div} \boldsymbol{u} - \beta \operatorname{grad} \tau + \boldsymbol{F} = \rho \ddot{\boldsymbol{u}} \tag{10.94$'$}$$

$$\dot{\tau} = \kappa \nabla^2 \tau + \frac{Q^*}{c_\epsilon} - \kappa \eta \operatorname{div} \dot{\boldsymbol{u}} \tag{10.95$'$}$$

Let us resolve the displacement and body force vectors into their potential and solenoidal parts

$$\boldsymbol{u} = \operatorname{grad} \phi + \operatorname{rot} \boldsymbol{\psi} \qquad \operatorname{div} \boldsymbol{\psi} = 0 \tag{10.101}$$

$$\boldsymbol{F} = \operatorname{grad} \xi + \operatorname{rot} \boldsymbol{\chi} \qquad \operatorname{div} \boldsymbol{\chi} = 0 \tag{10.102}$$

Introducing Eqs. (10.101) and (10.102) into (10.94) we receive

$$\operatorname{grad}\,[(\lambda+2\mu)\nabla^2\phi - \rho\ddot{\phi} - \beta\tau + \xi] + \operatorname{rot}\,(\mu\nabla^2\boldsymbol{\psi} - \rho\ddot{\boldsymbol{\psi}} + \boldsymbol{\chi}) = 0 \qquad \text{(a)}$$

Equation (a) is satisfied if

$$\left(\nabla^2 - \frac{1}{c_1^2}\frac{\partial^2}{\partial t^2}\right)\phi = K\tau - \frac{1}{\rho c_1^2}\xi \qquad (10.103)$$

$$\left(\nabla^2 - \frac{1}{c_2^2}\frac{\partial^2}{\partial t^2}\right)\boldsymbol{\psi} = -\frac{1}{\rho c_2^2}\boldsymbol{\chi} \qquad (10.104)$$

where

$$K = \frac{\beta}{\lambda+2\mu} = \frac{1-\nu}{1+\nu}\alpha, \qquad c_1 = \sqrt{\frac{\lambda+2\mu}{\rho}}, \qquad c_2 = \sqrt{\frac{\mu}{\rho}} \qquad (10.105)$$

Equations (10.103) and (10.104) mean that propagation of mechanical disturbance is a wave phenomenon. Introducing Eqs. (10.101) and (10.102) into (10.95) we receive

$$\left(\nabla^2 - \frac{1}{\kappa}\frac{\partial}{\partial t}\right)\tau - \eta\nabla^2\frac{\partial\phi}{\partial t} = -\frac{Q^*}{\lambda} \qquad (10.106)$$

Equation (10.106) means that heat transport is still based upon diffusion. The governing equations of thermoelasticity which take coupling effect into account are given by Eqs. (10.103), (10.104), and (10.106). Eliminating from the governing equations the temperature change τ or the potential function ϕ we obtain the system of differential equations

$$\left[\left(\nabla^2 - \frac{1}{c_1^2}\frac{\partial^2}{\partial t^2}\right)\left(\nabla^2 - \frac{1}{\kappa}\frac{\partial}{\partial t}\right) - \eta K\nabla^2\frac{\partial}{\partial t}\right]\phi = -\frac{1}{\rho c_1^2}\left(\nabla^2 - \frac{1}{\kappa}\frac{\partial}{\partial t}\right)\xi - \frac{KQ^*}{\lambda} \qquad (10.107)$$

or

$$\left[\left(\nabla^2 - \frac{1}{c_1^2}\frac{\partial^2}{\partial t^2}\right)\left(\nabla^2 - \frac{1}{\kappa}\frac{\partial}{\partial t}\right) - \eta K\nabla^2\frac{\partial}{\partial t}\right]\tau = -\left(\nabla^2 - \frac{1}{c_1^2}\frac{\partial^2}{\partial t^2}\right)\frac{Q^*}{\lambda} - \frac{\eta}{\rho c_1^2}\nabla^2\frac{\partial}{\partial t}\xi \qquad (10.107')$$

and

$$\left(\nabla^2 - \frac{1}{c_2^2}\frac{\partial^2}{\partial t^2}\right)\boldsymbol{\psi} = -\frac{1}{\rho c_2^2}\boldsymbol{\chi} \qquad (10.108)$$

Equations (10.107) and (10.107′) describe the longitudinal dilatational wave of the potential function ϕ and the temperature change τ, respectively, while Eq. (10.108) describes the transverse wave. Applying the divergence operation to Eq. (10.94′) and using the definition of dilatation $e = u_{k,k}$ we get

$$\left(\nabla^2 - \frac{1}{c_1^2}\frac{\partial^2}{\partial t^2}\right) e - K\nabla^2\tau + \frac{1}{\lambda + 2\mu} F_{i,i} = 0 \tag{10.109}$$

Eliminating the temperature change τ from Eqs. (10.95) and (10.109), we obtain the wave equation of dilatation e

$$\left[\left(\nabla^2 - \frac{1}{c_1^2}\frac{\partial^2}{\partial t^2}\right)\left(\nabla^2 - \frac{1}{\kappa}\frac{\partial}{\partial t}\right) - \eta K\nabla^2\frac{\partial}{\partial t}\right] e$$
$$= -\frac{K\nabla^2 Q^*}{\lambda} - \frac{1}{\lambda + 2\mu}\left(\nabla^2 - \frac{1}{\kappa}\frac{\partial}{\partial t}\right) F_{i,i} \tag{10.110}$$

10.5 The Variational Theorems of Thermoelasticity

Let us consider a variational principle for coupled thermoelasticity based upon a derivation by Biot with the basic thermodynamic laws of irreversible process taken into consideration. Under the assumption of small deviations of the thermodynamic system from the state of equilibrium we will consider the statement of virtual external work:

$$\int_D F_i \delta u_i \, dV + \int_B p_i \delta u_i \, dA \tag{a}$$

where F_i denotes the components of body forces and p_i the components of surface tractions applied to the surface B. Using Eq. (4.5′) and Gauss' divergence theorem in the second term of Eq. (a) we obtain

$$\int_B p_i \delta u_i \, dA = \int_D \sigma_{ji,j}\delta u_i \, dV + \int_D \sigma_{ji}\delta u_{i,j} \, dV$$
$$= \int_D \sigma_{ji,j}\delta u_i \, dV + \int_D \sigma_{ji}\delta \epsilon_{ij} \, dV \tag{b}$$

Taking into account the equation of motion (10.89), the virtual external work (a) reduces to

$$\int_D F_i \delta u_i \, dV + \int_B p_i \delta u_i \, dA = \int_D \rho \ddot{u}_i \delta u_i \, dV + \int_D \sigma_{ji}\delta\epsilon_{ij} \, dV \tag{10.111}$$

Substitution of the Duhamel-Neumann relation (4.18′) into the second term on the right-hand side of Eq. (10.111) yields

$$\int_D \sigma_{ji}\delta\epsilon_{ij} \, dV = \int_D (2\mu\epsilon_{ij} + \lambda\epsilon_{kk}\delta_{ij})\delta\epsilon_{ij} \, dV - \int_D \beta\tau\delta\epsilon_{kk} \, dV \tag{c}$$

We arrive at the theorem of virtual work from Eqs. (10.111) and (c)

$$\int_D F_i \delta u_i \, dV + \int_B p_i \delta u_i \, dA - \int_D \rho \ddot{u}_i \delta u_i \, dV = \delta W - \int_D \beta \tau \delta \epsilon_{kk} \, dV \qquad (10.112)$$

where

$$\begin{aligned} \delta W &= \int_D (2\mu \epsilon_{ij} \delta \epsilon_{ij} + \lambda \epsilon_{kk} \delta \epsilon_{mm}) \, dV \\ &= \int_D \sigma_{ij} \delta \epsilon_{ij} \, dV + \beta \int_D \tau \delta \epsilon_{kk} \, dV \end{aligned} \qquad (10.113)$$

The function W implies the work of the deformation may be expressed by

$$W = \int_D \left(\mu \epsilon_{ij} \epsilon_{ij} + \frac{\lambda}{2} \epsilon_{kk} \epsilon_{mm} \right) dV \qquad (10.114)$$

The three terms on the left-hand side of Eq. (10.112) express the virtual external work of the body forces, of tractions on the boundary, and of inertia forces, respectively, while the right-hand side expresses the virtual internal work. If the temperature change τ appearing in the last term of the right-hand side of Eq. (10.112) were known, that is, if the uncoupled thermoelastic problem were treated, the theorem of virtual work (10.112) would be complete. When the coupled thermoelastic problem is treated, the temperature change τ appearing in the last term of the right-hand side of Eq. (10.112) is unknown. In this case, an additional relation for heat conduction must be introduced. The entropy balance in the body without internal heat generation is

$$q_{i,i} = -T \dot{s} \approx -T_0 \dot{s} \qquad (10.115)$$

We introduce an entropy flux, H, which is related to the heat flux through the equation

$$q_i = T_0 \dot{H}_i, \qquad s = -H_{i,i} \qquad (10.116)$$

Multiplying δH_i by the Fourier law of heat conduction for an isotropic body (10.78) and integrating over the region D of the body we find, by use of Eq. (10.116)

$$\int_D \left(\tau_{,i} + \frac{T_0}{\lambda} \dot{H}_i \right) \delta H_i \, dV = 0 \qquad (10.117)$$

The first term of Eq. (10.117) reduces to

$$\begin{aligned} \int_D \tau_{,i} \delta H_i \, dV &= \int_D (\tau \delta H_i)_{,i} \, dV - \int_D \tau \delta H_{i,i} \, dV \\ &= \int_B \tau n_i \delta H_i \, dA - \int_D \tau \delta H_{i,i} \, dV \end{aligned} \qquad \text{(d)}$$

From the constitutive equation for entropy (10.79) and Eq. (10.116) we have

$$\delta H_{i,i} = -\beta \delta \epsilon_{kk} - \frac{c_\epsilon}{T_0} \delta \tau \qquad (10.118)$$

Substitution of Eq. (10.118) into Eq. (d) yields

$$\int_D \tau_{,i}\delta H_i \, dV = \int_B \tau n_i \delta H_i \, dA + \beta \int_D \tau \delta \epsilon_{kk} \, dV + \frac{c_\epsilon}{T_0} \int_D \tau \delta \tau \, dV \tag{e}$$

Introducing Eq. (e) into Eq. (10.117) yields

$$\int_B \tau n_i \delta H_i \, dA + \beta \int_D \tau \delta \epsilon_{kk} \, dV + \frac{c_\epsilon}{T_0} \int_D \tau \delta \tau \, dV + \frac{T_0}{\lambda} \int_D \dot{H}_i \delta H_i \, dV = 0 \tag{10.119}$$

We introduce the heat potential P

$$P = \frac{c_\epsilon}{2T_0} \int_D \tau^2 \, dV \tag{10.120}$$

where

$$\delta P = \frac{c_\epsilon}{T_0} \int_D \tau \delta \tau \, dV \tag{10.121}$$

and the dissipation function D where

$$\delta D = \frac{T_0}{\lambda} \int_D \dot{H}_i \delta H_i \, dV \tag{10.122}$$

Introducing Eqs. (10.121) and (10.122) into Eq. (10.119) we obtain the variational equation for heat conduction

$$\delta P + \delta D + \beta \int_D \tau \delta \epsilon_{kk} \, dV = - \int_B \tau n_i \delta H_i \, dA \tag{10.123}$$

Since the theorem of virtual work (10.112) and the variational equation of heat conduction (10.123) must be treated simultaneously in coupled thermoelasticity, elimination of the term $\beta \int_D \tau \delta \epsilon_{kk} \, dV$ in Eqs. (10.112) and (10.123) leads to

$$\delta W + \delta P + \delta D = \int_D F_i \delta u_i \, dV + \int_B p_i \delta u_i \, dA - \int_D \rho \ddot{u}_i \delta u_i \, dV - \int_B \tau n_i \delta H_i \, dA \tag{10.124}$$

The right-hand side in Eq. (10.124) expresses the virtual external work of the body forces, of tractions on the boundary, of inertia forces, and of heating of the boundary, respectively, while the left-hand side represents the virtual internal work of deformation, the variation of heat potential, and the variation of the dissipation function, respectively. Introducing the Biot thermoelastic potential Φ

$$\Phi = W + P = \int_D \left(\mu \epsilon_{ij} \epsilon_{ij} + \frac{\lambda}{2} \epsilon_{kk} \epsilon_{mm} + \frac{c_\epsilon}{2T_0} \tau^2 \right) dV \tag{10.125}$$

we obtain the variational principle for the coupled problem of thermoelasticity

$$\delta(\Phi + D) = \int_D (F_i - \rho\ddot{u}_i)\,\delta u_i\,dV + \int_B (p_i\delta u_i - \tau n_i \delta H_i)\,dA \tag{10.126}$$

10.6 Uniqueness Theorem

We need to discuss whether the solution of linear thermoelasticity leads to a unique solution of a given problem. We consider the time derivative of the sum of the work of the deformation (10.114), the heat potential (10.120) and the kinetic energy (10.2)

$$\begin{aligned}\frac{d}{dt}(W + P + K) &= \frac{d}{dt}\int_D \left(\mu\epsilon_{ij}\epsilon_{ij} + \frac{\lambda}{2}\epsilon_{kk}\epsilon_{mm}\right) dV \\ &\quad + \frac{d}{dt}\int_D \frac{c_\epsilon}{2T_0}\tau^2\,dV + \frac{d}{dt}\int_D \frac{1}{2}\rho v_i v_i\,dV \\ &= \int_D (2\mu\epsilon_{ij}\dot{\epsilon}_{ij} + \lambda\epsilon_{kk}\dot{\epsilon}_{mm})\,dV + \frac{c_\epsilon}{T_0}\int_D \tau\dot{\tau}\,dV + \int_D \rho v_i \dot{v}_i\,dV\end{aligned} \tag{10.127}$$

Using the Duhamel-Neumann relation (10.60) and Gauss' divergence theorem in the first integral of Eq. (10.127) we obtain

$$\begin{aligned}\int_D (2\mu\epsilon_{ij}\dot{\epsilon}_{ij} + \lambda\epsilon_{kk}\dot{\epsilon}_{mm})\,dV &= \int_D (\sigma_{ij}\dot{\epsilon}_{ij} + \beta\tau\dot{\epsilon}_{mm})\,dV \\ &= \int_D (\sigma_{ij}v_{i,j} + \beta\tau\dot{\epsilon}_{mm})\,dV \\ &= \int_D [(\sigma_{ij}v_i)_{,j} - \sigma_{ij,j}v_i + \beta\tau\dot{\epsilon}_{mm}]\,dV \\ &= \int_B \sigma_{ij}v_i n_j\,dA + \int_D [(F_i - \rho\dot{v}_i)v_i + \beta\tau\dot{\epsilon}_{mm}]\,dV \\ &= \int_B p_i v_i\,dA + \int_D [(F_i - \rho\dot{v}_i)v_i + \beta\tau\dot{\epsilon}_{mm}]\,dV\end{aligned}$$

Thus, Eq. (10.127) reduces to

$$\frac{d}{dt}(W + P + K) = \int_B p_i v_i\,dA + \int_D F_i v_i\,dV + \int_D \left(\frac{c_\epsilon}{T_0}\dot{\tau} + \beta\dot{\epsilon}_{mm}\right)\tau\,dV \tag{10.128}$$

Next, multiplying $\dot{H}_i$ by the Fourier law of heat conduction for an isotropic body (10.78) and integrating over the region D of the body we have, by use of Eq. (10.116)

$$\int_D \left(\tau_{,i} + \frac{T_0}{\lambda}\dot{H}_i\right)\dot{H}_i\,dV = 0 \tag{10.129}$$

The first term of Eq. (10.129) reduces to

$$\int_D \tau_{,i}\dot{H}_i \, dV = \int_D (\tau \dot{H}_i)_{,i} \, dV - \int_D \tau \dot{H}_{i,i} \, dV$$
$$= \int_B \tau n_i \dot{H}_i \, dA - \int_D \tau \dot{H}_{i,i} \, dV \tag{a}$$

From the constitutive equation for the entropy (10.79) and Eq. (10.116) we obtain

$$\dot{H}_{i,i} = -\beta \dot{\epsilon}_{kk} - \frac{c_\epsilon}{T_0}\dot{\tau} \tag{10.130}$$

Substitution of Eq. (10.130) into Eq. (a) yields

$$\int_D \tau_{,i}\dot{H}_i \, dV = \int_B \tau n_i \dot{H}_i \, dA + \int_D \left(\frac{c_\epsilon}{T_0}\dot{\tau} + \beta \dot{\epsilon}_{kk} \right) \tau \, dV \tag{b}$$

From Eqs. (10.78) and (10.116), we have

$$\dot{H}_i = \frac{q_i}{T_0} = -\frac{\lambda}{T_0}\tau_{,i} \tag{10.131}$$

Introducing Eq. (10.131) to Eq. (b) we obtain

$$\int_D \left(\frac{c_\epsilon}{T_0}\dot{\tau} + \beta \dot{\epsilon}_{kk} \right) \tau \, dV = \frac{\lambda}{T_0}\int_B \tau \frac{\partial \tau}{\partial n} \, dA - \frac{\lambda}{T_0}\int_D (\tau_{,i})^2 \, dV \tag{10.132}$$

Substitution of Eq. (10.132) into Eq. (10.128) yields the theorem of energy conservation

$$\frac{d}{dt}(W + P + K) + \frac{\lambda}{T_0}\int_D (\tau_{,i})^2 \, dV = \int_B p_i v_i \, dA + \int_D F_i v_i \, dV + \frac{\lambda}{T_0}\int_B \tau \frac{\partial \tau}{\partial n} \, dA \tag{10.133}$$

Let us consider a body V with boundary B subjected to mechanical and thermal loads. We assume that all physical quantities and their first derivatives are continuous, and that the second derivatives of the displacements and of the temperature change are also continuous for all $\boldsymbol{x} \in V + B$ and $t > 0$. When the linear thermoelastic body is subjected to mechanical and thermal loads, the governing equations are

Fundamental equations:

$$\sigma_{ji,j} + F_i = \rho \ddot{u}_i$$
$$\dot{\tau} = \kappa \tau_{,ii} + \frac{Q^*}{c_\epsilon} - \kappa \eta \dot{\epsilon}_{kk} \tag{c}$$

Boundary conditions:

$$\sigma_{ji}n_j = p_i$$
$$-\lambda\frac{\partial \tau}{\partial n} + q_b = h(\tau - \Theta) \tag{d}$$

Initial conditions:

$$u_i(\boldsymbol{x},0) = u_{0i}(\boldsymbol{x})$$
$$\dot{u}_i(\boldsymbol{x},0) = \dot{u}_{0i}(\boldsymbol{x}) \tag{e}$$
$$\tau(\boldsymbol{x},0) = \tau_0(\boldsymbol{x})$$

Duhamel-Neumann relations:

$$\sigma_{ij} = 2\mu\epsilon_{ij} + (\lambda\epsilon_{kk} - \beta\tau)\delta_{ij} \tag{f}$$

To prove the uniqueness of the solution of linear thermoelasticity, we assume that there exist two sets of solutions $u_i'(x,t)$, $T_i'(x,t)$, and $u_i''(x,t)$, $T_i''(x,t)$, both satisfying the governing equations (c) to (f). Introducing the difference functions

$$\hat{u}_i\,(x,t) = u'_i\,(x,t) - u''_i\,(x,t)\,, \qquad \widehat{T}\,(x,t) = T'\,(x,t) - T''\,(x,t) \tag{10.134}$$

the governing equations for the difference functions should be satisfied:

Fundamental equations:

$$\hat{\sigma}_{ji,j} = \rho\ddot{\hat{u}}_i$$
$$\dot{\hat{\tau}} = \kappa\hat{\tau}_{,ii} - \kappa\eta\dot{\hat{\epsilon}}_{kk} \tag{g}$$

Boundary conditions:

$$\hat{\sigma}_{ji}n_j = 0$$
$$-\lambda\frac{\partial\hat{\tau}}{\partial n} = h\hat{\tau} \tag{h}$$

Initial conditions:

$$\hat{u}_i(\boldsymbol{x},0) = 0$$
$$\dot{\hat{u}}_i(\boldsymbol{x},0) = 0 \tag{i}$$
$$\hat{\tau}(\boldsymbol{x},0) = 0$$

$$\hat{\sigma}_{ij} = 2\mu\hat{\epsilon}_{ij} + \left(\lambda\hat{\epsilon}_{kk} - \beta\hat{\tau}\right)\delta_{ij} \tag{j}$$

The theorem of energy conservation for the system of the forces with the difference functions is written as

$$\frac{d}{dt}(\widehat{W} + \widehat{P} + \widehat{K}) + \frac{\lambda}{T_0}\int_D (\hat{\tau}_{,i})^2\,dV = \frac{\lambda}{T_0}\int_B \hat{\tau}\frac{\partial\hat{\tau}}{\partial n}\,dA \tag{10.135}$$

Using the definitions of the functions $\widehat{W}$, $\widehat{P}$, $\widehat{K}$ and the boundary condition for the temperature, Eq. (10.135) has the form

$$\frac{d}{dt}\int_D\left[\left(\mu\hat{\epsilon}_{ij}\hat{\epsilon}_{ij}+\frac{\lambda}{2}\hat{\epsilon}_{kk}\hat{\epsilon}_{mm}\right)+\frac{c_\epsilon}{2T_0}\hat{\tau}^{\,2}+\frac{1}{2}\rho\hat{v}_i\hat{v}_i\right]dV$$

$$=-\left[\frac{\lambda}{T_0}\int_D(\hat{\tau}_{,i})^2\,dV+\frac{h}{T_0}\int_B\hat{\tau}^{\,2}\,dA\right] \tag{10.136}$$

The integrands on both sides in Eq. (10.136) are positive or zero at any time because the integrands are the sums of squares. The integrand of the left-hand side of Eq. (10.136) is zero at the initial time, since the difference functions satisfy the homogeneous initial conditions. On the other hand, the time derivative of the integral on the left-hand side is negative or zero because the integral on the right-hand side is negative or zero. Hence

$$\mu\hat{\epsilon}_{ij}\hat{\epsilon}_{ij}+\frac{\lambda}{2}\hat{\epsilon}_{kk}\hat{\epsilon}_{mm}+\frac{c_\epsilon}{2T_0}\hat{\tau}^{\,2}+\frac{1}{2}\rho\hat{v}_i\hat{v}_i=0 \qquad \text{for } t\geq 0 \tag{10.137}$$

Since Eq. (10.137) is the sum of squares, we get

$$\hat{v}_i=0,\quad \hat{\epsilon}_{ij}=0,\quad \hat{\tau}=0 \qquad \text{for } t\geq 0$$

or

$$v'_i=v''_i,\quad \epsilon'_{ij}=\epsilon''_{ij},\quad \sigma'_{ij}=\sigma''_{ij},\quad \tau'=\tau'' \qquad \text{for } t\geq 0 \tag{10.138}$$

From the first equation in the initial conditions (e) and the first equation in Eq. (10.138), it follows:

$$u'_i=u''_i \qquad \text{for } t\geq 0 \tag{10.139}$$

The uniqueness of the solution of the given problem of the linear thermoelasticity has been proved.

10.7 Reciprocal Theorem

10.7.1 Reciprocal Theorem

Let us consider a body V with boundary B subjected to mechanical and thermal loads. When external loads which are called the first state of forces are applied to the body V with the boundary B, the displacements $u'_i(x,t)$ and the temperature $T'(x,t)$ and, therefore, stresses $\sigma'_{ij}(x,t)$ and strains $\epsilon'_{ij}(x,t)$ are produced in the body. When another set of external loads, which is called the second state of forces, independent of the first state of forces, is applied to the same body V with the same boundary B, the displacements $u''_i(x,t)$ and the temperature $T''_i(x,t)$ and, therefore, stresses $\sigma''_{ij}(x,t)$ and strains $\epsilon''_{ij}(x,t)$ are independently produced in the body. We assume that all physical quantities and their first derivatives are continuous, and second derivatives of the displacements and of the temperatures are also continuous for all $\boldsymbol{x} \in V + B$ and $t > 0$. When the homogeneous initial conditions are considered, the governing equations for the first and the second state of the forces are:

Fundamental equation:

$$\sigma_{ji,j} + F_i = \rho \ddot{u}_i$$
$$\dot{\tau} = \kappa \tau_{,ii} + \frac{Q^*}{c_\epsilon} - \kappa \eta \dot{\epsilon}_{kk} \tag{a}$$

Boundary conditions:

$$\sigma_{ji} n_j = p_i \qquad -\lambda \frac{\partial \tau}{\partial n} + q_b = h(\tau - \Theta) \tag{b}$$

Initial conditions:

$$u_i(\boldsymbol{x}, 0) = 0$$
$$\dot{u}_i(\boldsymbol{x}, 0) = 0 \tag{c}$$
$$\tau(\boldsymbol{x}, 0) = 0$$

Duhamel-Neumann relations:

$$\sigma_{ij} = 2\mu \epsilon_{ij} + (\lambda e - \beta \tau)\,\delta_{ij}, \qquad e = \epsilon_{kk} \tag{d}$$

in which the first and the second state of the forces will be indicated by single and double primes. We introduce the Laplace transform

$$\bar{u}_i(\boldsymbol{x}, p) = L\{u_i(\boldsymbol{x}, t)\} = \int_0^\infty u_i(\boldsymbol{x}, t)\, e^{-pt} dt \tag{10.140}$$

Performing the Laplace transform on the above governing equations, we obtain

$$\bar{\sigma}_{ji,j} + \overline{F}_i = \rho p^2 \bar{u}_i \qquad \boldsymbol{x} \in V \tag{e}$$

$$\bar{\tau}_{,ii} - \frac{p}{\kappa}\bar{\tau} - \eta p \bar{\epsilon}_{kk} = -\frac{\overline{Q}^*}{\lambda} \qquad \boldsymbol{x} \in V \tag{f}$$

$$\bar{\sigma}_{ji} n_j = \bar{p}_i \qquad \boldsymbol{x} \in B \tag{g}$$

$$\lambda \bar{\tau}_{,n} + h\bar{\tau} = \bar{q}_b + h\overline{\Theta} \qquad \boldsymbol{x} \in B \tag{h}$$

$$\bar{\sigma}_{ij} = 2\mu \bar{\epsilon}_{ij} + (\lambda \bar{e} - \beta \bar{\tau})\delta_{ij} \tag{i}$$

Multiplying the Duhamel-Neumann relations on which the Laplace transform has been performed for the two states of the forces by $\bar{\epsilon}''_{ij}$ and $\bar{\epsilon}'_{ij}$, respectively, and subtracting, we get

$$\bar{\sigma}'_{ij}\bar{\epsilon}''_{ij} - \bar{\sigma}''_{ij}\bar{\epsilon}'_{ij} = \beta(\bar{\tau}''\,\bar{e}' - \bar{\tau}'\,\bar{e}'') \tag{j}$$

Integrating Eq. (j) over the volume of the body V yields

$$\int_V \left(\bar{\sigma}'_{ij}\bar{\epsilon}''_{ij} - \bar{\sigma}''_{ij}\bar{\epsilon}'_{ij}\right) dV = \beta \int_V (\bar{\tau}''\,\bar{e}' - \bar{\tau}'\,\bar{e}'')\, dV \tag{10.141}$$

The first integral on the left-hand side in Eq. (10.141) may be reduced as follows:

$$\begin{aligned}
\int_V \bar{\sigma}'_{ij}\bar{\epsilon}''_{ij}\, dV &= \frac{1}{2}\int_V \bar{\sigma}'_{ij}\left(\bar{u}''_{i,j} + \bar{u}''_{j,i}\right) dV = \int_V \bar{\sigma}'_{ij}\bar{u}''_{i,j}\, dV \\
&= \int_V \left[\left(\bar{\sigma}'_{ji}\bar{u}''_i\right)_{,j} - \bar{\sigma}'_{ij,j}\bar{u}''_i\right] dV \\
&= \int_B \bar{\sigma}'_{ji}\bar{u}''_i n_j\, dA - \int_V \bar{\sigma}'_{ji,j}\bar{u}''_i\, dV \\
&= \int_B \bar{p}'_i\bar{u}''_i\, dA + \int_V \overline{F}'_i\bar{u}''_i\, dV - \rho\int_V p^2\bar{u}'_i\bar{u}''_i\, dV
\end{aligned} \tag{k}$$

The second integral on the left-hand side in Eq. (10.141) may be reduced as follows:

$$\int_V \bar{\sigma}''_{ij}\bar{\epsilon}'_{ij}\, dV = \int_B \bar{p}''_i\bar{u}'_i\, dA + \int_V \overline{F}''_i\bar{u}'_i\, dV - \rho\int_V p^2\bar{u}'_i\bar{u}''_i\, dV \tag{l}$$

Substitution of Eqs. (k) and (l) into Eq. (10.141) yields

$$\int_B \left(\bar{p}'_i\bar{u}''_i - \bar{p}''_{ni}\bar{u}'_i\right) dA + \int_V \left(\overline{F}'_i\bar{u}''_i - \overline{F}''_i\bar{u}'_i\right) dV + \beta\int_V (\bar{\tau}'\,\bar{e}'' - \bar{\tau}''\,\bar{e}')\, dV = 0 \tag{10.142}$$

Equation (10.142) constitutes the first part of the reciprocal theorem.

Next, multiplying the heat conduction equations in the first and second states of the forces by $\bar{\tau}''$ and $\bar{\tau}'$, respectively, and subtracting the results and integrating over the body V, we obtain

$$\lambda\int_V \left(\bar{\tau}'_{,ii}\bar{\tau}'' - \bar{\tau}''_{,ii}\bar{\tau}'\right) dV - \lambda\eta p\int_V (\bar{e}'\,\bar{\tau}'' - \bar{e}''\,\bar{\tau}')\, dV + \int_V \left(\overline{Q}^{*\prime}\,\bar{\tau}'' - \overline{Q}^{*\prime\prime}\,\bar{\tau}'\right) dV = 0 \tag{10.143}$$

The first integral on the left-hand side in Eq. (10.143) may be reduced as follows:

$$\begin{aligned}
\int_V \bar{\tau}'_{,ii}\bar{\tau}''\, dV &= \int_V \left[(\bar{\tau}'_{,i}\bar{\tau}'')_{,i} - \bar{\tau}'_{,i}\bar{\tau}''_{,i}\right] dV \\
&= \int_B \bar{\tau}'_{,i}\bar{\tau}'' n_i\, dA - \int_V \bar{\tau}'_{,i}\bar{\tau}''_{,i}\, dV \\
&= \int_B \frac{\partial\bar{\tau}'}{\partial n}\bar{\tau}''\, dA - \int_V \bar{\tau}'_{,i}\bar{\tau}''_{,i}\, dV
\end{aligned} \tag{m}$$

The second integral on the left-hand side in Eq. (10.143) may be reduced as follows:

$$\int_V \bar{\tau}''_{,ii}\bar{\tau}' dV = \int_B \frac{\partial \bar{\tau}''}{\partial n}\bar{\tau}' dA - \int_V \bar{\tau}'_{,i}\bar{\tau}''_{,i} dV \tag{n}$$

Substitution of Eqs. (m) and (n) into Eq. (10.143) yields

$$\lambda \int_B \left(\frac{\partial \bar{\tau}'}{\partial n}\bar{\tau}'' - \frac{\partial \bar{\tau}''}{\partial n}\bar{\tau}'\right) dA - \lambda\eta p \int_V (\bar{e}'\,\bar{\tau}'' - \bar{e}''\,\bar{\tau}')\, dV$$
$$+ \int_V (\overline{Q}^{*\prime}\,\bar{\tau}'' - \overline{Q}^{*\prime\prime}\,\bar{\tau}')\, dV = 0 \tag{10.144}$$

Equation (10.144) constitutes the second part of the reciprocal theorem. Eliminating the term $\int_V (\bar{e}'\,\bar{\tau}'' - \bar{e}''\,\bar{\tau}')\, dV$ from Eqs. (10.142) and (10.144) leads to the reciprocal theorem

$$\lambda\eta p \int_B (\bar{p}'_i\bar{u}''_i - \bar{p}''_i\bar{u}'_i)\, dA + \lambda\eta p \int_V (\overline{F}'{}_i\bar{u}''_i - \overline{F}''{}_i\bar{u}'_i)\, dV$$
$$= \lambda\beta \int_B \left(\frac{\partial \bar{\tau}'}{\partial n}\bar{\tau}'' - \frac{\partial \bar{\tau}''}{\partial n}\bar{\tau}'\right) dA + \beta \int_V (\overline{Q}^{*\prime}\,\bar{\tau}'' - \overline{Q}^{*\prime\prime}\,\bar{\tau}')\, dV \tag{10.145}$$

Applying the following relations for the inverse Laplace transform

$$L^{-1}[\bar{f}(p)] = f(t)$$
$$L^{-1}[p\bar{f}(\boldsymbol{x}, p)] = \frac{\partial f(\boldsymbol{x}, t)}{\partial t} \qquad \text{for } f(\boldsymbol{x}, 0) = 0$$
$$L^{-1}[\bar{f}_1(\boldsymbol{x}, p)\cdot \bar{f}_2(\boldsymbol{x}, p)] = \int_0^t f_1(\boldsymbol{x}, s) f_2(\boldsymbol{x}, t-s)\, ds$$
$$= \int_0^t f_1(\boldsymbol{x}, t-s) f_2(\boldsymbol{x}, s)\, ds$$
$$L^{-1}[p\bar{f}_1(\boldsymbol{x}, p)\cdot \bar{f}_2(\boldsymbol{x}, p)] = \int_0^t \frac{\partial f_1(\boldsymbol{x}, s)}{\partial s} f_2(\boldsymbol{x}, t-s)\, ds$$
$$= \int_0^t \frac{\partial f_2(\boldsymbol{x}, s)}{\partial s} f_1(\boldsymbol{x}, t-s)\, ds$$
$$\text{for } f_1(\boldsymbol{x}, 0) = f_2(\boldsymbol{x}, 0) = 0$$

to the reciprocal theorem (10.145) in the Laplace transform domain, we arrive at the reciprocal theorem in the real space domain

$$\lambda\eta \int_V \left\{\int_0^t \left[F'{}_i(x, t-s)\frac{\partial u''{}_i(x, s)}{\partial s} - F''{}_i(x, t-s)\frac{\partial u'{}_i(x, s)}{\partial s}\right] ds\right\} dV$$
$$+ \lambda\eta \int_B \left\{\int_0^t \left[p'{}_i(x, t-s)\frac{\partial u''{}_i(x, s)}{\partial s} - p''{}_i(x, t-s)\frac{\partial u'{}_i(x, s)}{\partial s}\right] ds\right\} dA$$

$$= \beta \int_V \left\{ \int_0^t \left[Q^{*\prime}(x, t-s)\tau''(x, s) - Q^{*\prime\prime}(x, t-s)\tau'(x, s) \right] ds \right\} dV$$

$$+ \lambda\beta \int_B \left\{ \int_0^t \left[\frac{\partial \tau'}{\partial n}(x, t-s)\tau''(\boldsymbol{x}, s) - \frac{\partial \tau''}{\partial n}(x, t-s)\tau'(x, s) \right] ds \right\} dA$$

(10.146)

Next, we consider the reciprocal theorem in stationary and uncoupled thermoelasticity. In this case, the governing equations are for the first and the second state of the forces

Equation of motion:

$$\sigma_{ji,j} + F_i = 0 \tag{o}$$

Heat conduction equation:

$$\dot{\tau} = \kappa \tau_{,ii} + \frac{Q^*}{c_\epsilon} \tag{p}$$

Boundary conditions:

$$\sigma_{ji} n_j = p_i \tag{q}$$

$$-\lambda \frac{\partial \tau}{\partial n} + q_b = h(\tau - \Theta) \tag{r}$$

Initial condition:

$$\tau(\boldsymbol{x}, 0) = \tau_0(\boldsymbol{x}) \tag{s}$$

Duhamel-Neumann relations:

$$\sigma_{ij} = 2\mu\epsilon_{ij} + (\lambda e - \beta\tau)\delta_{ij}, \qquad e = \epsilon_{kk} \tag{t}$$

in which the first and second states of the forces will be indicated by single and double primes. Multiplying the Duhamel-Neumann relations for the two states of the forces by ϵ''_{ij} and ϵ'_{ij}, respectively, and subtracting, and integrating over the volume of the body V, we receive

$$\int_V \left(\sigma'_{ij}\epsilon''_{ij} - \sigma''_{ij}\epsilon'_{ij} \right) dV = \beta \int_V (\tau'' e' - \tau' e'')\, dV \tag{10.147}$$

The first integral on the left-hand side in Eq. (10.147) can be reduced as follows:

$$\int_V \sigma'_{ij}\epsilon''_{ij}\, dV = \int_V \sigma'_{ij} u''_{i,j}\, dV$$

$$= \int_V \left[(\sigma'_{ji} u''_i)_{,j} - \sigma'_{ij,j} u''_i \right] dV$$

$$= \int_B \sigma'_{ji} u''_i n_j \, dA - \int_V \sigma'_{ji,j} u''_i \, dV$$

$$= \int_B p'_i u''_i \, dA + \int_V F'_i u''_i \, dV \tag{u}$$

The second integral on the left-hand side in Eq. (10.147) can be reduced as follows:

$$\int_V \sigma''_{ij} \epsilon'_{ij} \, dV = \int_B p''_{ni} u'_i \, dA + \int_V F''_i u'_i \, dV \tag{v}$$

Substitution of Eqs. (u) and (v) into Eq. (10.147) yields the reciprocal theorem in stationary and uncoupled thermoelasticity.

$$\int_B p'_{ni} u''_i \, dA + \int_V F'_i u''_i \, dV + \beta \int_V \tau' e'' \, dV$$
$$= \int_B p''_{ni} u'_i \, dA + \int_V F''_i u'_i \, dV + \beta \int_V \tau'' e' \, dV \tag{10.148}$$

We consider a special case in stationary and uncoupled thermoelasticity. The first state of forces is subjected to only thermal loads and produces the displacements $u'_i(\boldsymbol{x}, t)$ and the temperature change $\tau'_i(\boldsymbol{x}, t)$. On the other hand, the second state of forces is subjected to only a unit body force $F''_i = \delta(\boldsymbol{x} - \boldsymbol{\xi})\delta_{ik}$ in the direction x_k at the point $\boldsymbol{\xi}$ in the body without surface tractions and independently produces the displacements $u''_i = u_i^{(k)}(\boldsymbol{x}, \boldsymbol{\xi})$. The displacements $u''_i = u_i^{(k)}(\boldsymbol{x}, \boldsymbol{\xi})$ in the second state of forces may be determined from the governing equation

$$\mu u^{(k)}_{i,jj} + (\lambda + \mu) u^{(k)}_{j,ij} + \delta(\boldsymbol{x} - \boldsymbol{\xi})\delta_{ik} = 0 \qquad (i, j, k = 1, 2, 3) \tag{10.149}$$

Under these assumptions, the reciprocal theorem (10.148) has the form

$$\beta \int_V \tau' u^{(k)}_{j,j}(\boldsymbol{x}, \boldsymbol{\xi}) \, dV(\boldsymbol{x}) = \int_V \delta(\boldsymbol{x} - \boldsymbol{\xi})\delta_{ik} u'_i(\boldsymbol{x}) \, dV(\boldsymbol{x}) \tag{10.150}$$

Using the shifting property of the Dirac function $\delta(\boldsymbol{x} - \boldsymbol{\xi})$

$$\int_V f(\boldsymbol{x})\delta(\boldsymbol{x} - \boldsymbol{\xi}) \, dV(\boldsymbol{x}) = f(\boldsymbol{\xi}) \tag{w}$$

Equation (10.150) becomes

$$u'_k(\boldsymbol{\xi}) = \beta \int_V \tau'(\boldsymbol{x}) u^{(k)}_{j,j}(\boldsymbol{x}, \boldsymbol{\xi}) \, dV(\boldsymbol{x})$$
$$= \beta \int_V \tau'(\boldsymbol{x}) \epsilon^{(k)}_{jj}(\boldsymbol{x}, \boldsymbol{\xi}) \, dV(\boldsymbol{x}) \qquad (i, j, k = 1, 2, 3) \tag{10.151}$$

and using the relationship $\sigma_{jj}^{(k)} = (3\lambda + 2\mu)\epsilon_{jj}^{(k)}$ in Eq. (10.151) we obtain

$$u'_k(\boldsymbol{\xi}) = \alpha \int_V \tau'(\boldsymbol{x})\sigma_{jj}^{(k)}(\boldsymbol{x}, \boldsymbol{\xi})\, dV(\boldsymbol{x}) \qquad (i, j, k = 1, 2, 3) \tag{10.152}$$

If the stress field in the second state of forces and the temperature field in the first state of forces are known, the displacement $u'_k(\boldsymbol{\xi})$ in the first state of forces may be obtained by Eq. (10.152). As many isothermal elastic problems have already been solved, thermoelastic problems with arbitrary temperature fields may be obtained from Eq. (10.152) using these solutions.

10.7.2 Practical Problem

We consider a one-dimensional problem of a circular cylinder subjected to thermal loading by use of the reciprocal theory discussed in Subsection 10.7.1.

The displacement u'_i in the Cartesian coordinate system in Eq. (10.151) can be reduced to the radial displacement u'_r of the cylindrical coordinate system for axisymmetric plane problems. The first state of forces consists only of thermal loading and produces the radial displacement $u'_r(r, t)$ and the temperature change $\tau'(r, t)$, while the second state of forces consists only of a unit body force $F''_r = \delta(r - \rho)$ in the direction r at the radius ρ in the body with traction free boundary surface and independently produces the radial displacement u''_r. In this case, Eq. (10.148) reduces to

$$\beta^* \int \tau' e'' 2\pi r\, dr = \int \delta(r - \rho) u'_r 2\pi r\, dr$$

Then,

$$u'_r = \frac{\beta^*}{\rho} \int \tau'(r, t) e''(r, t) r\, dr$$

After changing variables, we have

$$u'_r = \frac{\beta^*}{r} \int \tau'(\rho, t) e''(\rho, t) \rho\, d\rho \tag{10.153}$$

We consider the problem of a hollow circular cylinder with an inner radius a and an outer radius b. The second state of forces is defined as the radial unit body force uniformly distributed on a radius r in the hollow cylinder. The governing equation of the second state of forces is

$$\frac{d\sigma''_{rr}}{d\rho} + \frac{\sigma''_{rr} - \sigma''_{\theta\theta}}{\rho} + \delta(\rho - r) = 0 \tag{10.154}$$

By use of the Hooke's law, Eq. (10.154) reduces to

$$\frac{E^*}{1 - \nu^{*2}} \frac{d}{d\rho} \left[\frac{1}{\rho} \frac{d(\rho u'')}{d\rho} \right] + \delta(\rho - r) = 0 \tag{10.155}$$

The solution of Eq. (10.155) is

$$u_1'' = A_1\rho + B_1\rho^{-1} \qquad \text{for } a \le \rho < r$$
$$u_2'' = A_2\rho + B_2\rho^{-1} \qquad \text{for } r < \rho \le b$$

The boundary conditions are

$$\sigma_{rr1}'' = 0 \qquad \text{on } \rho = a$$
$$\sigma_{rr2}'' = 0 \qquad \text{on } \rho = b$$
$$u_1'' = u_2'', \quad \sigma_{rr2}'' - \sigma_{rr1}'' = -1 \qquad \text{on } \rho = r$$

The coefficients can be determined as

$$A_1 = \frac{(1-\nu^{*2})b^2}{2E^*(b^2-a^2)}\left(1 + \frac{1-\nu^*}{1+\nu^*}\frac{r^2}{b^2}\right)$$
$$B_1 = \frac{(1+\nu^*)^2 a^2 b^2}{2E^*(b^2-a^2)}\left(1 + \frac{1-\nu^*}{1+\nu^*}\frac{r^2}{b^2}\right)$$
$$A_2 = \frac{(1-\nu^{*2})a^2}{2E^*(b^2-a^2)}\left(1 + \frac{1-\nu^*}{1+\nu^*}\frac{r^2}{a^2}\right)$$
$$B_2 = \frac{(1+\nu^*)^2 a^2 b^2}{2E^*(b^2-a^2)}\left(1 + \frac{1-\nu^*}{1+\nu^*}\frac{r^2}{a^2}\right)$$

so that the displacements in the second state have the form

$$\begin{aligned} u_1'' &= \frac{(1+\nu^*)b^2}{2E^*(b^2-a^2)}\left(1 + \frac{1-\nu^*}{1+\nu^*}\frac{r^2}{b^2}\right)\left[(1-\nu^*)\rho + (1+\nu^*)\frac{a^2}{\rho}\right] \\ u_2'' &= \frac{(1+\nu^*)a^2}{2E^*(b^2-a^2)}\left(1 + \frac{1-\nu^*}{1+\nu^*}\frac{r^2}{a^2}\right)\left[(1-\nu^*)\rho + (1+\nu^*)\frac{b^2}{\rho}\right] \end{aligned} \tag{10.156}$$

The dilatation e'' is

$$e'' = \frac{du''}{d\rho} + \frac{u''}{\rho} = \begin{cases} \dfrac{(1-\nu^{*2})b^2}{E^*(b^2-a^2)}\left(1 + \dfrac{1-\nu^*}{1+\nu^*}\dfrac{r^2}{b^2}\right) & \text{for } a \le \rho < r \\ \dfrac{(1-\nu^{*2})a^2}{E^*(b^2-a^2)}\left(1 + \dfrac{1-\nu^*}{1+\nu^*}\dfrac{r^2}{a^2}\right) & \text{for } r < \rho \le b \end{cases} \tag{10.157}$$

Substitution of Eq. (10.157) into Eq. (10.153) yields

$$\begin{aligned} u_r' &= \frac{2\beta^*}{r}\left[A_1\int_a^r \tau'(\rho)\rho\,d\rho + A_2\int_r^b \tau'(\rho)\rho\,d\rho\right] \\ &= (1+\nu^*)\alpha^*\left[\frac{1}{r}\int_a^r \tau' r\,dr + \frac{r}{b^2-a^2}\left(\frac{a^2}{r^2} + \frac{1-\nu^*}{1+\nu^*}\right)\int_a^b \tau' r\,dr\right] \end{aligned} \tag{10.158}$$

The stresses are given by

$$\sigma'_{rr} = \alpha^* E^* \left[-\frac{1}{r^2} \int_a^r \tau' r \, dr + \frac{r^2 - a^2}{r^2(b^2 - a^2)} \int_a^b \tau' r \, dr \right]$$

$$\sigma'_{\theta\theta} = \alpha^* E^* \left[\frac{1}{r^2} \int_a^r \tau' r \, dr + \frac{r^2 + a^2}{r^2(b^2 - a^2)} \int_a^b \tau' r \, dr - \tau' \right] \tag{10.159}$$

$$\sigma'_{zz} = \begin{cases} 0 & \text{for plane stress} \\ \dfrac{\alpha E}{1 - \nu} \left(\dfrac{2\nu}{b^2 - a^2} \displaystyle\int_a^b \tau' r \, dr - \tau' \right) & \text{for plane strain with } \epsilon_0 = 0 \end{cases}$$

Example 10.1 Show that volume change, ΔV, due to an arbitrary distributed temperature change τ is given by

$$\Delta V = 3\alpha \int_V \tau \, dV \tag{10.160}$$

Solution Let us consider that the first equilibrium state corresponds to the uniform tension in all the direction without temperature change, so that

$$\sigma'_{ij} = p\delta_{ij}, \qquad \tau' = 0 \tag{10.161}$$

Furthermore, let us consider that the second equilibrium state corresponds to the thermoelastic deformation due to the arbitrarily distributed temperature change τ with the traction free boundary condition, so that

$$\sigma''_{ji} n_j = 0, \qquad \tau'' = \tau \tag{10.162}$$

Substituting Eqs. (10.161) and (10.162) into Eq. (10.147), we have

$$\int_V (p\epsilon''_{kk} - \sigma''_{ij}\epsilon'_{ij}) \, dV = \beta \int_V \tau \epsilon'_{kk} \, dV \tag{10.163}$$

Now, it can be found that the second term on the left-hand side of Eq. (10.163) vanishes when body force for the body is neglected under the condition of traction free, i.e.,

$$\begin{aligned} \int_V \sigma''_{ij}\epsilon'_{ij} \, dV &= \int_V \sigma''_{ij} u'_{i,j} \, dV \\ &= \int_S \sigma''_{ij} n_j u'_i \, dS - \int_V \sigma''_{ij,j} u'_i \, dV = 0 \end{aligned} \tag{10.164}$$

where the first equation of Eqs. (10.162) and the equilibrium equation

$$\sigma''_{ij,j} = 0 \tag{10.165}$$

are taken into account. Moreover, from Eq. (10.63), we obtain

$$\epsilon''_{kk} = \frac{3p}{3\lambda + 2\mu} \tag{10.166}$$

Substituting Eqs. (10.164) and (10.166) into Eq. (10.163) and considering Eq. (10.66), we have the following relation:

$$\int_V \epsilon''_{kk}\, dV = 3\alpha \int_V \tau\, dV \tag{10.167}$$

The left-hand side of Eq. (10.167) means volume change ΔV for the body. Thus the volume change of the body is given by Eq. (10.160).

Problems

10.1 Derive Eq. (10.32).

10.2 Derive Eq. (10.62).

10.3 Show that the equations of motion (10.89) reduce to the equations of equilibrium (4.1″) when motion of the body can be neglected.

10.4 Show that the equations of motion in terms of displacement components (10.93) reduce to the equations of equilibrium (4.23a) when motion of the elastic body can be neglected.

10.5 When the body force is zero and the internal heat generation in an elastic body does not exist, show that the equations (10.107), (10.107′), and (10.108) reduce to

$$\left[\left(\nabla^2 - \frac{1}{c_1^2}\frac{\partial^2}{\partial t^2}\right)\left(\nabla^2 - \frac{1}{\kappa}\frac{\partial}{\partial t}\right) - \eta K \nabla^2 \frac{\partial}{\partial t}\right]\phi = 0 \tag{10.168}$$

$$\left[\left(\nabla^2 - \frac{1}{c_1^2}\frac{\partial^2}{\partial t^2}\right)\left(\nabla^2 - \frac{1}{\kappa}\frac{\partial}{\partial t}\right) - \eta K \nabla^2 \frac{\partial}{\partial t}\right]\tau = 0 \tag{10.169}$$

$$\left(\nabla^2 - \frac{1}{c_2^2}\frac{\partial^2}{\partial t^2}\right)\boldsymbol{\psi} = 0 \tag{10.170}$$

10.6 When the initial motion and the initial temperature are zero, show that equations (10.168) through (10.170) reduce to, in the Laplace transform domain,

$$\left[\left(\nabla^2 - \frac{p^2}{c_1^2}\right)\left(\nabla^2 - \frac{p}{\kappa}\right) - \eta K p \nabla^2\right]\bar{\phi} = 0 \tag{10.171}$$

$$\left[\left(\nabla^2 - \frac{p^2}{c_1^2}\right)\left(\nabla^2 - \frac{p}{\kappa}\right) - \eta K p \nabla^2\right]\bar{\tau} = 0 \tag{10.172}$$

$$\left(\nabla^2 - \frac{p^2}{c_2^2}\right)\bar{\boldsymbol{\psi}} = 0 \tag{10.173}$$

in which $(\bar{\phi}, \bar{\tau}, \bar{\boldsymbol{\psi}}) = \int_0^\infty (\phi, \tau, \boldsymbol{\psi}) e^{-pt}\, dp$.

10.7 Derive the following equations from Eqs. (10.171) and (10.172):

$$(\nabla^2 - a_i)(\bar{\phi}_i, \bar{\tau}_i) = 0 \qquad i = 1, 2$$

where

$$\bar{\phi} = \bar{\phi}_1 + \bar{\phi}_2, \qquad \bar{\tau} = \bar{\tau}_1 + \bar{\tau}_2$$

$$a_1 = \frac{1}{2}\left\{\left[\frac{p^2}{c_1^2} + \left(\frac{1}{\kappa} + \eta K\right) p\right] + \sqrt{\left[\frac{p^2}{c_1^2} + \left(\frac{1}{\kappa} + \eta K\right) p\right]^2 - 4\frac{p^3}{c_1^2 \kappa}}\right\}$$

$$a_2 = \frac{1}{2}\left\{\left[\frac{p^2}{c_1^2} + \left(\frac{1}{\kappa} + \eta K\right) p\right] - \sqrt{\left[\frac{p^2}{c_1^2} + \left(\frac{1}{\kappa} + \eta K\right) p\right]^2 - 4\frac{p^3}{c_1^2 \kappa}}\right\}$$

10.8 When a hollow sphere with inner and outer radii a and b is subject to a temperature change $\tau(r) = C_0 + (C_1/r)$, calculate the volume change ΔV from Eq. (10.160).

Answer

$$\Delta V = 12\pi\alpha\left[\frac{1}{3}C_0(b^3 - a^3) + \frac{1}{2}C_1(b^2 - a^2)\right]$$

Appendix A

INVERSE LAPLACE TRANSFORM AND CONTOUR INTEGRALS

Let us consider a complex function $f(z)$ of the complex variable $z = x + iy$, where $i = \sqrt{-1}$, which is regular in the region S.

Cauchy's integral theorem If a complex function $f(z)$ of the complex variable $z = x + iy$ is regular in a simply connected region S, the integral

$$\int_L f(z)\,dz = 0 \tag{A.1}$$

is valid for any simple smooth closed regular contour L in S.

The complex variable t denotes any point on any simple smooth closed contour L in the simply connected region S. When the complex variable t moves from a start point a to an end point b on the contour L so that the region S is on the left side while moving along the contour L, the direction on the contour L is called positive. The integral of the form

$$\frac{1}{2\pi i}\int_L \frac{f(t)}{t-z}\,dt$$

taken over L in the region S with some point z which does not lie on L is called a Cauchy integral.

Theorem 1 If a complex function $f(z)$ is regular in the simply connected region S, a contour L is any simple smooth closed contour in the region S and the direction of the contour L is taken to be positive, the integral

$$f(z) = \frac{1}{2\pi i} \int_L \frac{f(t)}{t - z} dt \tag{A.2}$$

is valid for any point z in an inner region of L.

When a complex function $f(z)$ is regular in a region S: $0 < |z - a| < R$ but is not regular in the circular region $|z - a| < R$ which consists of the point a and the region S, the point a in the region S is called an isolated singularity of the complex function $f(z)$. When the point a is the isolated singularity of the complex function $f(z)$, the complex function $f(z)$ can be expressed for the region S: $0 < |z - a| < R$ by use of the Laurent expansion

$$f(z) = \sum_{n=1}^{\infty} \frac{c_{-n}}{(z - a)^n} + \sum_{n=0}^{\infty} c_n (z - a)^n \tag{A.3}$$

where

$$c_n = \frac{1}{2\pi i} \int_{|z-a|=r} \frac{f(z)}{(z - a)^{n+1}} dz \qquad (n\text{: integer, } 0 < r < R)$$

The second term on the right-hand side of Eq. (A.3) is convergent in the circular region $|z - a| < R$, is a regular function and is called the regular part of $f(z)$. On the other hand, the first term on the right-hand side of Eq. (A.3) is called the singular part of $f(z)$ or the principal part of the Laurent expansion at the singular point a. If the principal part of the Laurent expansion at the singular point a is expressed by a finite series, $f(z)$ can be expressed by

$$f(z) = \sum_{n=1}^{k} \frac{c_{-n}}{(z - a)^n} + \sum_{n=0}^{\infty} c_n (z - a)^n \tag{A.4}$$

In this case, $z = a$ is called the pole of order k of $f(z)$. The coefficient c_{-1} of the term $(z - a)^{-1}$ in the Laurent expansion (A.3) is called the residue of $f(z)$ at a, and the residue is expressed as $R(a)$:

$$R(a) = c_{-1} = \frac{1}{2\pi i} \int_{|z-a|=r} f(z)\, dz, \qquad 0 < r < R \tag{A.5}$$

If a complex function $f(z)$ has a pole of order one at the point a in the region S, the residue of $f(z)$ at a is given by

$$R(a) = \lim_{z \to a} (z - a) f(z) \tag{A.6}$$

Residue Theorem Let L denote a simple smooth closed contour in a region S, and all points in the inner region of regular contour L belong to the region S, and let a complex function $f(z)$ be a regular function inside L and on L, except for a finite number of points $a_1, a_2, \ldots, a_m$ inside L. Then

$$\frac{1}{2\pi i}\int_L f(z)\,dz = \sum_{n=1}^{m} R(a_n) \tag{A.7}$$

in which the direction of L is positive.

The inversion theorem for the Laplace transform (3.61) may be written as

$$f(t) = L^{-1}[\bar{f}(p)] = \frac{1}{2\pi i}\int_{\gamma-i\infty}^{\gamma+i\infty} \bar{f}(p)e^{pt}\,dp = \lim_{\beta\to\infty}\int_{\gamma-i\beta}^{\gamma+i\beta} \bar{f}(p)e^{pt}\,dp \tag{A.8}$$

Let us consider the inverse Laplace transform of the temperature $\overline{T}$. If $\overline{T}$ is a single-valued function of complex variable p in the left side region of integral path from $\gamma - i\infty$ to $\gamma + i\infty$, we consider the integral along the contour given in Fig. 3.6:

$$\lim_{R\to\infty}\oint e^{pt}\overline{T}(p)\,dp = \lim_{R\to\infty}\int_{BCDC'B'} e^{pt}\overline{T}(p)\,dp + \lim_{\beta\to\infty}\int_{\gamma-i\beta}^{\gamma+i\beta} e^{pt}\overline{T}(p)\,dp \tag{A.9}$$

If $|\overline{T}(p)|$ uniformly tends to zero at $R \to \infty$, the first integral on the right-hand side vanishes at the limit

$$\lim_{R\to\infty}\int_{BCDC'B'} e^{pt}\overline{T}(p)\,dp \le \lim_{R\to\infty}\int_{BCDC'B'} e^{\gamma t}\overline{T}(p)\,dp \to 0 \tag{a}$$

so that Eq. (A.9) reduces to

$$\lim_{R\to\infty}\oint e^{pt}\overline{T}(p)\,dp = \lim_{\beta\to\infty}\int_{\gamma-i\beta}^{\gamma+i\beta} \bar{f}(p)e^{pt}\,dp \tag{b}$$

First we proof Eq. (a). The integral $\int_{BCDC'B'} e^{pt}\overline{T}(p)\,dp$ can be expressed by a sum of three terms

$$\int_{BCDC'B'} e^{pt}\overline{T}(p)\,dp = \int_{BC} e^{pt}\overline{T}(p)\,dp + \int_{CDC'} e^{pt}\overline{T}(p)\,dp + \int_{C'B'} e^{pt}\overline{T}(p)\,dp \tag{c}$$

If an arbitray small constant ϵ is selected, $|\overline{T}(p)|$ uniformly tends to zero at $R \to \infty$ so that a radius R which satisfies the inequality $|\overline{T}(p)| < \epsilon$ can be found. On the circle of radius R, a relationship

$$|e^{pt}| = e^{Re(pt)} = e^{tR\cos\varphi}$$

is satisfied. Then the second term on the right-hand side in Eq. (c) reduces to

$$\left|\int_{CDC'} e^{pt}\overline{T}(p)\,dp\right| \le R\int_{\pi/2}^{3\pi/2} |\overline{T}(p)| e^{tR\cos\varphi}\,d\varphi \le R\epsilon\int_{\pi/2}^{3\pi/2} e^{tR\cos\varphi}\,d\varphi$$

$$= 2R\epsilon\int_{0}^{\pi/2} e^{-tR\sin\varphi}\,d\varphi \tag{d}$$

Using $t > 0$ and the Jordan's inequality

$$\frac{2}{\pi}\varphi \le \sin\varphi \qquad \left(0 \le \varphi \le \frac{\pi}{2}\right) \tag{A.10}$$

Equation (d) yields

$$\left|\int_{CDC'} e^{pt}\overline{T}(p)\,dp\right| \le 2\epsilon\int_{0}^{\pi/2} e^{-2tR\varphi/\pi}\,d\varphi = \frac{\pi}{t}\epsilon(1 - e^{-tR}) < \frac{\pi}{t}\epsilon$$

As $\epsilon \to 0$ for $R \to \infty$, we get

$$\lim_{R\to\infty}\int_{CDC'} e^{pt}\overline{T}(p)\,dp = 0 \qquad (t > 0) \tag{e}$$

Next we consider the first term of Eq. (c). The first term of Eq. (c) reduces to

$$\left|\int_{BC} e^{pt}\overline{T}(p)\,dp\right| \le R\epsilon\int_{\pi/2-\theta}^{\pi/2} e^{tR\cos\varphi}\,d\varphi \le R\epsilon\int_{0}^{\theta} e^{tR\sin\varphi}\,d\varphi \le R\epsilon\theta e^{tR\sin\varphi} \tag{f}$$

Using $\gamma = R\sin\theta$, Eq. (f) yields

$$\left|\int_{BC} e^{pt}\overline{T}(p)\,dp\right| \le \gamma\epsilon\frac{\theta}{\sin\theta}e^{t\gamma}$$

When $R \to \infty$, $\epsilon \to 0$ so that $\theta/\sin\theta \to 1$. Then we obtain

$$\lim_{R\to\infty}\int_{BC} e^{pt}\overline{T}(p)\,dp = 0 \qquad (t > 0) \tag{g}$$

The third term of Eq. (a) yields by the same way

$$\lim_{R\to\infty}\int_{C'B'} e^{pt}\overline{T}(p)\,dp = 0 \qquad (t > 0) \tag{h}$$

Now substitution of Eqs. (f), (g), (h) into (c) yields Eq. (a). This result shows that the inverse Laplace transform of the temperature $\overline{T}(p)$ reduces to a complex integral

$$T(t) = L^{-1}[\overline{T}(p)] = \frac{1}{2\pi i}\int_{\gamma-i\infty}^{\gamma+i\infty}\overline{T}(p)e^{pt}\,dp = \lim_{\beta\to\infty}\int_{\gamma-i\beta}^{\gamma+i\beta}\overline{T}(p)e^{pt}\,dp$$

$$= \lim_{R\to\infty}\oint \overline{T}(p)e^{pt}\,dp \tag{A.11}$$

Appendix B

BESSEL FUNCTIONS

The differential equation

$$\frac{d^2 f(z)}{dz^2} + \frac{1}{z}\frac{df(z)}{dz} + \left(1 - \frac{\nu^2}{z^2}\right) f(z) = 0 \tag{B.1}$$

is called the Bessel's differential equation of order ν. The first solution of the Bessel's equation is expressed by

$$J_\nu(z) = \sum_{m=0}^{\infty} \frac{(-1)^m}{m!\Gamma(\nu + m + 1)} \left(\frac{z}{2}\right)^{\nu+2m} \tag{B.2}$$

and is called the Bessel function of the first kind of order ν. The second solution of Bessel's equation which is available for all values of ν is expressed by

$$Y_\nu(z) = \frac{J_\nu(z)\cos\nu\pi - J_{-\nu}(z)}{\sin\nu\pi} \tag{B.3}$$

and is called the Bessel function of the second kind of order ν. If ν is not an integer, $J_\nu(z)$ may be used instead of $Y_\nu(z)$ as a second solution. Functions $J_\nu(z)$ and $J_{-\nu}(z)$ are independent solutions of Eq. (B.1). When ν is an integer n, $J_{-\nu}(z)$ is dependent on $J_\nu(z)$

$$J_n(z) = (-1)^n J_{-n}(z) \tag{B.4}$$

We adopt $Y_n(z)$ being defined as $\lim_{\nu\to n} Y_\nu(z)$ as the Bessel function of the second kind of an integral order n:

$$Y_0(z) = \frac{2}{\pi}\left(\gamma + \ln\frac{z}{2}\right) J_0(z) - \frac{2}{\pi}\sum_{r=1}^{\infty}\frac{(-1)^r}{(r!)^2}\left(\frac{z}{2}\right)^{2r}\sum_{k=1}^{r}\frac{1}{k} \qquad \text{for } n=0$$

$$Y_n(z) = \frac{2}{\pi}\left(\gamma + \ln\frac{z}{2}\right) J_n(z) - \frac{1}{\pi}\sum_{r=0}^{n-1}\frac{(n-r-1)!}{r!}\left(\frac{2}{z}\right)^{n-2r} \tag{B.5}$$

$$-\frac{1}{\pi}\sum_{r=0}^{\infty}\frac{(-1)^r}{r!(n+r)!}\left(\frac{z}{2}\right)^{n+2r}\left(\sum_{k=1}^{r}\frac{1}{k}+\sum_{k=1}^{n+r}\frac{1}{k}\right) \qquad \text{for } n \le 1$$

where γ is the Euler's constant which is defined by $\gamma = \lim_{m\to\infty}(\sum_{k=1}^{m}(1/k) - \ln m) = 0.5772\ldots$, and the last term $\sum_{k=1}^{r}(1/k) + \sum_{k=1}^{n+r}(1/k)$ is replaced by $\sum_{k=1}^{n}(1/k)$ when $r = 0$. The complete solution of Eq. (B.1) is

$$\begin{aligned} f(z) &= AJ_\nu(z) + BY_\nu(z) && \text{for all values of } \nu \\ f(z) &= AJ_n(z) + BY_n(z) && \text{for an integer } n \\ f(z) &= AJ_\nu(z) + BJ_{-\nu}(z) && \text{for non-integer } \nu \end{aligned} \tag{B.6}$$

We list the formulas of the Bessel functions:

$$\frac{2\nu}{z}J_\nu(z) = J_{\nu-1}(z) + J_{\nu+1}(z) \tag{B.7}$$

$$\frac{dJ_\nu(z)}{dz} = \frac{1}{2}[J_{\nu-1}(z) - J_{\nu+1}(z)] \tag{B.8}$$

$$= J_{\nu-1}(z) - \frac{\nu}{z}J_\nu(z) = -J_{\nu+1}(z) + \frac{\nu}{z}J_\nu(z) \tag{B.9}$$

$$\frac{d}{dz}[z^\nu J_\nu(az)] = az^\nu J_{\nu-1}(az), \qquad \frac{d}{dz}[z^{-\nu}J_\nu(az)] = -az^{-\nu}J_{\nu+1}(az) \tag{B.10}$$

$$J_n(z) = (-1)^n J_{-n}(z) = (-1)^n J_n(-z) \tag{B.11}$$

$$J_{\nu+1}(z)Y_\nu(z) - J_\nu(z)Y_{\nu+1}(z) = \frac{2}{\pi z} \tag{B.12}$$

Above formulas except for the last equation are satisfied by $Y_\nu(z)$ instead of $J_\nu(z)$.

The differential equation

$$\frac{d^2 f(z)}{dz^2} + \frac{1}{z}\frac{df(z)}{dz} - \left(1 + \frac{\nu^2}{z^2}\right) f(z) = 0 \tag{B.13}$$

is called the modified Bessel's differential equation of order ν. The complete solution of Eq. (B.13) is

$$f(z) = AI_\nu(z) + BK_\nu(z) \qquad \text{for all values of } \nu$$

$$f(z) = AI_n(z) + BK_n(z) \qquad \text{for an integer } n \tag{B.14}$$
$$f(z) = AI_\nu(z) + BI_{-\nu}(z) \qquad \text{for non-integer } \nu$$

The function $I_\nu(z)$ is called the modified Bessel function of the first kind of order ν, and is defined by

$$I_\nu(z) = \sum_{m=0}^{\infty} \frac{1}{m!\Gamma(\nu+m+1)} \left(\frac{z}{2}\right)^{\nu+2m} \tag{B.15}$$

where Γ stands for the gamma function defined by Eq. (C.12), the function $K_\nu(z)$ is called the modified Bessel function of the second kind of order ν, and is defined by

$$K_\nu(z) = \frac{\pi}{2} \frac{I_{-\nu}(z) - I_\nu(z)}{\sin \nu\pi} \tag{B.16}$$

We adopt $K_\nu(z)$ being defined as $\lim_{\nu \to n} K_\nu(z)$ as the Bessel function of the second kind of integer order n:

$$K_0(z) = -\left(\gamma + \ln\frac{z}{2}\right) I_0(z) + \sum_{r=1}^{\infty} \frac{1}{(r!)^2} \left(\frac{z}{2}\right)^{2r} \sum_{k=1}^{r} \frac{1}{k} \qquad \text{for } n = 0$$

$$K_n(z) = (-1)^{n+1}\left(\gamma + \ln\frac{z}{2}\right) I_n(z) + \frac{1}{2}\sum_{r=0}^{n-1}(-1)^r \frac{(n-r-1)!}{r!}\left(\frac{z}{2}\right)^{-n+2r} \tag{B.17}$$
$$\times \frac{(-1)^n}{2} \sum_{r=0}^{\infty} \frac{1}{r!(n+r)!}\left(\frac{z}{2}\right)^{n+2r} \left(\sum_{k=1}^{r}\frac{1}{k} + \sum_{k=1}^{n+r}\frac{1}{k}\right) \qquad \text{for } n \leq 1$$

where γ is the Euler's constant and the last term $\sum_{k=1}^{r}(1/k) + \sum_{k=1}^{n+r}(1/k)$ is replaced by $\sum_{k=1}^{n}(1/k)$ when $r = 0$.

We list the formulas for the modified Bessel functions:

$$\frac{2\nu}{z} I_\nu(z) = I_{\nu-1}(z) - I_{\nu+1}(z) \tag{B.18}$$
$$\frac{dI_\nu(z)}{dz} = \frac{1}{2}[I_{\nu-1}(z) + I_{\nu+1}(z)] \tag{B.19}$$
$$= I_{\nu-1}(z) - \frac{\nu}{z} I_\nu(z) = I_{\nu+1}(z) + \frac{\nu}{z} I_\nu(z) \tag{B.20}$$
$$\frac{d}{dz}\left[z^\nu I_\nu(az)\right] = az^\nu I_{\nu-1}(az), \qquad \frac{d}{dz}\left[z^{-\nu} I_\nu(az)\right] = az^{-\nu} I_{\nu+1}(az) \tag{B.21}$$
$$I_n(z) = I_{-n}(z) = (-1)^n I_n(-z) \tag{B.22}$$
$$\frac{2\nu}{z} K_\nu(z) = K_{\nu+1}(z) - K_{\nu-1}(z) \tag{B.23}$$

$$\frac{dK_\nu(z)}{dz} = -\frac{1}{2}\left[K_{\nu-1}(z) + K_{\nu+1}(z)\right] \tag{B.24}$$

$$= -K_{\nu-1}(z) - \frac{\nu}{z}K_\nu(z) = -K_{\nu+1}(z) + \frac{\nu}{z}K_\nu(z) \tag{B.25}$$

$$\frac{d}{dz}\left[z^\nu K_\nu(az)\right] = -az^\nu K_{\nu-1}(az), \qquad \frac{d}{dz}\left[z^{-\nu} K_\nu(az)\right] = -az^{-\nu} K_{\nu+1}(az) \tag{B.26}$$

$$K_n(z) = K_{-n}(z) \tag{B.27}$$

$$I_{\nu+1}(z)K_\nu(z) + I_\nu(z)K_{\nu+1}(z) = \frac{1}{z} \qquad \text{for } z \neq 0 \tag{B.28}$$

$$I_\nu(ze^{\pm(1/2)\pi i}) = e^{\pm(1/2)\nu\pi i} J_\nu(z) \tag{B.29}$$

$$K_\nu(ze^{\pm(1/2)\pi i}) = \mp\frac{1}{2}\pi e^{\mp(1/2)\pi(\nu\mp1)i}\left[J_\nu(z) \mp iY_\nu(z)\right] \tag{B.30}$$

$$\int_0^z J_\nu(kz)J_\nu(lz)z\,dz = \frac{z}{k^2 - l^2}[kJ_\nu(lz)J_{\nu+1}(kz) - lJ_\nu(kz)J_{\nu+1}(lz)]$$

$$= \frac{z}{k^2 - l^2}[lJ_{\nu-1}(lz)J_\nu(kz) - kJ_{\nu-1}(kz)J_\nu(lz)] \qquad k \neq l \tag{B.31}$$

$$\int_0^z J_\nu^2(kz)z\,dz = \frac{z^2}{2}\left[J_\nu^2(kz) - J_{\nu-1}(kz)J_{\nu+1}(kz)\right] \tag{B.32}$$

$$\int^z Y_\nu(kz)Y_\nu(lz)z\,dz = \frac{z}{k^2 - l^2}[kY_\nu(lz)Y_{\nu+1}(kz) - lY_\nu(kz)Y_{\nu+1}(lz)]$$

$$= \frac{z}{k^2 - l^2}[lY_{\nu-1}(lz)Y_\nu(kz) - kY_{\nu-1}(kz)Y_\nu(lz)] \qquad k \neq l \tag{B.33}$$

$$\int^z Y_\nu^2(kz)z\,dz = \frac{z^2}{2}\left[Y_\nu^2(kz) - Y_{\nu-1}(kz)Y_{\nu+1}(kz)\right] \tag{B.34}$$

$$\int^z J_\nu(kz)Y_\nu(lz)z\,dz = \frac{z}{k^2 - l^2}[kY_\nu(lz)J_{\nu+1}(kz) - lJ_\nu(kz)Y_{\nu+1}(lz)]$$

$$= \frac{z}{k^2 - l^2}[lY_{\nu-1}(lz)J_\nu(kz) - kJ_{\nu-1}(kz)Y_\nu(lz)] \qquad k \neq l \tag{B.35}$$

$$\int^z J_\nu(kz)Y_\nu(kz)z\,dz = \frac{z^2}{4}[2J_\nu(kz)Y_\nu(kz) - J_{\nu-1}(kz)Y_{\nu+1}(kz) - J_{\nu+1}(kz)Y_{\nu-1}(kz)] \tag{B.36}$$

The last four equations have a restriction on ν, if the lower limit is zero.

$$\int_0^z I_\nu(kz)I_\nu(lz)z\,dz = \frac{z}{k^2-l^2}[kI_\nu(lz)I_{\nu+1}(kz) - lI_\nu(kz)I_{\nu+1}(lz)]$$
$$= \frac{z}{k^2-l^2}[kI_\nu(lz)I_{\nu-1}(kz) - lI_\nu(kz)I_{\nu-1}(lz)]$$
$$\text{Re}(\nu) > -1, \quad k \neq l \tag{B.37}$$

$$\int_0^z I_\nu^2(kz)z\,dz = \frac{z^2}{2}\left[I_\nu^2(kz) - I_{\nu-1}(kz)I_{\nu+1}(kz)\right], \quad \text{Re}(\nu) > -1 \tag{B.38}$$

$$\int_z^\infty K_\nu(kz)K_\nu(lz)z\,dz = \frac{z}{k^2-l^2}[kK_\nu(lz)K_{\nu+1}(kz) - lK_\nu(kz)K_{\nu+1}(lz)]$$
$$= \frac{z}{k^2-l^2}[kK_{\nu-1}(lz)K_\nu(kz) - lK_{\nu-1}(kz)K_\nu(lz)]$$
$$z > 0, \quad \text{Re}(k+l) > 0, \quad k \neq l \tag{B.39}$$

$$\int_z^\infty K_\nu^2(kz)z\,dz = -\frac{z^2}{4}\left[K_\nu^2(kz) - K_{\nu-1}(kz)K_{\nu+1}(kz)\right]$$
$$z > 0, \quad \text{Re}(k) > 0 \tag{B.40}$$

$$\int_0^z K_\nu(kz)I_\nu(lz)z\,dz = -\frac{z}{k^2-l^2}[kI_\nu(lz)K_{\nu+1}(kz) + lK_\nu(kz)I_{\nu+1}(lz)] + \frac{(l/k)^\nu}{k^2-l^2}$$
$$= -\frac{z}{k^2-l^2}[kK_{\nu-1}(kz)I_\nu(lz) + lI_{\nu-1}(lz)K_\nu(kz)] + \frac{(l/k)^\nu}{k^2-l^2}$$
$$\text{Re}(\nu) > -1, \quad k^2 \neq l^2 \tag{B.41}$$

$$\int_0^z I_\nu(kz)K_\nu(kz)z\,dz = -\frac{z^2}{4}[2I_\nu(kz)K_\nu(kz) + I_{\nu-1}(kz)K_{\nu+1}(kz)$$
$$+ I_{\nu+1}(kz)K_{\nu-1}(kz)] - \frac{\nu}{2k^2}, \qquad \text{Re}(\nu) > -1, \quad k^2 \neq l^2 \tag{B.42}$$

$$\int_0^z K_\nu(kz)J_\nu(lz)z\,dz = \frac{z}{k^2+l^2}[lK_\nu(kz)J_{\nu+1}(lz) - kJ_\nu(lz)K_{\nu+1}(kz)] + \frac{(l/k)^\nu}{k^2+l^2}$$
$$\text{Re}(\nu) > -1, \quad k^2 \neq -l^2 \tag{B.43}$$

Spherical Bessel Functions

The differential equation

$$\frac{d^2 f}{dz^2} + \frac{2}{z}\frac{df}{dz} + \left[1 - \frac{n(n+1)}{z^2}\right] = 0 \tag{B.44}$$

is called the spherical Bessel's differential equation of order n, in which n is zero or positive integer. The first and second solutions of this differential equation is expressed by

$$j_n(z) = (2z)^n \sum_{m=0}^{\infty} \frac{(-1)^m (n+m)!}{m!(2n+2m+1)!} z^{2m} \tag{B.45}$$

$$y_n(z) = -\frac{1}{2^n z^{n+1}} \left[\sum_{m=0}^{n} \frac{(2n-2m)!}{m!(n-m)!} z^{2m} + (-1)^n \sum_{m=n+1}^{\infty} \frac{(-1)^m (n-m)!}{m!(2n-2m)!} z^{2m} \right] \tag{B.46}$$

and are called the spherical Bessel functions of the first and second kind of order n. They are related to the Bessel's functions $J_n(z)$ and $Y_n(z)$ by

$$j_n(z) = \sqrt{\frac{\pi}{2z}} J_{n+\frac{1}{2}}(z), \quad y_n(z) = \sqrt{\frac{\pi}{2z}} Y_{n+\frac{1}{2}}(z) \tag{B.47}$$

We list the formulas of the spherical Bessel functions:

$$\frac{2n+1}{z} j_n(z) = j_{n-1}(z) + j_{n+1}(z) \tag{B.48}$$

$$\frac{dj_n(z)}{dz} = \frac{1}{2n+1} [n j_{n-1}(z) - (n+1) j_{n+1}(z)] \tag{B.49}$$

$$= \frac{n}{z} j_n(z) - j_{n+1}(z) = -\frac{n+1}{z} j_n(z) + j_{n-1}(z) \tag{B.50}$$

$$\frac{d}{dz}\left[z^{n+1} j_n(az)\right] = az^{n+1} j_{n-1}(az), \quad \frac{d}{dz}\left[z^{-n} j_n(az)\right] = -az^{-n} j_{n+1}(az) \tag{B.51}$$

$$j_{n+1}(z) y_n(z) - j_n(z) y_{n+1}(z) = \frac{1}{z^2} \tag{B.52}$$

Above formulas from Eqs. (B.48) through (B.51) are satisfied also by $y_n(z)$. The spherical Bessel functions of order n are related to the trigonometric functions by

$$j_n(z) = (-1)^n (2z)^n \frac{d^n}{d(z^2)^n} \left[\frac{\sin z}{z}\right] = (-1)^n z^n \left(\frac{1}{z}\frac{d}{dz}\right)^n \left[\frac{\sin z}{z}\right] \tag{B.53}$$

$$y_n(z) = (-1)^{n+1} (2z)^n \frac{d^n}{d(z^2)^n} \left[\frac{\cos z}{z}\right] = (-1)^{n+1} z^n \left(\frac{1}{z}\frac{d}{dz}\right)^n \left[\frac{\cos z}{z}\right] \tag{B.54}$$

It follows from these formulas that

$$j_0(z) = \frac{1}{z}\sin z, \quad j_1(z) = \frac{1}{z^2}[\sin z - z\cos z], \quad j_2(z) = \frac{1}{z^3}[(3-z^2)\sin z - 3z\cos z], \ldots \tag{B.55}$$

$$y_0(z) = -\frac{1}{z}\cos z, \quad y_1(z) = -\frac{1}{z^2}[\cos z + z\sin z], \quad y_2(z) = -\frac{1}{z^3}[(3-z^2)\cos z + 3z\sin z], \ldots \tag{B.56}$$

Appendix C

LEGENDRE FUNCTIONS

The Laplace differential equation for a spherical coordinate system (r, θ, ϕ) has the form

$$\frac{\partial}{\partial r}\left(r^2\frac{\partial V}{\partial r}\right)+\frac{1}{\sin\theta}\frac{\partial}{\partial\theta}\left(\sin\theta\frac{\partial V}{\partial\theta}\right)+\frac{1}{\sin^2\theta}\frac{\partial^2 V}{\partial\phi^2}=0 \tag{C.1}$$

We apply the method of separation of variables, so that

$$V(r,\theta,\phi)=R(r)\Theta(\theta)\Phi(\phi) \tag{C.2}$$

Then, the following differential equation system is obtained:

$$\frac{d^2\Phi}{d\phi^2}+\mu^2\Phi=0 \tag{C.3}$$

$$\frac{d^2R}{dr^2}+\frac{2}{r}\frac{dR}{dr}-\frac{\nu(1+\nu)}{r^2}R=0 \tag{C.4}$$

$$\frac{d^2\Theta}{d\theta^2}+\frac{\cos\theta}{\sin\theta}\frac{d\Theta}{d\theta}+\left[\nu(1+\nu)-\frac{\mu^2}{\sin^2\theta}\right]\Theta=0 \tag{C.5}$$

in which, μ^2 and $\nu(1+\nu)$ are real constants.

General solutions for $R(r)$ and $\Phi(\phi)$ are given by (with constants A, B, C, D)

$$\begin{aligned} R(r)&=Ar^{\nu}+Br^{-(1+\nu)}\\ \Phi(\phi)&=C\sin\mu\phi+D\cos\mu\phi \end{aligned} \tag{C.6}$$

Now, we apply the transformation of variable

$$x = \cos\theta \tag{C.7}$$

and Eq. (C.5) can be rewritten as

$$(1-x^2)\frac{d^2\Theta}{dx^2} - 2x\frac{d\Theta}{dx} + \left[\nu(1+\nu) - \frac{\mu^2}{1-x^2}\right]\Theta = 0 \tag{C.8}$$

which is called the Legendre's associated differential equation, and the two solutions independent of each other are given by

$$\Theta(x) = P_\nu^\mu(x), \qquad Q_\nu^\mu(x) \tag{C.9}$$

where $P_\nu^\mu(x)$ and $Q_\nu^\mu(x)$ are called associated Legendre function of the first and second kinds, respectively, and are defined for real x by

$$P_\nu^\mu(x) = \frac{e^{\mu\pi i}}{\Gamma(1-\mu)}\left(\frac{1+x}{1-x}\right)^{\mu/2}\Gamma\left(-\nu, \nu+1, 1-\mu; \frac{1-x}{2}\right) \tag{C.10}$$

$$\begin{aligned} Q_\nu^\mu(x) &= \frac{\pi}{2\sin\mu\pi}\left[\cos\mu\pi\, P_\nu^\mu(x) - \frac{\Gamma(\nu+\mu+1)}{\Gamma(\nu-\mu+1)}P_\nu^{-\mu}(x)\right] \qquad \mu \neq \text{integer} \\ &= \frac{\pi}{2\sin(\nu+\mu)\pi}\left[\cos(\nu+\mu)\pi\, P_\nu^\mu(x) - P_\nu^\mu(-x)\right] \qquad \nu+\mu \neq \text{integer} \end{aligned} \tag{C.11}$$

where Γ means the gamma function, F is the Gauss hypergeometric function, and they are defined by

$$\Gamma(x) = \int_0^\infty e^{-t}t^{x-1}\,dt \qquad x > 0 \tag{C.12}$$

$$F(\alpha, \beta, \gamma; x) = \frac{\Gamma(\gamma)}{\Gamma(\alpha)\Gamma(\beta)}\sum_{n=0}^{\infty}\frac{\Gamma(\alpha+n)\Gamma(\beta+n)}{\Gamma(\gamma+n)}\frac{x^n}{n!} \tag{C.13}$$

Furthermore, $\Gamma(x)$ for $x < 0$ is evaluated from the formula

$$\Gamma(x)\Gamma(1-x) = \frac{\pi}{\sin\pi x} \tag{C.14}$$

Then, we have

$$\Gamma(-n) = \pm\infty \qquad n = 0, 1, 2, \ldots \tag{C.15}$$

General solution $\Theta(\theta)$ available to the three-dimensional problems of the sphere region defined by $0 \leq \theta \leq \pi$ becomes

$$\Theta(\theta) = P_n^m(x) \qquad n = 0, 1, 2, \ldots \tag{C.16}$$

where $x = \cos\theta$, and $P_n^m(x)$ is called the associated Legendre polynomial. Therefore, the general solution $V(r, \theta, \phi)$ of Eq. (C.1) in the region $0 \le \theta \le \pi, 0 \le \phi \le 2\pi$ is

$$V(r,\theta,\phi) = \begin{pmatrix} r^n \\ r^{-(n+1)} \end{pmatrix} P_n^m(x) \begin{pmatrix} \sin m\phi \\ \cos m\phi \end{pmatrix} \qquad n, m = 0, 1, 2, \ldots \tag{C.17}$$

Furthermore, the general solution $V(r, \theta)$ of Eq. (C.1) for axisymmetric problems is

$$V(r,\theta) = \begin{pmatrix} r^n \\ r^{-(n+1)} \end{pmatrix} P_n(x) \qquad n = 0, 1, 2, \ldots \tag{C.18}$$

where

$$P_\nu(x) = P_\nu^m(x)\big|_{m=0}, \qquad Q_\nu(x) = Q_\nu^m(x)\big|_{m=0} \tag{C.19}$$

are called, respectively, the Legendre functions of the first and second kind for real ν, and, especially, $P_n(x)$ for zero or integer of n is called the Legendre polynomial. The functions $P_n(x)$ and $P_n^m(x)$ are normalized and orthogonal. These orthogonal functions satisfy the following relations:

$$\int_{-1}^{1} P_n(x) P_{n'}(x)\, dx = \begin{cases} \dfrac{2}{2n+1} & \text{for } n = n' \\ 0 & \text{for } n \neq n' \end{cases} \tag{C.20}$$

$$\int_{-1}^{1} P_n^m(x) P_{n'}^m(x)\, dx = \begin{cases} \dfrac{2}{2n+1} \dfrac{(n+m)!}{(n-m)!} & \text{for } n = n' \\ 0 & \text{for } n \neq n' \end{cases} \tag{C.21}$$

Thus, the series expansions of an arbitrary function $f(x)$ in terms of $P_n(x)$ or $P_n^m(x)$ are given by

$$\begin{aligned} f(x) &= \sum_{n=0}^{\infty} A_n P_n(x) \\ A_n &= \frac{2n+1}{2} \int_{-1}^{1} f(x) P_n(x)\, dx \end{aligned} \tag{C.22}$$

or

$$\begin{aligned} f(x) &= \sum_{n=0}^{\infty} A_n^m P_n^m(x) \\ A_n^m &= \frac{2n+1}{2} \frac{(n-m)!}{(n+m)!} \int_{-1}^{1} f(x) P_n^m(x)\, dx \end{aligned} \tag{C.23}$$

We list the formulas for the Legendre functions. In the following expressions, ν and μ are real numbers, n and m are integers, and x is a real variable.

[1] Associated Legendre functions $P_\nu^\mu(x)$ and $Q_\nu^\mu(x)$

$$P_\nu^\mu(x) = P_{-(1+\nu)}^\mu(x) \tag{C.24}$$

$$Q_\nu^\mu(x)\sin(\nu+\mu)\pi - Q_{-(1+\nu)}^\mu(x)\sin(\nu-\mu)\pi = \pi\cos\nu\pi\cos\mu\pi\, P_\nu^\mu(x) \tag{C.25}$$

Denoting $P_\nu^\mu(x)$ or $Q_\nu^\mu(x)$ by $R_\nu^\mu(x)$, recurrence formulas for $R_\nu^\mu(x)$ are given as

$$(\nu-\mu+1)R_{\nu+1}^\mu(x) - (2\nu+1)xR_\nu^\mu(x) + (\nu+\mu)R_{\nu-1}^\mu(x) = 0 \tag{C.26}$$

$$R_{\nu+1}^\mu(x) - R_{\nu-1}^\mu(x) - (2\nu+1)(1-x^2)^{1/2}R_\nu^{\mu-1}(x) = 0 \tag{C.27}$$

$$R_{\nu-1}^\mu(x) - xR_\nu^\mu(x) + (\nu-\mu+1)(1-x^2)^{1/2}R_\nu^{\mu-1}(x) = 0 \tag{C.28}$$

$$xR_\nu^\mu(x) - R_{\nu+1}^\mu(x) + (\nu+\mu)(1-x^2)^{1/2}R_\nu^{\mu-1}(x) = 0 \tag{C.29}$$

$$(\nu-\mu)xR_\nu^\mu(x) - (\nu+\mu)R_{\nu-1}^\mu(x) + (1-x^2)^{1/2}R_\nu^{\mu+1}(x) = 0 \tag{C.30}$$

$$(\nu-\mu+1)R_{\nu+1}^\mu(x) - (\nu+\mu+1)xR_\nu^\mu(x) + (1-x^2)^{1/2}R_\nu^{\mu+1}(x) = 0 \tag{C.31}$$

$$R_\nu^{\mu+2}(x) - 2(\mu+1)x(1-x^2)^{-1/2} + (\nu-\mu)(\nu+\mu+1)R_\nu^\mu(x) = 0 \tag{C.32}$$

$$\begin{aligned}(1-x^2)\frac{dR_\nu^\mu(x)}{dx} &= (\nu+1)xR_\nu^\mu(x) - (\nu-\mu+1)R_{\nu+1}^\mu(x)\\ &= (\nu+\mu)R_{\nu-1}^\mu(x) - \nu xR_\nu^\mu(x)\\ &= (1-x^2)^{1/2}R_\nu^{\mu+1}(x) - \mu xR_\nu^\mu(x)\\ &= \mu xR_\nu^\mu(x) - (\nu+\mu)(\nu-\mu+1)(1-x^2)^{1/2}R_\nu^{\mu-1}(x)\end{aligned} \tag{C.33}$$

[2] Associated Legendre polynomials $P_n^m(x)$ and $Q_n^m(x)$
The formulas (C.24) to (C.33) are valid for $P_n^m(x)$ and $Q_n^m(x)$ when ν and μ are replaced by integers n and m. Furthermore

$$P_n^m(x) = 0 \qquad \text{for } m > n \tag{C.34}$$

$$P_n^m(-x) = (-1)^{n+m}P_n^m(x) \tag{C.35}$$

$$Q_n^m(-x) = (-1)^{n+m+1}P_n^m(x) \tag{C.36}$$

$$P_n^m(x) = \frac{(1-x^2)^{m/2}}{2^m}\sum_{r=0}^{n-m}\frac{(2n-r)!}{r!(n-r)!(n-m-r)!}\left(\frac{x-1}{2}\right)^{n-m-1} \tag{C.37}$$

From Eq. (C.37), we have

$$P_1^1(x) = (1-x^2)^{1/2}, \qquad P_2^1(x) = 3(1-x^2)^{1/2}x,$$

$$P_2^2(x) = 3(1-x^2), \qquad P_3^1(x) = \frac{3}{2}(1-x^2)^{1/2}(5x^2-1),$$
$$P_3^2(x) = 15(1-x^2)x, \qquad P_3^3(x) = 15(1-x^2)^{3/2}, \tag{C.38}$$
$$P_4^1(x) = \frac{5}{2}(1-x^2)^{1/2}x(7x^2-3), \ldots$$

[3] Legendre functions $P_\nu(x)$ and $Q_\nu(x)$
From Eqs. (C.24) to (C.33), we obtain recurrence formulas for $P_\nu(x)$ and $Q_\nu(x)$ by putting $\mu = 0$

$$P_\nu(x) = P_{-(1+\nu)}(x) \tag{C.39}$$

$$Q_\nu(x)\sin\nu\pi - Q_{-(1+\nu)}(x)\sin\nu\pi = \pi\cos\nu\pi\, P_\nu(x) \tag{C.40}$$

Denoting $P_\nu(x)$ or $Q_\nu(x)$ by $R_\nu(x)$, recurrence formulas for $R_\nu(x)$ are given as

$$(\nu+1)R_{\nu+1}(x) - (2\nu+1)xR_\nu(x) + \nu R_{\nu-1}(x) = 0 \tag{C.41}$$

$$\begin{aligned}(1-x^2)\frac{dR_\nu(x)}{dx} &= (\nu+1)\,[xR_\nu(x) - R_{\nu+1}(x)] \\ &= \nu\,[R_{\nu-1}(x) - xR_\nu(x)]\end{aligned} \tag{C.42}$$

[4] Legendre polynomials $P_n(x)$

$$P_n(x) = \frac{1}{2^n n!}\frac{d^n}{dx^n}(x^2-1)^n = F\left(-n, n+1, 1; \frac{1-x}{2}\right) \tag{C.43}$$

From Eq. (C.43), we receive the following relations:

$$\begin{aligned} P_0(x) &= 1, \\ P_1(x) &= x, \\ P_2(x) &= \frac{1}{2}(3x^2-1), \\ P_3(x) &= \frac{1}{2}(5x^3-3x), \\ P_4(x) &= \frac{1}{8}(35x^4-30x^2+3), \\ &\ldots \end{aligned} \tag{C.44}$$

From Eqs. (C.39) to (C.42), we obtain recurrence formulas for $P_n(x)$ by putting $\nu = n$

$$P_n(x) = P_{-(1+n)}(x) \tag{C.45}$$

$$P_n(-x) = (-1)^n P_n(x) \tag{C.46}$$

$$(n+1)P_{n+1}(x) - (2n+1)xP_n(x) + nP_{n-1}(x) = 0 \tag{C.47}$$

$$
\begin{aligned}
(1-x^2)\frac{dP_n(x)}{dx} &= (n+1)\,[xP_n(x) - P_{n+1}(x)] \\
&= n\,[P_{n-1}(x) - xP_n(x)]
\end{aligned} \tag{C.48}
$$

Furthermore, we have the following formulas:

$$
\frac{d}{dx}\,[xP_n(x) - P_{n+1}(x)] = -nP_n(x) \tag{C.49}
$$

$$
\int P_n(x)\,dx = \begin{cases} -\dfrac{1}{n}\,[xP_n(x) - P_{n+1}(x)] & \text{for } n \neq 0 \\ x & \text{for } n = 0 \end{cases} \tag{C.50}
$$

$$
\int xP_n(x)\,dx = \begin{cases} -\dfrac{1}{2-n(n+1)} \\ \quad \times \left\{(n+1)x\,[xP_n(x) - P_{n+1}(x)] \right. \\ \quad \left. -(1-x^2)P_n(x)\right\} & \text{for } n \neq 0, 1 \\ \dfrac{1}{3}x^3 & \text{for } n = 1 \\ \dfrac{1}{2}x^2 & \text{for } n = 0 \end{cases} \tag{C.51}
$$

$$
\int (1-x^2)\frac{dP_n(x)}{dx}\,dx = \begin{cases} (1-x^2)P_n(x) + \dfrac{2}{2-n(n+1)} \\ \quad \times \left\{(n+1)x\,[xP_n(x) - P_{n+1}(x)] \right. \\ \quad \left. -(1-x^2)P_n(x)\right\} & \text{for } n \neq 0, 1 \\ x - \dfrac{x^3}{3} & \text{for } n = 1 \\ 0 & \text{for } n = 0 \end{cases} \tag{C.52}
$$

BOOKS ON THERMAL STRESSES

The list includes most, if not all, books published originally in English. It also includes most books translated into English. There exists a multitude of books in other languages; of those, an insignificant number published in German, Japanese, Polish, Rumanian, Russian, and Serbian are listed.

1. P. P. Benham and R. Hoyle (Editors), Thermal stress, Sir Isaac Pitman & Sons, London (1964).
2. I. A. Birger and B. F. Shorr (Editors), Thermal strength of machine elements (in Russian), Mashinostoyenie, Moscow (1975).
3. B. A. Boley (Editor), Thermoinelasticity (IUTAM Symposium East Kilbride), Springer-Verlag, Wien (1970).
4. B. A. Boley and J. H. Weiner, Theory of thermal stresses, John Wiley & Sons, Inc., New York (1960).
5. D. Burgreen, Elements of thermal stress analysis, C. P. Press, Jamaica, N.Y. (1971).
6. R. S. Čukić, Rešenja Nekih Zadataka Thermoelastičnosti Primenom Integralnih Transformacija, Naučna Knjiga, Beograd (1979).
7. R. E. Czarnota-Bojarski, M. Sokolowski, and H. Zorski (Editors), Trends in elasticity and thermoelasticity, Wolters-Noordhoff Publishing, Netherlands (1971).
8. R. S. Dhaliwal and A. Singh, Dynamic coupled thermoelasticity, Hindustan Pub., Delhi (1980).
9. A. J. Fletcher, Thermal stress and strain generation in heat treatment, Elsevier Applied Science, London (1989).
10. Y. B. Fridman, Strength and deformation in nonuniform temperature fields, Consultants Bureau, New York (1964).
11. K. V. Frolov, Yu. L. Izrailev, N. A. Makhutov, E. M. Morozov, and V. Z. Parton, Thermal stresses and strength of turbins: calculation and design, Second augmented edition, English edition editor: R. B. Hetnarski, translated by N. Wadwha and R. Wadwha, Hemisphere Publ., New York (1990).
12. B. E. Gatewood, Thermal stresses, McGraw-Hill, New York (1957).

13. I. I. Goldenblat and N. A. Nikolaenko, Calculation of thermal stresses in nuclear reactors, Consultants Bureau, New York (1964).
14. N. Harwood and W. M. Cummings (Editors), Thermoelastic stress analysis, Adam Hilger, Bristol (1991).
15. D. P. H. Hasselman and R. A. Heller (Editors), Thermal stresses in severe environments, Proc. of an international conference held at Virginia Polytechnic Institute and State Univ., Blacksburg, March 19–21, 1980, Plenum Press, New York (1980).
16. R. B. Hetnarski (Editor), Thermal stresses I, North-Holland, Amsterdam (1986).
17. R. B. Hetnarski (Editor), Thermal stresses II, North-Holland, Amsterdam (1987).
18. R. B. Hetnarski (Editor), Thermal stresses III, North-Holland, Amsterdam (1989).
19. R. B. Hetnarski (Editor), Thermal stresses IV, North-Holland, Amsterdam (1996).
20. R. B. Hetnarski (Editor), Thermal stresses V, Lastran, Rochester (1999).
21. R. B. Hetnarski and J. Ignaczak, Mathematical theory of elasticity, to be published.
22. T. R. Hsu, The finite element method in thermomechanics, Allen & Unwin, Boston (1986).
23. D. Iesan, Theory of thermoelasticity (in Rumanian), Editura Academiei Republicii Socialiste Romania, Bucuresti (1979).
24. D. J. Johns, Thermal stress analyses, Pergamon Press, London (1965).
25. G. S. Kit and M. G. Krivtsun, Plane problems of thermoelasticity for bodies with cracks (in Russian), Naukova Dumka, Kiev (1983).
26. A. D. Kovalenko, Thermoelasticity (Basic theory and applications), Wolters-Noordhoff Publishing, Groningen (1969).
27. V. D. Kupradze (Editor), Three-dimensional problems of the mathematical theory of elasticity and thermoelasticity, North-Holland, Amsterdam (1979).
28. D. J. Littler (Editor), Thermal stresses and thermal fatigue, Butterworths, London (1971).
29. S. S. Manson, Thermal stress and low-cycle fatigue, McGraw-Hill, New York (1966).
30. E. Melan and H. Parkus, Wärmespannungen infolge stationärer Temperaturfelder, Springer, Wien (1953).
31. N. Naerlović-Veljković, Uvod U Termoelastičnost, Naučna Knjiga, Beograd (1977).
32. V. S. Nikishin, Thermal stresses in a composite cylinder with an arbitrary temperature distribution along its length, Plenum, New York (1966).
33. N. Noda, R. B. Hetnarski, and Y. Tanigawa, Thermal Stresses, First Edition, Lastran, Rochester (2000).
34. W. Nowacki, Problems of thermoelasticity (in Polish), P. W. N.—Polish Scientific Publishers, Warsaw (1959).
35. W. Nowacki, Thermoelasticity, Pergamon, Oxford (1962).
36. W. Nowacki, Thermoelasticity, First edition, Pergamon Press, Oxford, Gauthier Villars, Paris, and P. W. N.—Polish Scientific Publishers, Warsaw (1962).
37. W. Nowacki, Dynamic problems of thermoelasticity, Translation from Polish was edited by P. H. Francis and R. B. Hetnarski, Noordhoff Int. Pub., Leyden (1975).
38. W. Nowacki, Thermoelasticity, Second edition, Pergamon Press, Oxford and P. W. N.—Polish Scientific Publishers, Warsaw (1986).
39. W. K. Nowacki (Editor), Progress in thermoelasticity, VIIIth European Mechanics Colloquium, Warsaw 1967, P. W. N.—Polish Scientific Publishers, Warsaw (1969).
40. J. L. Nowinski, Theory of thermoelasticity with applications, Sijthoff and Noordhoff, Alphen aan den Rijn (1978).
41. P. M. Ogibalov, V. F. Gribanov, Thermal stability of plates and shells (in Russian), Publication of the Moscow University, Moscow (1968).
42. Z. Orłoś (Editor), Thermal stresses (in Polish), P. W. N.—Polish Scientific Publishers, Warsaw (1991).
43. H. Parkus, Instationäre Wärmespannungen, Springer, Wien (1959).
44. H. Parkus, Thermoelasticity, Blaisdell Pub., Waltham (1968).
45. Ya. S. Podstrihach, W. A. Lomakin, and Io. M. Kolyano, Thermoelasticity of bodies of inhomogeneous structure (in Russian), Nauka, Moscow (1981).
46. A. A. Pomarantsev, Thermal stresses in solids of revolution of arbitrary-shape, Gordon & Breach Science, New York (1970).
47. Proc. of the First Int. Symp. on Thermal Stresses and Related Topics, Hamamatsu, June 5–7, 1995, Editor: N. Noda, Shizuoka University, Hamamatsu (1995).

48. Proc. of the Second Int. Symp. on Thermal Stresses and Related Topics, Rochester, June 8–11, 1997, Editors: R. B. Hetnarski, N. Noda, and H. Ghoneim, Rochester Institute of Technology, Rochester (1997).
49. Proc. of the Third Int. Congress on Thermal Stresses, Cracow, June 13–17, 1999, Editors: J. J. Skrzypek and R. B. Hetnarski, Cracow University of Technology, Cracow (1999).
50. Proc. of the Fourth Int. Congress on Thermal Stresses, Osaka, June 8–11, 2001, Editors: Y. Tanigawa, R. B. Hetnarski, and N. Noda, Osaka Prefecture Univ., Sakai (2001).
51. T. Rożnowski, Moving heat sources in thermoelasticity, Ellis Horwood, Chichester (1989). Originally published in Polish by P. W. N.—Polish Scientific Publishers, Warsaw (1988).
52. I. N. Sneddon, The linear theory of thermoelasticity, Springer-Verlag, Wien (1974).
53. R. Stojanovic, Nonlinear thermoelasticity, Springer, Wien (1972).
54. S. Taira, Thermal stress and thermal fatigue (in Japanese), Nikkan Kougyo, Tokyo (1974).
55. Y. Takeuti, Analyses of thermal stresses (in Japanese), Nisshin, Tokyo (1971).
56. Y. Takeuti and N. Noda, Analyses of thermal stresses (in Japanese), Nisshin, Tokyo (1981).
57. E. A. Thornton, Thermal structures for aerospace applications, AIAA Education Series, Editor: J. S. Przemieniencki, Reston, VA (1996).
58. V. M. Vihak, Management of thermal stresses and displacements (in Russian), Naukova Dumka, Kiev (1988).
59. Z. Zudans, T. C. Yen, and W. H. Steigelmann, Thermal stress techniques in the nuclear industry, Elsevier, New York (1965).

INDEX